ARTHROPOD VECTOR: CONTROLLER OF DISEASE TRANSMISSION

VECTOR SALIVA-HOST-PATHOGEN INTERACTIONS

VOLUME 2

Edited by

STEPHEN K. WIKEL

Emeritus Professor and Chair of Medical Sciences, Sch   ,
Quinnipiac University, CT, United States

SERAP AKSOY

Professor of Epidemiology, Yale School of Public Health, Yale University School of Medicine,
New Haven, CT, United States

GEORGE DIMOPOULOS

Professor, Department of Molecular Microbiology and Immunology,
Johns Hopkins Bloomberg School of Public Health, Baltimore, MD, United States

Academic Press is an imprint of Elsevier
125 London Wall, London EC2Y 5AS, United Kingdom
525 B Street, Suite 1800, San Diego, CA 92101-4495, United States
50 Hampshire Street, 5th Floor, Cambridge, MA 02139, United States
The Boulevard, Langford Lane, Kidlington, Oxford OX5 1GB, United Kingdom

Library of Congress Cataloging-in-Publication Data
A catalog record for this book is available from the Library of Congress

British Library Cataloguing-in-Publication Data
A catalogue record for this book is available from the British Library

ISBN: 978-0-12-805360-7

For information on all Academic Press publications visit our website at
https://www.elsevier.com/books-and-journals

  Working together
to grow libraries in
developing countries

www.elsevier.com • www.bookaid.org

Publisher: Sara Tenney
Acquisition Editor: Kristi Gomez
Editorial Project Manager: Pat Gonzalez
Production Project Manager: Julia Haynes
Designer: Matthew Limbert

Typeset by TNQ Books and Journals

Contents—Volume 2

Contents—Volume 1

List of Contributors

John F. Andersen NIH/NIAID Laboratory of Malaria and Vector Research, Rockville, MD, United States

Sarah Bonnet UMR BIPAR 956 INRA-ANSES-ENVA, Maisons-Alfort, France

Nathalie Boulanger Université de Strasbourg, Strasbourg, France; Centre National de Référence Borrelia, Strasbourg, France

Guy Caljon University of Antwerp, Antwerp, Belgium

Adriana Costero-Saint Denis National Institute of Allergy and Infectious Diseases, NIH Rockville, Maryland, United States

Iliano V. Coutinho-Abreu National Institutes of Health, Rockville, MD, United States

Waldionê de Castro National Institutes of Health, Rockville, MD, United States

Carl De Trez Vrije Universiteit Brussel (VUB), Brussels, Belgium; VIB Structural Biology Research Center (SBRC), Brussels, Belgium

Ranadhir Dey Center for Biologics Evaluation and Research, FDA, Silver Spring, MD, United States

Erol Fikrig Yale University School of Medicine, New Haven, CT, United States; Howard Hughes Medical Institute, Chevy Chase, MD, United States

Stephen Higgs Kansas State University, Manhattan, KS, United States

Yan-Jang S. Huang Kansas State University, Manhattan, KS, United States

Shaden Kamhawi National Institutes of Health, Rockville, MD, United States

Randall Kincaid National Institute of Allergy and Infectious Diseases, NIH Rockville, Maryland, United States

Michail Kotsyfakis Czech Academy of Sciences, Budweis, Czech Republic

Wolfgang W. Leitner National Institute of Allergy and Infectious Diseases, NIH Rockville, Maryland, United States

Erin E. McClure University of Maryland School of Medicine, Baltimore, MD, United States

Hira L. Nakhasi Center for Biologics Evaluation and Research, FDA, Silver Spring, MD, United States

Sukanya Narasimhan Yale University School of Medicine, New Haven, CT, United States

Fabiano Oliveira National Institutes of Health, Rockville, MD, United States

Joao H.F. Pedra University of Maryland School of Medicine, Baltimore, MD, United States

Anne Poinsignon Institute of Research for Development (IRD), MIVEGEC Unit, Montpellier, France; Institute Pierre Richet (IPR), Bouake, Ivory Coast

Franck Remoue Institute of Research for Development (IRD), MIVEGEC Unit, Montpellier, France; Institute Pierre Richet (IPR), Bouake, Ivory Coast

José M.C. Ribeiro NIH/NIAID Laboratory of Malaria and Vector Research, Rockville, MD, United States

Andre Sagna Institute of Research for Development (IRD), MIVEGEC Unit, Montpellier, France; Institute Pierre Richet (IPR), Bouake, Ivory Coast

Tyler R. Schleicher Yale University School of Medicine, New Haven, CT, United States

Tiago D. Serafim National Institutes of Health, Rockville, MD, United States

Dana K. Shaw University of Maryland School of Medicine, Baltimore, MD, United States

Benoît Stijlemans Vrije Universiteit Brussel (VUB), Brussels, Belgium; VIB Inflammation Research Center, Ghent, Belgium

Jesus G. Valenzuela National Institutes of Health, Rockville, MD, United States

Jan Van Den Abbeele Institute of Tropical Medicine Antwerp (ITM), Antwerp, Belgium

Dana L. Vanlandingham Kansas State University, Manhattan, KS, United States

Esther von Stebut Johannes Gutenberg-University, Mainz, Germany

Tonu Wali National Institute of Allergy and Infectious Diseases, NIH Rockville, Maryland, United States

Stephen Wikel Quinnipiac University, Hamden, CT, United States

Preface

These two volumes bring together in one place an up-to-date, multidisciplinary examination, by leading authorities, of factors that make the vector arthropod the controller of pathogen transmission. Arthropod vector ability to transmit infectious agents is increasingly recognized as being impacted by the themes addressed in these two volumes: vector microbiome, arthropod innate immunity, and vector saliva stimulation and modulation of host defenses. The three areas addressed in these two volumes are increasingly active areas of investigation that are resulting in significant new insights for understanding vector competence of a variety of arthropod vector species. These research areas increasingly represent opportunities for translation of basic findings into novel approaches for controlling arthropod vectors and vector-borne diseases.

The first volume examines arthropod factors that determine vector competence in the context of gut and reproductive organ–associated microbiomes and complex interactions of vector arthropod immune defenses with the microbiome and vector-borne infectious agents. The introductory chapter identifies knowledge gaps in the biology of vector transmission of disease causing microbes that are addressed in the chapters of these volumes. First volume vector microbiome–related topics encompass: vector microbiome–mediated immune inductions; microbiome influence on shaping arthropod immunity; impact of vector microbiota on disease transmission; and engineered microbiome in control applications. Arthropod vector immune defenses include signaling pathways; priming of innate defenses during vector infection; and developing vector immunity-based novel control strategies for arthropod transmitted pathogens.

The second chapter of the first volume explores the relationships of vector arthropod immune signal transduction pathways with mitochondrial regulation. Microbiome vector immune responses are reviewed in chapters that focus on *Wolbachia*-mediated immune inductions; *Wolbachia* in host and microbe dialogue; role of microbiota in development of the arthropod immune system; and interactions of vector and microbe in mosquito immunity and development. Chapter themes focusing on novel vector-borne disease control approaches include use of symbionts to control dengue; modulation of mosquito immune defenses as a control strategy; targeting dengue virus replication in the mosquito vector; paratransgenesis-mediated pathogen control utilizing engineered symbionts; mosquitocidal activity of antibodies that target chloride channels; and modulation of mosquito immune defenses to control dengue. The first volume closes with a chapter addressing the implication of insulin-like peptides in modulating *Plasmodium* infection in the mosquito.

The second volume provides expert reviews of the complex interactions occurring between hosts, disease vectors, and vector-borne pathogens. Topics addressed in this volume include: vector saliva composition; saliva stimulation, modulation, and suppression of host defenses; influence of vector saliva on transmission and establishment of infectious agents; use of host antibody responses to vector saliva as biomarkers of exposure and risk of vector-borne pathogen infection; and development of saliva molecule–based disease transmission blocking vaccines. This volume concludes with a chapter examining multiple considerations that must be addressed in translating basic arthropod disease vector research described in these two volumes into commercial products to control vectors and the diseases they transmit.

The scope of second volume chapter themes provides a comprehensive examination of the multiple, interrelated, and complex interactions occurring at the host interface with the vector arthropod. The first chapter explores the cutaneous innate and adaptive immune defenses that confront the blood-feeding vector arthropod. The next three chapters focus on how arthropod disease vectors counteract the challenges posed by the host defenses of hemostasis, itch and pain responses, and wound healing, respectively. The fifth chapter examines the genomics and proteomics of vector saliva to provide the underpinnings for subsequent chapters that address the interactions of sandflies, tsetse flies, mosquitoes, and ticks with host defenses and how those interactions create environments favorable for pathogen transmission and establishment of infections agents. These chapters describe both conserved and unique host defense countermeasures across the range of arthropod disease vectors. A subsequent chapter is an excellent example of increasingly detailed molecular characterizations of the interplay between vectors and hosts by examining how tick saliva interacting with Nod-like receptors modulates host innate immune response signaling. Additional chapters describe the use of saliva biomarkers as indicators of vector exposure and epidemiological role in assessing potential risk of infection with malaria and African trypanosomiasis. Multiple authors provide insights as to how basic knowledge of the interactions between vectors and hosts can be used to develop immunologically based strategies to control vector-borne protozoa, bacteria, and arboviruses.

The idea for this two-volume work emerged from the 2015 Keystone Symposium entitled, "The Arthropod Vector: The Controller of Transmission." Our hope is that information contained in these two volumes will stimulate further research and encourage both interdisciplinary collaborations and new avenues of research on arthropod vectors of disease.

Stephen K. Wikel
Serap Aksoy
George Dimopoulos

Network of Cells and Mediators of Innate and Adaptive Cutaneous Immunity: Challenges for an Arthropod Vector

Esther von Stebut

Johannes Gutenberg-University, Mainz, Germany

THE CUTANEOUS IMMUNE SYSTEM

Together with the lung and gut, the skin is the one of the largest organs and represents the host's external barrier to the environment. In addition to its well-characterized mechanical function, the skin forms an immunological barrier to the outside that provides the first line of defense against infection. It is now clear that understanding the mechanisms that regulate immune responses requires the study of interactions between different immune and nonimmune cellular compartments of specific tissues. Maintenance of immune homeostasis in the skin relies on a balanced equilibrium of interactions between different cellular, molecular, and microbial components. Dysregulation such as that induced by vectors or pathogens introduced into skin contributes to the pathogenesis of inflammatory responses of the skin.

As a first-line defense organ, the skin evaluates immunologically relevant signals to orchestrate the appropriate immune response (Di Meglio et al., 2011). To fulfill its role as an immunological barrier, the skin performs several, sequential tasks. First, the skin is the initial line of defense against exogenous challenges such as pathogenic bacteria, parasites, and other agents (toxins, haptens, or allergens). Second, the skin initiates and shapes immune responses through a dense network of immunologically active cells involved in induction of either tolerance or inflammation-associated immunity, both under steady-state conditions and upon challenge. Third, once activated, a network of resident and immune cells that migrate into the skin actively participate in the orchestration of local as well as systemic immune reactions. Finally, controlling unwanted excessive inflammation is also a task performed by various immune cells in skin.

The main task of the skin-resident immune cells is the containment of the invading pathogen to the skin to prevent spreading to inner organs and/or the elimination of the pathogen if possible. Several pathogens developed strategies to circumvent these functions. In addition, alerting circulating immune cells being recruited to the skin to migrate along gradients of pathogen-induced proinflammatory cytokines and/or chemokines is a main role of these skin cells.

In this chapter, various elements of the cutaneous immune system will be introduced by using the example of *Leishmania major* infection and focusing on initiators of adaptive immunity.

SKIN-RESIDENT IMMUNE COMPONENTS

The skin harbors resident immune cells in both the epidermis and the dermis. Most cells in the epidermis are epithelial keratinocytes, but ~5% of epidermal cells are Langerhans cells (LCs), belonging to the family of dendritic cells (DCs), and gamma/delta (γδ) T cells. The dermis consists of fibroblasts, as the main stromal cell–type, skin-resident macrophages (MΦ), dermal DCs, and mast cells (MCs) that are found in the upper papillary dermal compartment, and they are long-lived.

Complement Activation

Serum complement is quickly activated after the vector contact induced skin damage and pathogen inoculation. The complement system is designed to help clear pathogens from the organism by disrupting the target pathogen's plasma membrane. For example, *L. major* inoculation into skin leads to rapid activation of complement by both the classical and the alternative pathways (Dominguez et al., 2003). Pathogen opsonization is generally fast and occurs within seconds/minutes after pathogen contact resulting in lysis via the membrane attack complex (C5b–C9 complex) and may contribute to efficient killing of arthropod vector–inoculated pathogens. However, several pathogens have developed strategies to circumvent lysis by, e.g., membrane alterations that prevent the insertion of the C5b–C9 complex into parasites' outer membranes, by producing protein kinases able to phosphorylate complement, or by expressing elongated, complement-binding surface molecules, such as LPG and gp63 on *L. major*, that impede lysis (Mosser and Brittingham, 1997). On the other hand, some pathogens utilize complement activation and opsonization to quickly invade various host cells, such as neutrophils and MΦ, and to evade a hostile environment (Sacks and Sher, 2002). In addition, complement components such as C3a and C5a are potent chemoattractants for various immune cells. Both MΦ and neutrophils migrate along gradients of C3a and C5a to the site of complement activation to prevent skin invasion by pathogens (Teixeira et al., 2006).

Arthropod vector components such as saliva are known to activate complement; the exact contribution of these components and how this may alter subsequent complement functions need to be clarified.

Mast Cells

Both MCs and MC-derived products are key factors for the induction of early local inflammation against pathogens in the skin. Preferentially localized at the borders of the organism (the skin, lung, gut), MCs contribute as sentinels of the immune system to local defense mechanisms. They produce a large variety of mediators and cytokines, which are prestored and can be released within seconds after cell activation (Galli et al., 1999), and they play an important role in the regulation of protective adaptive immune responses against pathogens (Marshall, 2004). The skin and peritoneum mainly contain connective tissue–type MCs, whereas mucosal MCs are predominantly found in the lung and gut.

Previously, various studies showed that MCs are responsible for survival in various models of acute bacterial and parasitic infections (Maurer et al., 1998; Galli et al., 2008); however, an investigation using mice devoid of MC in a c-kit-independent fashion has shown that MC may not be essential for these responses (Rodewald and Feyerabend, 2012). More studies are currently underway investigating this phenomenon.

In earlier studies, several authors demonstrated that *Leishmania*-infected MCs upon activation degranulate and release preformed TNFα both in vitro and in vivo (Bidri et al., 1997; Saha et al., 2004). MC degranulation after arthropod vector contact with skin is a known fact, and one of the main mediators responsible for itching induced by arthropod bites is the saliva-induced release of preformed histamine from skin MC granules. Previously, we demonstrated that MCs also mediate recruitment of neutrophils and MΦ to the inflamed tissue. Release of TNFα from MC promotes the influx of neutrophils, which release chemokines (such as MIP-1α/β, MIP-2) that in turn results in MΦ recruitment (von Stebut et al., 2003b; Maurer et al., 2006; Dudeck et al., 2011).

Macrophages

MΦs are of myeloid origin and reside in various tissues including the skin. In inflammatory situations, monocytes move from peripheral blood into the skin, where transmigration through epithelial structures contributes to monocyte differentiation into MΦ. Immigration of MΦ to skin occurs as early as 3–4 days poststimulus, thus following that of neutrophils (von Stebut et al., 2003b; von Stebut, 2007b). However, immigration of inflammatory MΦ during physiological low-dose infection with *L. major* is delayed by several weeks (Belkaid et al., 2000).

MΦs are part of the mononuclear phagocyte system, and one of their main functions is to engulf and digest cellular debris, foreign substances, pathogens, or tumor cells. They also play an important role in the innate immune response of the skin and serve as antigen presenting cells (APCs), especially in the case of restimulation of already primed T cells. Their priming capacity of naïve T cells is limited compared to that of DC. In the skin, they are located in the various layers of the dermis and in the subcutaneous fat; however, a perivascular accumulation in the upper dermis can be observed (von Stebut, 2007a; von Stebut et al., 1998).

Depending on the micromilieu of the surrounding skin tissue, MΦs can exhibit two different functions. They can exert proinflammatory functions with the production of various cytokines including interleukin (IL)-1, IL-6, IL-12, and TNFα. As M1 MΦ, they are also able to synthesize nitric oxide from arginine, which enables them to kill invading pathogens, such as *L. major*. Under certain circumstances, such as wound healing or sclerosis of the skin, they can be M2 MΦ capable of producing regulatory cytokines such as TGFβ and also synthesize ornithine from arginine responsible for tissue repair (Sindrilaru and Scharffetter-Kochanek, 2013). Pathogen-associated modulation of MΦ function into either the M1 or M2 phenotype is an important factor for disease control.

Dendritic Cells

DCs belong to the myeloid lineage and are APCs important for the orchestration of adaptive T and B cell immunity as well as immune tolerance. In general, they are positioned at potential pathogen entry sites such as the skin. Data show that DCs that internalized a pathogen are the critical APCs responsible for naïve T cell priming against pathogen antigens. On pathogen uptake, DCs become activated and process the antigen. DCs upregulate MHC class I and II and costimulatory molecules and migrate to the draining lymph node where they encounter and present antigen to naïve T cells (von Stebut et al., 1998). Very recently, it was shown that *L. major* parasite releases a soluble Mincle ligand capable of dampening DC activation and subsequent immune activation, indicating that arthropod vector and pathogen-derived factors may attempt to inhibit DC function (Iborra et al., 2016).

DCs contribute to T cell priming and education by producing specific cytokines. Under some circumstances, DCs primarily induce T-helper (Th)1 immunity characterized by high levels of IFNγ (so-called DC1 cells), in other settings so-called DC2 preferentially induce Th2 immune responses with predominant IL-4, IL-5, and IL-13 production. In the presence of IL-1, IL-6, IL-23, and/or TGFβ, either regulatory T cells or Th17 cells develop (Kautz-Neu et al., 2012). Differences in the production of infection-induced proinflammatory cytokines produced by DCs from various inbred mouse strains are genetically determined and contribute to disease outcome. As such, DCs from *Leishmania*-resistant C57BL/6 mice produce IL-12, IL-1α, IL-27, and IL-23, whereas *Leishmania*-susceptible BALB/c DCs produce similar amounts of IL-12 and IL-27, but less IL-1α, and more inhibitory IL-12p80 and IL-23 (von Stebut et al., 2003a; Nigg et al., 2007; Lopez Kostka et al., 2009; Kautz-Neu et al., 2011). As a result, immunity against *Leishmania* in C57BL/6 mice is mainly IFNγ-driven and leads to NO-associated parasite killing, whereas BALB/c mice T cell responses are Th2/Th17 and regulatory T cells (Tregs) predominant creating an environment responsible for disease susceptibility.

Recently, Naik et al. (2012, 2015) demonstrated that skin DCs not only induce immunity against pathogens dangerous for the host, but also initiate immune responses against commensals of the skin. In this context, CD103+ dermal DCs (dDCs) produce IL-1 and stimulate CD8+ T cells to produce IL-17 (so-called Tc17 cells) in the skin recognizing *Staphylococcus epidermidis*. How anticommensal immunity may be important for skin defense mechanisms against vector-borne pathogens will be discussed further.

For a number of years, several DC subtypes have been characterized. In the skin, at least five different DC subsets can be found: epidermal LCs, dDCs, and DC subsets that migrate into the skin on inflammation that are referred to as inflammatory DCs. LCs in transit to the epidermis can also be found in the dermis. The dDC subset can be divided into Langerin+ CD103+ dDC and Langerin[neg] CD103[neg] dDC. In addition, plasmocytoid DC as well as CD8a+ DC are situated in the draining lymph node with the ability to translocate to the skin during inflammation (Fig. 1.1).

Only recently, several groups discovered that these different DC subsets not only represent DCs with different surface marker expression, but they have, in fact, also different roles for resulting immune responses in skin (Durai and Murphy, 2016). In cutaneous leishmaniasis, we and others have demonstrated that epidermal LCs preferentially induce regulatory T cells that serve to control adaptive immunity in the skin (Kautz-Neu et al., 2011; Brewig et al., 2009). Because LCs represent the most superficial DC subsets that encounter antigen without indicating a breach in the barrier function of the skin, regulation of immunity is the result. If antigen enters the skin more deeply and the host senses, as a result, "danger," other DC subsets are recruited. Several groups demonstrated that CD103+ dDC may be primarily responsible for the induction of protective immunity against *L. major* (Ashok et al., 2014; Martínez-López et al., 2015; Ritter et al., 2004). The precise roles of the other DC subsets are unknown so far; however, cross-presentation of antigen from skin origin DCs in the lymph node to lymph node resident DCs may also occur and shape the resulting immune responses. Finally, active or passive transport of pathogens from skin to draining lymph nodes may allow lymph node–resident DCs to pick up foreign antigen and be responsible for T cell priming in this setting (Kautz-Neu et al., 2010).

Epidermal Gamma/Delta T Cells

The majority of T cells have T cell receptors (TCRs) composed of two glycoprotein chains called α and β. In contrast, a much less frequently encountered T cell subset has a TCR comprised

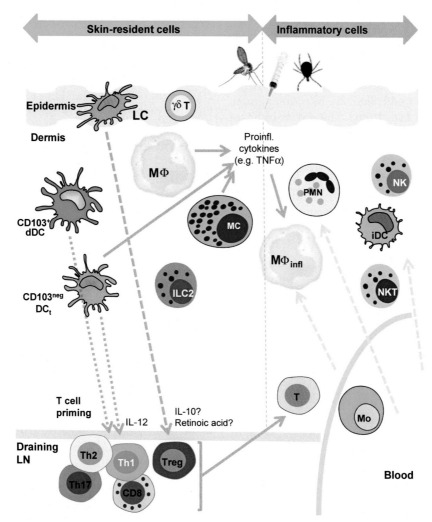

FIGURE 1.1 **Skin immune system: skin-resident immune cells and those immigrating upon inflammation.** Epidermal Langerhans cells (LCs) as well as CD103[+] and CD103[neg] dermal DC (dDC) reside in the skin and can be activated to migrate to the draining lymph node (LN). There, T cell priming and education occurs, where LCs preferentially induce regulatory T cells (Tregs), whereas the other DC subsets appear to induce other T-helper (Th) subsets capable of producing IFNγ (Th1), IL-4/IL-13/IL-10 (Th2), and IL-17 (Th17). In addition, activated DCs also cross-present antigen in an MHC class I-dependent context to CD8[+] T cells. Type-2 innate lymphoid cells (ILCs) as well as epidermal gamma/delta (γδ) T cells are present in skin under steady-state conditions. On activation through the epidermis, skin-resident macrophages (MΦs), mast cells (MCs), and DCs release various proinflammatory mediators such as IL-1 and TNFα, which then induce migration of various inflammatory cells into skin along chemokine gradients. Among the first cells to arrive in the skin after irritation are neutrophils [polymorphonuclear cells (PMN)], followed by inflammatory macrophages, inflammatory DCs, natural killer (NK) cells, T cell subsets as well as NKT cells.

of one γ and one δ chain. γδ T cells have their highest abundance in gut mucosa, but they are also found in the epidermis (Satoskar et al., 1997). While TCRα$^{-/-}$ developed nonhealing lesions with high parasite burdens, TCRδ$^{-/-}$ C57BL/6 mice efficiently controlled the infection similar to wild-type mouse strains. These data indicate that alpha/beta T cells are required for protection against infection, whereas γδ T cells are dispensable. Recently, several authors reported that γδ T cells contribute to skin immunity against pathogens or in inflammatory situations, as in psoriasis, by producing certain cytokines, such as IL-17A (Adami et al., 2014). How an arthropod vector modulates γδ T cell function is currently not understood.

Innate Lymphoid Cells

Recently, NK1.1 positive innate lymphoid cells (ILCs) capable of releasing significant amounts of T cell–related cytokines have been characterized. They are now designated ILC1 (capable of producing IFNγ), ILC2 (producing IL-4, IL-5, and IL-13), and ILC3 (characterized by IL-17/IL-22 synthesis) (Gronke et al., 2016). Under steady-state conditions, the skin appears to contain substantial numbers of ILC2, whereas only a few ILC1 and even fewer ILC3 are present (Kim et al., 2016). Additionally, ILC2 cells are capable of producing amphiregulin, a protein suggested to play an important role in wound healing (Sonnenberg and Artis, 2015). Amphiregulin serves to initiate epithelial cell proliferation and tissue repair through the activation of its receptor, EGFR (Wills-Karp and Finkelman, 2011).

The respective roles of these ILC subsets for pathogen control in skin have not yet been investigated in detail. Because these subsets are capable of releasing substantial amounts of proinflammatory cytokines into the tissue within a short time and can shape DC function, they may be important contributors to skin-associated pathogen control.

Stromal Cells (Keratinocytes, Fibroblasts)

Antimicrobial peptides (AMPs) with strong effects against various bacteria, viruses, and fungi are present in the skin (Schauber and Gallo, 2008). Cutaneous production of AMPs is important for pathogen control on the skin surface. Cathelicidin is one main AMPs with the ability to directly exert antimicrobial toxicity. Additionally, it initiates host responses with enhancement of cell influx resulting in proinflammatory responses in the skin. Cathelicidins are involved in the pathogenesis of various inflammatory skin diseases, with high levels found in psoriatic skin, whereas low levels characterize atopic dermatitis. Vitamin D3 is a major regulator of cathelicidin production. The other well-defined AMP in skin is β-defensin, but more than 20 proteins with antimicrobial activity are known to occur in the skin (Schauber and Gallo, 2008). Many cells in skin are capable of producing AMPs, such as keratinocytes, and MCs. In addition, rapid influx of AMP-producing neutrophils after skin irritation enhances AMP levels in the skin. On infection or other barrier disruption of the skin, cathelicidin is strongly upregulated. Thus, inoculation of various pathogens by arthropod vectors will likely alter AMP production by skin cells, such as keratinocytes.

Keratinocytes also produce several proinflammatory cytokines on stimulation to trigger subsequent immune activation. They were identified as an important source of early chemokine and cytokine expression in the skin during various inflammatory processes, such as delayed-type hypersensitivity reactions and infections (Ehrchen et al., 2010). The relevance for keratinocyte-derived mediators in pathogen defense mechanisms has been shown in several models. For example, *L. major* infection leads to strong induction of several cytokines, such as IL-1α/β, IL-12, osteopontin, IL-4, and IL-6, while local application of anti-IL-4, or the absence of IL-6 in stromal keratinocytes, resulted in worsening of disease outcome because of preferential Th2 induction. Induction of the chemokine CXCL11 (I-TAC) in keratinocytes upon *L. major* infection is important for ultimate control of the pathogen via modulation of DC function (Roebrock et al., 2014).

Dermal fibroblasts also contribute to the skin immune response. In wound healing, chemokine release and migration of fibroblasts are of critical importance (Sakthianandeswaren et al., 2005). On infection with *L. major*, fibroblasts directly interact with the pathogen leading to parasite uptake (Bogdan et al., 2000). Fibroblasts appear to release CCL2, thus modulating monocytes/MΦ recruitment in skin (Goncalves et al., 2011).

RECRUITMENT OF INFLAMMATORY CELLS TO THE SKIN

On stimulation, a variety of immune cells are rapidly recruited to the skin from the blood or lymphatic tissue, leading to the resulting immune responses. Depending on the nature of the trigger, different immune cells may be recruited, such as monocytes/MΦ or neutrophils in response to distinct pathogens. In addition, the sequence of events is often critical for disease outcome or pathogen control as with recruitment of neutrophils prior to MΦ/DC and then T cells.

Neutrophils

On perturbation of the skin barrier by arthropod bites, needle injection, or tape stripping, neutrophils are rapidly recruited to the skin via the blood (Beil et al., 1992; Peters et al., 2008). Skin immigration by neutrophils occurs as early as 60 min after skin barrier disruption (Ribeiro-Gomes and Sacks, 2012). Interestingly, when parasites are injected into the ear pinna, neutrophil numbers decrease in both strains of mice 3 days after needle injection. This recruitment is independent of pathogen inoculation and a result of

sensing damage of the skin. Thus, even though they are not skin-resident immune cells, neutrophils strongly contribute to the early immune response of the skin. Later on, a second wave of neutrophils immigrates into the infection site after *L. major* inoculation within 7–10 days (Ribeiro-Gomes and Sacks, 2012), indicating that pathogen-specific factors also contribute to neutrophil recruitment within skin.

Neutrophils represent the immune systems frontline of innate defense against infection. The rapid recruitment to skin ensures their early presence at the site of entry. Neutrophils use three strategies to kill invading pathogens: (1) phagocytosis and lysis of microbes, (2) secretion of AMPs, and (3) formation of neutrophil extracellular traps (NETs) used for killing of extracellular bacteria (Brinkmann et al., 2010; Brinkmann and Zychlinsky, 2007). Neutrophil NETs consist of released granule proteins containing a high concentration of antimicrobial components and chromatin, which bind and disarm extracellular bacteria NETs independent of phagocytic uptake.

However, neutrophils entering skin are also exploited by pathogens, as they represent an important immune evasion strategy for *Leishmania* parasites. Even though MΦs are the final host cells for the *Leishmania* parasite, neutrophils are among the first leukocytes infected, and on ingestion of *L. major*, they secrete high levels of MIP-1β that attracts MΦ. Within neutrophils, the parasites survive and, in contrast, infection of neutrophils prolongs survival of the latter (van Zandbergen et al., 2004). MΦs, in turn, readily phagocytose infected neutrophils, resulting in release of the antiinflammatory cytokine TGF-β. Thus, within the skin, *Leishmania* uses neutrophils to silently enter their final host cell.

Natural Killer and Natural Killer T Cells

Natural killer (NK) cells are effector cells of the innate immune system that are part of the protective immune response against several types of tumors and microbial infections. In the skin, they rapidly accumulate after pathogen inoculation and represent a major source of early IFNγ production in an antigen-independent fashion (Müller et al., 2001; Bajénoff et al., 2006). TLR9-dependent NK cell activation via IL-12 promoted early control of parasite replication in *L. major* infections (Liese et al., 2007).

Natural Killer T (NKT) cells are a rare population of innate-like lymphocytes recognizing glycolipid antigens presented by surface CD1d (Bendelac, 1995). These cells have an effector phenotype and on stimulation can rapidly secrete various cytokines, including both IL-4 and IFNγ (Bendelac et al., 2007). NKT cell activation occurs in various infectious diseases. On cutaneous challenge, NKT cells slowly accumulate in the skin over weeks post infection (Griewank et al., 2014). A potentially promising approach that showed beneficial effects is αGalCer stimulation of NKT cells in a BALB/c LACK-based vaccination trial against *L. major* (Dondji et al., 2008).

Adaptive Immune Cells

Accumulation of B cells in the form of plasma cells in human skin and under experimental conditions is indicative for infections with *Borrelia spp.* or *Treponema spp.* that are the causative agents of syphilis or leprosy. Acute skin reactions to arthropod bites may also contain numerous plasma cells. Their exact role is not fully understood. Antibody production by B cells that are not necessarily residing in skin is important for control of various pathogens inoculated through the skin, including defense against Lyme disease.

T cell priming by DC migrating from the skin after activation by pathogen uptake occurs primarily in skin-draining lymph nodes. Only T cells that are primed by skin-derived DC are capable of migrating to skin upon antigen challenge, whereas T cells primed against the same antigen via the intestine cannot (Dudda et al., 2004). While aiming for induction of adaptive immunity against skin-invading pathogens, as in vaccination trials, it is important to consider the route of vaccination to obtain best efficacy. This effect is also responsible for the eradication of the pox virus in humans with only immunization trough injured skin, but not other tissues, inducing the observed solid protection (Liu et al., 2010). Immune memory response responsible for protection against pox virus is transmitted by virus-reactive T cells that reside in the skin (Jiang et al., 2012). Skin as the initial trigger of antiviral immunity also serves as the reservoir for memory T cells and lifelong immunity.

One special feature of the skin is its ability to direct T cell differentiation into one of the four major subsets, Th1, Th2, Th17, and Tregs (Di Meglio et al., 2011). This feature distinguishes the skin from other organs with epithelial barriers and contact to the environment. For example, mainly Tregs are induced in the intestine while the lung favors induction of Th2 responses. This ability of skin is mainly promoted by antigen/pathogen/adjuvant-stimulated DC. Two factors may be responsible for induction of different Th cell subsets via skin: (1) the DC subset (Kautz-Neu et al., 2011) and (2) the cytokines released by DC upon stimulation (von Stebut et al., 1998; von Stebut et al., 2003a; Nigg et al., 2007; Lopez Kostka et al., 2009).

THE SKIN MICROBIOME AIDS ANTIPATHOGEN IMMUNITY

Extensive cross talk between epithelial, stromal and immune cells regulates immune responses in the skin and beyond to ensure effective host defense and to maintain or restore tissue homeostasis. In addition, the diverse microbial communities that colonize the surface of the skin constantly interact with host epithelial and immune cells, thereby influencing local and systemic immunity.

In various barrier tissues, such as the skin, recent studies highlighted a crucial role of communication between host cells and the microbiota in the regulation of immune responses. Skin microbiota, such as *S. epidermidis*, were critically important for protective host defense against the arthropod-transmitted infection *L. major* (Naik et al., 2012, 2015). Commensal-specific T cell responses are induced by dDC and do not mediate inflammation, which is a unique property. These findings indicate that the skin immune system is dynamic and shaped in part by commensals present on the skin. These commensal-specific responses are important for efficient control of protection against invasive pathogens.

OUTLOOK: HOW VECTOR AND PATHOGEN MODULATE SKIN IMMUNITY

As described earlier, pathogens developed strategies that allow them to modulate subsequent immune activating events in skin to enable or improve their survival after invading the host. For example, *Leishmania* parasites express proteins that protect them from immediate complement-mediated lysis (Mosser and Brittingham, 1997) or express soluble proteins, e.g., Mincle, that suppress subsequent DC activation to delay immune activation (Iborra et al., 2016).

The effect of arthropod vector–derived immune modulators is currently being investigated in many laboratories. As such, it is known that sand fly saliva modulates DC stimulation, initiates rapid neutrophil recruitment to skin, and induces the priming of antigen-specific T cells that shape the resulting immune responses (Loeuillet et al., 2016; Reed et al., 2016; Oliveira et al., 2013). Knowledge about these vector-derived factors may enable a better understanding of resulting antipathogen immunity and will allow for development of novel therapeutic strategies. Translational models using human skin transplanted onto immunodeficient mice will enable studying these factors with relevance for medical approaches.

Acknowledgment

This work was funded by research grants by the Deutsche Forschungsgemeinschaft (DFG).

References

Adami, S., Cavani, A., Rossi, F., Girolomoni, G., 2014. The role of interleukin-17A in psoriatic disease. BioDrugs 28, 487–497.

Ashok, D., Schuster, S., Ronet, C., Rosa, M., Mack, V., Lavanchy, C., Marraco, S.F., Fasel, N., Murphy, K.M., Tacchini-Cottier, F., Acha-Orbea, H., 2014. Cross-presenting dendritic cells are required for control of *Leishmania major* infection. Eur. J. Immunol. 44, 1422–1432.

Bajénoff, M., Breart, B., Huang, A.Y., Qi, H., Cazareth, J., Braud, V.M., Germain, R.N., Glaichenhaus, N., 2006. Natural killer cell behavior in lymph nodes revealed by static and real-time imaging. J. Exp. Med. 203, 619–631.

Beil, W.J., Meinardus-Hager, G., Neugebauer, D.C., Sorg, C., 1992. Differences in the onset of the inflammatory response to cutaneous leishmaniasis in resistant and susceptible mice. J. Leukoc. Biol. 52, 135–142.

Belkaid, Y., Mendez, S., Lira, R., Kadambi, N., Milon, G., Sacks, D., 2000. A natural model of *Leishmania major* infection reveals a prolonged "silent" phase of parasite amplification in the skin before the onset of lesion formation and immunity. J. Immunol. 165, 969–977.

Bendelac, A., 1995. CD1: presenting unusual antigens to unusual T lymphocytes. Science 269, 185–186.

Bendelac, A., Savage, P.B., Teyton, L., 2007. The biology of NKT cells. Annu. Rev. Immunol. 25, 297–336.

Bidri, M., Vouldoukis, I., Mossalayi, M.D., Debre, P., Guillosson, J.J., Mazier, D., Arock, M., 1997. Evidence for direct interaction between mast cells and Leishmania parasites. Parasite Immunol. 19, 475–483.

Bogdan, C., Donhauser, N., Döring, R., Röllinghoff, M., Diefenbach, A., Rittig, M.G., 2000. Fibroblasts as host cells in latent leishmaniosis. J. Exp. Med. 191, 2121–2130.

Brewig, N., Kissenpfennig, A., Malissen, B., Veit, A., Bickert, T., Fleischer, B., Mostböck, S., Ritter, U., 2009. Priming of CD8+ and CD4+ T cells in experimental leishmaniasis is initiated by different dendritic cell subtypes. J. Immunol. 182, 774–783.

Brinkmann, V., Zychlinsky, A., 2007. Beneficial suicide: why neutrophils die to make NETs. Nat. Rev. Microbiol. 5, 577–582.

Brinkmann, V., Laube, B., Abu Abed, U., Goosmann, C., Zychlinsky, A., 2010. Neutrophil extracellular traps: how to generate and visualize them. J. Vis. Exp. 24, 36.

Dominguez, M., Moreno, I., Aizpurua, C., Torano, A., 2003. Early mechanisms of Leishmania infection in human blood. Microbes Infect. 5, 507–513.

Dondji, B., Deak, E., Goldsmith-Pestana, K., Perez-Jimenez, E., Esteban, M., Miyake, S., Yamamura, T., McMahon-Pratt, D., 2008. Intradermal NKT cell activation during DNA priming in heterologous prime-boost vaccination enhances T cell responses and protection against Leishmania. Eur. J. Immunol. 38, 706–719.

Dudda, J.C., Simon, J.C., Martin, S., 2004. Dendritic cell immunization route determines CD8+ T cell trafficking to inflamed skin: role for tissue microenvironment and dendritic cells in establishment of T cell-homing subsets. J. Immunol. 172, 857–863.

Dudeck, A., Suender, C.A., Lopez Kostka, S., von Stebut, E., Maurer, M., 2011. Mast cells promote Th1 and Th17 responses by modulating dendritic cell maturation and function. Eur. J. Immunol. 41, 1883–1893.

Durai, V., Murphy, K.M., 2016. Functions of murine dendritic cells. Immunity 45, 719–736.

Ehrchen, J.M., Roebrock, K., Foell, D., Nippe, N., von Stebut, E., Weiss, J.M., Münck, N.A., Viemann, D., Varga, G., Müller-Tidow, C., Schuberth, H.J., Roth, J., Sunderkötter, C., 2010. Keratinocytes determine Th1 immunity during early experimental leishmaniasis. PLoS Pathog. 6, e1000871.

Galli, S.J., Maurer, M., Lantz, C.S., 1999. Mast cells as sentinels of innate immunity. Curr. Opin. Immunol. 11, 53–59.

Galli, S.J., Grimbaldeston, M., Tsai, M., 2008. Immunomodulatory mast cells: negative, as well as positive, regulators of immunity. Nat. Rev. Immunol. 8, 478–486.

Goncalves, R., Zhang, X., Cohen, H., Debrabant, A., Mosser, D.M., 2011. Platelet activation attracts a subpopulation of effector monocytes to sites of Leishmania major infection. J. Exp. Med. 208, 1253–1265.

Griewank, K.G., Lorenz, B., Fischer, M.R., Boon, L., Lopez Kostka, S., von Stebut, E., 2014. Immune modulating effects of NKT cells in a physiologically low dose Leishmania major infection model after αGalCer analog PBS57 stimulation. PLoS Negl. Trop. Dis. 8, e2917.

Gronke, K., Kofoed-Nielsen, M., Diefenbach, A., 2016. Innate lymphoid cells, precursors and plasticity. Immunol. Lett. 179, 9–18.

Iborra, S., Martínez-López, M., Cueto, F.J., Conde-Garrosa, R., Del Fresno, C., Izquierdo, H.M., Abram, C.L., Mori, D., Campos-Martín, Y., Reguera, R.M., Kemp, B., Yamasaki, S., Robinson, M.J., Soto, M., Lowell, C.A., Sancho, D., 2016. Leishmania uses mincle to target an inhibitory ITAM signaling pathway in dendritic cells that dampens adaptive immunity to infection. Immunity 2016 (45), 788–801.

Jiang, X., Clark, R.A., Liu, L., Wagers, A.J., Fuhlbrigge, R.C., Kupper, T.S., 2012. Skin infection generates non-migratory memory CD8+ T(RM) cells providing global skin immunity. Nature 483, 227–231.

Kautz-Neu, K., Meyer, R.G., Clausen, B.E., von Stebut, E., 2010. Leishmaniasis, contact hypersensitivity and graftversus-host disease: understanding the role of dendritic cell subsets in balancing skin immunity and tolerance. Exp. Dermatol. 19, 760–771.

Kautz-Neu, K., Noordegraaf, M., Dinges, S., Bennett, C.L., John, D., Clausen, B.E., von Stebut, E., 2011. Langerhans cells are negative regulators of the anti-Leishmania response. J. Exp. Med. 208, 885–891.

Kautz-Neu, K., Schwonberg, K., Fischer, M.R., Schermann, A.I., von Stebut, E., 2012. Dendritic cells in Leishmania major infections: mechanisms of parasite uptake, cell activation and evidence for physiological relevance. Med. Microbiol. Immunol. 201, 581–592.

Kim, C.H., Hashimoto-Hill, S., Kim, M., 2016. Migration and tissue tropism of innate lymphoid cells. Trends Immunol. 37, 68–79.

Liese, J., Schleicher, U., Bogdan, C., 2007. TLR9 signaling is essential for the innate NK cell response in murine cutaneous leishmaniasis. Eur. J. Immunol. 37, 3424–3434.

Liu, L., Zhong, Q., Tian, T., Dubin, K., Athale, S.K., Kupper, T.S., 2010. Epidermal injury and infection during poxvirus immunization is crucial for the generation of highly protective T cell-mediated immunity. Nat. Med. 16, 224–227.

Loeuillet, C., Bañuls, A.L., Hide, M., 2016. Study of Leishmania pathogenesis in mice: experimental considerations. Parasit. Vectors 9, 144.

Lopez Kostka, S., Dinges, S., Griewank, K., Iwakura, Y., Udey, M.C., von Stebut, E., 2009. IL-17 promotes progression of cutaneous leishmaniasis in susceptible mice. J. Immunol. 182, 3039–3046.

Marshall, J.S., 2004. Mast-cell responses to pathogens. Nat. Rev. Immunol. 4, 787–799.

Martínez-López, M., Iborra, S., Conde-Garrosa, R., Sancho, D., 2015. Batf3-dependent CD103+ dendritic cells are major producers of IL-12 that drive local Th1 immunity against Leishmania major infection in mice. Eur. J. Immunol. 45, 119–129.

Maurer, M., Echtenacher, B., Hultner, L., Kollias, G., Mannel, D.N., Langley, K.E., Galli, S.J., 1998. The c-kit ligand, stem cell factor, can enhance innate immunity through effects on mast cells. J. Exp. Med. 188, 2343–2348.

Maurer, M., Lopez Kostka, S., Siebenhaar, F., Moelle, K., Metz, M., Knop, J., von Stebut, E., 2006. Skin mast cells control T cell-dependent host defense in Leishmania major infections. FASEB J. 20, 2460–2467.

Di Meglio, P., Perera, G.K., Nestle, F.O., 2011. The multitasking organ: recent insights into skin immune function. Immunity 35, 857–869.

Mosser, D.M., Brittingham, A., 1997. Leishmania, macrophages and complement: a tale of subversion and exploitation. Parasitology 115, S9–S23.

Müller, K., van Zandbergen, G., Hansen, B., Laufs, H., Jahnke, N., Solbach, W., Laskay, T., 2001. Chemokines, natural killer cells and granulocytes in the early course of *Leishmania major* infection in mice. Med. Microbiol. Immunol. 190, 73–76.

Naik, S., Bouladoux, N., Wilhelm, C., Molloy, M.J., Salcedo, R., Kastenmuller, W., Deming, C., Quinones, M., Koo, L., Conlan, S., Spencer, S., Hall, J.A., Dzutsev, A., Kong, H., Campbell, D.J., Trinchieri, G., Segre, J.A., Belkaid, Y., 2012. Compartmentalized control of skin immunity by resident commensals. Science 337, 1115–1119.

Naik, S., Bouladoux, N., Linehan, J.L., Han, S.J., Harrison, O.J., Wilhelm, C., Conlan, S., Himmelfarb, S., Byrd, A.L., Deming, C., Quinones, M., Brenchley, J.M., Kong, H.H., Tussiwand, R., Murphy, K.M., Merad, M., Segre, J.A., Belkaid, Y., 2015. Commensal-dendritic-cell interaction specifies a unique protective skin immune signature. Nature 520, 104–108.

Nigg, A.P., Zahn, S., Rückerl, D., Hölscher, C., Yoshimoto, T., Ehrchen, J.M., Wölbing, F., Udey, M.C., von Stebut, E., 2007. Dendritic cell-derived IL-12p40 homodimer contributes to susceptibility in cutaneous leishmaniasis in BALB/c mice. J. Immunol. 178, 7251–7258.

Oliveira, F., de Carvalho, A.M., de Oliveira, C.I., 2013. Sand-fly saliva-leishmania-man: the trigger trio. Front Immunol. 4, 375.

Peters, N.C., Egen, J.G., Secundino, N., Debrabant, A., Kimblin, N., Kamhawi, S., Lawyer, P., Fay, M.P., Germain, R.N., Sacks, D., 2008. In vivo imaging reveals an essential role for neutrophils in leishmaniasis transmitted by sand flies. Science 321, 970–974.

Reed, S.G., Coler, R.N., Mondal, D., Kamhawi, S., Valenzuela, J.G., 2016. Leishmania vaccine development: exploiting the host-vector-parasite interface. Expert Rev. Vaccines 15, 81–90.

Ribeiro-Gomes, F.L., Sacks, D., 2012. The influence of early neutrophil-Leishmania interactions on the host immune response to infection. Front. Cell. Infect. Microbiol. 2, 59.

Ritter, U., Meissner, A., Scheidig, C., Körner, H., 2004. CD8α- and Langerin-negative dendritic cells, but not Langerhans cells, act as principal antigen-presenting cells in leishmaniasis. Eur. J. Immunol. 34, 1542–1550.

Rodewald, H.R., Feyerabend, T.B., 2012. Widespread immunological functions of mast cells: fact or fiction? Immunity 37, 13–24.

Roebrock, K., Sunderkötter, C., Münck, N.A., Wolf, M., Nippe, N., Barczyk, K., Varga, G., Vogl, T., Roth, J., Ehrchen, J., 2014. Epidermal expression of I-TAC (Cxcl11) instructs adaptive Th2-type immunity. FASEB J. 28, 1724–1734.

Sacks, D., Sher, A., 2002. Evasion of innate immunity by parasitic protozoa. Nat. Immunol. 3, 1041–1047.

Saha, B., Tonkal, A.M., Croft, S., Roy, S., 2004. Mast cells at the host-pathogen interface: host-protection versus immune evasion in leishmaniasis. Clin. Exp. Immunol. 137, 19–23.

Sakthianandeswaren, A., Elso, C.M., Simpson, K., Curtis, J.M., Kumar, B., Speed, T.P., Handman, E., Foote, S.J., 2005. The wound repair response controls outcome to cutaneous leishmaniasis. Proc. Natl. Acad. Sci. U.S.A. 102, 15551–15556.

Satoskar, A., Okano, M., David, J.R., 1997. γδ T cells are not essential for control of cutaneous *Leishmania major* infection in genetically resistant C57BL/6 mice. J. Infect Dis. 176, 1649–1652.

Schauber, J., Gallo, R.L., 2008. Antimicrobial peptides and the skin immune defense system. J. Allergy Clin. Immunol. 122, 261–266.

Sindrilaru, A., Scharffetter-Kochanek, K., 2013. Disclosure of the culprits: macrophages-versatile regulators of wound healing. Adv. Wound Care (New Rochelle) 2, 357–368.

Sonnenberg, G.F., Artis, D., 2015. Innate lymphoid cells in the initiation, regulation and resolution of inflammation. Nat. Med. 21, 698–708.

von Stebut, E., 2007a. Cutaneous Leishmania infection: progress in pathogenesis research and experimental therapy. Exp. Dermatol. 16, 340–346.

von Stebut, E., 2007b. Immunology of cutaneous leishmaniasis: the role of mast cells, phagocytes and dendritic cells for protective immunity. Eur. J. Dermatol. 17, 115–122.

von Stebut, E., Belkaid, Y., Jakob, T., Sacks, D.L., Udey, M.C., 1998. Uptake of *Leishmania major* amastigotes results in activation and interleukin 12 release from murine skin-derived dendritic cells: implications for the initiation of anti-Leishmania immunity. J. Exp. Med. 188, 1547–1552.

von Stebut, E., Ehrchen, J.M., Belkaid, Y., Kostka, S.L., Molle, K., Knop, J., Sunderkotter, C., Udey, M.C., 2003a. Interleukin 1α promotes Th$_1$ differentiation and inhibits disease progression in *Leishmania major*-susceptible BALB/c mice. J. Exp. Med. 198, 191–199.

von Stebut, E., Metz, M., Milon, G., Knop, J., Maurer, M., 2003b. Early macrophage influx to sites of cutaneous granuloma formation is dependent on MIP-1α/β released from neutrophils recruited by mast cell-derived TNFα. Blood 101, 210–215.

Teixeira, M.J., Teixeira, C.R., Andrade, B.B., Barral-Netto, M., Barral, A., 2006. Chemokines in host-parasite interactions in leishmaniasis. Trends Parasitol. 22, 32–40.

Wills-Karp, M., Finkelman, F.D., 2011. Innate lymphoid cells wield a double-edged sword. Nat. Immunol. 12, 1025–1027.

van Zandbergen, G., Klinger, M., Mueller, A., Dannenberg, S., Gebert, A., Solbach, W., Laskay, T., 2004. Cutting edge: neutrophil granulocyte serves as a vector for Leishmania entry into macrophages. J. Immunol. 173, 6521–6525.

Vector Arthropods and Host Pain and Itch Responses

Stephen Wikel

Quinnipiac University, Hamden, CT, United States

INTRODUCTION

Itch and pain sensations are closely related phenomena due to shared mediators, receptors, and neuronal pathways that induce protective responses of scratching or withdrawal from a painful stimulus (Liu and Ji, 2013). Significant advances are occurring in identifying pruriceptors, itch receptors, of primary sensory neurons that transduce cutaneous itch sensations and in characterizing molecular mechanisms that initiate and mediate signaling of the itch response (Potenzieri and Undem, 2011; Liu and Ji, 2013; Ringkamp and Meyer, 2014; Schmelz, 2015; Taves and Ji, 2015). Although itch and pain share some common elements, specific itch and pain pathways exist (Schmelz, 2015).

This chapter provides a review of itch and pain responses with emphasis on recent advances characterizing mediators and mechanisms involved in stimulation and responses of the diverse array of cutaneous pruriceptors. Itch and pain mechanisms are examined in the context of skin damage from vector arthropod probing, attachment, and feeding. Vector saliva molecules that inhibit itch and pain are described as well as other potential suppressors of those responses identified by analyses of salivary gland transcriptomes and proteomes.

Host awareness of arthropod probing and feeding results in behaviors that reduce both infestation and pathogen transmission. Cutaneous hypersensitivity to arthropod saliva develops as a result of repeated bites, and that reactivity can involve both immediate (type I) and delayed (type IV) hypersensitivities (Mellanby, 1946). Heightened awareness of the feeding arthropod induces host behavior that is associated with interruption of blood feeding, removal of the potential vector, and a reduced likelihood of transmission of vector-borne infectious agents (Burke et al., 2005). Acquired resistance to ticks involves cutaneous inflammation at the tick bite site and increased grooming that reduces tick burden (Hart, 2000). These phenomena are examined in the context of the mechanisms of host defenses, vector countermeasures to those defenses, and the dynamics of vector-borne pathogen transmission. Should antivector and transmission blocking vaccine strategies incorporate antigens that induce an acceptable level of cutaneous hypersensitivity to alert a host to the presence of a probing or feeding vector?

MECHANISMS OF ITCH AND PAIN

This section focuses primarily on the mediators of itch and their interactions with pruriceptors. Nociceptors, pain receptors, and the neuroanatomy of pathways responsible for peripheral and central processing of pain and itch signals overlap in part, but independent pathways arc identified for itch and pain. Pruriceptive itch is an acute response that arises in the skin; generally last from several minutes to several days; and, it is relieved by scratching that may remove the stimulus (Davidson et al., 2009; Metz et al., 2011; Schmelz, 2015). Readers interested in descriptions of the peripheral and central neuroanatomy and neurophysiology of itch and pain as well as the interactions of itch and pain pathways are referred to reviews by Davidson and Giesler (2010), Dhand and Aminoff (2014), and Schmelz (2015).

Histamine is the most commonly recognized mediator and receptor family associated with itch (McNeil and Dong, 2012). Antihistamines are the long-standing preferred treatment for pruritus (Liu and Ji, 2013). The role of histamine–receptor interactions is complex, since histamine can also elicit a pain response and antihistamines cannot always alleviate itching (McNeil and Dong, 2012). In recent years, expanded mediators and receptors that mediate itch by primary sensory neuron stimulation were described (Liu and Ji, 2013). Itch sensations are also mediated by signaling pathways involving protease-activated receptors (PARs) of polymodal C-fibers (Reddy et al., 2008); mediators derived from skin cells that bind to toll-like receptors (TLRs) on primary sensory neurons (Taves and Ji, 2015); interleukin-31 (IL-31) cytokine receptors reacting to T lymphocyte cytokine secretion (Cevikbas et al., 2014); transient receptor potential cation vanilloid channel subfamily V1/A1 (TRPV1/A1) are histaminergic (TRPV1) and responsive to tissue-derived reactive oxygen species (TRPA1) transducers of itch and pain (Nakagawa and Hiura, 2013; Ji, 2015);

serotonin binding 5-hydroxytryptamine receptors (5-HTR) (McNeil and Dong, 2012); tumor necrosis factor receptor (TNFR) binding the pro-inflammatory cytokine tumor necrosis factor (Ji, 2015); endothelin A receptor (ETAR) stimulated by keratinocyte-derived endothelian-1 peptide (Kido-Nakahara et al., 2014); and Mas-related G protein coupled receptors (Mrgprs) family transducers activated by multiple nonhistaminergic compounds (Liu and Dong, 2015). In addition to inducing pruriceptors, histamine, endothelian-1, and serotonin can stimulate pain responses (Liu and Ji, 2013). Relative to our interest in vector arthropods, itch directs the scratching response to the site of the inducing cutaneous stimulus (Schmelz, 2015). These activating mediators and neurophysiology of these receptors will be examined because they can potentially be involved in acute itch responses induced and/or modulated by probing, attachment, and blood feeding of vector arthropods, such as mosquitoes, bugs, biting flies, and ticks.

HISTAMINE

Importance of histamine as a mediator of itch is well established. In the context of this chapter, vector arthropods induce host cutaneous responses that involve histamine derived from immediate hypersensitivity reactions, cutaneous basophil hypersensitivity responses, and mediators of inflammation, such as complement-derived anaphylatoxins, and from direct damage to mast cells in the dermis. Histamine-binding molecules occur in the saliva of several ixodid tick species (Valdés, 2014).

Histamine induction of itch responses is mediated by stimulation of receptors on C-fibers that are unmyelinated and mechano-insensitive (Schmelz et al., 1997) while mechano-heat sensitive pain receptors are unresponsive or weakly responsive to histamine (Schmelz, 2015). Four G-coupled histamine receptors (H1, H2, H3, H4) are identified (Huang and Thurmond, 2008). Stimulation of histamine H1 and H4 receptors present on dorsal root ganglions induce itching sensation and scratching (Han et al., 2006; Strakhova et al., 2009). Redundancy exists in histamine sensitive itch receptors. Histamine also induces itch by stimulation of TRPV1-mediated activation of the phospholipase A2 and 12-lipoxygenase pathway (Kim et al., 2004; Shim et al., 2007).

SEROTONIN

Serotonin, 5-hydroxytrytamine, is taken up by platelets and subsequently released when they are activated as at a site of vascular injury (Cerrito et al., 1993). Human mast cells are also reported to synthesize and release serotonin (Kushnir-Sukhov et al., 2007) and application of serotonin to skin induces a flare response and itch (Hosogi et al., 2006). The 5-HT2 receptors are considered to be more dominant than 5-HT1 receptors in direct itch perception of serotonin (Yamaguchi et al., 1999; Kim et al., 2008). Histamine and serotonin are considered here because they are both amines. Another consideration is the tick saliva lipocalin of *Dermacentor reticulatus* that possesses two high affinity–binding sites that bind histamine in the one site and serotonin in the other site (Sangamnatdej et al., 2002).

PROTEASE ACTIVATED RECEPTORS

Skin PARs are G-protein coupled receptors of polymodal C fibers directly activated by cysteine proteases (Reddy et al., 2008; Johanek et al., 2008; Soh et al., 2010). In a unique activation sequence, cysteine proteases cleave the neuron receptor resulting in the N-terminal fragment binding to its receptor (Soh et al., 2010). The PAR2 receptor is considered the most effective of the PAR receptors mediating acute itch (Steinhoff et al., 2003). PAR2 receptors are involved in regulating activity of cells in the skin as well as in itch

response activation. Both keratinocytes and neutrophils possess PAR2 receptors that when activated alter expression of adhesion molecules with PAR2 receptor engagement contributing to development of cutaneous inflammation (Seeliger et al., 2003; Shpacovitch et al., 2004; Buddenkotte et al., 2005). Other proteases that induce itch responses by activating PARs include cathepsin S, mast cell–derived tryptase, and kallikreins (Costa et al., 2008, 2010). Inhibitors of PAR2 are identified in sand fly saliva that inhibit cutaneous inflammation and possibly induction of an itch response (Collin et al., 2012).

TOLL-LIKE RECEPTORS

TLRs are present on primary sensory neurons (Taves and Ji, 2015). TLRs are central recognition elements of cells of the innate immune system, including keratinocytes, of pathogen-associated molecular patterns (PAMPs) and damage-associated molecular patterns (DAMPs) of self-cells, resulting in orchestration of innate and adaptive immune responses (Liu and Ding, 2016). Itch and pain primary sensory neurons express functional TLR3, TLR4, and TLR7 that are linked to sensing cutaneous danger signals, such as ectoparasites (Taves and Ji, 2015). This cutaneous sensory role is a logical extension of the established sensing by PAMPs and DAMPs and provides a rapid alert system for eliciting the avoidance behavior of scratching. TLRs add to the diverse array of itch receptors that sense the cutaneous environment from keratinocytes (endothelin 1, nerve growth factor, RNA) and from mast cells (histamine, tryptase, serotonin).

TRPV AND TRPA

Stimulation of both itch and pain signals are mediated by TRPV1 and TRPA1 channels acting as receptors and major transducers of the itch response (Nakagawa and Hiura, 2013; Ji, 2015).

As occurs for IL-31–mediated itch with coexpression of TRPV1, histamine stimulation of itch requires TRPV1 ion channel involvement (Shim and Oh, 2008). Mediators TLR4, TLR5, and TLR7 act upon TRPV1 and/or TRPA1 receptors on pruriceptive sensory neurons, and TRPV1 and TRPVA1 are effectors for a wide variety of itch mediators (Ji, 2015). Reactive oxygen species generated by an inflammatory cellular influx induce itch by stimulating TRPA1 as an activating receptor (Liu and Ji, 2012).

ENDOTHELIN 1

Endothelin 1 is a peptide produced by keratinocytes that acts in an autocrine manner to promote DNA synthesis and regeneration of normal function of keratinocytes and contributes to cutaneous inflammatory responses (Tsuboi et al., 1994). Endothelial cells, dendritic cells, and mast cells also produce endothelin 1 that acts as a regulator of cellular functions, as a vasoconstrictor, and an inducer of pain in addition to inducing itch (Guruli et al., 2004; Metz et al., 2006; Kido-Nakahara et al., 2014). Endothelin A receptor and endothelin B receptor both bind to endothelin 1 with the endothelian A receptor engagement, the more significant regulator of both pain and itch phenomena (McQueen et al., 2007).

INTERLEUKIN-31 AND INTERLEUKIN-13

IL-31 and IL-13 are tentatively linked to the induction of itch (McNeil and Dong, 2012; Cevikbas et al., 2014). The IL-31 receptor is coexpressed on dorsal root ganglia with TRPV1 receptors (Cevikbas et al., 2014). In addition to definitively determining the sites of action of these cytokines, it remains to be determined if their itch-stimulating activity is mediated directly by these cytokines or through other molecules that are present in inflammatory responses (McNeil and Dong, 2012). Both IL-31 and IL-13 are released from Th2 CD4+ T lymphocytes and contribute broadly to what can be described as allergic inflammation (Brombacher, 2000; Maier et al., 2014). Interleukin-31 cytokine signaling plays a role in maintaining an intact cutaneous barrier (Hänel et al., 2016).

TUMOR NECROSIS FACTOR

Tumor necrosis factor alpha (TNF-α) is a proinflammatory cytokine of vast importance in the pathophysiology of numerous diseases (Vassalli, 1992). Primary sources of TNF-α are macrophages, dendritic cells, natural killer cells, and T lymphocytes. Among its biological activities, TNF-α activates neutrophils and endothelial cells. Drug therapies that modulate TNF-α are extensively used to treat conditions with an inflammatory component, including skin diseases (Campa et al., 2015; Jinesh, 2015). In light of their importance in orchestrating inflammatory and immune responses, the fact that vector arthropods inhibit proinflammatory cytokines, including TNF-α, is not surprising (Fontaine et al., 2011; Wikel, 2013; Chen et al., 2014). TNF-α acting upon TNF receptor 1 (TNFR1) mediates pain responses (Storan et al., 2015). Likewise, TNFR1 stimulation by TNF-α is a receptor for induction of itch responses (Ji, 2015). In addition to TNF-α, both IL-2 and IL-6 are reported to induce itch (Steinhoff et al., 2011).

Mas-RELATED G PROTEIN COUPLED RECEPTORS

The Mas-related G protein coupled receptors (Mrgprs) family is comprised of 10 members in humans; however, a common ligand is not identified for this receptor family (McNeil and Dong, 2012; Liu and Dong, 2015). Some members of the Mrgpr family of receptors utilize TRPA1 as a pruriceptive receptor for neuronal activation

(Sikand et al., 2011). Activators of itch response by members of the Mrgpr family include β-alanine, chloroguine, and several other chemicals (Liu and Dong, 2015). Roles of these receptors as mediators of itch responses are unclear.

VECTOR ARTHROPOD STIMULATION AND MODULATION OF HOST ITCH AND PAIN RESPONSES

Successful arthropod blood feeding involves mouthpart insertion or laceration of host skin; disruption of capillaries from which blood is obtained; and, introduction during this process of a complex saliva that modulates host defenses, including for some vectors host itch and pain responses. The following sections of this chapter examine general patterns of vector arthropod blood feeding; cutaneous responses induced by saliva of vector arthropods; actions of mediators of pain and itch on the feeding arthropod; vector saliva molecules that counteract itch and pain mediators; and how itch and pain responses impact vector-borne infectious agent transmission.

Although many vector arthropod–host relationships remain to be studied, cutaneous immediate and delayed hypersensitivity immune responses to vector saliva result in cutaneous irritation that induce behaviors that can interrupt vector feeding and reduce transmission and establishment of vector-borne infectious agents by rapidly feeding vectors, such as mosquitoes, and long-duration blood feeders, such as ticks (Hart, 2000; Sorokin, 2003; Burke et al., 2005).

BLOOD FEEDING

Arthropod blood feeding is divided into the two general patterns of solenophages that use stylet-like mouthparts to feed from venules or small hemorrhages within tissues created by their probing, and telmophages that use slashing mouthparts to create blood pools from which to feed. (Lavoipierre, 1965; Bergman, 1996). Pharmacologically active saliva is introduced into host skin by the arthropod during the processes of probing, creating the feeding site, and blood feeding. Vessel-feeding solenophages include mosquitoes, triatomids, sucking lice, and bed bugs. Blood pool–feeding telmophages include fleas, biting flies, and ticks. Due to blood feeding and attachment to the host for days, tick countermeasures to host itch and pain responses are the most extensively studied (Wikel, 2013).

HOST RESPONSES TO ARTHROPOD BITES

Cutaneous reactions to vector arthropod bites vary within a population depending on genetic differences and frequency of bites from members of different arthropod species. Development of type I immunoglobulin E (IgE)-mediated hypersensitivity reactions to stinging insects, including anaphylactic reactions that are fatal, is well characterized as to basic pathogenesis and clinical outcomes with venom allergens characterized for many species (Tankersley and Ledford, 2015). Cutaneous hypersensitivity to biting arthropods involves both immediate, type I, and delayed, type IV, immune hypersensitivity reactions that are both characterized in part by presence of a significant itch component.

In a seminal study, Mellanby (1946) reported that human volunteers develop a predictable sequential cutaneous response pattern to insect bites. Human volunteers were experimentally exposed to bites of *Aedes aegypti* several times over the course of approximately one month. Initial bite sensitization (stage I) resulted in only a delayed response at approximately 24h after the bite with a central wheal and erythematous flare accompanied by itching. Repeated *A. aegypti* bites (stage II) resulted in immediate development of a wheal at the bite followed by a delayed reaction, as just described. Additional bites (stage III) induced an immediate response with no subsequent delayed reaction. Continued exposure to bites (stage IV) induced a state of specific unresponsiveness, tolerance, to bites of *A. aegypti* with neither immediate nor delayed reaction development. An individual reacting at the first stage to one species of mosquito can be reactive at the third stage to a different species of mosquito, and the sequential pattern of skin responses appears to apply to a variety of blood-feeding arthropods (Mellanby, 1946).

Cutaneous hypersensitivity reactions accompanied by itch are induced by feeding of the sucking louse, *Pediculus humanus humanus* (Peck et al., 1943), fleas (Feingold et al., 1968), scabies mite, *Sarcoptes scabiei* (Mellanby, 1944); Triatoma bites (Moffitt et al., 2003), Diptera (black flies, deer flies, tsetse flies, stable flies, horse flies) (Klotz et al., 2009; Wilson, 2014), and ticks (Beaudouin et al., 1997). Allergens that induce host IgE are identified for several biting arthropod species (Hoffman, 2008). Cell-mediated immune responses with Th2 CD4+ T lymphocyte polarization are linked to cutaneous basophil hypersensitivity responses (Otsuka and Kabashima, 2015). Basophils have an established role in expression of acquired resistance to tick infestation (Allen, 1973; Wada et al., 2010).

ARTHROPODS AND ITCH MEDIATORS

Cutaneous hypersensitivity reactions to biting insects and ticks involve proinflammatory cytokines, mast cells, basophils, eosinophils, and mediators released from these cells (Wikel, 2013; Cantillo et al., 2014). Itch-inducing mediators released by arthropod bites include histamine, serotonin, leukotrienes, mast cell tryptase, and cysteine proteases (Wilson, 2014). Mosquito bites activate mast cells resulting in an inflammatory cell influx to the bite site and changes in draining lymph nodes that help shape the adaptive immune response (Demeure et al., 2005).

ARTHROPODS: HISTAMINE AND SEROTONIN

Histamine is arguably an important mediator of host cutaneous itch responses to arthropod bites, and serotonin is another amine that contributes as well to arthropod induced itch. Members of the *Anopheles gambiae* D7 family of saliva proteins bind both serotonin and histamine, reducing itch/pain responses, vasoconstriction, and aggregation of platelets (Calvo et al., 2006). D7 proteins occur in saliva or their transcripts are present in salivary gland transcriptomes of a variety of sand flies (Martín-Martín et al., 2013; Kato et al., 2013), midges (Lehiy and Drolet, 2014), and additional mosquito species (Girard et al., 2010; Malafronte Rdos et al., 2003). The broad distribution of D7 proteins in saliva of vector insects provides a mechanism that could potentially limit itch and pain responses during blood feeding due to histamine and serotonin release at the bite site. Reduction of cutaneous itch due to D7 proteins would likely be most effective if the bitten host is at stage 1 of the sequence of reactivity described by Mellanby (1946). During stage 1 of cutaneous hypersensitivity, an itch response does not occur until approximately 24h after the bite.

Interactions between ticks and histamine are the most extensively characterized for a blood-feeding arthropod vector–host relationship. Skin histamine content is linked to expression of acquired resistance to tick feeding (Willadsen et al., 1979; Wikel, 1982a) and a factor responsible for disruption of normal tick-feeding behavior (Paine et al., 1983). Presence of histamine blocking agents are found in ixodid tick saliva (Chinery and Ayitey-Smith, 1977; Paesen et al., 1999, 2000; Valdés, 2014). Although histamine binders are present in tick saliva that inhibit the action of histamine, a histamine releasing factor also occurs in tick saliva during the rapid engorgement phase of tick feeding (Dai et al., 2010). Adult female tick feeding is divided into an initial phase that occurs during the first week or more in which tick weigh increases approximately 10-fold followed by a rapid engorgement phase of 12–24 h during which weight increases to 100-fold over unfed weight (Kaufman, 1989).

Basophil-rich skin reactions are associated with acquired resistance that limits tick feeding resulting from repeated infestations of guinea pigs with *Dermacentor andersoni* larvae (Allen, 1973). Cattle exposed to multiple infestations with female *Ixodes holocyclus* develop bite site basophil accumulations that are not evident during exposures of previously uninfested cattle (Allen et al., 1977). Accumulation of basophils at tick-feeding sites and their association with acquired resistance suggests that basophil derived mediators, such as histamine, play a role in host rejection of feeding ticks. Similar to basophils, mast cells degranulated by tick feeding either through anaphylatoxins or IgE-mediated mechanisms contribute to bite site inflammation and induction of itch and pain responses (van Nunen, 2015).

A positive relationship exists between bovine skin histamine concentration and bovine host resistance to infestation with *Rhipicephalus (Boophilus) microplus* (Willadsen et al., 1979). Treatment with the antihistamine mepyramine maleate diminished the skin hypersensitivity response (Willadsen et al., 1979). Administration of an H1 receptor antagonist reduced expression of rabbit acquired resistance to female *Ixodes ricinus* (Brossard, 1982). Histamine content was determined for *D. andersoni* attachment sites during both an initial exposure to 100 larvae and second infestation with a similar number of larvae 7 days later on a different skin site (Wikel, 1982a,b). During the second infestation, larval engorgement was reduced by 92%. When compared to normal skin of an uninfested guinea pig, histamine content of a similar site upon which larvae were feeding, during an initial infestation, was reduced by 53% and histamine content was increased 500% during a second exposure in which acquired resistance was expressed (Wikel, 1982a). Concurrent

administration of H1 and H2 histamine receptor antagonists during a second infestation with *D. andersoni* larvae significantly blocked expression of acquired resistance (Wikel, 1982a). These findings indicated that histamine was reduced during an initial infestation, potentially by a saliva factor, and that histamine, likely attributable to the basophil hypersensitivity response at attachment sites, reduced tick feeding.

Histamine has a direct negative impact on physiology of the feeding tick. Injection of histamine under attachment sites of *R. microplus* larvae caused detachment from the skin of the bovine host when administered during the first 3 days of infestation (Kemp and Bourne, 1980). Using an in vitro membrane blood-feeding model and electrophysiological recording of tick feeding, histamine or serotonin caused diminished salivation and uptake of blood by *D. andersoni* females (Paine et al., 1983). The amount of histamine needed to disrupt feeding was 10 mM (Paine et al., 1983), which was consistent with that reported for skin of guinea pigs expressing acquired resistance to infestation (Wikel, 1982a). Therefore, it is advantageous for the feeding tick to suppress host histamine and serotonin to reduce direct physiological effects on the tick, modulate inflammation, and to reduce itch and pain responses that could stimulate host grooming and removal of the feeding tick.

Saliva of multiple ixodid tick species contain histamine binding proteins that protect the feeding tick by inhibiting the direct impact of this amine on tick physiology and also by reducing bite site histamine induction of itch responses. First tick salivary gland histamine blocking agent was described for *Rhipicephalus sanguineus sanguineus* (Chinery and Ayitey-Smith, 1977). Three histamine binding proteins occur in the saliva of *Rhipicephalus appendiculatus* and each histamine binding protein was reported to have two binding sites for the same ligand (Paesen et al., 1999). The two histamine binding sites were found to differ in their affinity for histamine (Paesen et al., 2000). One binding site had high affinity for the ligand while the second binding site had low affinity for histamine. These differences were subsequently resolved when it was observed that the ligand for the high affinity binding site is histamine and the ligand for the site with low affinity for histamine is serotonin for the histamine-binding protein from the salivary glands of *Dermacentor reticulatus* (Sangamnatdej et al., 2002). This lipocalin structure capable of binding histamine and serotonin to two active sites on the same molecule provides the tick with an effective way to block or reduce the action of these molecules to induce an inflammatory reaction at the bite and to reduce itch and pain responses, since receptors for both molecules are present on cutaneous neurons. This duality of binding also can be an advantage when blood feeding on a range of diverse host species. Lipocalins are expressed in salivary glands of *Ixodes scapularis* (McNally et al., 2012), *Ixodes persulcatus* (Konnai et al., 2011), *Ornithodoros moubata* (Díaz-Martín et al., 2011), *R. microplus* (Tirloni et al., 2014), *Amblyomma americanum* (Radulović et al., 2014), and *Haemaphysalis longicornis* (Tirloni et al., 2015). Although saliva lipocalins may contribute to inhibition of inflammation and potentially bind histamine and/or serotonin, the functional properties of lipocalins derived from these tick species remain to be defined.

The role of histamine at the tick–host interface becomes more complex with identification of a histamine-releasing factor secreted in the saliva of *Dermacentor variabilis* (Mulenga et al., 2003) and *I. scapularis* (Dai et al., 2010). Expression of a histamine-releasing factor late in the blood meal acquisition process is advantageous for increasing blood flow to the bite site during the rapid engorgement phase. Histamine-binding factors present early in the multiple day feeding cycle reduce inflammation and itch at the bite site while histamine-releasing factor expressed late during engorgement and close to time of detachment from the host facilitates achieving complete engorgement.

Saliva of the triatomid *Rhodnius prolixus* contains a lipocalin nitrophorin that transports nitric oxide to the bite site, where it acts as a vasodilator, and upon release of nitric oxide binds histamine (Ribeiro and Walker, 1994; Montfort et al., 2000; Andersen et al., 2005). Lipocalins are identified in salivary gland transcriptome of *Triatoma rubida* that have 40% identity of primary sequence with *Triatoma protracta lipocalins*; however, functional roles of these lipocalins are not known, but they are not acting as nitrophorins (Ribeiro et al., 2012).

ARTHROPODS: PROTEASES

Itch responses are also stimulated in the absence of histamine by cysteine proteases (Soh et al., 2010) as well as other proteases that activate PARs, including cathepsin S, mast cell–derived tryptase, and kallikreins (Costa et al., 2008, 2010). Vector arthropod salivary glands contain inhibitors of protease-activated receptors (PARs) that diminish both inflammatory and itch responses. *Lutzomyia longipalpis* salivary gland contains a novel protein, Lufaxin that prevents activation of PAR2 receptors by activated coagulation factor Xa (FXa), thus inhibiting edema associated with inflammation and thrombus formation (Collin et al., 2012). Ability of Lufaxin to inhibit pain and itch responses is unknown. Saliva of the midge, *Culicoides sonorensis* contains protease inhibitors (Lehiy and Drolet, 2014).

PAR-mediated itch is induced by cysteine proteases (Reddy et al., 2008; Soh et al., 2010) and by mast cell–derived tryptase, trypsin, cathepsin S, and kallikreins that are endogenous proteases (Costa et al., 2008, 2010). Vector salivary glands produce cystatins that inhibit cysteine proteases (Ribeiro et al., 2007; Schwarz et al., 2012) and serpins that are a large superfamily of broadly acting protease inhibitors (Law et al., 2006; Tirloni et al., 2016).

Cystatins identified in tick salivary secretions are linked largely to modulation of host immune defenses (Schwarz et al., 2012). *Ixodes scapularis* cystatin, sialostatin L, inhibits caspase-1–mediated inflammation, blocks dendritic cell interferon-β responses, and, suppresses proliferation of CD8+ T lymphocytes (Kotsyfakis et al., 2006; Chen et al., 2014; Lieskovská et al., 2015). Salivary gland cystatins occur in *R. sanguineus* (Anatriello et al., 2010), *A. americanum* (Karim et al., 2005), and *Amblyomma variegatum* (Ribeiro et al., 2011). Although their presence suggests a potential function, ability of these salivary cystatins to modulate host itch responses remains to be determined.

Serpins are a large superfamily of protease inhibitors with broad functional diversity including inhibition of neutrophil elastase, cathepsin (G, L and K), thrombin, trypsin, and C1 esterase (Law et al., 2006). *Aedes aegypti* female salivary gland serpin is a coagulation factor Xa inhibitor (Stark and James, 1998), and multiple serpin transcripts occur in the *A. aegypti* female salivary gland transcriptome (Ribeiro et al., 2007). Salivary glands of numerous tick species produce serpins. *R. microplus* salivary gland produces three serpins with apparent different function profiles that include inhibiting cathepsin G–activated platelet aggregation (Tirloni et al., 2016). An *I. ricinus* salivary serpin reduces IL-6–dependent differentiation of Th17 CD4+ T lymphocytes (Páleníková et al., 2015), resulting in suppression of innate and adaptive immune response components. An *A. americanum* salivary serpin expressed early is speculated to be important in establishing a successful feeding lesion (Chalaire et al., 2011). Proteases present in vector saliva have established roles in modulating hemostasis, inflammation, and innate and adaptive immunity. Examining vector arthropod saliva protease suppression of pain and itch responses is a fertile area of investigation with practical potential for contributing to novel control strategies.

Salivary kininases are present in both *I. scapularis* (Ribeiro and Mather, 1998) and *R. microplus* (Bastiani et al., 2002), which could con-

tribute to suppression of pain and itch responses. *I. scapularis* kininase is a metallo dipeptidyl carboxypeptidase with properties similar to angiotensin converting enzymes common to many eukaryotic species (Ribeiro and Mather, 1988). *R. microplus* salivary kininase is a thiol-activated metalloendopeptidase that does not possess angiotensin-converting enzyme activity (Bastiani et al., 2002).

ARTHROPODS: ADDITIONAL ITCH AND PAIN RECEPTORS

TNF-α acting upon TNFR1 mediates both pain (Storan et al., 2015) and itch responses (Ji, 2015). TNF-α is an important proinflammatory cytokine expressed early in the innate immune/inflammatory response to injury and infection. Therefore, it is not surprising that salivary glands of vector arthropods have molecules that downregulate the actions of TNF-α and other proinflammatory cytokines. Multiple ixodid tick species produce salivary inhibitors of TNF-α and IL-1 (Wikel, 2013; Kotál et al., 2015). *Aedes aegypti* salivary gland extract downregulated expression of TNF-α along with Th1 CD4+ T lymphocyte cytokines (Wasserman et al., 2004). Salivary glands of the tsetse fly, *Glossina morsitans morsitans*, contains a peptide inhibitor of TNF-α (Bai et al., 2015). *Simulium bannaense* salivary gland antimicrobial peptide inhibits production of TNF-α, IL-1, and IL-6 (Wu et al., 2015). *L. longipalpis* salivary gland maxadilan downregulates TNF-α, IL-12, and nitric oxide while upregulating Th2 cytokines (Brodie et al., 2007). Whether or not vector arthropod downregulation of TNF-α contributes to suppression of pain and itch responses remains to be determined.

Vector arthropod interactions with other itch and pain receptors also remain to be defined; however, salivary gland molecules are identified that could potentially modulate responses that could alert a host to probing and feeding by an arthropod. Two families of peptides occurring in salivary glands of *Hyalomma asiaticum asiaticum* suppress TNF-α as well as acting as scavengers of free radicals (Wu et al., 2010). Although TRPA1 is responsive to reactive oxygen species (Nakagawa and Hiura, 2013; Ji, 2015), the ability of the *Hyalomma* peptide families to modulate pain and or itch signaling is unknown. *I. scapularis* salivary protein Salp 25D reduces inflammation at the bite site by detoxifying reactive oxygen species (Narasimhan et al., 2007).

The description of new itch and pain receptors (Liu and Ji, 2013) expands opportunities for research on the ability of vector arthropod saliva molecules to manipulate stimulation of itch and pain. Development of novel control strategies that incorporate induction of acceptable levels of itch responses should be considered as potential components of immunological strategies to reduce vector feeding and vector-targeted infectious agent transmission blocking vaccines.

IMPACT OF ITCH AND PAIN ON VECTOR FEEDING AND PATHOGEN TRANSMISSION

Development of acquired resistance to vector feeding is most thoroughly examined for ixodid ticks infesting laboratory hosts and cattle; however, host resistance to myiasis, midges, biting flies, mosquitoes, lice, fleas, and mites also occurs (Allen, 1989, 1994; Wikel, 1982b, 1996; Wada et al., 2010). Acquired resistance reduces feeding success and infestation burden of the host due to development of immune responses resulting from prior infestation. Cutaneous hypersensitivities can be allergic, type I, or cell-mediated, type IV, reactions that are stronger than the ability of vector arthropod saliva to block host cell–released mediators that negatively impact vector feeding. Host mediators released by hypersensitivity reactions can act directly upon the vector or induce host grooming and avoidance behaviors that limit infestation. Host immune responses to vector saliva

molecules can reach levels that eventually neutralize the saliva biological activities allowing for itch responses to be expressed that result in host awareness of infestation and initiation of grooming behavior.

The role of grooming behavior in reducing ectoparasite infestation was confirmed when restricting self-grooming resulted in increased numbers of *R. microplus* infesting cattle (Snowball, 1956; Riek, 1962; Bennett, 1969). Self-grooming behavior limits tick infestations of wild bovids (Hart, 2000). Infestation-induced itching is linked to host grooming behavior that limits sucking louse, *Linognathus vituli*, infestation of calves (Weeks et al., 1995), and feline infestations with fleas (Eckstein and Hart, 2000).

Bovine resistance to tick infestation has a genetic basis with cattle of *Bos indicus* genetic composition more highly resistant to ticks than cattle of *Bos taurus* genetic background (Jonsson et al., 2014). Tick resistance of *B. indicus* breeds is recently linked in part to greater density and longer tongue papillae that allow more efficient grooming to remove ticks than by the tongue structure of less resistant *B. taurus* breeds (Veríssimo et al., 2015). When animals are restricted from grooming, acquired tick resistance is still expressed as fewer feeding ticks, reduced engorgement weight, impaired molting, reduced oviposition, and abnormal blood meals due to feeding on numerous leukocytes at bite site inflammatory infiltrates (Allen, 1989, 1994; Wikel, 1982a,b, 1996; Wikel and Allen, 1976; Schoeler and Wikel, 2001).

Humans develop itch responses to ectoparasitic arthropods (Klotz et al., 2009). *Sarcoptes scabiei* is well recognized as a cause of itch in individuals with primary infestations and heightened itch upon reinfestation (Mellanby, 1944; Mounsey et al., 2013). Salivary secretions of multiple mosquito species stimulate local allergic, type I, hypersensitivity (Peng and Simons, 2007; Cantillo et al., 2014). Individuals residing on Heron Island, Australia during seabird breeding season reported development of pruritis to bites of the argasid tick *Ornithodoros capensis* that were characterized as cell-mediated, type IV, hypersensitivity reactions (Humphery-Smith et al., 1991). Development of pruritic responses are reported by people infested with larval ticks that remain attached to the skin for days (Lee et al., 2011).

Repeated experimental exposures to bites of *I. scapularis* nymphs do not induce expression of acquired resistance to infestation (Schoeler et al., 1999). Tick researchers have anecdotally noted that frequent tick bite acquired in the field induces cutaneous hypersensitivity that alerts them to a tick attaching to the skin. The question of whether or not development of cutaneous hypersensitivity induced by repeated exposures to *I. scapularis* bites induced cutaneous hypersensitivity and thus a lower risk for acquiring Lyme disease was determined by an analysis of multiple years of data from twice yearly serosurveys and questionnaire responses of residents of a Lyme disease endemic area (Burke et al., 2005). Individuals reporting itch associated with tick bite had a statistically significant lower likelihood of infection with *Borrelia burgdorferi*. The value of a pruritic response to *I. scapularis* bites in reducing the likelihood of acquiring Lyme disease is linked to the requirement for the tick to be attached at least 24 h with maximum transmission of *B. burgdorferi* between 48 and 72 h after attachment (des Vignes et al., 2001). However, it is important to note that *Ehrlichia phagocytophilum* was frequently transmitted within 24 h of *I. scapularis* nymph attachment (des Vignes et al., 2001). Powassan virus is transmitted by *I. scapularis* nymphs as rapidly as 15 min after attachment to the host (Ebel and Kramer, 2004). Itch induced by tick probing and attachment might not be as effective in preventing Powassan virus infection as it might be for disrupting the feeding tick and reducing transmission of *B. burgdorferi*.

Histologic examination of *I. scapularis* bite sites on humans and experimentally infested mice provide clues to hypersensitivity to tick infestation (Krause et al., 2009). Experimental first murine infestation was characterized by an inflammatory cellular influx surrounding the bite site but not actually at the tick mouthparts. Repeated infestations were histologically different with an intense accumulation of inflammatory cells at the tick mouthparts. Tick-bite site biopsies of *I. scapularis*–infested people were histologically similar to those observed for experimentally infested mice. The hypothesis put forward is that the inflammatory cellular influx at tick attachment sites on repeated exposed individuals create a hostile environment for tick-transmitted infectious agents and contribute to cutaneous hypersensitivity (Krause et al., 2009). Studies linked acquired resistance to tick feeding to cutaneous basophil-rich influxes at bite sites of repeatedly infested hosts (Allen, 1973, 1989; Wikel, 1982a,b; Wada et al., 2010).

Mosquito bite–induced itch is reported to be linked to reduced transmission of malaria (Kumar, 1996) and Ross River virus (Dugdale, 2002). Repeated host tick infestation result in development of acquired resistance limits both subsequent tick feeding and transmission of infectious agents. The first report of transmission blocking immunity described acquired resistance induced by infestation of rabbits with pathogen-free adult *D. andersoni* that resulted in significant protection from acquiring infection with highly virulent type A *Francisella tularensis* when infested with infected *D. andersoni* nymphs (Bell et al., 1979). Infestation of tick resistant guinea pigs with Thogoto virus–infected *R. appendiculatus* reduced transmission of virus to cofeeding uninfected ticks, indicating that tick resistance created a host environment unfavorable for the virus (Jones and Nuttall, 1990). Infestation of guinea pigs with uninfected *I. scapularis* nymphs induced acquired resistance that reduced transmission of *B. burgdorferi* by infestation with infected nymphs (Nazario et al., 1998). In the absence of acquired resistance, repeated infestations with uninfected *I. scapularis* nymphs induced resistance to *I. scapularis* nymph transmission of *B. burgdorferi* (Wikel et al., 1997).

WHAT ARE THE NEXT STEPS?

Recent advances in understanding receptors and signaling pathways of itch and pain responses have increased the array of molecules that stimulate these responses and linked itch to multiple components of inflammatory and immune responses. The use of next-generation sequencing and proteomics is providing more complete analyses of salivary gland bioactive molecules and provides a database for selection and analysis of protection inducing immunogens and identification of molecules that counteract host itch and pain defenses. Interest is increasing in development of vaccines that block vector feeding and/or inhibit vector-borne infectious agent transmission as standalone vaccines or components of infectious agent-specific vaccines. Incorporation of immunogens that enhance cutaneous irritation, without significant pathology, could be a valuable component in alerting human or veterinary hosts of vector probing and attempted feeding.

References

Allen, J.R., 1973. Tick resistance: basophils in skin reactions of resistant guinea pigs. Int. J. Parasitol. 3, 195–200.

Allen, J.R., 1989. Immunology of interactions between ticks and laboratory animals. Exp. Appl. Acarol. 7, 5–13.

Allen, J.R., 1994. Host resistance to ectoparasites. Rev. Sci. Tech. 13, 1287–1303.

Allen, J.R., Doube, B.M., Kemp, D.H., 1977. Histology of bovine skin reactions to Ixodes holocyclus Neumann. Can. J. Comp. Med. 41, 26–35.

Anatriello, E., Ribeiro, J.M., de Miranda-Santos, I.K., Brandão, L.G., Anderson, J.M., Valenzuela, J.G., Maruyama, S.R., Silva, J.S., Ferreira, B.R., 2010. An insight into the sialotranscriptome of the brown dog tick, Rhipicephalus sanguineus. BMC Genomics 11, 450.

Andersen, J.F., Gudderra, N.P., Francischetti, I.M., Ribeiro, J.M., 2005. The role of salivary lipocalins in blood feeding by Rhodnius prolixus. Arch. Insect Biochem. Physiol. 58, 97–105.

Bai, X., Yao, H., Du, C., Chen, Y., Lai, R., Rong, M., 2015. An immunoregulatory peptide from tsetse fly salivary glands of Glossina morsitans morsitans. Biochimie 118, 123–128.

Bastiani, M., Hillebrand, S., Horn, F., Kist, T.B., Guimarães, J.A., Termignoni, C., 2002. Cattle tick Boophilus microplus salivary gland contains a thiol-activated metalloendopeptidase displaying kininase activity. Insect Biochem. Mol. Biol. 32, 1439–1446.

Beaudouin, M.D., Kanny, G., Guerin, B., Guerin, L., Plenat, F., Moneret-Vautrin, D.A., 1997. Unusual manifestations of hypersensitivity after a tick bite: report of two cases. Ann. Allergy Asthma Immunol. 79, 43–46.

Bell, J.F., Stewart, S.J., Wikel, S.K., 1979. Resistance to tick-borne Francisella tularensis by tick-sensitized rabbits: allergic klendusity. Am. J. Trop. Med. Hyg. 28, 876–880.

Bennett, G.F., 1969. Boophilus microplus (Acarina: Ixodidae): experimental infestations on cattle restrained from grooming. Exp. Parasitol. 26, 323–328.

Bergman, D.K., 1996. Mouthparts and feeding mechanisms of haematophagous arthropods. In: Wikel, S.K. (Ed.), The Immunology of Host-ectoparasitic Arthropod Relationships. CAB International, Wallingford, UK, pp. 30–61.

Brodie, T.M., Smith, M.C., Morris, R.V., Titus, R.G., 2007. Immunomodulatory effects of the Lutzomyia longipalpis salivary gland protein maxadilan on mouse macrophages. Infect. Immun. 75, 2359–2365.

Brombacher, F., 2000. The role of interleukin-13 in infectious diseases and allergy. Bioessays 22, 646–656.

Brossard, M., 1982. Rabbits infested with adult Ixodes ricinus L.: effects of mepyramine on acquired resistance. Experientia 38, 702–704.

Buddenkotte, J., Stroh, C., Engels, I.H., Moormann, C., Shpacovitch, V.M., Seeliger, S., Vergnolle, N., Vestweber, D., Luger, T.A., Schulze-Osthoff, K., Steinhoff, M., 2005. Agonists of proteinase-activated receptor-2 stimulate upregulation of intercellular cell adhesion molecule-1 in primary human keratinocytes via activation of NF-kappa B. J. Invest. Dermatol. 124, 38–45.

Burke, G., Wikel, S.K., Spielman, A., Telford, S.R., McKay, K., Krause, P.J., Tick-Borne Infection Study Group, 2005. Hypersensitivity to ticks and Lyme disease risk. Emerg. Infect. Dis. 11, 36–41.

Calvo, E., Mans, B.J., Andersen, J.F., Ribeiro, J.M., 2006. Function and evolution of a mosquito salivary protein family. J. Biol. Chem. 281, 1935–1942.

Campa, M., Ryan, C., Menter, A., 2015. An overview of developing TNF-α targeted therapy for the treatment of psoriasis. Expert. Opin. Investig. Drugs. 24, 1343–1354.

Cantillo, J.F., Fernández-Caldas, E., Puerta, L., 2014. Immunological aspects of the immune response induced by mosquito allergens. Int. Arch. Allergy Immunol. 165, 271–282.

Cerrito, F., Lazzaro, M.P., Gaudio, E., Arminio, P., Aloisi, G., 1993. 5HT2-receptors and serotonin release: their role in human platelet aggregation. Life. Sci. 53, 209–215.

Cevikbas, F., Wang, X., Akiyama, T., Kempkes, C., Savinko, T., Antal, A., Kukova, G., Buhl, T., Ikoma, A., Buddenkotte, J., Soumelis, V., Feld, M., Alenius, H., Dillon, S.R., Carstens, E., Homey, B., Basbaum, A., Steinhoff, M., 2014. A sensory neuron-expressed IL-31 receptor mediates T helper cell-dependent itch: involvement of TRPV1 and TRPA1. J. Allergy Clin. Immunol. 133, 448–460.

Chalaire, K.C., Kim, T.K., Garcia-Rodriguez, H., Mulenga, A., 2011. *Amblyomma americanum* (L.) (Acari: Ixodidae) tick salivary gland serine protease inhibitor (serpin) 6 is secreted into tick saliva during tick feeding. J. Exp. Biol. 214, 665–673.

Chen, G., Wang, X., Severo, M.S., Sakhon, O.S., Sohail, M., Brown, L.J., Sircar, M., Snyder, G.A., Sundberg, E.J., Ulland, T.K., Olivier, A.K., Andersen, J.F., Zhou, Y., Shi, G.P., Sutterwala, F.S., Kotsyfakis, M., Pedra, J.H., 2014. The tick salivary protein sialostatin L2 inhibits caspase-1-mediated inflammation during *Anaplasma phagocyto-philum* infection. Infect. Immun. 82, 2553–2564.

Chinery, W.A., Ayitey-Smith, E., 1977. Histamine blocking agent in the salivary gland homogenate of the tick, *Rhipicephalus sanguineus sanguineus*. Nature 265, 366–367.

Collin, N., Assumpção, T.C., Mizurini, D.M., Gilmore, D.C., Dutra-Oliveira, A., Kotsyfakis, M., Sá-Nunes, A., Teixeira, C., Ribeiro, J.M., Monteiro, R.Q., Valenzuela, J.G., Francischetti, I.M., 2012. Lufaxin, a novel factor Xa inhibitor from the salivary gland of the sand fly *Lutzomyia longipalpis* blocks protease-activated receptor 2 activation and inhibits inflammation and thrombosis in vivo. Arterioscler. Thromb. Vasc. Biol. 32, 2185–2198.

Costa, R., Manjavachi, M.N., Motta, E.M., Marotta, D.M., Juliano, L., Torres, H.A., Pesquero, J.B., Calixto, J.B., 2010. The role of kinin B1 and B2 receptors in the scratching behaviour induced by proteinase-activated receptor-2 agonists in mice. Br. J. Pharmacol. 159, 888–897.

Costa, R., Marotta, D.M., Manjavachi, M.N., Fernandes, E.S., Lima-Garcia, J.F., Paszcuk, A.F., Quintão, N.L., Juliano, L., Brain, S.D., Calixto, J.B., 2008. Evidence for the role of neurogenic inflammation components in trypsin-elic-ited scratching behaviour in mice. Br. J. Pharmacol. 154, 1094–1103.

Dai, J., Narasimhan, S., Zhang, L., Liu, L., Wang, P., Fikrig, E., 2010. Tick histamine release factor is critical for *Ixodes scapularis* engorgement and transmission of the Lyme disease agent. PLoS Pathog. 6, 1001205.

Davidson, S., Giesler, G.J., 2010. The multiple pathways for itch and their interactions with pain. Trends Neurosci. 33, 550–558.

Davidson, S., Zhang, X., Khasabov, S.G., Simone, D.A., Giesler Jr., G.J., 2009. Relief of itch by scratching: state-dependent inhibition of primate spinothalamic tract neurons. Nat. Neurosci. 12, 544–546.

Demeure, C.E., Brahimi, K., Hacini, F., Marchand, F., Péronet, R., Huerre, M., St-Mezard, P., Nicolas, J.F., Brey, P., Delespesse, G., Mécheri, S., 2005. *Anopheles* mosquito bites activate cutaneous mast cells leading to a local inflammatory response and lymph node hyperplasia. J. Immunol. 174, 3932–3940.

des Vignes, F., Piesman, J., Heffernan, R., Schulze, T.L., Stafford 3rd, K.C., Fish, D., 2001. Effect of tick removal on transmission of *Borrelia burgdorferi* and *Ehrlichia phago-cytophila* by *Ixodes scapularis* nymphs. J. Infect. Dis. 183, 773–778.

Dhand, A., Aminoff, M.J., 2014. The neurology of itch. Brain 137, 313–322.

Díaz-Martín, V., Manzano-Román, R., Siles-Lucas, M., Oleaga, A., Pérez-Sánchez, R., 2011. Cloning, charac-terization and diagnostic performance of the salivary lipocalin protein TSGP1 from *Ornithodoros moubata*. Vet. Parasitol. 178, 163–172.

Dugdale, A.E., 2002. Itching bites may limit Ross River virus infection. Med. J. Aust. 177, 399–400.

Ebel, G.D., Kramer, L.D., 2004. Short report: duration of tick attachment required for transmission of powassan virus by deer ticks. Am. J. Trop. Med. Hyg. 71, 268–271.

Eckstein, R.A., Hart, B.L., 2000. Grooming and control of fleas in cats. Appl. Anim. Behav. Sci. 68, 141–150.

Feingold, B.F., Benjamini, E., Michaeli, D., 1968. The aller-gic response to insect bites. Annu. Rev. Entomol. 13, 137–158.

Fontaine, A., Diouf, I., Bakkali, N., Missé, D., Pagès, F., Fusai, T., Rogier, C., Almeras, L., 2011. Implication of haema-tophagous arthropod salivary proteins in host-vector interactions. Parasit. Vectors 4, 187.

Girard, Y.A., Mayhew, G.F., Fuchs, J.F., Li, H., Schneider, B.S., McGee, C.E., Rocheleau, T.A., Helmy, H., Christensen, B.M., Higgs, S., Bartholomay, L.C., 2010. Transcriptome changes in *Culex quinquefasciatus* (Diptera: Culicidae) salivary glands during West Nile virus infection. J. Med. Entomol. 4, 421–435.

Guruli, G., Pflug, B.R., Pecher, S., Makarenkova, V., Shurin, M.R., Nelson, J.B., 2004. Function and survival of den-dritic cells depend on endothelin-1 and endothelin recep-tor autocrine loops. Blood 104, 2107–2115.

Han, S.K., Mancino, V., Simon, M.I., 2006. Phospholipase Cbeta 3 mediates the scratching response activated by the histamine H1 receptor on C-fiber nociceptive neu-rons. Neuron 52, 691–703.

Hänel, K.H., Pfaff, C.M., Cornelissen, C., Amann, P.M., Marquardt, Y., Czaja, K., Kim, A., Lüscher, B., Baron, J.M., 2016. Control of the physical and antimicrobial skin bar-rier by an IL-31-IL-1 signaling network. J. Immunol. 196, 3233–3244.

Hart, B.L., 2000. Role of grooming in biological control of ticks. Ann. N.Y. Acad. Sci. 916, 565–569.

Hoffman, D.R., 2008. Biting insect allergens. Clin. Allergy Immunol. 21, 251–260.

Hosogi, M., Schmelz, M., Miyachi, Y., Ikoma, A., 2006. Bradykinin is a potent pruritogen in atopic dermatitis: a switch from pain to itch. Pain 126, 16–23.

Huang, J.F., Thurmond, R.L., 2008. The new biology of histamine receptors. Curr. Allergy Asthma Rep. 8, 21–27.

Humphery-Smith, I., Thong, Y.H., Moorhouse, D., Creevey, C., Gauci, M., Stone, B., 1991. Reactions to argasid tick bites by island residents on the Great Barrier Reef. Med. J. Aust. 155, 181–186.

Ji, R.R., 2015. Neuroimmune interactions in itch: Do chronic itch, chronic pain, and chronic cough share similar mechanisms? Pulm. Pharmacol. Ther. 35, 81–86.

Jinesh, S., 2015. Pharmaceutical aspects of anti-inflammatory TNF-blocking drugs. Inflammopharmacology 23, 71–77.

Johanek, L.M., Meyer, R.A., Friedman, R.M., Greenquist, K.W., Shim, B., Borzan, J., Hartke, T., LaMotte, R.H., Ringkamp, M., 2008. A role for polymodal C-fiber afferents in nonhistaminergic itch. J. Neurosci. 28, 7659–7669.

Jones, L.D., Nuttall, P.A., 1990. The effect of host resistance to tick infestation on the transmission of Thogoto virus by ticks. J. Gen. Virol. 71, 1039–1043.

Jonsson, N.N., Piper, E.K., Constantinoiu, C.C., 2014. Host resistance in cattle to infestation with the cattle tick *Rhipicephalus microplus*. Parasite. Immunol. 36, 553–559.

Karim, S., Miller, N.J., Valenzuela, J., Sauer, J.R., Mather, T.N., 2005. RNAi-mediated gene silencing to assess the role of synaptobrevin and cystatin in tick blood feeding. Biochem. Biophys. Res. Commun. 334, 1336–1342.

Kato, H., Jochim, R.C., Gomez, E.A., Uezato, H., Mimori, T., Korenaga, M., Sakurai, T., Katakura, K., Valenzuela, J.G., Hashiguchi, Y., 2013. Analysis of salivary gland transcripts of the sand fly *Lutzomyia ayacuchensis*, a vector of Andean-type cutaneous leishmaniasis. Infect. Genet. Evol. 13, 56–66.

Kaufman, W.R., 1989. Tick-host interaction: a synthesis of current concepts. Parasitol. Today. 5, 47–56.

Kemp, D.H., Bourne, A., 1980. *Boophilus microplus*: the effect of histamine on the attachment of cattle-tick larvae-studies in vivo and in vitro. Parasitology 80, 487–496.

Kido-Nakahara, M., Buddenkotte, J., Kempkes, C., Ikoma, A., Cevikbas, F., Akiyama, T., Nunes, F., Seeliger, S., Hasdemir, B., Mess, C., Buhl, T., Sulk, M., Müller, F.U., Metze, D., Bunnett, N.W., Bhargava, A., Carstens, E., Furue, M., Steinhoff, M., 2014. Neural peptidase endothelin-converting enzyme 1 regulates endothelin 1-induced pruritus. J. Clin. Invest. 124, 2683–2695.

Kim, B.M., Lee, S.H., Shim, W.S., Oh, U., 2004. Histamine-induced Ca2+ influx via the PLA(2)/lipoxygenase/TRPV1 pathway in rat sensory neurons. Neurosci. Lett. 361, 159–162.

Kim, D.K., Kim, H.J., Kim, H., Koh, J.Y., Kim, K.M., Noh, M.S., Kim, J.J., Lee, C.H., 2008. Involvement of serotonin receptors 5-HT1 and 5-HT2 in 12(S)-HPETE-induced scratching in mice. Eur. J. Pharmacol. 579, 390–394.

Klotz, J.H., Pinnas, J.L., Klotz, S.A., Schmidt, J.O., 2009. Anaphylactic reactions to arthropod bites and stings. Am. Entomol. 55, 134–139.

Konnai, S., Nishikado, H., Yamada, S., Imamura, S., Ito, T., Onuma, M., Murata, S., Ohashi, K., 2011. Molecular identification and expression analysis of lipocalins from blood feeding taiga tick, *Ixodes persulcatus* Schulze. Exp. Parasitol. 127, 467–474.

Kotál, J., Langhansová, H., Lieskovská, J., Andersen, J.F., Francischetti, I.M., Chavakis, T., Kopecký, J., Pedra, J.H., Kotsyfakis, M., Chmelař, J., 2015. Modulation of host immunity by tick saliva. J. Proteomics 128, 58–68.

Kotsyfakis, M., Sá-Nunes, A., Francischetti, I.M., Mather, T.N., Andersen, J.F., Ribeiro, J.M., 2006. Antiinflammatory and immunosuppressive activity of sialostatin L, a salivary cystatin from the tick *Ixodes scapularis*. J. Biol. Chem. 281, 26298–26307.

Krause, P.J., Grant-Kels, J.M., Tahan, S.R., Dardick, K.R., Alarcon-Chaidez, F., Bouchard, K., Visini, C., Deriso, C., Foppa, I.M., Wikel, S., 2009. Dermatologic changes induced by repeated *Ixodes scapularis* bites and implications for prevention of tick-borne infection. Vector Borne Zoonotic Dis. 9, 603–610.

Kumar, A., 1996. Itching and immunity. Lancet 348, 1383.

Kushnir-Sukhov, N.M., Brown, J.M., Wu, Y., Kirshenbaum, A., Metcalfe, D.D., 2007. Human mast cells are capable of serotonin synthesis and release. J. Allergy. Clin. Immunol. 119, 498–499.

Lavoipierre, M.M.J., 1965. Feeding mechanisms of blood-sucking arthropods. Nature 208, 302–303 London.

Law, R.H., Zhang, Q., McGowan, S., Buckle, A.M., Silverman, G.A., Wong, W., Rosado, C.J., Langendorf, C.G., Pike, R.N., Bird, P.I., Whisstock, J.C., 2006. An overview of the serpin superfamily. Genome. Biol. 7, 216.

Lee, Y.B., Jun, J.B., Kim, J.Y., Cho, B.K., Park, H.J., 2011. Multiple bites from the larvae of *Haemaphysalis longicornis*. Arch. Dermatol. 147, 1333–1334.

Lehiy, C.J., Drolet, B.S., 2014. The salivary secretome of the biting midge, *Culicoides sonorensis*. PeerJ 2, e426.

Lieskovská, J., Páleníková, J., Širmarová, J., Elsterová, J., Kotsyfakis, M., Campos Chagas, A., Calvo, E., Růžek, D., Kopecký, J., 2015. Tick salivary cystatin sialostatin L2 suppresses IFN responses in mouse dendritic cells. Parasite. Immunol. 37, 70–78.

Liu, L., Ji, R.-R., 2013. New insights into the mechanisms of itch: are pain and itch controlled by distinct mechanisms? Eur. J. Physiol. 465, 1671–1685.

Liu, Q., Ding, J.L., 2016. The molecular mechanisms of TLR-signaling cooperation in cytokine regulation. Immunol. Cell. Biol. http://dx.doi.org/10.1038/icb.2016.18.

Liu, Q., Dong, X., 2015. The role of the Mrgpr receptor family in itch. In: Cowan, A., Yosipovitch, G. (Eds.), Pharmacology of Itch, Handbook of Experimental Pharmacology 226. Springer-Verlag, Berlin, pp. 71–88.

Liu, T., Ji, R.-R., 2012. Oxidative stress induces itch via activation of transient receptor potential subtype ankyrin 1 in mice. Neurosci. Bull. 28, 145–154.

Maier, E., Werner, D., Duschl, A., Bohle, B., Horejs-Hoeck, J., 2014. Human Th2 but not Th9 cells release IL-31 in a STAT6/NF-κB-dependent way. J. Immunol. 193, 645–654.

Malafronte Rdos, S., Calvo, E., James, A.A., Marinotti, O., 2003. The major salivary gland antigens of *Culex quinquefasciatus* are D7-related proteins. Insect Biochem. Mol. Biol. 33, 63–71.

Martín-Martín, I., Molina, R., Jiménez, M., 2013. Molecular and immunogenic properties of apyrase SP01B and D7-related SP04 recombinant salivary proteins of *Phlebotomus perniciosus* from Madrid, Spain. BioMed Res. Int. 2013, 526069.

McNally, K.L., Mitzel, D.N., Anderson, J.M., Ribeiro, J.M., Valenzuela, J.G., Myers, T.G., Godinez, A., Wolfinbarger, J.B., Best, S.M., Bloom, M.E., 2012. Differential salivary gland transcript expression profile in *Ixodes scapularis* nymphs upon feeding or flavivirus infection. Ticks Tick Borne Dis. 3, 18–26.

McNeil, B., Dong, X., 2012. Peripheral mechanisms of itch. Neurosci. Bull. 28, 100–110.

McQueen, D.S., Noble, M.A., Bond, S.M., 2007. Endothelin-1 activates ETA receptors to cause reflex scratching in BALB/c mice. Br. J. Pharmacol. 151, 278–284.

Mellanby, K., 1944. The development of symptoms, parasitic infection and immunity in human scabies. Parasitology 35, 197–206.

Mellanby, K., 1946. Man's reaction to mosquito bites. Nature 158, 554.

Metz, M., Grundmann, S., Stander, S., 2011. Pruritus: an overview of current concepts. Vet. Dermatol. 22, 121–131.

Metz, M., Lammel, V., Gibbs, B.F., Maurer, M., 2006. Inflammatory murine skin responses to UV-B light are partially dependent on endothelin-1 and mast cells. Am. J. Pathol. 169, 815–822.

Moffitt, J.E., Venarske, D., Goddard, J., Yates, A.B., deShazo, R.D., 2003. Allergic reactions to *Triatoma* bites. Ann. Allergy Asthma Immunol. 91, 122–128.

Montfort, W.R., Weichsel, A., Andersen, J.F., 2000. Nitrophorins and related antihemostatic lipocalins from *Rhodnius prolixus* and other blood-sucking arthropods. Biochim. Biophys. Acta 1482, 110–118.

Mounsey, K.E., McCarthy, J.S., Walton, S.F., 2013. Scratching the itch: new tools to advance understanding of scabies. Trends. Parasitol. 29, 35–42.

Mulenga, A., Macaluso, K.R., Simser, J.A., Azad, A.F., 2003. The American dog tick, *Dermacentor variabilis*, encodes a functional histamine release factor homolog. Insect Biochem. Mol. Biol. 33, 911–919.

Nakagawa, H., Hiura, A., 2013. Four possible itching pathways related to the TRPV1 channel, histamine, PAR-2 and serotonin. Malays. J. Med. Sci. 20, 5–12.

Narasimhan, S., Sukumaran, B., Bozdogan, U., Thomas, V., Liang, X., DePonte, K., Marcantonio, N., Koski, R.A., Anderson, J.F., Kantor, F., Fikrig, E., 2007. A tick antioxidant facilitates the Lyme disease agent's successful migration from the mammalian host to the arthropod vector. Cell Host Microbe 2, 7–18.

Nazario, S., Das, S., de Silva, A.M., Deponte, K., Marcantonio, N., Anderson, J.F., Fish, D., Fikrig, E., Kantor, F.S., 1998. Prevention of *Borrelia burgdorferi* transmission in Guinea pigs by tick immunity. Am. J. Trop. Med. Hyg. 58, 780–785.

Otsuka, A., Kabashima, K., 2015. Mast cells and basophils in cutaneous immune responses. Allergy 70, 131–140.

Paesen, G.C., Adams, P.L., Harlos, K., Nuttall, P.A., Stuart, D.I., 1999. Tick histamine-binding proteins: isolation, cloning, and three-dimensional structure. Mol. Cell. 3, 661–671.

Paesen, G.C., Adams, P.L., Harlos, K., Nuttall, P.A., Stuart, D.I., 2000. Tick histamine-binding proteins: lipocalins with a second binding cavity. Biochim. Biophys. Acta 1482, 92–101.

Paine, S.H., Kemp, D.H., Allen, J.R., 1983. In vitro feeding of *Dermacentor andersoni* (Stiles): effects of histamine and other mediators. Parasitology 86, 419–428.

Páleníková, J., Lieskovská, J., Langhansová, H., Kotsyfakis, M., Chmelař, J., Kopecký, J., 2015. *Ixodes ricinus* salivary serpin IRS-2 affects Th17 differentiation via inhibition of the interleukin-6/STAT-3 signaling pathway. Infect. Immun. 83, 1949–1956.

Peck, S.M., Wright, W.H., Gant Jr., J.Q., 1943. Cutaneous reactions due to the body louse (*Pediculus humanus*). J. Am. Med. Assoc. 123, 821–825.

Peng, Z., Simons, F.E.R., 2007. Advances in mosquito allergy. Curr. Opin. Allergy Clin. Immunol. 7, 350–354.

Potenzieri, C., Undem, B.J., 2011. Basic mechanisms of itch. Clin. Exp. Allergy 42, 8–19.

Radulović, Ž.M., Kim, T.K., Porter, L.M., Sze, S.H., Lewis, L., Mulenga, A., 2014. A 24-48 h fed *Amblyomma americanum* tick saliva immuno-proteome. BMC Genomics 15, 518.

Reddy, V.B., Iuga, A.O., Shimada, S.G., LaMotte, R.H., Lerner, E.A., 2008. Cowhage-evoked itch is mediated by a novel cysteine protease: a ligand of protease-activated receptors. J. Neurosci. 28, 4331–4335.

Ribeiro, J.M., Anderson, J.M., Manoukis, N.C., Meng, Z., Francischetti, I.M., 2011. A further insight into the sialome of the tropical bont tick, *Amblyomma variegatum*. BMC Genomics 12, 136.

Ribeiro, J.M., Arcà, B., Lombardo, F., Calvo, E., Phan, V.M., Chandra, P.K., Wikel, S.K., 2007. An annotated catalogue of salivary gland transcripts in the adult female mosquito, *Aedes aegypti*. BMC Genomics 8, 6.

Ribeiro, J.M., Assumpção, T.C., Pham, V.M., Francischetti, I.M., Reisenman, C.E., 2012. An insight into the sialotranscriptome of *Triatoma rubida* (Hemiptera: Heteroptera). J. Med. Entomol. 49, 563–572.

Ribeiro, J.M.C., Mather, T.N., 1998. *Ixodes scapularis*: salivary kininase activity is a metallo dipeptidyl carboxypeptidase. Exp. Parasitol. 89, 213–221.

Ribeiro, J.M., Walker, F.A., 1994. High affinity histamine-binding and antihistaminic activity of the salivary nitric oxide-carrying heme protein (nitrophorin) of *Rhodnius prolixus*. J. Exp. Med. 180, 2251–2257.

Riek, R.F., 1962. Studies on the reactions of animals to infestation with ticks. VI. Resistance of cattle to infestation with the tick, *Boophilus microplus* (Canestrini). Aust. J. Agric. Res. 13, 532–550.

Ringkamp, M., Meyer, R., 2014. Pruriceptors. In: Carstens, E., Akiyama, T. (Eds.), Itch: Mechanisms and Treatment. CRC Press, Boca Raton, FL, pp. 129–139.

Sangamnatdej, S., Paesen, G.C., Slovak, M., Nuttall, P.A., 2002. A high affinity serotonin- and histamine-binding lipocalin from tick saliva. Insect. Mol. Biol. 11, 79–86.

Schmelz, M., 2015. Itch and pain differences and commonalities. In: Schaible, H.-G. (Ed.), Pain Control, Handbook of Experimental Pharmacology 227. Springer-Verlag, Berlin, pp. 285–301.

Schoeler, G.B., Manweiler, S.A., Wikel, S.K., 1999. *Ixodes scapularis*: effects of repeated infestations with pathogen-free nymphs on macrophage and T lymphocyte cytokine responses of BALB/c and C3H/HeN mice. Exp. Parasitol. 92, 239–248.

Schoeler, G.B., Wikel, S.K., 2001. Modulation of host immunity by haematophagous arthropods. Ann. Trop. Med. Parasitol. 95, 755–771.

Schmelz, M., Schmidt, R., Bickel, A., Handwerker, H.O., Torebjörk, H.E., 1997. Specific C-receptors for itch in human skin. J. Neurosci. 17, 8003–8008.

Schwarz, A., Valdés, J.J., Kotsyfakis, M., 2012. The role of cystatins in tick physiology and blood feeding. Ticks Tick Borne Dis. 3, 117–127.

Seeliger, S., Derian, C.K., Vergnolle, N., Bunnett, N.W., Nawroth, R., Schmelz, M., Von Der Weid, P.Y., Buddenkotte, J., Sunderkötter, C., Metze, D., Andrade-Gordon, P., Harms, E., Vestweber, D., Luger, T.A., Steinhoff, M., 2003. Proinflammatory role of proteinase-activated receptor-2 in humans and mice during cutaneous inflammation in vivo. FASEB J. 17, 1871–1885.

Shim, W.S., Oh, U., 2008. Histamine-induced itch and its relationship with pain. Mol. Pain 4, 29.

Shim, W.S., Tak, M.H., Lee, M.H., Kim, M., Kim, M., Koo, J.Y., Lee, C.H., Kim, M., Oh, U., 2007. TRPV1 mediates histamine-induced itching via the activation of phospholipase A2 and 12-lipoxygenase. J. Neurosci. 27, 2331–2337.

Shpacovitch, V.M., Varga, G., Strey, A., Gunzer, M., Mooren, F., Buddenkotte, J., Vergnolle, N., Sommerhoff, C.P., Grabbe, S., Gerke, V., Homey, B., Hollenberg, M., Luger, T.A., Steinhoff, M., 2004. Agonists of proteinase-activated receptor-2 modulate human neutrophil cytokine secretion, expression of cell adhesion molecules, and migration within 3-D collagen lattices. J. Leukoc. Biol. 76, 388–398.

Sikand, P., Dong, X., LaMotte, R.H., 2011. BAM8-22 peptide produces itch and nociceptive sensations in humans independent of histamine release. J. Neurosci. 31, 7563–7567.

Snowball, G.J., 1956. The effect of self-licking by cattle on infestation of cattle tick, *Boophilus microplus* (Canestrini). Aust. J. Agric. Res. 7, 227–232.

Soh, U.J., Dores, M.R., Chen, B., Trejo, J., 2010. Signal transduction by protease-activated receptors. Br. J. Pharmacol. 160, 191–203.

Sorokin, M., 2003. Itching bites may limit Ross River virus infection. Med. J. Aust. 178, 143–144.

Stark, K.R., James, A.A., 1998. Isolation and characterization of the gene encoding a novel factor Xa-directed anticoagulant from the yellow fever mosquito, *Aedes aegypti*. J. Biol. Chem. 273, 20802–20809.

Steinhoff, M., Cevikbas, F., Ikoma, A., Berger, T.G., 2011. Pruritus: management algorithms and experimental therapies. Semin. Cutan. Med. Surg. 30, 127–137.

Steinhoff, M., Neisius, U., Ikoma, A., Fartasch, M., Heyer, G., Skov, P.S., Luger, T.A., Schmelz, M., 2003. Proteinase-activated receptor-2 mediates itch: a novel pathway for pruritus in human skin. J. Neurosci. 23, 6176–6180.

Storan, E.R., O'Gorman, S.M., McDonald, I.D., Steinhoff, M., 2015. Role of cytokines and chemokines in itch. In: Cowan, A., Yosipovitch, G. (Eds.), Pharmacology of Itch, Handbook of Experimental Pharmacology 226. Springer-Verlag, Berlin, pp. 163–176.

Strakhova, M.I., Nikkel, A.L., Manelli, A.M., Hsieh, G.C., Esbenshade, T.A., Brioni, J.D., Bitner, R.S., 2009. Localization of histamine H4 receptors in the central nervous system of human and rat. Brain. Res. 1250, 41–48.

Tankersley, M.S., Ledford, D.K., 2015. Stinging insect allergy: state of the art 2015. J. Allergy Clin. Immunol. Pract. 3, 315–322.

Taves, S., Ji, R.-R., 2015. Itch control by Toll-like receptors. In: Cowan, A., Yosipovitch, G. (Eds.), Pharmacology of Itch. Handbook of Experimental Pharmacology 226. Springer-Verlag, Berlin, pp. 135–150.

Tirloni, L., Islam, M.S., Kim, T.K., Diedrich, J.K., Yates 3rd, J.R., Pinto, A.F., Mulenga, A., You, M.J., Da Silva Vaz Jr., I., 2015. Saliva from nymph and adult females of *Haemaphysalis longicornis*: a proteomic study. Parasit. Vectors 8, 338.

Tirloni, L., Kim, T.K., Coutinho, M.L., Ali, A., Seixas, A., Termignoni, C., Mulenga, A., da Silva Vaz Jr., I., 2016. The putative role of *Rhipicephalus microplus* salivary serpins in the tick-host relationship. Insect. Biochem. Mol. Biol. 71, 12–28.

Tirloni, L., Reck, J., Terra, R.M., Martins, J.R., Mulenga, A., Sherman, N.E., Fox, J.W., Yates 3rd, J.R., Termignoni, C., Pinto, A.F., Vaz Ida Jr., S., 2014. Proteomic analysis of cattle tick *Rhipicephalus (Boophilus) microplus* saliva: a comparison between partially and fully engorged females. PLoS One 9, e94831.

Tsuboi, R., Sato, C., Shi, C.M., Nakamura, T., Sakurai, T., Ogawa, H., 1994. Endothelin-1 acts as an autocrine growth factor for normal human keratinocytes. J. Cell. Physiol. 159, 213–220.

Valdés, J.J., 2014. Antihistamine response: a dynamically refined function at the host-tick interface. Parasit. Vectors 7, 491.

van Nunen, S., 2015. Tick-induced allergies: mammalian meat allergy, tick anaphylaxis and their significance. Asia Pac. Allergy 5, 3–16.

Vassalli, P., 1992. The pathophysiology of tumor necrosis factors. Annu. Rev. Immunol. 10, 411–452.

Veríssimo, C.J., D'Agostino, S.M., Pessoa, F.F., de Toledo, L.M., Santos, I.K., 2015. Length and density of filiform tongue papillae: differences between tick-susceptible and resistant cattle may affect tick loads. Parasit. Vectors 8, 594.

Wada, T., Ishiwata, K., Koseki, H., Ishikura, T., Ugajin, T., Ohnuma, N., Obata, K., Ishikawa, R., Yoshikawa, S., Mukai, K., Kawano, Y., Minegishi, Y., Yokozeki, H., Watanabe, N., Karasuyama, H., 2010. Selective ablation of basophils in mice reveals their nonredundant role in acquired immunity against ticks. J. Clin. Invest. 120, 2867–2875.

Wasserman, H.A., Singh, S., Champagne, D.E., 2004. Saliva of the Yellow Fever mosquito, *Aedes aegypti*, modulates murine lymphocyte function. Parasite. Immunol. 26, 295–306.

Weeks, C.A., Nicol, C.J., Titchener, R.N., 1995. Effects of the sucking louse (*Linognathus vituli*) on the grooming behaviour of housed calves. Vet. Rec. 137, 33–35.

Wikel, S.K., 1982a. Histamine content of tick attachment sites and the effects of H1 and H2 histamine antagonists on the expression of resistance. Ann. Trop. Med. Parasitol. 76, 179–185.

Wikel, S.K., 1982b. Immune responses to arthropods and their products. Annu. Rev. Entomol. 27, 21–48.

Wikel, S.K., 1996. Host immunity to ticks. Annu. Rev. Entomol. 41, 1–22.

Wikel, S., 2013. Ticks and tick-borne pathogens at the cutaneous interface: host defenses, tick countermeasures, and a suitable environment for pathogen establishment. Front. Microbiol. 4, 337.

Wikel, S.K., Allen, J.R., 1976. Acquired resistance to ticks. I. Passive transfer of resistance. Immunology 30, 311–316.

Wikel, S.K., Ramachandra, R.N., Bergman, D.K., Burkot, T.R., Piesman, J., 1997. Infestation with pathogen-free nymphs of the tick *Ixodes scapularis* induces host resistance to transmission of *Borrelia burgdorferi* by ticks. Infect. Immun. 65, 335–338.

Willadsen, P., Wood, G.M., Riding, G.A., 1979. The relation between skin histamine concentration, histamine sensitivity, and the resistance of cattle to the tick, *Boophilus microplus*. Zeitschrift fur Parasitenkunde 59, 87–93.

Wilson, A.D., 2014. Immune responses to ectoparasites of horses, with a focus on insect bite hypersensitivity. Parasite. Immunol. 36, 560–572.

Wu, J., Mu, L., Zhuang, L., Han, Y., Liu, T., Li, J., Yang, Y., Yang, H., Wei, L., 2015. A cecropin-like antimicrobial peptide with anti-inflammatory activity from the black fly salivary glands. Parasit. Vectors 8, 561.

Wu, J., Wang, Y., Liu, H., Yang, H., Ma, D., Li, J., Li, D., Lai, R., Yu, H., 2010. Two immunoregulatory peptides with antioxidant activity from tick salivary glands. J. Biol. Chem. 285, 16606–16613.

Yamaguchi, T., Nagasawa, T., Satoh, M., Kuraishi, Y., 1999. Itch-associated response induced by intradermal serotonin through 5-HT2 receptors in mice. Neurosci. Res. 35, 77–83.

3

Arthropod Modulation of Wound Healing

Stephen Wikel

Quinnipiac University, Hamden, CT, United States

INTRODUCTION

Vector arthropods expose host cells, molecules, and multiple defense mechanisms to both mechanical injury from insertion of mouthparts and biological actions of saliva molecules that facilitate both blood meal acquisition and infectious agent transmission (Bernard et al., 2014, 2015; Wikel, 2013). Sequential, overlapping, and functionally coordinated phases of acute cutaneous wound healing are activated within seconds of injury and evolve over days, weeks, and months until skin integrity is restored (Shaw and Martin, 2009; Greaves et al., 2013; Eming et al., 2014). Phases of wound healing include hemostasis, inflammation, proliferation, extracellular matrix formation, and tissue remodeling with the earliest aspects of the final phase activated ~3 days after injury (Greaves et al., 2013).

Blood-feeding arthropod vectors modulate the phases of wound healing largely based on their duration of attachment to the host, which ranges from minutes for mosquitoes to more than a week for adult ixodid ticks. Feeding lesion structure also impacts how wound healing components and pathways are downregulated or deviated by vessel feeders and those vector

species that create pool-like lesions to obtain a blood meal. Hematophagy arose independently multiple times among the Arthropoda (Adams, 1999). Therefore, it is not surprising that saliva of different arthropod families, genera, and species evolved diverse repertoires of bioactive molecules that modulate host defenses, including wound healing.

This chapter examines vector arthropod interactions with host acute wound healing initiated by mouthpart insertion into the skin followed by tissue disruption and blood feeding. Specific components of each phase of wound healing are described. Wound healing begins within seconds of injury. Key cellular and molecular events, interactions, and interrelationships are described for each phase of wound healing. Vector arthropod saliva modulation of specific wound healing mechanisms, cells, and molecules is examined in the context of each phase and relative to the timeline of initiation and interactions with other phases. Significant advances are occurring in dissecting the events of wound healing. Likewise, rapid advances in characterizing saliva transcriptomes and proteomes of vector arthropod species facilitate increasingly well-characterized interactions of vectors with host wound healing pathways. Here, we examine those relationships.

In addition to specific content provided in this chapter on wound healing, readers are encouraged to consult second-volume chapters that address aspects of specific themes that relate to wound healing: vector arthropod modulation of skin cells and mediators (Chapter 1), vectors and host hemostasis (Chapter 2), vector saliomes (Chapter 5), and host inflammatory and immune responses induced and/or modulated by specific vector arthropods (Chapters 6, 8, 9, and 10).

VECTOR ARTHROPOD FEEDING

Patterns of vector arthropod blood meal acquisition are divided into solenophagy, "vessel feeders," and telmophagy, "pool feeders" (Lavoipierre, 1965; Bergman, 1996). Solenophage mouthparts are styletlike for piercing and cannulating small venules and for creating small hemorrhagic foci within host dermis for blood feeding. Vector arthropods evolved mouthpart structures suitable for their ectoparasitic lifestyle (Bergman, 1996). Examples of solenophages are mosquitoes, triatomids, bed bugs, and sucking lice. In contrast, telmophages have slashing mouthparts that create open wounds or pools of blood within the skin. Telmophages include biting flies, fleas, and ticks. Salivary glands of some ixodid tick species also produce attachment cement to help anchor the tick to the host for the long duration of blood feeding (Moorhouse, 1969).

Vector arthropod interactions with wound healing and other host defenses should be considered in the context of duration of blood feeding by a vector and by the fact that an individual host may be fed on by the same vector species many times within a day and during the life span of the host. Influence of vector modulation of specific phases of wound healing has direct immediate benefits for the arthropod, whether it blood feeds for only seconds or for days.

Long-term benefit to the vector of blocking or deviating host defenses is reducing the likelihood that a host will develop responses that inhibit the ability of members of that species to obtain a blood meal. Maintaining feeding site integrity by altering sequential events of wound healing is particularly important for ixodid ticks. All blood-feeding vectors face challenges from host hemostasis, inflammation, and initiation of immune responses that could impair subsequent blood feeding.

Duration of blood feeding differs among both solenophage and telmophage species. Among solenophages, female mosquitoes complete blood meal uptake in less than 2 min (Chadee and Beier, 1997); both adult male and female triatomines blood feed for 20–30 min (Martinez-Ibarra et al., 2001); and bed bugs engorge on blood for 5–10 min (Thomas et al., 2004). Telmophages feed to repletion within seconds to minutes in the cases of biting flies and fleas and hours to days for argasid and ixodid ticks. Cat flea, *Ctenocephalides felis felis*, creates a feeding pool and engorges in 10 min (Cadiergues et al., 2000); female sandy fly, *Lutzomyia longipalpis*, requires ~20 s to feed, but infection with *Leishmania mexicana* prolongs blood feeding to several minutes (Rogers and Bates, 2007); argasid ticks repeatedly consume blood for as long as 2 h (Cooley and Kohls, 1944); and ixodid ticks remain attached to the host and can engorge for more than a week (Kaufman, 1989).

Complexity of vector arthropod salivary gland transcriptomes and proteomes correlates with the repertoire of bioactive molecules needed to modulate host defenses for successful blood feeding. As general estimates, salivary gland transcriptomes of *Phlebotomus papatasi* predict ~49 secreted proteins (Abdeladhim et al., 2012) and those of female *Aedes aegypti* mosquito predict 55 secreted proteins (Ribeiro et al., 2007). Salivary gland transcriptome complexity increases greatly for ixodid ticks to ~700 potentially secreted proteins for *Dermacentor andersoni* (Alarcon-Chaidez et al., 2007). Next-generation sequencing and proteome analyses reveal that these salivary gland transcriptome estimates likely significantly underestimate the true complexity of repertoires of salivary gland–secreted bioactive molecules (Kotál et al., 2015).

WOUND HEALING: CELLS, MOLECULES, MECHANISMS, AND PHASES

What are the events initiated by cutaneous injury? Wound healing consists of a series of complex interactions among cells, molecules, and pathways that are divided into the temporally interconnected phases of hemostasis, inflammation, proliferation of cellular and structural components, and extracellular matrix formation and remodeling (Shaw and Martin, 2009; Greaves et al., 2013; Eming et al., 2014). To understand the relationships between vector arthropods and wound healing, an overview is provided for specific components and events of each phase of normal wound healing and the relationships of how these phases are interrelated over the timeline of wound healing.

HEMOSTASIS: FIRST PHASE OF WOUND HEALING

Disruption of endothelial cells exposes both extracellular matrix, resulting in platelet activation and aggregation, and tissue factor (thromboplastin) that complexes with and activates coagulation cascade factor VII that in turn activates factor X and prothrombin to produce a limited amount of thrombin, which enhances the coagulation cascade and results in fibrin stabilization of the platelet thrombus (Versteeg et al., 2013; Golebiewska and Poole, 2015). Thrombin amplifies the coagulation pathway by activation of factors VIII and V, which are, respectively, cofactors for factors IX and X, and greatly increases the amount of fibrin generated to

stabilize the platelet plug (Versteeg et al., 2013). Platelet-derived mediators ADP, serotonin, GDP, and others from dense granules along with α-granule-derived von Willebrand factor amplify platelet activation and aggregation to stop blood loss (Blair and Flaumenhaft, 2009; Golebiewska and Poole, 2015). Activation of the platelet integrin αIIbβ3, which binds fibrinogen, along with glycoprotein VI receptor interaction with collagen results in further platelet aggregation, activation, and mediator secretion (Du et al., 1991; Watson et al., 2005).

Activated platelet-derived growth factors, cytokines, and chemokines regulate all downstream phases of wound healing by contributing to fibroblast and endothelial cell recruitment and activation, inflammatory cell influx starting ~1 h postinjury and persisting for days to more than a week, fibroblast recruitment and activation, initiation of signaling events resulting in angiogenesis, and extracellular matrix remodeling (Golebiewska and Poole, 2015). Activated platelets secrete platelet-derived growth factor (PDGF), transforming growth factor-β (TGF-β), epidermal growth factor (EGF), interleukin-1 (IL-1), vascular endothelial growth factor (VEGF), and fibroblast growth factor (FGF) (Barrientos et al., 2008). Actions of these molecules include PDGF and IL-1 recruiting of neutrophils to the wound; VEGF and FGF recruitment of endothelial cells and initiating angiogenesis; fibroplasia and extracellular matrix remodeling by activities of PDGF, FGF, and TGF-β; and reepithelization orchestrated by EGF and FGF (Barrientos et al., 2008). In addition, within seconds of cutaneous injury, damaged cells undergo changes resulting in expression of damage-associated molecular patterns (DAMPs) that activate cells of the innate immune system contributing to initiation of the inflammatory response (Shaw and Martin, 2009; Greaves et al., 2013; Strbo et al., 2014).

In addition to platelet aggregation and activation, thromboxane A2 released from platelets immediately upon injury induces vasoconstriction that lasts ~10 min before a transition to vasodilation mediated by histamine occurs (Sinno and Prakash, 2013). Histamine release at this point is attributed to activation of the alternative pathway of complement with the generation of anaphylatoxins that cause mast cell degranulation (Sinno and Prakash, 2013). Direct injury-induced damage to mast cells could also induce mediator release. Vasodilation contributes to the entry of serum factors and leukocytes into the wound site. Vasodilation decreases ~1 h postinjury. These events provide a continuum from the hemostasis phase to the overlapping inflammatory phase of wound healing.

INFLAMMATION: SECOND PHASE OF WOUND HEALING

Mediators produced during the hemostasis phase of wound healing provide signals that result in the inflammatory phase (Shaw and Martin, 2009; Greaves et al., 2013). Inflammatory phase involves elements of the innate immune response, specifically an initial influx of neutrophils entering the wound beginning ~1 h postinjury followed within several hours by monocytes that become macrophages on entering the tissues from the vasculature compartment (Eming et al., 2007, 2009; Greaves et al., 2013). Wound-infiltrating neutrophils and macrophages kill microbes, elaborate connective tissue growth factors to enhance subsequent fibrosis and extracellular matrix production, and promote keratinocyte migration, blood vessel formation, and reepithelization (Shaw and Martin, 2009; Greaves et al., 2013; Sinno and Prakash, 2013). Pattern recognition receptors of neutrophils, macrophages, dendritic cells, keratinocytes, and other cells bind to pathogen associated molecular patterns and DAMPs to initiate both antimicrobial and injury repair responses (Strbo et al., 2014). Additional innate immune response components that contribute to the inflammatory phase are complement (Sinno and

Prakash, 2013), mast cells (Eming et al., 2009), dendritic cells (Cumberbatch et al., 2000), and keratinocytes (Strbo et al., 2014).

Adhesion molecules and chemokines are essential for leukocyte migration from the vascular compartment into tissues (Carlos and Harlan, 1994). Endothelial cell expression of E-selectin, P-selectin, and ICAM-1 serves as important regulatory elements of cutaneous wound healing (Yukami et al., 2007). Both neutrophils and macrophages that migrated into the wound site are actively engaged in phagocytosis of cellular debris, cytokine stimulation of fibroblasts and extracellular matrix formation, and regulation of new blood vessel formation. Alternative activation of macrophages is recognized as an important part of healing with production of antiinflammatory cytokines and stimulation of tissue repair (Eming et al., 2007, 2009; Greaves et al., 2013; Strbo et al., 2014). Macrophages produce VEGF and PDGF that contribute to the formation of granulation tissue and the transition into the proliferation phase of wound healing (Barrientos et al., 2008). Although less well defined, adaptive immune responses are linked to wound healing through T and B lymphocyte activation and production of soluble mediators and direct interactions with keratinocytes that play an important role in the proliferation phase of wound healing (Eming et al., 2009; Strbo et al., 2014). These cells, molecules, and pathways constitute the inflammatory phase of the continuum of interconnected stages of acute wound healing.

PROLIFERATION: THIRD PHASE OF WOUND HEALING

This phase of wound healing is characterized by reduced numbers of neutrophils and macrophages accompanied by influx and proliferation of fibroblasts, endothelial cells, and keratinocytes into the wound (Diegelmann and Evans, 2004; Olczyk et al., 2014). Proliferation phase is evident at 2–3 days postinjury and overlaps with the inflammatory phase initiated within hours after cutaneous injury by elaboration of PDGF, a variety of cytokines, and FGF generated during hemostasis and inflammatory phases (Greaves et al., 2013; Olczyk et al., 2014; Seifert and Maden, 2014). Macrophage-derived mediators contributing to the proliferation phase include TGF-β, FGF, IL-1, and VEGF that are also produced by platelets, keratinocytes, fibroblasts, neutrophils, and smooth muscle cells (Barrientos et al., 2008; Greaves et al., 2013; Olczyk et al., 2014).

Proliferative phase involves wound site infiltration by fibroblasts, endothelial cells, and keratinocytes that orchestrate synthesis and remodeling of extracellular matrix, angiogenesis, and formation of granulation tissue to restore normal dermis and epidermis (Gurtner et al., 2008; Schultz and Wysocki, 2009; Olczyk et al., 2014; Pastar et al., 2014). Growth factors, cytokines, chemokines, and integrins regulate collagen deposition to replace the earlier deposited fibrin meshwork with more tightly organized and stronger collagen fibrils (Schultz and Wysocki, 2009; Greaves et al., 2013). Fibroblasts are predominant in this phase because of their secretion of extracellular matrix proteins, particularly collagens, in response to growth factors (Clark, 2001; Schultz and Wysocki, 2009).

Angiogenesis takes place during the proliferative phase with endothelial cell migration from capillaries occurring at the wound margins in response to VEGF, FGF, and other mediators derived from activated platelets as well as the subsequent infiltration of macrophages, keratinocytes, endothelial cells, and fibroblasts (Greaves et al., 2013; Walsh et al., 2015). Matrix metalloprotease remodeling of collagen-rich extracellular matrix contributes to migration of proliferating endothelial cells to establish a new capillary network and formation of granulation tissue (Greaves et al., 2013; Olczyk et al., 2014). Significantly, inhibition of angiogenesis delays wound healing (Clark, 2001).

TISSUE REMODELING: FOURTH PHASE OF WOUND HEALING

This final phase of acute cutaneous wound healing begins approximately 2–3 weeks postinjury with transformation of abundant fibroblasts into myofibroblasts, containing microfilaments of α-smooth muscle actin (Olczyk et al., 2014). Matrix metalloproteases are present in granulation tissue (Madlener et al., 1998), and they remodel collagen (Seifert and Maden, 2014). Extracellular matrix composition changes with an increase in cross-linked collagens that strengthen tissue and more closely resemble normal skin (Diegelmann and Evans, 2004; Hinz, 2007). Granulation tissue transitions into mechanically stronger scar tissue accompanied by decreased numbers of fibroblasts and myofibrils and reduced density of capillaries as they transition into larger vessels (Diegelmann and Evans, 2004). Matrix metalloproteases enable keratinocytes to migrate and proliferate during epithelialization of the wound (Pastar et al., 2014).

VECTOR ARTHROPOD MODULATION OF WOUND HEALING

Complex nature of disease vector countermeasures to host defenses is evident in the ability of saliva molecules of short- and long-duration blood-feeding arthropods to modulate specific components of the four overlapping and integrated phases of acute cutaneous wound healing. Modulation of wound healing is the ideal topic to integrate the known countermeasures of vector arthropods to hemostasis, inflammation, innate immunity, and adaptive immunity as well as to characterize both pluripotency and redundancy of biological activities of vector saliva. Temporal relationships are best illustrated for host defenses and vector countermeasures in the wound healing cascade of events. For additional information on vector arthropod manipulation of specific defenses refer to chapters addressing vectors and host hemostasis (Chapter 2), vector saliomes (Chapter 5), and host inflammatory and immune responses (Chapters 6, 8, 9, and 10).

Two examples of solenophage blood feeding are female mosquitoes that obtain a blood meal within 2 min (Chadee and Beier, 1997), and triatomids for which all five instars blood feed with adult females and males requiring 20–30 min to engorge (Martinez-Ibarra et al., 2001). Primary challenge confronting rapidly feeding solenophages is the hemostasis phase of wound healing: platelet aggregation, coagulation cascade initiation, vasoconstriction, fibrin meshwork deposition, thrombus formation, and vasodilation. Saliva of solenophages also contains molecules that modulate inflammatory responses and innate and adaptive immune defenses (Assumpção et al., 2008; Schneider and Higgs, 2008; Fontaine et al., 2011; Conway et al., 2014).

Head lice are long-duration blood-feeding solenophages that take 4–10 blood meals per day over their life span of ~27 days (Bonilla et al., 2013). Lice have the potential to sensitize a host because of continuous association for approximately a month combined with multiple introductions of saliva while blood feeding each day. Very little is known about the relationships of head and body lice saliva with host inflammatory and immune responses.

Biting flies and ticks are telmophages whose laceration of skin to create a feeding pool provides strong activating signals for wound repair by disrupting blood vessels and releasing DAMPs. Female sand flies consume blood over several minutes from a pool created by their slashing mouthpart disruption of skin capillaries and lymphatic vessels (Rogers and Bates, 2007). Female sand flies feed during the period when hemostasis poses a threat to successful blood meal acquisition. In addition, sand fly saliva effectively suppresses or deviates host inflammatory and immune responses (Abdeladhim et al., 2014). Argasid ticks feeding from 15 min to 2 h (Cooley and Kohls, 1944) encounter host hemostasis, initiation of a neutrophil influx, and soluble mediators of the early inflammatory phase of wound repair at the bite site. In contrast, adult ixodid ticks remain attached to the host and blood feed over a period of several days to more than a week resulting in saliva that inhibits, deviates, or modulates aspects of each of the cutaneous wound repair phases of hemostasis, inflammation, proliferation, and tissue remodeling.

Complexity of salivary gland transcriptomes and proteomes of mosquitoes, triatomids, sand flies, and ticks is consistent with blood feeding biology and complexity of host defenses encountered (Ribeiro et al., 2007; Assumpção et al., 2008; Abdeladhim et al., 2012; Kotál et al., 2015). Diverse saliva compositions evolved across the spectrum of vector arthropods to modulate host defenses (Ribeiro and Francischetti, 2003). Significant differences occur in the repertoire of saliva bioactive molecules between ixodid tick species (Ribeiro et al., 2006; Alarcon-Chaidez et al., 2007).

Salivary gland–secreted molecules of rapid blood feeders that inhibit inflammation and innate and adaptive immune responses would appear to be unnecessary for successful blood feeding requiring only seconds to minutes to complete. The value of inhibitors of these host defenses is likely found in long-term relationships between a vector species and vertebrate hosts that will be individually fed on many times every day over a life span of years. The objective is to keep a susceptible population of hosts available for blood meals. Insertion of vector mouthparts and salivation while blood feeding introduce molecules that can induce host immune responses capable of impairing the ability to obtain a blood meal, a phenomenon known as acquired resistance to infestation (Trager, 1939; Allen, 1973; Wikel, 1982).

Range of host species is an evolutionary consideration in driving complexity and redundant nature of biological activities commonly found in saliva of many vector species (Fontaine et al., 2011). Need to down-regulate or deviate inflammatory and immune responses and suppress later stages of wound healing is important for those vectors, ixodid ticks, and sucking lice that feed for prolonged periods in continuous association with a single host. In the case of ixodid ticks, simultaneous saliva introduction of very small concentrations of multiple biologically active proteins combined with sequential introduction or shuttling among members of a family of related proteins during the course of feeding is the way to reduce potential immunogenicity of these molecules (Chmelař et al., 2016b). Immunomodulatory molecules in saliva also act as a shield to reduce the likelihood of development of host immune responses that will significantly diminish the overall biological activity of other saliva molecules.

FIRST PHASE: HEMOSTASIS AND VECTORS

Solenophage and telmophage arthropod saliva inhibits the hemostasis phase of acute wound healing that is activated immediately on vector mouthpart-induced cutaneous injury: platelet activation and aggregation, vasoconstriction, and coagulation pathway activation (Francischetti, 2010; Fontaine et al., 2011; Chmelař et al., 2012). Because blood feeding arose multiple times among the arthropods (Adams, 1999), different vector species evolved diverse saliva compositions that target the same host molecular and cellular defense mechanisms, such as platelet activation and aggregation, in multiple different ways (Francischetti, 2010; Fontaine et al., 2011). Next-generation sequencing and proteome analyses of salivary glands are revealing an increasingly higher degree of functional redundancies of saliva molecule families (Kotál et al., 2015; Chmelař et al., 2016a).

Platelet Aggregation

Saliva of both rapid- and long-duration blood-feeding arthropods inhibits platelet aggregation (Francischetti, 2010). Phylogenetically diverse vector species may utilize similar molecules, such as apyrases, to inhibit platelet aggregation as well as possessing other saliva modulators of platelet activation and aggregation (Fontaine et al., 2011). Multiple cellular and molecular components of hemostasis may be counteracted by a single saliva molecule. A saliva lipocalin of the triatomine, *Dipetalogaster maxima*, simultaneously blocks platelet aggregation by binding TXA_2, vasoconstriction by inhibiting $PGF_2\alpha$, and angiogenesis by binding 15(S)-HETE (Assumpção et al., 2010). Vector arthropod saliva may inhibit a single host protein with multiple key host defense functions, such as thrombin, a serine protease that converts soluble fibrinogen to fibrin, activates coagulation factors, and activates and aggregates platelets (Berliner, 1992). Saliva inhibitors of thrombin are present in mosquitoes (Watanabe et al., 2011), triatomines (Campos et al., 2004), and ticks (Francischetti, 2010).

Apyrases are platelet aggregation inhibitors present in saliva of multiple vector species, including triatomine bugs, mosquitoes, fleas, deer flies, sand flies, argasid ticks, and ixodid ticks (Francischetti, 2010). Apyrases inhibit platelet aggregation by hydrolyzing ATP and ADP to AMP and phosphate (Komozsynski and Wojtczak, 1996; Gachet, 2001). Female and male bed bugs, *Cimex lectularius*, blood feed to repletion in 5–10 min (Thomas et al., 2004), and their saliva contains apyrase (Valenzuela et al., 1998). *A. aegypti* saliva also inhibits platelet aggregation with apyrase (Champagne et al., 1995), but that is only one of the inhibitors. *A. aegypti* saliva also contains the protein aegyptin that blocks von Willibrand factor binding sequence on collagen to inhibit platelet aggregation and concurrently blocks collagen initiation of coagulation by activation of factor XII (Calvo et al., 2007, 2010; Mizurini et al., 2013).

Saliva of both *A. aegypti* and *Anopheles gambiae* contains apyrases and D7 proteins that bind serotonin, norepinephrine, and histamine that inhibit platelet aggregation, vasoconstriction, and pain/itch responses (Calvo et al., 2006). The D7 family of proteins is an example of molecules with broad activities that facilitate successful blood feeding. Similar to the action of the saliva of *A. aegypti*, collagen binding–induced platelet aggregation is blocked by the *Anopheles stephensi* saliva protein, anopheline antiplatelet protein (Hayashi et al., 2012).

Among pool blood feeders, saliva of the flea, *Xenopsylla cheopis*, and the cattle tick, *Rhipicephalus* (*Boophilus*) *microplus*, contains apyrases (Liyou et al., 1999; Andersen et al., 2007). Saliva of Argasids (*Ornithodorus savignyi*, *Ornithodoros moubata*, *Ornithodoros coriaceus*) and ixodid ticks (*Dermacentor variabilis*, *Ixodes scapularis*, *Ixodes pacificus*) contains disintegrins that block platelet aggregation and activation by acting as antagonists of fibrinogen binding to platelet integrin α2bβ3 (Francischetti, 2010; Assumpção et al., 2012; Chmelař et al., 2012). *I. scapularis* saliva also contains metalloprotease activity that degrades fibrinogen and fibrin to inhibit platelet aggregation (Francischetti et al., 2003). *Ixodes ricinus* saliva serpin inhibits inflammation and platelet aggregation induced by thrombin and cathepsin G (Chmelař et al., 2011). Saliva of *Amblyomma americanum* saliva contains a serinelike and papainlike cysteine protease inhibitor that reduces platelet aggregation by ~47% and inhibits plasmin (Mulenga et al., 2013).

Vasodilation

Vasoconstriction is an immediate response to cutaneous injury acting in concert with platelet aggregation to reduce blood loss. Successful blood meal acquisition requires that vector arthropods that feed either rapidly or for long durations facilitate blood flow by direct inhibition of mediators of vasoconstriction or by introduction to the host of vasodilators. Telmophages require potent saliva vasodilators because of the nature of cutaneous wounds they create and the need to maintain a pool of blood for feeding. This is a particular challenge for ixodid ticks that blood feed for days. In addition to reducing blood loss directly, vasodilation also reduces efficiency of platelet aggregation at the wound site (Sneddon and Vane, 1988).

First vector-associated vasodilator reported was a tachykinin peptide from saliva of *A. aegypti* (Ribeiro, 1992). This activity was subsequently attributed to two *A. aegypti* saliva sialokinin peptides (Champagne and Ribeiro, 1994). *Aedes albimanus* saliva catechol oxidase/peroxidase maintains blood flow at the bite site by inhibiting vasoconstriction function (Ribeiro and Nussenzveig, 1993). Host inhibition of vasoconstriction is achieved in an essentially similar manner by related mosquito species. Saliva of *Aedes triseriatus* contains a tachykinin, and *A. gambiae* saliva catechol oxidase/peroxidase activity counteracts vasoconstriction (Ribeiro et al., 1994).

Biting flies feed from pools of blood created by their slashing mouthparts that disrupt the skin. The black fly *Simulium vittatum* produces *Simulium vittatum* erythema protein, a powerful vasodilator (Cupp et al., 1998). Maxadilan isolated from saliva of *L. longipalpis*, the New World sand fly, is one of the most potent vasodilators ever characterized (Lerner et al., 1991). Maxadilan is a selective inhibitor of the PAC1 receptor that inhibits arterial vasodilation and leakage of plasma proteins from venules (Banki et al., 2014). In contrast to *Lutzomyia*, Old World sand fly, *P. papatasi*, does not produce maxadilan, but instead induces vasodilation with saliva adenosine and 5′-AMP (Ribeiro et al., 1999).

Saliva of the bed bug, *C. lectularius*, and the triatomine bug, *Rhodnius prolixus*, contains nitric oxide storing and transporting salivary nitrophorins that induce vasodilation and inhibit platelet aggregation at the bite site (Valenzuela et al., 1995; Yuda et al., 1997). Another example of multiple functionality of a saliva molecule is dipetalodipin, a lipocalin inhibitor of vasoconstriction, platelet aggregation, and angiogenesis present in saliva of the triatomine, *D. maxima* (Assumpção et al., 2010).

To date, tick saliva does not appear to contain proteins with specific vasodilator activity. Both adenosine and prostaglandin E2 are present in saliva of *Rhipicephalus sanguineus* (Oliveira et al., 2011). Although specific vasodilator properties were not determined in this tick–host relationship, these two compounds are known to act as vasodilators. *I. scapularis* saliva kininase could act upon kinins that induce vasodilation and smooth muscle contraction (Ribeiro and Mather, 1998).

Coagulation Pathways

Because blood feeding arose multiple times within the Arthropoda, it is not unexpected that different genera of hematophagous arthropods evolved diverse saliva molecules that inhibit coagulation pathway components (Francischetti et al., 2009; Fontaine et al., 2011; Assumpção et al., 2012). Coagulation factors are one of multiple aspects of hemostasis inhibited in the multifaceted ways used by vector species to maintain favorable blood-feeding sites, to disrupt the normal progression of wound healing, and to provide an environment that is exploited by vector-borne pathogens to establish host infection. Effective inhibition of multiple components of hemostasis is particularly important for solenophage vectors whose mouthparts are narrow fascicles through which blood must be drawn and saliva introduced. An example is the mosquito mouthpart diameter at end of the fascicle that is calculated to be 1 μm in diameter (Ramasubramanian et al., 2008).

More than one component of hemostasis can be inhibited by an individual saliva molecule. In addition to reducing platelet aggregation, *A. aegypti* saliva aegyptin inhibits factor XII activation and the factor XI–factor XII amplification steps initiated by the tissue factor exposed during mouthpart insertion and feeding (Mizurini

et al., 2013). Blocked factor XII activation reduces fibrin clot stability. *A. stephensi* saliva reduces factor XII–kallikrein reciprocal activation and bradykinin formation by binding both factor XII and kininogen that inhibit interactions on the activating surface (Isawa et al., 2002).

Thrombin converts soluble fibrinogen to insoluble fibrin to stabilize the clot. Thrombin also contributes to activation and aggregation of platelets by engaging platelet membrane protease-activated receptors. Therefore, it is not surprising that saliva of members of all blood-feeding arthropod families has saliva proteins that inhibit activated factor X (FXa) and activated factor V (FVa), which convert prothrombin (FII) to thrombin (FIIa) (Fontaine et al., 2011). Alboserpin is an *A. aegypti* saliva factor Xa inhibitor serpin that blocks prothrombinase (Calvo et al., 2011). *L. longipalpis* salivary glands contain Lufaxin, a novel factor Xa inhibitor that does not bind factor X, inhibits the enzyme prothrombinase, and also acts to reduce inflammation that is the next overlapping phase of wound healing (Collin et al., 2012). Multiple factor Xa inhibitors are present among argasid and ixodid tick species (Chmelař et al., 2012).

I. scapularis saliva proteins ixolaris and penthalaris inhibit factor VIIa–tissue factor complex–induced conversion of factor X to factor Xa (Francischetti et al., 2002, 2004).

All of these arthropod-mediated manipulations of host hemostasis can impair wound healing sufficiently to facilitate blood feeding. Can these saliva molecules be used to develop immunological-based strategies to impair blood feeding and transmission and establishment of infectious agents?

SECOND PHASE: INFLAMMATION AND VECTORS

All blood-feeding arthropods initiate the inflammatory response because of tissue damage induced by mouthpart probing or laceration of skin and blood meal uptake. Initiation of the inflammatory phase of cutaneous wound healing starts upon injury with the release of DAMPs from disrupted cells that in turn interact with receptors on keratinocytes, dermal macrophages, and dendritic cells to induce release of proinflammatory cytokines and chemokines, activate the alternative complement pathway, degranulate mast cells, and initiate an influx of neutrophils at ~1 h postinjury. These events are the underpinnings for stimulation of adaptive immune responses to saliva molecules. Vector modulation of host inflammatory and additional immune response effectors may also provide protection for the vector midgut as a blood meal is consumed.

Host–vector interactions involve diverse inflammatory features. An example of a unique response to vector feeding is the human cutaneous histopathologic reaction to *I. scapularis* nymphs characterized by an influx of inflammatory cells at the periphery of the feeding lesion but not adjacent to the mouthparts (Krause et al., 2009). *A. stephensi* bites induce mast cell degranulation that results in fluid extravasation and signaling, driving a neutrophil influx to the bite site accompanied by draining lymph node changes consistent with B and T lymphocyte priming (Demeure et al., 2005).

Pool feeder disruption of the skin provides a strong proinflammatory stimulus with the potential for exposure to cutaneous microflora in the case of open wounds. Arthropod vectors that require a longer time period to engorge to repletion (triatomines, argasid ticks, ixodid ticks) are more likely to be negatively impacted by the inflammatory response, other elements of innate immunity, and the adaptive immune response. Host innate and adaptive immunity is a particular challenge for ixodid ticks that remain attached to the host with introduction of saliva and uptake of blood for days (Wikel, 2013; Kotál et al., 2015). Vector saliva–induced immune sensitization, resulting in acquired resistance or cutaneous hypersensitivity, can

negatively impact repeated blood meal acquisition from the same host by direct effects on the vector or heightened cutaneous awareness resulting in grooming activity (Mellanby, 1946; Allen, 1973). Although it is advantageous for vector arthropods to downregulate host adaptive immune defenses, host sensitivity to vector blood feeding results in awareness of infestation, removal of the arthropod, and reduced infectious agent transmission (des Vignes et al., 2001; Burke et al., 2005).

Why would saliva of a rapidly blood-feeding arthropod impact cells and molecules of the adaptive immune response, which requires days to develop? For a rapid blood-feeding vector, the answer is most likely not primarily related to impacting bite site wound healing phases. An arthropod vector feeding for a few minutes to a few hours would not be negatively impacted by adaptive immunity while obtaining an initial blood meal from a previously uninfested host. However, that individual host may become the subsequent source of blood for many mosquitoes or biting flies over days, weeks, months, and years. Multiple bites may occur daily from mosquitoes or flies of the same and different species. Therefore, modulation of host adaptive immunity by rapid blood feeders likely evolved to avoid development of responses that would impair the ability of a species to successfully obtain blood meals over the life span of a host.

Female *A. aegypti* salivary gland extract suppresses both B and T lymphocyte proliferation, reduces production of the Th1 cytokines IL-2 and IFN-γ and the Th2 cytokines IL-4 and IL-10, and dendritic cell secretion of the Th1 polarizing cytokine IL-12 (Wasserman et al., 2004). While *A. aegypti* salivary gland extract suppressed Th1 and Th2 cytokines, a similar response was not observed for *Culex quinquefasciatus* salivary gland extract, indicating different mosquito strategies for modulating host immune defenses (Wanasen et al., 2004). *A. aegypti* salivary gland extract does not directly inhibit dendritic cell maturation or function; however, it induces caspases-mediated apoptosis of both naïve CD4[+], CD8[+] T and B lymphocytes, but not memory lymphocytes (Bizzarro et al., 2013). Although these responses likely minimally impact for these rapid blood feeders, mosquito modulation of lymphocyte function reduces host ability to induce an antimosquito response.

Mosquito modulation of host immune defenses has obvious implications for transmission of vector-borne pathogens. Feeding by either *A. aegypti* or *Culex pipiens* significantly downregulated Th1 cytokines and upregulated Th2 cytokines of virus-susceptible, but not flavivirus-resistant mice (Zeidner et al., 1999). A novel protein, SAAG-4, of female *A. aegypti* saliva programs CD4[+] T lymphocytes to suppress IFN-γ production and polarizes to a Th2 response producing IL-4 (Boppana et al., 2009). Saliva of *A. aegypti* induces a strong cutaneous Th2 cytokine profile with significantly reduced Th1 cytokines for feeding by both uninfected and chikungunya-infected mosquitoes, while needle inoculation of virus induces a strong Th1 with an undetectable Th2 cytokine profile (Thangamani et al., 2010). Mosquito salivary gland–derived molecules enhance infectivity of numerous mosquito-borne arboviruses (Conway et al., 2014).

Sand flies directly alter neutrophil function, thus impacting the initial cellular influx of the inflammatory phase and providing a more favorable host environment for depositing *Leishmania* parasites (Abdeladhim et al., 2014). Saliva of *P. papatasi* and *Phlebotomus duboscqi* inhibits host production of neutrophil migration signals MIP-1α, tumor necrosis factor (TNF), and leukotriene B4 by enhancing dendritic cell expression of the antiinflammatory cytokine IL-10 and PGE2 (Carregaro et al., 2008). Concomitantly, saliva of these two sand fly species reduces dendritic cell MHC II and costimulatory molecule B7.2 expression, thus diminishing antigen-presenting capability (Carregaro et al., 2008). Neutrophils exposed to saliva of *L. longipalpis* undergo apoptosis mediated by expression of Fas-ligand and

action of caspases (Prates et al., 2011). Saliva of *L. longipalpis* inhibits both alternative and classical complement pathway activation, thus downregulating, in the case of the alternative pathway, an important factor contributing to the early inflammatory response (Mendes-Sousa et al., 2013). Neutrophils play a role in wound healing of the sand fly bite site. They uptake *Leishmania* parasites within neutrophil phagolysosomes, and they orchestrate the cytokine-mediated influx of macrophages that phagocytose apoptotic neutrophils and become parasite infected (Peters and Sacks, 2009; Ribeiro-Gomes and Sacks, 2012). Saliva of *L. longipalpis* induces a chemokine response chemotactic for macrophages (Teixeira et al., 2005). In leishmaniasis, neutrophils associated with wound healing are exploited by the sand fly–transmitted parasite to help establish macrophage infection. Further impacting macrophage function at the bite site, saliva of sand fly downregulates Th1 and enhances Th2 cytokine polarization (Gomes and Oliveira, 2012).

Saliva of the triatomine, *D. maxima*, inhibits neutrophil oxidative burst (Assumpção et al., 2013). The salivary gland transcriptome of *Triatoma infestans* encodes serine protease and serpins that potentially inhibit complement activation, allergens that induce cutaneous hypersensitivity to bites, and other potential immunomodulators (Assumpção et al., 2008).

Countermeasures of ixodid ticks that delay host wound healing have direct implications for individual feeding ticks as well as a population of ticks that might infest an individual host. Modulation of the inflammatory phase of wound healing is particularly important, because ixodid ticks remain attached, salivate, and consume blood for approximately 4 or 5 days for larvae and nymphs to more than a week for adults. As in the case of rapidly blood-feeding vectors, suppression of host immune defenses by ticks reduces the likelihood that subsequent infestations of a host by the same tick species will result in the expression of acquired resistance that rejects the feeding tick.

Ticks modulate essentially all elements of inflammation and innate and adaptive immunity including complement, neutrophils, macrophages, cytokines, chemokines, adhesion molecules, mast cells, reactive oxygen species, nitric oxide, B and T lymphocyte activation, polarization of T lymphocyte responses, and other immune defenses (Kazimírová and Štibrániová, 2013; Wikel, 2013; Kotál et al., 2015; Chmelař et al., 2016a,b). The combination of next-generation sequencing and proteomics is expanding insights into the diversity redundancy of complex families of tick salivary gland–secreted proteins (Schwarz et al., 2014; Chmelař et al., 2016a). In addition, nonprotein tick saliva molecules contribute significantly to modification of host defenses (Oliveira et al., 2011).

Ticks evolved effective countermeasures to the earliest elements of the inflammatory response. Leukocyte migration from the vascular compartment to the tissues at the site of tick feeding can be reduced by downregulation of proinflammatory cytokine and chemokine production combined with suppression of expression of endothelial cell adhesion molecules and leukocyte integrin. Saliva of multiple tick species downregulates expression of the proinflammatory cytokines TNF and IL-1 and chemokines, including IL-8 and MIP-1α (Vancová et al., 2010; Kazimírová and Štibrániová, 2013; Stibrániová et al., 2013; Wikel, 2013; Kotál et al., 2015). Saliva inhibition of alternative pathway of complement activation by introduced microbes or damaged tissue would further reduce chemotaxis by reducing generation of anaphylatoxins (Wikel, 2013). Salivary gland extract of *D. andersoni* reduces cutaneous microvascular endothelial cell expression of ICAM-1 and that of *I. scapularis* inhibits endothelial cell expression of VCAM-1 and P-selectin (Maxwell et al., 2005). *I. scapularis* saliva disintegrin metalloprotease-like molecules downregulate neutrophil surface expression of β2 integrin (Guo et al., 2009). Cellular influxes to tick bite sites on previously unexposed hosts are characterized by inflammatory cell influx to the periphery of the feeding lesion and not adjacent to the tick mouthparts (Krause et al., 2009).

Reduction of leukocyte accumulation at the bite site combined with downregulation of functions of neutrophils, macrophages, and dendritic cells modifies inflammatory and innate immune responses and deviates or diminishes adaptive immune defenses. Inflammatory response and tissue damage are suppressed by *I. scapularis* salivary inhibition of neutrophil aggregation, phagocytosis, granule enzyme secretion, and reactive oxygen species activity (Ribeiro et al., 1990). A saliva lipocalin of *I. ricinus* binds leukotriene B4, resulting in both a reduction in number and activation of neutrophils at the bite site (Beaufays et al., 2008). Salp 16–like saliva proteins of *Ixodes persulcatus* suppress both activated neutrophil reactive oxygen species production and their chemotactic responsiveness to IL-8 (Hidano et al., 2014). *D. variabilis* saliva contains PGE2 that stimulates macrophage PGE2 secretion that in turn inhibits macrophage secretion of TNF and reduces fibroblast migration to the tick bite site, which negatively impacts wound healing (Poole et al., 2013). In addition to inhibiting TNF and IL-12 secretion, macrophages exposed for 24h to *R. microplus* salivary gland extract expressed fewer CD69, CD80 (B7-1), and CD86 (B7-2) costimulatory molecules, which implies impair development of T cell responses (Brake and Pérez de León, 2012).

Tick saliva of multiple species impacts chemotaxis, differentiation, maturation, cytokine secretion, and antigen presentation by dendritic cells that are a bridge between innate and adaptive immunity (Kazimírová and Štibrániová, 2013; Wikel, 2013; Kotál et al., 2015). These saliva-induced changes impact innate and adaptive immune function and modulate the healing response for maintaining a functioning feeding site. *I. scapularis* saliva PGE2 inhibits dendritic cell expression of the costimulatory molecule CD40 that provides important activating signals at the immunologic synapse when interacting with

T lymphocyte–expressed CD40 ligand (Sá-Nunes et al., 2007). A lipocalin in *Rhipicephalus appendiculatus* saliva reprograms dendritic cells by inhibiting upregulation of the costimulatory molecule CD86 and the maturation marker CD83; reducing secretion of IFN-α, IFN-γ, IL-1, IL-6, IL-12, and TNF; diminishing human dendritic cell Th1 and Th17 polarizing ability; and inhibiting dendritic cell differentiation from monocytes (Preston et al., 2013). Tick saliva alters dendritic cell phagocytosis, cytokine expression, uptake and presentation of *Borrelia* antigens, and costimulatory molecule expression that alter the host response to the spirochete and facilitate establishment of infection (Mason et al., 2013).

Adaptive immune responses that have implications for altering wound healing to maintain integrity of the feeding site are suppressed during tick feeding. Numerous tick–host associations are characterized by saliva-induced Th2 polarization with downregulation of the Th1 cytokines, IL-2, and IFN-γ and upregulation of the Th2 cytokines IL-4, IL-5, IL-6, and IL-10 (Kazimírová and Štibrániová, 2013; Stibrániová et al., 2013; Wikel, 2013; Kotál et al., 2015). Transcriptional profiling of *I. scapularis* attachment sites during initial and second infestations reveals inhibition of Th17 responses (Heinze et al., 2012). Tick saliva impairs T lymphocyte cytokine signaling, traps important growth cytokines, and alters T cell receptor signal transduction pathways (Kazimírová and Štibrániová, 2013; Wikel, 2013; Kotál et al., 2015). Antibody responses can be inhibited by defective T lymphocyte help for B lymphocytes and by direct tick saliva inhibition of B lymphocyte proliferation (Yu et al., 2006).

Altering or delaying wound healing is particularly important for ixodid ticks. Reducing the influx of inflammatory cells and mediators facilitates blood feeding. However, the reduced influx must not be so significant locally or systemically to increase the likelihood that a host will succumb to opportunistic infections. Timing, magnitude, and duration of host modulation are important parameters that remain to be thoroughly characterized.

THIRD PHASE: PROLIFERATION AND VECTORS

Proliferation phase of wound healing begins within several hours of injury with initiation linked to events of the inflammatory phase with which it overlaps (Greaves et al., 2013; Olczyk et al., 2014; Seifert and Maden, 2014). This phase of wound healing is characterized by the action of growth factors (PDGF, FGF, TGF-β, VEGF) produced by numerous cells including platelets, macrophages, and keratinocytes and fibroblast migration into the site with deposition of extracellular matrix, collagen fibril formation, angiogenesis, and emergence of granulation tissue (Gurtner et al., 2008; Schultz and Wysocki, 2009; Olczyk et al., 2014; Pastar et al., 2014). Temporal emergence of the proliferation phase means that long-duration blood-feeding ixodid ticks could be most directly impacted by this phase. However, saliva enzymes that can modify extracellular matrix and inhibitors of angiogenesis also occur in the saliva of more rapidly feeding vectors. Modification of extracellular matrix may be universally important for mouthpart probing and maintaining an effective feeding site for both short- and long-duration blood feeders. Role of angiogenesis inhibitors in saliva of rapidly blood-feeding vectors remains to be determined.

Saliva lipocalin, dipetalodipin, of the triatomine *D. maxima* inhibits host angiogenesis by binding 15(S)-HETE to block endothelial tube formation that may inhibit granulation tissue formation at the bite site (Assumpção et al., 2010). Disintegrin present in horsefly saliva is an angiogenesis inhibitor (Ma et al., 2010, 2011). Angiogenesis inhibitors have not been identified to date in mosquitoes and sand flies (Assumpção et al., 2012). First tick inhibitor of angiogenesis was detected in *Haemaphysalis longicornis* saliva as a troponin I–like inhibitor of human vascular endothelial cell capillary formation (Fukumoto et al., 2006). Haemangin is a novel Kunitz-type inhibitor of angiogenesis and inducer of apoptosis from the salivary glands of *H. longicornis* (Islam et al., 2009). Microvascular endothelial cell proliferation is inhibited and apoptosis is induced by saliva of both *I. scapularis* and (*Rhipicephalus*) *Boophilus microplus* (Francischetti et al., 2005). Genes encoding disintegrins with potential antiangiogenesis properties occur in salivary gland transcripts of several tick species (Assumpção et al., 2012).

Growth factors are important signaling and regulating molecules with multiple actions contributing to the orchestration of wound healing (Schultz and Wysocki, 2009; Greaves et al., 2013; Seifert and Maden, 2014). Tick salivary glands of several ixodid species produce inhibitors of wound healing growth factors. Salivary gland transcriptome of *D. andersoni* provided important clues for existence of saliva molecules modulating host wound healing (Alarcon-Chaidez et al., 2007). A search for tick saliva molecules inhibiting growth factors revealed that *I. scapularis* and *I. ricinus* bind PDGF; *Dermacentor reticulatus* and *R. appendiculatus* salivary gland extracts bind TGF-β, FGF, and hepatocyte growth factor (HGF); and *Amblyomma variegatum* binds FGF, HGF, PDGF, and TGF-β (Hajnická et al., 2011). Disruption of signaling during this later stage of wound healing contributes to maintaining a blood pool for the tick as it enters the rapid engorgement phase of blood feeding. Suppression of growth factors reduces revascularization, impairs development of a collagen meshwork, and suppresses formation of granulation tissue at the feeding site.

FOURTH PHASE: REMODELING AND VECTORS

Ixodid tick feeding would delay completion of the normal phases of acute wound healing until after the tick has detached. Many ixodid species produce an attachment cement cone that extends from the epidermal surface into the dermis in a pattern that varies among tick species (Moorhouse, 1969). Attachment cement is frequently left behind in the skin where it acts as a foreign body and it is eventually shed during healing of the skin. Tick saliva proteins are trapped in attachment cement where they have the potential to provide a prolonged antigenic stimulus (Allen et al., 1979).

During this final phase of wound healing, collagen is remodeled by matrix metalloproteases yielding collagen bundles that increase tissue tensile strength and result in extracellular matrix more closely resembling that of normal skin (Seifert and Maden, 2014). Ixodid ticks requiring a functioning feeding site for more than a week would be anticipated to have evolved saliva molecules to delay all phases of wound healing until detachment from the host. Modulating the final two phases of wound healing would not appear to be as important for a vector that obtains a blood meal for a few minutes to a few hours.

Matrix metalloproteases are important in regulation of tissue repair and cellular migration (Parks et al., 2004). Substrates for matrix metalloproteases include collagens, laminin, and fibrin. Salivary gland transcriptomes of numerous tick species contain transcripts encoding metalloproteases (Chmelař et al., 2016b). These saliva enzymes may contribute to modulation of host hemostasis, inflammation, and tissue repair (Decrem et al., 2008). Remodeling and disruption of extracellular matrix would contribute to maintaining a functioning feeding site.

CONCLUDING STATEMENT

Examination of the phases of wound healing provides an excellent platform for understanding the overlapping, complex, and integrated ways in which host defenses are suppressed and manipulated by blood-feeding vector arthropods. These vector-induced changes create environments that are favorable for introduction and establishment of infectious agents. This author believes that the clades of infectious agents transmitted by an individual vector species are dependent evolutionarily on how the vector modifies host defenses.

Our understanding of the relationships of vector arthropods with the molecules, cells, and pathways of wound healing has progressed at an increasing rate over the past few decades. There remain many unanswered questions and new tools are emerging to be used by an increasingly larger body of highly skilled and innovative researchers. Embedded in these complex associations are the underpinnings of novel immunological control approaches to block vectors and the infectious agents they transmit.

References

Abdeladhim, M., Jochim, R.C., Ben Ahmed, M., Zhioua, E., Chelbi, I., Cherni, S., Louzir, H., Ribeiro, J.M., Valenzuela, J.G., 2012. Updating the salivary gland transcriptome of *Phlebotomus papatasi* (Tunisian strain): the search for sand fly-secreted immunogenic proteins for humans. PLoS One 7, e47347.

Abdeladhim, M., Kamhawi, S., Valenzuela, J., 2014. What's behind a sand fly bite? The profound effect of sand fly saliva on host hemostasis, inflammation and immunity. Infect. Genet. Evol. 28, 691–703.

Adams, T.S., 1999. Hematophagy and hormone release. Ann. Entomol. Soc. Am. 92, 1–13.

Alarcon-Chaidez, F.J., Sun, J., Wikel, S.K., 2007. Construction and characterization of a cDNA library from the salivary glands of *Dermacentor andersoni* Stiles (Acari: Ixodidae). Insect Biochem. Mol. Biol. 37, 48–71.

Allen, J.R., 1973. Tick resistance: basophils in skin reactions of resistant guinea pigs. Int. J. Parasitol. 3, 195–200.

Allen, J.R., Khalil, H.M., Graham, J.E., 1979. The location of tick salivary antigens, complement and immunoglobulin in the skin of guinea-pigs infested with *Dermacentor andersoni*. Immunology 38, 467–472.

Andersen, J.F., Hinnebusch, B.J., Lucas, D.A., Conrads, T.P., Veenstra, T.D., Pham, V.M., Ribeiro, J.M., 2007. An insight into the sialome of the oriental rat flea, *Xenopsylla cheopis* (Rots). BMC Genomics 8, 102.

Assumpção, T.C., Alvarenga, P.H., Ribeiro, J.M., Andersen, J.F., Francischetti, I.M., 2010. Dipetalodipin, a novel multifunctional salivary lipocalin that inhibits platelet aggregation, vasoconstriction, and angiogenesis through unique binding specificity for TXA2, PGF2alpha, and 15(S)-HETE. J. Biol. Chem. 285, 39001–39012.

Assumpção, T.C., Francischetti, I.M., Andersen, J.F., Schwarz, A., Santana, J.M., Ribeiro, J.M., 2008. An insight into the sialome of the blood-sucking bug *Triatoma infestans*, a vector of Chagas' disease. Insect Biochem. Mol. Biol. 38, 213–232.

Assumpção, T.C., Ma, D., Schwarz, A., Reiter, K., Santana, J.M., Andersen, J.F., Ribeiro, J.M., Nardone, G., Yu, L.L., Francischetti, I.M., 2013. Salivary antigen-5/CAP family members are Cu^{2+}-dependent antioxidant enzymes that scavenge O_2 and inhibit collagen-induced platelet aggregation and neutrophil oxidative burst. J. Biol. Chem. 288, 14341–14361.

Assumpção, T.C., Ribeiro, J.M., Francischetti, I.M., 2012. Disintegrins from hematophagous sources. Toxins 4, 296–322.

Banki, E., Hajna, Z., Kemeny, A., Botz, B., Nagy, P., Bolcskei, K., Toth, G., Reglodi, D., Helyes, Z., 2014. The selective PAC1 receptor agonist maxadilan inhibits neurogenic vasodilation and edema formation in the mouse skin. Neuropharmacology 85, 538–547.

Barrientos, S., Stojadinovic, O., Golinko, M.S., Brem, H., Tomic-Canic, M., 2008. Growth factors and cytokines in wound healing. Wound Repair Regen. 16, 585–601.

Beaufays, J., Adam, B., Mentsen-Dedoyart, C., Fievez, L., Grosjean, A., Decrem, Y., Prevot, P.-P., Santini, S., Brasseur, R., Brossard, M., Vanhaeverbeek, M., Bureau, F., Heinen, E., Lins, L., Vanhamme, L., Godfroid, E., 2008. Ir-LBP, an *Ixodes ricinus* tick salivary LTB4-binding lipocalin, interferes with host neutrophil function. PLoS One 3, e3987.

Bergman, D.K., 1996. Mouthparts and feeding mechanisms of haematophagous arthropods. In: Wikel, S.K. (Ed.), The Immunology of Host-ectoparasitic Arthropod Relationships. CAB International, Wallingford, UK, pp. 30–61.

Berliner, L.J., 1992. Thrombin Structure and Function. Springer, New York.

Bernard, Q., Jaulhac, B., Boulanger, N., 2014. Smuggling across the border: how arthropod-borne pathogens evade and exploit the host defense system of the skin. J. Invest. Dermatol. 134, 1211–1219.

Bernard, Q., Jaulhac, B., Boulanger, N., 2015. Skin and arthropods: an effective interaction used by pathogens in vector-borne diseases. Eur. J. Dermatol. 25 (Suppl. 1), 18–22.

Bizzarro, B., Barros, M.S., Maciel, C., Gueroni, D.I., Lino, C.N., Campopiano, J., Kotsyfakis, M., Amarante-Mendes, G.P., Calvo, E., Capurro, M.L., Sá-Nunes, A., 2013. Effects of *Aedes aegypti* salivary components on dendritic cell and lymphocyte biology. Parasites Vectors 6, 329.

Blair, P., Flaumenhaft, R., 2009. Platelet alpha-granules: basic biology and clinical correlates. Blood Rev. 23, 177–189.

Bonilla, D.L., Durden, L.A., Eremeeva, M.E., Dasch, G.A., 2013. The biology and taxonomy of head and body lice-implications for louse-borne disease prevention. PLoS Pathog. 9, e1003724.

Boppana, V.D., Thangamkani, S., Adler, A.J., Wikel, S.K., 2009. SAAG-4 is a novel mosquito salivary protein that programs host CD4+ T cells to express IL-4. Parasite Immunol. 31, 287–295.

Brake, D.K., Pérez de León, A.A., 2012. Immunoregulation of bovine macrophages by factors in the salivary glands of *Rhipicephalus microplus*. Parasites Vectors 5, 38.

Burke, G., Wikel, S.K., Spielman, A., Pollack, R., McKay, K., Krause, P.J., Tick-borne Disease Study Group, 2005. Cutaneous tick hypersensitivity in humans is associated with decreased Lyme disease risk. Emerging Infect. Dis. 11, 36–41.

Cadiergues, M.C., Hourcq, P., Cantaloube, B., Franc, M., 2000. First bloodmeal of *Ctenocephalides felis felis* (Siphonaptera: Pulicidae) on cats: time to initiation and duration of feeding. J. Med. Entomol. 37, 634–636.

Calvo, E., Mans, B.J., Andersen, J.F., Ribeiro, J.M., 2006. Function and evolution of a mosquito salivary protein family. J. Biol. Chem. 281, 1935–1942.

Calvo, E., Mizurini, D.M., Sá-Nunes, A., Ribeiro, J.M., Andersen, J.F., Mans, B.J., Monteiro, R.Q., Kotsyfakis, M., Francischetti, I.M., 2011. Alboserpin, a factor Xa inhibitor from the mosquito vector of yellow fever, binds heparin and membrane phospholipids and exhibits antithrombotic activity. J. Biol. Chem. 286, 27998–28010.

Calvo, E., Tokumasu, F., Marinotti, O., Villeval, J.L., Ribeiro, J.M., Francischetti, I.M., 2007. Aegyptin, a novel mosquito salivary gland protein, specifically binds to collagen and prevents its interaction with platelet glycoprotein VI, integrin alpha2beta1, and von Willebrand factor. J. Biol. Chem. 282, 26928–26938.

Calvo, E., Tokumasu, F., Mizurini, D.M., McPhie, P., Narum, D.L., Ribeiro, J.M., Monteiro, R.Q., Francischetti, I.M., 2010. Aegyptin displays high-affinity for the von Willebrand factor binding site (RGQOGVMGF) in collagen and inhibits carotid thrombus formation in vivo. FEBS J. 277, 413–427.

Campos, I.T., Tanaka-Azevedo, A.M., Tanaka, A.S., 2004. Identification and characterization of a novel factor XIIa inhibitor in the hematophagous insect, *Triatoma infestans* (Hemiptera: Reduviidae). FEBS Lett. 577, 512–516.

Carlos, T.M., Harlan, J.M., 1994. Leukocyte-endothelial adhesion molecules. Blood 84, 2068–2101.

Carregaro, V., Valenzuela, J.G., Cunha, T.M., Verri Jr., W.A., Grespan, R., Matsumura, G., Ribeiro, J.M., Elnaiem, D.E., Silva, J.S., Cunha, F.Q., 2008. *Phlebotomine salivas* inhibit immune inflammation-induced neutrophil migration via an autocrine DC-derived PGE2/IL-10 sequential pathway. J. Leukocyte Biol. 84, 104–114.

Chadee, D.D., Beier, J.C., 1997. Factors influencing the duration of blood-feeding by laboratory reared and wild *Aedes aegypti* (Diptera: Culicidae) from Trinidad, West Indies. Ann. Trop. Med. Parasitol. 91, 199–207.

Champagne, D.E., Ribeiro, J.M., 1994. Sialokinin I and II: vasodilatory tachykinins from the yellow fever mosquito *Aedes aegypti*. Proc. Nat. Acad. Sci. U.S.A. 91, 138–142.

Champagne, D.E., Smartt, C.T., Ribeiro, J.M., James, A.A., 1995. The salivary gland specific apyrase of the mosquito Aedes aegypti is a member of the 5′-nucleotidase family. Proc. Nat. Acad. Sci. U.S.A. 92, 694–698.

Chmelař, J., Oliveira, C.J., Rezacova, P., Francischetti, I.M., Kovarova, Z., Pejler, G., Kopacek, P., Ribeiro, J.M., Mares, M., Kopecky, J., Kotsyfakis, M., 2011. A tick salivary protein targets cathepsin G and chymase and inhibits host inflammation and platelet aggregation. Blood 117, 736–744.

Chmelař, J., Calvo, E., Pedra, J.H., Francischetti, I.M., Kotsyfakis, M., 2012. Tick salivary secretion as a source of antihemostatics. J. Proteomics 75, 3842–3854.

Chmelař, J., Kotál, J., Karim, S., Kopacek, P., Francischetti, I.M., Pedra, J.H., Kotsyfakis, M., 2016a. Sialomes and Mialomes: a systems-biology view of tick tissues and tick-host interactions. Trends Parasitol. 32, 242–254.

Chmelař, J., Kotál, J., Kopecký, J., Pedra, J.H., Kotsyfakis, M., 2016b. All for one and one for all on the tick-host battlefield. Trends Parasitol. 32, 368–377.

Clark, R.A.F., 2001. Fibrin and wound healing. Ann. N. Y. Acad. Sci. 936, 355–367.

Collin, N., Assumpção, T.C., Mizurini, D.M., Gilmore, D.C., Dutra-Oliveira, A., Kotsyfakis, M., Sá-Nunes, A., Teixeira, C., Ribeiro, J.M., Monteiro, R.Q., Valenzuela, J.G., Francischetti, I.M., 2012. Lufaxin, a novel factor Xa inhibitor from the salivary gland of the sand fly *Lutzomyia longipalpis* blocks protease-activated receptor 2 activation and inhibits inflammation and thrombosis in vivo. Arterioscler. Thromb. Vasc. Biol. 32, 2185–2198.

Conway, M.J., Colpitts, T.M., Fikrig, E., 2014. Role of the vector in arbovirus transmission. Annu. Rev. Virol. 1, 71–88.

Cooley, R.A., Kohls, G.M., 1944. The Argasidae of North America, Central America and Cuba. Am. Midl. Nat. Monogr. 1, 1–152.

Cumberbatch, M., Dearman, R.J., Griffiths, C.E., Kimber, I., 2000. Langerhans cell migration. Clin. Exp. Dermatol. 25, 413–418.

Cupp, M.S., Ribeiro, J.M., Champagne, D.E., Cupp, E.W., 1998. Analyses of cDNA and recombinant protein for a potent vasoactive protein in saliva of a blood-feeding black fly, Simulium vittatum. J. Exp. Biol. 201, 1553–1561.

Decrem, Y., Beaufays, J., Blasioli, V., Lahaye, K., Brossard, M., Vanhamme, L., Godgroid, E., 2008. A family of putative metalloproteases in the salivary glands of the tick *Ixodes ricinus*. FEBS J. 275, 1485–1499.

Demeure, C.E., Brahimi, K., Hacini, F., Marchand, F., Péronet, R., Huerre, M., St-Mezard, P., Nicolas, J.F., Brey, P., Delespesse, G., Mécheri, S., 2005. *Anopheles* mosquito bites activate cutaneous mast cells leading to a local inflammatory response and lymph node hyperplasia. J. Immunol. 174, 3932–3940.

des Vignes, F., Piesman, J., Heffernan, R., Schulze, T.L., Stafford 3rd, K.C., Fish, D., 2001. Effect of tick removal on transmission of *Borrelia burgdorferi* and *Ehrlichia phagocytophila* by *Ixodes scapularis* nymphs. J. Infect. Dis. 183, 773–778.

Diegelmann, R.F., Evans, M.C., 2004. Wound healing: an overview of acute, fibrotic and delayed healing. Front. Biosci. 9, 283–289.

Du, X., Plow, E.F., Frelinger III, A.L., O'Toole, T.E., Loftus, J., Ginsberg, M.H., 1991. Ligands "activate" integrin $\alpha_{IIb}\beta_3$ (platelet GPIIb-IIIa). Cell 65, 409–416.

Eming, S.A., Hammerschmidt, M., Krieg, T., Roers, A., 2009. Interrelation of immunity and tissue repair or regeneration. Semin. Cell Dev. Biol. 20, 517–527.

Eming, S.A., Krieg, T., Davidson, J.M., 2007. Inflammation in wound repair: molecular and cellular mechanisms. J. Invest. Dermatol. 127, 514–525.

Eming, S.A., Martin, P., Tomic-Canic, M., 2014. Wound repair and regeneration: mechanisms, signaling and translation. Sci. Transl. Med. 6, 265.

Fontaine, A., Diouf, I., Bakkali, N., Missé, D., Pagès, F., Fusai, T., Rogier, C., Almeras, L., 2011. Implication of haematophagous arthropod salivary proteins in host-vector interactions. Parasites Vectors 4, 187.

Francischetti, I.M., 2010. Platelet aggregation inhibitors from hematophagous animals. Toxicon 56, 1130–1144.

Francischetti, I.M., Mather, T.N., Ribeiro, J.M., 2003. Cloning of a salivary gland metalloprotease and characterization of gelatinase and fibrin(ogen)lytic activities in the saliva of the Lyme disease tick vector *Ixodes scapularis*. Biochem. Biophys. Res. Commun. 305, 869–875.

Francischetti, I.M., Mather, T.N., Ribeiro, J.M., 2004. Penthalaris, a novel recombinant five-Kunitz tissue factor pathway inhibitor (TFPI) from the salivary gland of the tick vector of Lyme disease, *Ixodes scapularis*. Thromb. Haemostasis 91, 886–898.

Francischetti, I.M., Mather, T.N., Ribeiro, J.M., 2005. Tick saliva is a potent inhibitor of endothelial cell proliferation and angiogenesis. Thromb. Haemostasis 94, 167–174.

Francischetti, I.M.B., Sa-Nunes, A., Mans, B.J., Santos, I.M., Ribeiro, J.M.C., 2009. The role of saliva in tick feeding. Front. Biosci. 14, 2051–2088.

Francischetti, I.M.B., Valenzuela, J.G., Andersen, J.F., Mather, T.N., Ribeiro, J.M.C., 2002. Ixolaris, a novel recombinant tissue factor pathway inhibitor (TFPI) from the salivary gland of the tick, *Ixodes scapularis*: identification of factor X and factor Xa as scaffolds for the inhibition of factor VIIa/tissue factor complex. Blood 99, 3602–3612.

Fukumoto, S., Sakaguchi, T., You, M., Xuan, X., Fujisaki, K., 2006. Tick troponin I-like molecule is a potent inhibitor for angiogenesis. Microvasc. Res. 71, 218–221.

Gachet, C., 2001. ADP receptors of platelets and their inhibition. Thromb. Haemostasis 86, 222–232.

Golebiewska, E.M., Poole, A.W., 2015. Platelet secretion: from hemostasis to wound healing and beyond. Blood Rev. 29, 153–162.

Gomes, R., Oliveira, F., 2012. The immune response to sand fly salivary proteins and its influence on *Leishmania* immunity. Front. Immunol. 3, 110.

Greaves, N.S., Ashcroft, K.J., Baguneid, M., Bayat, A., 2013. Current understanding of molecular and cellular mechanisms in fibroplasia and angiogenesis during acute wound healing. J. Dermatol. Sci. 72, 206–217.

Guo, X., Booth, C.J., Paley, M.A., Wang, X., DePonte, K., Fikrig, E., Narasimhan, S., Montogomery, R.R., 2009. Inhibition of neutrophil function by two tick salivary proteins. Infect. Immun. 77, 2320–2329.

Gurtner, G.C., Werner, S., Barrandon, Y., Longaker, M.T., 2008. Wound repair and regeneration. Nature 453, 314–321.

Hajnická, V., Vančová-Stibrániová, I., Slovák, M., Kocáková, P., Nuttall, P.A., 2011. Ixodid tick salivary gland products target host wound healing growth factors. Int. J. Parasitol. 41, 213–223.

Hayashi, H., Kyushiki, H., Nagano, K., Sudo, T., Iyori, M., Matsuoka, H., Yoshida, S., 2012. Anopheline anti-platelet protein from a malaria vector mosquito has anti-thrombotic effects in vivo without compromising hemostasis. Thromb. Res. 129, 169–175.

Heinze, D.M., Carmical, J.R., Aronson, J.F., Thangamani, S., 2012. Early immunologic events at the tick-host interface. PLoS One 7, e47301.

Hidano, A., Konnai, S., Yamada, S., Githaka, N., Isezaki, M., Higuchi, H., Nagahata, H., Ito, T., Takano, A., Ando, S., Kawabata, H., Murata, S., Ohahsi, K., 2014. Suppressive effects of neutrophil by Salp16-like salivary gland proteins from *Ixodes persulcatus* Schulze tick. Insect Mol. Biol. 23, 466–474.

Hinz, B., 2007. Formation and function of the myofibroblast during tissue repair. J. Invest. Dermatol. 127, 526–537.

Isawa, H., Yuda, M., Orito, Y., Chinzei, Y., 2002. A mosquito salivary protein inhibits activation of the plasma contact system by binding to factor XII and high molecular weight kininogen. J. Biol. Chem. 277, 27651–27658.

Islam, M.K., Tsuji, N., Miyoshi, T., Alim, M.A., Huang, X., Hatta, T., Fujisaki, K., 2009. The Kunitz-like modulatory protein haemangin is vital for hard tick blood-feeding success. PLoS Pathog. 5, e1000497.

Kaufman, W.R., 1989. Tick-host interaction: a synthesis of current concepts. Parasitol. Today 5, 47–56.

Kazimírová, M., Štibrániová, I., 2013. Tick salivary compounds: their role in modulation of host defences and pathogen transmission. Front. Cell. Infect. Microbiol. 3, 43.

Komozsynski, M.A., Wojtczak, A., 1996. Apyrases (ATP diphosphohydrolase, EC 3.6.1.5): function and relationship to ATPases. Biochem. Biophys. Acta 1310, 233–241.

Kotál, J., Langhansová, H., Lieskovská, J., Andersen, J.F., Francischetti, I.M., Chavakis, T., Kopecký, J., Pedra, J.H., Kotsyfakis, M., Chmelař, J., 2015. Modulation of host immunity by tick saliva. J. Proteomics 128, 58–68.

Krause, P.J., Grant-Kels, J.M., Tahan, S.R., Dardick, K.R., Alarcon-Chaidez, F., Bouchard, K., Visini, C., Deriso, C., Foppa, I.M., Wikel, S., 2009. Dermatologic changes induced by repeated *Ixodes scapularis* bites and implications for prevention of tick-borne infection. Vector-Borne Zoonotic Dis. 9, 603–610.

Lavoipierre, M.M.J., 1965. Feeding mechanisms of blood-sucking arthropods. Nature Lond. 208, 302–303.

Lerner, E.A., Ribeiro, J.M., Nelson, R.J., Lerner, M.R., 1991. Isolation of maxadilan, a potent vasodilatory peptide from the salivary glands of the sand fly *Lutzomyia longipalpis*. J. Biol. Chem. 266, 11234–11236.

Liyou, N., Hamilton, S., Elvin, C., Willadsen, P., 1999. Cloning and expression of ecto 5-nucleotidase from the cattle tick *Boophilus microplus*. Insect Mol. Biol. 8, 257–266.

Ma, D., Gao, L., An, S., Song, Y., Wu, J., Xu, X., Lai, R., 2010. A horsefly saliva antigen 5-like protein containing RTS motif is an angiogenesis inhibitor. Toxicon 55, 45–51.

Ma, D., Xu, X., An, S., Liu, H., Yang, X., Andersen, J.F., Wang, Y., Tokumasu, F., Ribeiro, J.M., Francischetti, I.M., Lai, R., 2011. A novel family of RGD-containing disintegrins (Tablysin-15) from the salivary gland of the horsefly *Tabanus yao* targets αIIbβ3 or αVβ3 and inhibits platelet aggregation and angiogenesis. Thromb. Haemostasis 105, 1032–1045.

Madlener, M., Parks, W.C., Werner, S., 1998. Matrix metalloproteinases (MMPs) and their physiological inhibitors (TIMPs) are differentially expressed during excisional skin wound repair. Exp. Cell Res. 242, 201–210.

Martinez-Ibarra, J.A., Miguel-Alvarez, A., Arredondo-Jimenez, J.I., Rodriguez-Lopez, M.H., 2001. Update on the biology of *Triatoma dimidiata* Latreille (Hemiptera: Reduviidae) under laboratory conditions. J. Am. Mosq. Control Assoc. 17, 209–210.

Mason, I.M., Veerman, C.C., Geijtenbeek, T.B., Hovius, J.W., 2013. Ménage à trois: *Borrelia*, dendritic cells, and tick saliva interactions. Trends Parasitol. 30, 95–103.

Maxwell, S.S., Stoklasek, T.A., Dashı, Y., Macaluso, K.R., Wikel, S.K., 2005. Tick modulation of the in vitro expression of adhesion molecules by skin-derived endothelial cells. Ann. Trop. Med. Parasitol. 99, 661–672.

Mellanby, K., 1946. Man's reaction to mosquito bites. Nature 158, 554.

Mendes-Sousa, A.F., Nascimento, A.A., Queiroz, D.C., Vale, V.F., Fujiwara, R.T., Araújo, R.N., Pereira, M.H., Gontijo, N.F., 2013. Different host complement systems and their interactions with saliva from *Lutzomyia longipalpis* (Diptera, Psychodidae) and *Leishmania infantum* promastigotes. PLoS One 8, e79787.

Mizurini, D.M., Francischetti, I.M., Monteiro, R.Q., 2013. Aegyptin inhibits collagen-induced coagulation activation in vitro and thromboembolism in vivo. Biochem. Biophys. Res. Commun. 436, 235–239.

Moorhouse, D.E., 1969. The attachment of some ixodid ticks to their natural hosts. In: Proceedings of the Second International Congress of Acarology. Hungary, Hungarian Academy of Sciences, Budapest, pp. 319–327.

Mulenga, A., Kim, T., Ibelli, A.M., 2013. *Amblyomma americanum* tick saliva serine protease inhibitor 6 is a cross-class inhibitor of serine proteases and papain-like cysteine proteases that delays plasma clotting and inhibits platelet aggregation. Insect Mol. Biol. 22, 306–319.

Olczyk, P., Mencner, L., Komosinska-Vassev, K., 2014. The role of extracellular matrix components in cutaneous wound healing. BioMed Res. Int. 2014:747584.

Oliveira, C.J., Sá-Nunes, A., Francischetti, I.M., Carregaro, V., Anatriello, E., Silva, J.S., Santos, I.K., Ribeiro, J.M., Ferreira, B.R., 2011. Deconstructing tick saliva: non-protein molecules with potent immunomodulatory properties. J. Biol. Chem. 286, 10960–10969.

Parks, W.C., Wilson, C.L., Lopez-Boado, S., 2004. Matrix metalloproteases as modulators of inflammation and innate immunity. Nature Rev. Immunol. 4, 617–629.

Pastar, I., Stojadinovic, O., Yin, N.C., Ramirez, H., Nusbaum, A.G., Sawaya, A., Patel, S.B., Khalid, L., Isseroff, R.R., Tomic-Canic, M., 2014. Epithelialization in wound healing: a comprehensive review. Adv. Wound Care 3, 445–464.

Peters, N.C., Sacks, D.L., 2009. The impact of vector-mediated neutrophil recruitment on cutaneous leishmaniasis. Cell. Microbiol. 11, 1290–1296.

Poole, N.M., Mamidanna, G., Smith, R.A., Coons, L.B., Cole, J.A., 2013. Prostaglandin E2 in tick saliva regulates macrophage cell migration and cytokine profile. Parasites Vectors 6, 261.

Prates, D.B., Araújo-Santos, T., Luz, N.F., Andrade, B.B., França-Costa, J., Afonso, L., Clarêncio, J., Miranda, J.C., Bozza, P.T., Dosreis, G.A., Brodskyn, C., Barral-Netto, M., Borges, V.M., Barral, A., 2011. *Lutzomyia longipalpis* saliva drives apoptosis and enhances parasite burden in neutrophils. J. Leukocyte Biol. 90, 575–582.

Preston, S.G., Majtán, J., Kouremenou, C., Rysnik, O., Burger, L.F., Cabezas Cruz, A., Chiong Guzman, M., Nunn, M.A., Paesen, G.C., Nuttall, P.A., Austyn, J.M., 2013. Novel immunomodulators from hard ticks selectively reprogramme human dendritic cell responses. PLoS Pathog. 9, e1003450.

Ramasubramanian, M.K., Barham, O.M., Swaninatham, V., 2008. Mechanics of a mosquito bite with applications to microneedle design. Bioinspiration Biomimetics 3, 046001.

Ribeiro, J.M., 1992. Characterization of a vasodilator from the salivary glands of the yellow fever mosquito *Aedes aegypti*. J. Exp. Biol. 165, 61–71.

Ribeiro, J., Alarcon-Chaidez, F., Francischetti, I.M.B., Mans, B., Mather, T.N., Valenzuela, J.G., Wikel, S.K., 2006. An annotated catalog of salivary gland transcripts from *Ixodes scapularis* ticks. Insect Biochem. Mol. Biol. 36, 111–129.

Ribeiro, J.M.C., Arca, B., Lombardo, F., Calvo, E., Phan, V.M., Chandra, P.K., Wikel, S.K., 2007. An annotated catalogue of salivary gland transcripts in the adult female mosquito, *Aedes aegypti*. BMC Genomics 8, 6.

Ribeiro, J.M.C., Francischetti, I.M.B., 2003. Role of arthropod saliva in blood feeding: sialome and post-sialome perspectives. Annu. Rev. Entomol. 48, 73–88.

Ribeiro, J.M., Katz, O., Pannell, L.K., Waitumbi, J., Warburg, A., 1999. Salivary glands of the sand fly *Phlebotomus papatasi* contain pharmacologically active amounts of adenosine and 5'-AMP. J. Exp. Biol. 202, 1551–1559.

Ribeiro, J.M.C., Mather, T.N., 1998. *Ixodes scapularis*: salivary kininase activity is a metallo dipeptidyl carboxypeptidase. Exp. Parasitol. 89, 213–221.

Ribeiro, J.M., Nussenzveig, R.H., 1993. The salivary catechol oxidase/peroxidase activities of the mosquito *Anopheles albimanus*. J. Exp. Biol. 179, 273–287.

Ribeiro, J.M., Nussenzveig, R.H., Tortorella, G., 1994. Salivary vasodilators of *Aedes triseriatus* and *Anopheles gambiae* (Diptera: Culicidae). J. Med. Entomol. 31, 747–753.

Ribeiro, J.M.C., Weis, J.J., Telford III, S.R., 1990. Saliva of the tick *Ixodes dammini* inhibits neutrophil function. Exp. Parasitol. 70, 382–388.

Ribeiro-Gomes, F.L., Sacks, D., 2012. The influence of early neutrophil-*Leishmania* interactions on the host immune response to infection. Cell. Infect. Microbiol. 2, 59.

Rogers, M.E., Bates, P.A., 2007. *Leishmania* manipulation of sand fly feeding behavior results in enhanced transmission. PLoS Pathog. 3, e91.

Sá-Nunes, A., Bafica, A., Lucas, D.A., Conrads, T.P., Veenstra, T.D., Andersen, J.F., Mather, T.N., Ribeiro, J.M.C., Francischetti, I.M.B., 2007. Prostaglandin E2 is a major inhibitor of dendritic cell maturation and function in *Ixodes scapularis* saliva. J. Immunol. 179, 1497–1505.

Schneider, B.S., Higgs, S., 2008. The enhancement of arbovirus transmission and disease by mosquito saliva is associated with modulation of the host immune response. Trans. R. Soc. Trop. Med. Hyg. 102, 400–408.

Schultz, G.S., Wysocki, A., 2009. Interactions between extracellular matrix and growth factors in wound healing. Wound Repair Regen. 17, 153–162.

Schwarz, A., Tenzer, S., Hackenberg, M., Erhart, J., Gerhold-Ay, A., Mazur, J., Kuharev, J., Ribeiro, J.M., Kotsyfakis, M., 2014. A systems level analysis reveals transcriptomic and proteomic complexity in *Ixodes ricinus* midgut and salivary glands during early attachment and feeding. Mol. Cell. Proteomics 13, 2725–2735.

Seifert, A.W., Maden, M., 2014. New insights into vertebrate skin regeneration. In: Kwang, W.J. (Ed.). Kwang, W.J. (Ed.), International Review of Cell and Molecular Biology, vol. 310. Elsevier, Amsterdam, pp. 129–169.

Shaw, T.J., Martin, P., 2009. Wound repair at a glance. J. Cell Sci. 122, 3209–3213.

Sinno, H., Prakash, S., 2013. Complements and the wound healing cascade: an updated review. Plast. Surg. Int. 2013:146764.

Sneddon, J.M., Vane, J.R., 1988. Endothelium derived relaxing factor reduces platelet adhesion to bovine endothelial cells. Proc. Nat. Acad. Sci. U.S.A. 85, 2800–2804.

Stibrániová, I., Lahová, M., Bartíková, P., 2013. Immunomodulators in tick saliva and their benefits. Acta Virol. 57, 200–216.

Strbo, N., Yin, N., Stojadinovic, O., 2014. Innate and adaptive immune responses in wound epithelization. Adv. Wound Care 3, 492–501.

Teixeira, C.R., Teixeira, M.J., Gomes, R.B., Santos, C.S., Andrade, B.B., Raffaele-Netto, I., Silva, J.S., Guglielmotti, A., Miranda, J.C., Barral, A., Brodskyn, C., Barral-Netto, M., 2005. Saliva from *Lutzomyia longipalpis* induces CC chemokine ligand 2/monocyte chemoattractant protein-1 expression and macrophage recruitment. J. Immunol. 175, 8346–8353.

Thangamani, S., Higgs, S., Ziegler, S., Vanlandingham, D., Tesh, R., Wikel, S., 2010. Host immune response to mosquito-transmitted chikungunya virus differs from that elicited by needle inoculated virus. PLoS One 5, 8.

Thomas, I., Kihiczak, G.G., Schwartz, R.A., 2004. Bedbug bites: a review. Int. J. Dermatol. 43, 430–433.

Trager, W., 1939. Acquired immunity to ticks. J. Parasitol. 25, 57–81.

Valenzuela, J.G., Charlab, R., Galperin, M.Y., Ribeiro, J.M., 1998. Purification, cloning, and expression of an apyrase from the bed bug *Cimex lectularius*. A new type of nucleotide-binding enzyme. J. Biol. Chem. 273, 30583–30590.

Valenzuela, J.G., Walker, F.A., Ribeiro, J.M., 1995. A salivary nitrophorin (nitric-oxide-carrying hemoprotein) in the bedbug *Cimex lectularius*. J. Exp. Biol. 198, 1519–1526.

Vancová, I., Hajincká, V., Slovák, M., Nuttall, P.A., 2010. Anti-chemokine activities of ixodid ticks depend on tick species, developmental stage, and duration of feeding. Vet. Parasitol. 167, 274–278.

Versteeg, H.H., Heemskerk, J.W.M., Levi, M., Reitsma, P.H., 2013. New fundamentals in hemostasis. Physiol. Rev. 93, 327–358.

Walsh, T.G., Metharom, P., Berndt, M.C., 2015. The functional role of platelets in the regulation of angiogenesis. Platelets 26, 199–211.

Wanasen, N., Nussenzveig, R.H., Champagne, D.E., Soong, L., Higgs, S., 2004. Differential modulation of murine host immune response by salivary gland extracts from the mosquitoes *Aedes aegypti* and *Culex quinquefasciatus*. Med. Vet. Entomol. 18, 191–199.

Wasserman, H.A., Singh, S., Champagne, D.E., 2004. Saliva of the yellow fever mosquito, *Aedes aegypti*, modulates murine lymphocyte function. Parasite Immunol. 26, 295–306.

Watanabe, R.M.O., Tanaka-Azevero, A.M., Araujo, M.S., Juliano, M.A., Tanaka, A.S., 2011. Characterization of thrombin inhibitory mechanism or rAaTI, a Kazal-type inhibitor from *Aedes aegypti* with anticoagulant activity. Biochimie 93, 618–623.

Watson, S.P., Auger, J.M., McCarty, O.J., Pearce, A.C., 2005. GPVI and integrin alphaIIb beta3 signaling in platelets. J. Thromb. Haemostasis 3, 1752–1762.

Wikel, S.K., 1982. Immune responses to arthropods and their products. Annu. Rev. Entomol. 27, 21–48.

Wikel, S., 2013. Ticks and tick-borne pathogens at the cutaneous interface: host defenses, tick countermeasures, and a suitable environment for pathogen establishment. Front. Microbiol. 4, 337.

Yu, D., Liang, J., Yu, H., Wu, H., Xu, C., Liu, J., Lai, R., 2006. A tick B-cell inhibitory protein from salivary glands of the hard tick, *Hyalomma asiaticum asiaticum*. Biochem. Biophys. Res. Commun. 343, 585–590.

Yuda, M., Higuchi, K., Sun, J., Kureishi, Y., Ito, M., Chinzei, Y., 1997. Expression, reconstitution and characterization of prolixin-S as a vasodilator–a salivary gland nitric-oxide-binding hemoprotein of *Rhodnius prolixus*. Eur. J. Biochem./FEBS 249, 337–342.

Yukami, T., Hasegawa, M., Matsushita, Y., Fujita, T., Matsushita, T., Horikawa, K., Yanaba, K., Hamaguchi, Y., Nagaoka, T., Ogawa, F., Fujimoto, M., Steeber, D.A., Tedder, T.F., Takehara, K., Sato, S., 2007. Endothelial selectins regulate skin wound healing in cooperation with L-selectin and ICAM-1. J. Leukocyte Biol. 82, 519–531.

Zeidner, N.S., Higgs, S., Happ, C.M., Beaty, B.J., Miller, B.R., 1999. Mosquito feeding modulates Th1 and Th2 cytokines in flavivirus susceptible mice: an effect mimicked by injection of sialokinins, but not demonstrated in flavivirus resistant mice. Parasite Immunol. 21, 35–44.

Salivary Kratagonists: Scavengers of Host Physiological Effectors During Blood Feeding

John F. Andersen, José M.C. Ribeiro

NIH/NIAID Laboratory of Malaria and Vector Research, Rockville, MD, United States

INTRODUCTION

The salivas of blood sucking arthropod species contain a large variety of pharmacologically active components (Ribeiro, 1995). Almost 30 years ago, antihistaminic activity was reported from the saliva of the hematophagous kissing bug, *Rhodnius prolixus* (Ribeiro, 1982). Somewhat surprisingly, in 1994, the nitrophorins, a novel class of salivary proteins from this insect were shown to tightly bind histamine (Ribeiro and Walker, 1994) and were therefore responsible for the observed antihistaminic activity. It may appear surprising that histamine, an amine of molecular weight 100, is antagonized by a 20,000 Da protein from the bug's saliva. At first glance it would appear an inefficient solution to a problem, but today we know that unrelated arthropods such as ticks, mosquitoes, sand flies and kissing bugs have independently evolved proteins that antagonize, through high-affinity binding, not only histamine, but also serotonin, norepinephrine, adenosine diphosphate (ADP), adenosine triphosphate, thromboxane

A2 (TXA2), leukotriene B4 (LTB4), cysteinyl leukotrienes, and even collagen, all agonists important in hemostasis and inflammation (Calvo et al., 2009, 2006, 2007; Mans et al., 2008a,b,c; Andersen et al., 2003; Francischetti et al., 2002; Paesen et al., 1999). More recently, mechanistically similar protein antagonists have also been found in tick saliva that recognize several cytokines and chemokines (Frauenschuh et al., 2007; Gillespie et al., 2001; Vancova et al., 2007; Konik et al., 2006). Many of these proteins have been crystallized and their binding sites and ligand specificities characterized (Paesen et al., 1999; Andersen et al., 1998; Mans et al., 2007).

Salivary proteins utilizing this sequestration mechanism of action tend to target agonists that elicit rapid host physiological responses, which significantly affect blood feeding. Thromboxane secretion and granule release by platelets occurs within seconds after activation by collagen exposed via wounding by insect mouthparts, leading to an amplification of platelet activation and subsequent aggregation (Swieringa et al., 2014). Binding of collagen, ADP and TXA2 by salivary proteins quickly reduces the local free concentrations of these agonists (Calvo et al., 2009; Alvarenga et al., 2010; Soter et al., 1983). Mast cells are also activated almost immediately during feeding by the formation of IgE complexes with salivary proteins bound to FcεR receptors, resulting in the release of histamine and serotonin-containing granules and the secretion of the cysteinyl leukotriene LTC4, which is subsequently converted to the active metabolites LTC4 and LTD4. These mediators cause vascular leakage, itching, and pain that can reduce the quality of the blood meal and elicit host defensive responses (Calvo et al., 2009; Alvarenga et al., 2010; Soter et al., 1983).

It is interesting to note that blood feeding arthropods are found in diverse phylogenetic groups that evolved independently to hematophagy and have acquired the agonist sequestration mechanism by convergence using a number of different protein families. The evolutionary recruitment of certain families suitable for this function into the salivary proteome suggests that proteins having similar functions may act in the endogenous physiology of the blood feeder. Although this type of antagonism could be included under the classification of Chemical Antagonism as defined by the IUPHAR (Neubig et al., 2003), it is not even considered in their Table 7 description of terms associated with antagonists. Because most chemical antagonists are small molecules that react chemically with the agonists or bind to their receptors, we believe that most scientists do not consider antagonism by sequestration as a possibility, and accordingly do not plan experiments toward their identification. We thus propose a novel general term to describe all these antagonists that inactivate their respective agonists by tightly binding to them; we consider the term kratagonists [composed of the Greek work κρατώ (krato) that means hold, and the term agonists] as more suitable to describe and identify this class of chemical antagonists (Ribeiro and Arca, 2009). In immunology, a similar function has been described in the case of "soluble receptors." Soluble receptors derive from membrane receptors that are clipped and solubilized by proteases. Their high affinity toward agonists can compete with the membrane receptors for the agonist, thus having a modulatory function (Levine, 2008; Pizzolo et al., 1987). They are, however, a special kratagonist subset deriving from the receptor gene product, unlike salivary kratagonists, which are unrelated to the agonists' receptors. Perhaps by "thinking kratagonist" a number of protein antagonists with unknown function could be mechanistically characterized, if the right experiments are conducted. We believe this class of antagonists could be acting in the immunological and neurobiological systems of animals.

MODES OF KRATAGONIST IDENTIFICATION

In most cases, the existence of a kratagonist has been suggested by bioassay results that are consistent with the removal of a receptor agonist from solution. This is followed up with physical measurements of binding affinity made using isothermal titration calorimetry (ITC) or surface plasmon resonance (SPR). For small molecules, ITC data allow for a complete solution thermodynamic analysis of the binding reaction. Highly enthalpic processes are indicative of the formation of electrostatic and hydrogen-bonding interactions between the ligand and protein, while reactions exhibiting favorable entropies are more likely due to hydrophobic contacts of the ligand with the protein. The equilibrium constant obtained from fitting the concentration dependence of the binding enthalpy per injection gives a measure of binding affinity, and the stoichiometry of binding provides an estimate of the number of ligand binding sites. SPR is most commonly used to measure interactions between macromolecules, but it also can be used for small-molecule–binding measurements and has the advantage of providing kinetic parameters for ligand binding and release mechanisms.

DIVERSITY OF KRATAGONIST STRUCTURE AND FUNCTION

As a general observation, kratagonists are derived from secreted, extracellular protein families whose structures are tolerant of a high degree of amino acid substitution. That kratagonists belong to a number of protein families is reflective of the independent evolution of blood feeding in different arthropod lineages. In the salivas of blood-feeding arthropods, members of four protein families have been identified as kratagonists of small-molecule agonists. In ticks and triatomine Hemiptera, lipocalins are used to bind biogenic amine, nucleotide, and eicosanoid effectors of hemostasis and inflammation. In mosquitoes, D7 proteins derived from the arthropod odorant binding protein family serve the same function as described for lipocalins above, and in the sand flies, members of the insect yellow/major royal jelly protein family have been recently identified as high-affinity biogenic amine–binding proteins. Most recently, in horse flies of the family Tabanidae, CAP domain/antigen V proteins have recently been found to be kratagonists of cysteinyl leukotrienes (Fig. 4.1).

BIOGENIC AMINE–BINDING LIPOCALINS

The lipocalins are a large family of proteins belonging to the calicin superfamily that serve a variety of ligand binding functions in invertebrates and vertebrates (Schiefner and Skerra, 2015). Two well-known examples from mammals are the retinol-binding protein and odorant-binding protein of the olfactory mucosa (Pelosi et al., 2014a).

The lipocalin structure is comprised of an eight-stranded antiparallel β-sheet folded to form a barrel that contains a binding pocket at its center. Surrounding the entry to the central pocket of the lipocalin β-barrel is a series of flexible loops connecting strands A-B, C-D, E-F, and G-H (Andersen et al., 2005). The pocket itself is lined by amino acid residues originating from elements of both the loops and the β-strands, giving the pockets a wide variety of shape, hydrophobicity, and charge characteristics.

In ticks, biogenic amine–binding lipocalins are widespread and appear to have arisen prior to the divergence of soft and hard forms (Mans et al., 2008b). Crystal structures have been determined of variants from the hard tick *Rhipicephalus appendiculatus* (HBP-1 and HBP-2) and the soft tick *Argas monolakensis* (monomine and monotonin) (Mans et al., 2008b; Paesen et al., 1999). They show a high degree of conservation in both the structure of the binding pocket and the ligand-binding mode. In the histamine-binding lipocalins, the charged amino group of the ligand is stabilized by hydrogen bond/salt bridge and cation pi interactions (Mans et al., 2008b; Paesen et al., 1999). The identities and positions of most of the residues involved in these interactions are conserved in the two groups of ticks, with an aspartate residue on β-strand G forming a salt bridge with the amino group, a serine near the N-terminus forming a hydrogen bond, and a tyrosine residue on β-strand A participating in a cation pi interaction. The imidazole portion of the molecule extends toward the entry to the pocket and is surrounded by both hydrophobic and polar residues with the latter forming hydrogen bonds with the imidazole nitrogen positions. The crystal structure of the histamine-binding protein HBP2 from *R. appendiculatus* revealed a second binding site having higher affinity for histamine (Paesen et al., 1999, 2000; Syme et al., 2010). Unlike the site described above, which lies deep within the β-barrel structure, this site

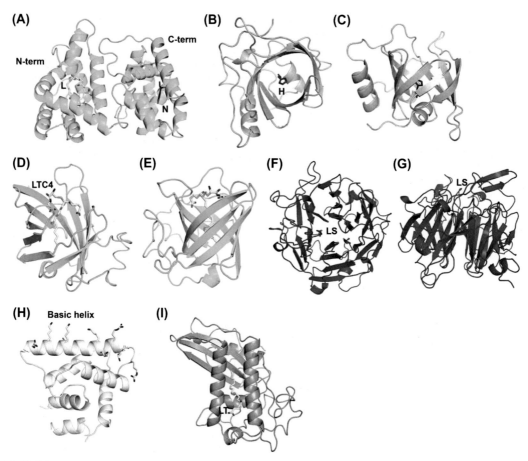

FIGURE 4.1 Structural diversity of arthropod salivary kratagonists. (A) Two-domain D7 protein (AeD7L1) from *Aedes aegypti* with ligands bound in the N- and C-terminal domains (Calvo et al., 2009). *L*, leukotriene C4, *N*, norepinephrine. (B) Histamine-binding lipocalin (monomine) from *Argas monolakensis* (top view) (Mans and Ribeiro, 2008a). *H*, histamine. (C) The complex shown in B viewed from the side. (D) Leukotriene-binding lipocalin from *Rhodnius prolixus* (LTBP1) with leukotriene C4 (LTC4) bound (top view) (Jablonka et al., 2016). (E) The structure in D viewed from the side. (F) The structure of LJM11 from *Lutzomyia longipalpis* viewed from the top (Xu et al., 2011). (G) The structure in panel F viewed from the side. (H) The structure of PdSP15b, a scavenger of polyphosphate and heparin from the sand fly *Phlebotomus duboscqi* (Alvarenga et al., 2013). The basic helix labeled on the figure with side chains highlighted is the likely point of ligand interaction. (I) The CAP domain protein tablysin-15 from the saliva of the horse fly *Tabanus yao* showing a bound molecule of leukotriene C4 (LT) in a hydrophobic binding pocket (Xu et al., 2012).

is formed in part by residues contained in the loops surrounding the entry to the barrel. At this site, the imidazole group of histamine is stabilized by stacking and edge-on interactions as well as hydrogen bonding with aspartate and glutamate residues on loop A-B and β-strand D, respectively.

Lipocalins related to the histamine-binding proteins, but showing strong selectivity for serotonin, are also present in the salivas of hard and soft ticks. The binding mode of the serotonin ligand is very similar to histamine in the histamine-binding proteins, with the stabilization of the charged amino group being due to equivalentresidues,withbothhydrogen-bonding and cation-pi interactions being conserved (Mans et al., 2008b; Sangamnatdej et al., 2002). Comparisons of histamine- and serotonin-binding forms show that specificity is imparted by changes in amino acid residue identity surrounding the indole and imidazole portions of serotonin and histamine in the two complexes, respectively. It seems clear from these similarities that the histamine- and serotonin-binding forms are derived from a common biogenic amine-binding ancestor. A second two-binding site lipocalin similar to HBP2 was later described from *R. appendiculatus*. Binding experiments demonstrated that this protein has a site that is selective for serotonin, with a second site that binds histamine (Sangamnatdej et al., 2002). A corresponding two-binding site lipocalin, TGSP-1, has also been found in the saliva of the soft tick *Ornithodoros savignyi* (Mans et al., 2008b). In this case, it was shown that the lower site favors serotonin, while the upper site binds histamine with higher affinity. Site-directed mutagenesis experiments suggested that both binding sites in TGSP-1 involve the same amino acid residues as the corresponding sites in HBP2 indicating conservation of both pockets.

In a remarkable bit of convergent evolution, triatomine bugs have independently evolved lipocalins that perform the same biogenic amine–binding function seen in ticks. Nitrophorins are lipocalins that contain a single heme molecule tethered to the protein via Fe(III) coordination with a histidine residue in the "proximal"-binding pocket (Andersen and Montfort, 2000; Ribeiro et al., 1993, 1995; Weichsel et al., 1998). In the salivary gland the protein is loaded with a single molecule of nitric oxide in the distal pocket that is released when the protein is injected into the host during feeding. In its place, histamine occupies the distal pocket and binds via a coordination bond with the heme iron, as well as a series of hydrogen bond/salt bridge interactions with polar and charged residues including an aspartate residue that forms an electrostatic interaction with the protonated amino group. One form of nitrophorin-like lipocalin in *R. prolixus* saliva, known as ABP, has lost the ability to bind heme due to structural changes that include substitution of the proximal histidine residue. Instead of heme, these proteins were found to bind catecholamines and serotonin with high affinity, thereby serving a different function in the saliva (Andersen et al., 2003). Although the two groups are clearly independently derived, the triatomine and tick biogenic amine–binding lipocalins share similarities in the nature of the binding pocket. This is particularly true in the stabilization of the amino group where both hydrogen bond/salt bridge and cation-pi interactions are utilized (Xu et al., 2013).

EICOSANOID-BINDING LIPOCALINS

In a study of complement inhibition by saliva of the soft tick *Ornithodoros moubata*, a salivary lipocalin given the name OMCI was found to inhibit the complement cascade via binding with component C5. Recombinant OMCI, expressed in yeast, showed the presence of an apparent fatty acid in the binding pocket located at the center of the β-barrel (Nunn et al., 2005; Roversi et al., 2007, 2013). The ligand was identified as ricinoleic acid, suggesting that this family of tick lipocalins functions as a binder of biologically active eicosanoids such as leukotrienes and thromboxanes, which are potent mediators of inflammation and platelet activation. In the OMCI-ricinoleic acid complex, a hydroxyl group at the ω-6 carbon forms a hydrogen bond with histidine and aspartate side chains located on β-strand G of the lipocalin barrel. It seemed likely that these residues form hydrogen bonds with the ω-5 hydroxyl group of thromboxane A2 or the ω-9 hydroxyl of LTB4, a 5-lipoxygenase derivative (Roversi et al., 2007). The bicyclic ring system of thromboxane A2 is not present in ricinoleic acid, but the positioning of glutamine and arginine residues on β-strand F of the lipocalin β-barrel made them candidates for possible hydrogen-bonding interactions with this portion of the ligand. Later, the protein was shown to specifically bind LTB4, a modulator that is essential for neutrophil chemotaxis in vivo, and the structure of the OMCI-LTB4 complex was determined (Roversi et al., 2013). In this structure, the side chains of His 119 and Asp 121 were found to form hydrogen bonds with the ω-9 hydroxyl as expected. A similar protein, Ir-LBP, has also been described from the hard tick, *Ixodes ricinus* and found to inhibit neutrophil function via the high affinity binding of LTB4, again showing functional conservation between hard and soft ticks (Beaufays et al., 2008). Moubatin, a lipocalin closely related to OMCI has been described as an inhibitor of collagen-mediated platelet aggregation, a property consistent with the scavenging of thromboxanes (Keller et al., 1993). Expression and assay of several members of the moubatin clade (moubatin, TSGP2 and TSGP3) revealed that thromboxane A2 (TXA2) analogs bind with high affinity to moubatin establishing this as the mechanism of action (Keller et al., 1993; Mans et al., 2003; Mans and Ribeiro, 2008a). The proteins are also able to bind LTB4 with high affinity. A second group of soft tick lipocalins has also been identified that bind the cysteinyl leukotrienes LTC4, LTD4, and LTE4, which contain a fatty acid conjugated with glutathione or modified peptides related to glutathione (Mans and Ribeiro, 2008b). No structures have been determined for these complexes.

Eicosanoid-binding lipocalins are also present in the saliva of triatomine bugs. Two related lipocalins from *Dipetologaster maximus* and *Triatoma infestans* are potent inhibitors of collagen-mediated platelet aggregation and were found to bind TXA2 analogs as well as a number of other hydroxylated eicosanoid fatty acids (Assumpcao et al., 2010). The

sequences of these two proteins are highly similar to that of pallidipin, a known platelet aggregation inhibitor from *Triatoma pallidipennis* (Haendler et al., 1995), making it likely that this protein also targets TXA2. In *R. prolixus*, a potent inhibitor of collagen-mediated platelet activation was isolated from salivary gland extracts and found to belong to the lipocalin family and given the name RPAI-1. The protein has general similarities to pallidipin but the degree of amino acid sequence identity with this protein was low. Experiments using platelet preparations and exogenously provided agonists showed that RPAI-1 does not bind TXA2 but functions by scavenging the agonist ADP, which is released from collagen-activated platelets as a component of the dense granules (Francischetti et al., 2000, 2002). Granule secretion is induced by TXA2 (Li et al., 2003), so binding of this ligand will prevent the release of ADP and subsequent aggregation. In *R. prolixus* (tribe Rodniini) evolution has apparently produced a secretion targeted at sequestering the granule product ADP, while *Triatoma* and *Dipetologaster* species (tribe Triatomini) aim to prevent ADP secretion by sequestering TXA2.

LTBP1, a lipocalin with moderate sequence identity to RPAI1 was isolated from the *R. prolixus* salivary gland extract and found not to bind ADP or TXA2 analogs nor inhibit platelet aggregation. Screening of potential ligands by isothermal titration calorimetry showed that this protein specifically bound cysteinyl leukotrienes, particularly LTC4, with very high affinity (~1 nM) (Jablonka et al., 2016). Cysteinyl leukotrienes are secreted by activated mast cells in the skin and cause rapid wheal and flare reactions similar to those caused by insect bites. Crystal structures of ligand-free LTBP1 and its complex with various cysteinyl leukotrienes showed that conformational changes in the G-H loop are involved in stabilizing the leukotriene molecule in the binding pocket. The hydrophobic fatty acid penetrates deeply into the interior of the protein while the peptide (glutathione in the case of LTC4) is bound at a surface site (Jablonka et al., 2016). No comparable protein has yet been identified from *Triatoma* or *Dipetologaster* species.

ODORANT-BINDING PROTEIN RELATIVES

The all-helical odorant binding protein (OBP) family is found in the sensory organs of arthropods, where its members are thought to mediate transport of ligands from the surface of a receptor sensillum through the receptor lymph to an olfactory receptor on the neuron membrane (Pelosi et al., 2014a). The role of these proteins in olfaction is not completely clear, however, and sequestration of ligands may be an important part of their function in addition to transport. In addition to the OPBs, a second family of soluble proteins known as chemosensory proteins is also present in chemoreceptors that also appear to have a ligand-binding function (Pelosi et al., 2014b).

In blood feeding Diptera, highly modified odorant-binding proteins known as D7s are found in the saliva and serve a kratagonist function (Calvo et al., 2006). Single-domain and two-domain forms of D7 are present in the salivas of mosquitoes. The single-domain proteins are made up of a bundle of eight α-helices surrounding a small, central ligand–binding pocket. The seventh and eighth helices G and H are uniquely positioned in the D7 proteins producing a binding pocket unlike the any of the olfactory/gustatory OBPs that have been structurally characterized (Calvo et al., 2009; Mans et al., 2007; Alvarenga et al., 2010). Single-domain D7 proteins from *Anopheles gambiae* have been shown to be kratagonists for biogenic amines, and the structure of one of these, D7R4, has been determined in complex with numerous ligands. AeD7 and AnStD7L1, two-domain D7 forms from *Aedes aegypti* and *A. stephensi*, respectively, have also

been characterized. AeD7 binds biogenic amines and eicosanoid compounds, and AnStD7L1 is only known to bind eicosanoids (Calvo et al., 2009; Alvarenga et al., 2010).

The crystal structure of the serotonin complex of D7R4 reveals an entry channel lying between helices B, G, and H, that is lined with hydrophobic side chains stabilizing the aromatic nucleus of the ligand. The amino group is oriented toward the opening of the pocket and forms hydrogen bond or salt bridge interactions with the side chains of aspartate and glutamate residues located on helix H. Glutamate and histidine residues on the back side of the pocket form hydrogen bonds with the hydroxyl groups of serotonin and the catecholamines epinephrine and norepinephrine (Mans et al., 2007).

The C-terminal domain of the two-domain form AeD7 has been shown to bind biogenic amines and serve an apparently identical function to the single domain D7 proteins of *A. gambiae* (Calvo et al., 2009). This portion of the protein exhibits a relatively high degree of sequence similarity to the single domain D7R4 of *A. gambiae*, with most residues involved with biogenic amine–binding being conserved. The binding mode of the ligand in this form is very similar to D7R4, but crystal structure evidence suggests that the binding mechanisms of the two forms differ. The ligand bound and free forms of D7R4 have similar structures containing eight helices, but in AeD7, the C-terminal helix is unwound in the ligand-free form and becomes helical on ligand binding. This corresponds to a rotation of Glu 268 and Arg 176 side chains to a position covering the ligand, thereby burying it in the interior of the C-terminal domain. It was suggested that the broader ligand selectivity exhibited by AeD7 relative to D7R4 was due to the facilitation of ligand entry as a result of the conformational change mechanism in this variant of the protein (Calvo et al., 2009).

The crystal structure of AeD7 revealed a second potential ligand-binding site in the N-terminal domain that takes the form of a long narrow hydrophobic channel having an entry point at the surface surrounded by a number of polar or charged residues. The shape of this pocket suggested that a linear hydrophobic molecule such as an eicosanoid fatty acid could be accommodated (Calvo et al., 2009). ITC experiments showed that this was indeed the case, with the protein binding LTC4, LTD4, and LTE4 with approximately equal affinity. The crystal structure of the complex with LTE4 showed that the fatty acid portion of the molecule is inserted into the pocket with the carboxyl group participating in hydrogen-bonding interactions near the entrance. The peptide portion of the leukotriene ligand is disordered in the structure suggesting a lesser influence on binding, consistent with the lack of discrimination by the protein for the different cysteinyl leukotrienes (Calvo et al., 2009).

Structurally, the N-terminal domain of AnStD7L1, the two-domain D7 protein from *A. stephensi* appears very similar to that of AeD7, and as expected, it was found to bind cysteinyl leukotrienes in a similar manner to AeD7 (Alvarenga et al., 2010). Screening of additional eicosanoids showed that it also bound the TXA2 analog U46619 with an affinity of approximately 100 nM. This compound shows no detectable binding with AeD7. AnStD7L1 was subsequently shown to be a potent inhibitor of collagen-mediated platelet aggregation, acting by scavenging TXA2. A structure of the AnStD7L1 complex with U46619 was obtained, revealing that a single amino acid difference likely accounts for the difference in TXA2 binding by AeD7 and AnStD7L1 (Alvarenga et al., 2010). In AnStD7L1 the ω-5 hydroxyl group of U46619 participates in a hydrogen bond with the hydroxyl group of Tyr-52. In AeD7, this position contains a phenylalanine (Phe 50) and is unable to form this hydrogen bond (Alvarenga et al., 2010). Since the binding sites are otherwise almost identical, it appears that selectivity for TXA2 is imparted by the presence of the tyrosine hydroxyl group at position 52.

"YELLOW" PROTEINS FROM SAND FLIES

Among the most abundant proteins in the salivary secretions of sand flies are members of the yellow/major royal jelly protein family. In metazoans, this family is restricted to insects where it is widely distributed (Drapeau et al., 2006). Similar proteins are known in bacteria and fungi as well, and the presence of yellow proteins in insects is apparently due to lateral gene transfer from a bacterium (Ferguson et al., 2011). In *Drosophila*, mutations in yellow protein genes cause defects in melanization, as well as neurological effects (Arakane et al., 2010; Han et al., 2002). Members of the yellow family have been identified as dopachrome conversion enzymes in both *Drosophila melanogaster* and mosquitoes. In the sand fly, *Lutzomyia longipalpis*, the salivary yellow protein LJM11 elicits strong delayed-type hypersensitivity responses in the skin of mice and shows potential as a vaccine candidate to be used against *Leishmania* parasites (Gomes et al., 2012; Xu et al., 2011). LJM111, a second yellow protein from *L. longipalpis* saliva strongly inhibits inflammatory responses in rodents, but the mechanism for this activity is not known. In ITC screening experiments designed to search for small-molecule ligands, three yellow protein family members from *L. longipalpis* were identified as high affinity binders of biogenic amines and exhibited subtle differences in ligand selectivity with LJM17 binding serotonin with very high affinity and LJM111 binding catecholamines extremely tightly (Xu et al., 2011). The role of catecholamine binding in the immunosuppressive activities of LJM111 is not known, but catecholamines are known to be modulators of various immune processes as well as potent vasoconstrictors.

The crystal structure of the yellow protein LJM11 has been determined, revealing a six-bladed β-propeller fold with a channel at its center that serves as a ligand-binding pocket

(Xu et al., 2011). The structure of the LJM11-serotonin complex shows that the ligand-binding site is located in the channel at the end of the molecule commonly referred to as the "top" side. The complex loop structures surrounding this binding site provide the amino acid side chains involved in specific interactions. In this protein, the amino group of the ligand is stabilized via hydrogen bonding interactions with the side chains of Asn-342 and Thr-327 (Xu et al., 2011).

CAP DOMAIN PROTEINS FROM TABANID FLIES

Members of the cysteine-rich secretory, antigen 5 and pathogenesis-related protein 1 (CAP) family are widely distributed in the salivas of blood-feeding insects as well as the venoms and secretions of a wide range of both invertebrates and vertebrates. Functionally, few of these have been characterized, but tablysin-15, from the saliva of the horse fly *Tabanus yao*, has been shown to have a disintegrin function mediated by an Arg-Gly-Asp motif contained in an external loop (Ma et al., 2011). In the crystal structure of tablysin-15 a deep hydrophobic channel was observed that opened to the surface of the protein between helices 1 and 3. Residual electron density in the channel suggested that the protein might bind a fatty acid in this position in a manner similar to the N-terminal domain of two-domain D7 proteins (Xu et al., 2012). Screening of potential ligands using ITC showed that the cysteinyl leukotrienes LTC4, LTD4, and LTE4 bound tablysin-15 with affinity constants of about 100 nM. Contraction of a guinea pig ileum preparation was prevented in the presence of tablysin-15, demonstrating that this protein can function to sequester proinflammatory leukotrienes during feeding (Xu et al., 2012). A crystal structure of the complex of tablysin-15 with a cysteinyl leukotriene showed that the lipid portion of the ligand

is inserted into the pocket in a similar manner to AeD7 (Xu et al., 2012). The carboxyl end of the fatty acid chain is stabilized at the surface of the protein by hydrogen bonding or salt bridge interactions with two histidine side chains (His 53 and His 130) and a tryptophan side chain (Trp 59). Like the complexes of D7 proteins with cysteinyl leukotrienes, the peptidyl portion of the ligand is disordered suggesting that it is not stabilized by contacts with the protein.

SALIVARY KRATAGONISTS OF MACROMOLECULAR EFFECTORS

In addition to the small-molecule agonists described above, various macromolecules are important in the inflammatory and hemostatic responses elicited by blood feeding. Collagen exposed by wounding of the skin serves as a primary stimulus for the activation of platelets (Manon-Jensen et al., 2016). Inorganic polyphosphate is secreted by platelets and mast cells and is thought to provide an endogenous contact surface for the activation of coagulation factor XII to XIIa, which goes on to activate factor XI (to factor XIa) and prekallikrein (to kallikrein). It is also an activator of the cleavage of factor XI by thrombin, a key alternative modulator of the intrinsic coagulation pathway (Bjorkqvist et al., 2014). Heparin is released by activated mast cells where it also can serve as a contact surface for FXII as well as a cofactor for mast cell proteases (Oschatz et al., 2011). Factor XIa is the activator of factor IX and kallikrein cleaves high molecular weight kininogen releasing the potent proinflammatory molecule bradykinin (Hofman et al., 2016). In mosquito saliva, aegyptin (from *A. aegypti*) and anopheline antiplatelet protein (from *A. stephensi*) bind tightly to collagen and inhibit the activation of platelets (Calvo et al., 2007; Yoshida et al., 2008; Sugiyama et al., 2014). pSP15 is an odorant-binding protein family member from the saliva of the sand fly *Phlebotomus papatasi* that has been shown to bind polyphosphate and heparin and act as an inhibitor of factor XII activation and the subsequent production of bradykinin (Alvarenga et al., 2013). Its structure reveals the presence of an α-helical segment studded with basic side chains that is thought to be the point of interaction with polyanionic ligands.

Due mainly to the length of time they spend attached to the host, hard ticks need to limit innate and acquired immunity during feeding. To accomplish this, hard tick saliva contains a variety of protein and small-molecule components directed at the inhibition of these systems. Among these are a group of chemokine-binding proteins that may be considered kratagonists. Three small proteins, given the names Evasin-1, -3, and -4, have been identified in the saliva of *Rhipicephalus sanguineus* and found to bind and neutralize a variety of chemokines (Bonvin et al., 2016; Deruaz et al., 2008). Evasin-1 was specific for CCL3, CCL4, and CCL18, while Evasin-3 was effective against CXCL8 and CXCL1, and Evasin-4 neutralized CCL5 and CCL11. Evasin-1 and -4 are related in sequence, while Evasin-3 shows no relationship to the others. In a murine model, Evasin-1 was found to inhibit leukocyte recruitment in vivo after administration of CCL3 or murine MIP-1α and also showed potent antiinflammatory activity against CCL3 and CCL4 in mouse models of psoriasis and pulmonary fibrosis (Deruaz et al., 2008). Evasin-3 was shown to inhibit neutrophil-mediated inflammation and produced significant effects on neutrophil recruitment in a mouse model of antigen-induced arthritis (Deruaz et al., 2008).

Activation of complement at the feeding site also occurs rapidly and can negatively affect feeding, presumably through the production of strongly proinflammatory anaphylatoxins. Because of this, complement inhibitors are present in the salivas of blood-feeding insects and ticks. In both hard and soft ticks, scavengers of complement factor C5 have been identified. These proteins bind the substrate of the C5

convertase rather than the enzyme complex, so this can be considered a sequestration mechanism and the proteins can be considered kratagonists. OmCI from *Ornithodoros moubata* and RaCI from *R. appendiculatus* have been shown to bind with high affinity to human C5 with affinities of less than 10nM (Nunn et al., 2005; Jore et al., 2016). The two proteins are structurally unrelated with OmCI belonging to the lipocalin family, and RaCI is a small protein stabilized by two disulfide bonds. Notably, the two tick proteins bind to different sites on the C5 surface, while the C5-binding therapeutic monoclonal antibody eculizumab binds to a third site (Jore et al., 2016). Therefore, the crystal structures of the two tick proteins in complex with C5 have revealed additional binding sites that may be useful for the development of new drugs for complement disorders.

CONCLUSIONS

The kratagonist mechanism has now been shown to operate in the salivas of all blood-feeding arthropods examined to date. The strategy of down-modulating or totally preventing an activated state by removing important effectors from the feeding area must be highly adaptive. It is noteworthy that in organisms such as the triatomine bugs, only a small fraction of the salivary lipocalins present have been functionally characterized, suggesting that kratagonist activities must extend well beyond what has been shown thus far. Perhaps, new aspects of vertebrate physiology will be revealed by understanding the ligand-binding functions of these proteins. In a practical sense, some of these molecules could be developed into pharmaceutically useful products, as has been attempted with a tick histamine-binding protein (Weston-Davies et al., 2005). More likely, understanding of the modes of interaction of kratagonists with their targets will provide insight into the development of molecular scavengers that are more suitable than the native molecules for administration to human hosts. Therapeutic monoclonal antibodies such as adalimumab for inflammatory diseases and eculizumab for complement disorders illustrate the effectiveness and potential of sequestration mechanisms in the modulation of physiological processes for the treatment of disease.

References

Alvarenga, P.H., Francischetti, I.M., Calvo, E., Sa-Nunes, A., Ribeiro, J.M., Andersen, J.F., 2010. The function and three-dimensional structure of a thromboxane A2/cysteinyl leukotriene-binding protein from the saliva of a mosquito vector of the malaria parasite. PLoS Biol. 8, e1000547.

Alvarenga, P.H., Xu, X., Oliveira, F., Chagas, A.C., Nascimento, C.R., Francischetti, I.M., Juliano, M.A., Juliano, L., Scharfstein, J., Valenzuela, J.G., Ribeiro, J.M., , Andersen, J.F., 2013. Novel family of insect salivary inhibitors blocks contact pathway activation by binding to polyphosphate, heparin, and dextran sulfate. Arterioscler. Vasc. Biol. 33, 2759–2770.

Andersen, J.F., Montfort, W.R., 2000. The crystal structure of nitrophorin 2. A trifunctional antihemostatic protein from the saliva of Rhodnius prolixus. J. Biol. Chem. 275, 30496–30503.

Andersen, J.F., Weichsel, A., Balfour, C.A., Champagne, D.E., Montfort, W.R., 1998. The crystal structure of nitrophorin 4 at 1.5 A resolution: transport of nitric oxide by a lipocalin-based heme protein. Structure 6, 1315–1327.

Andersen, J.F., Francischetti, I.M., Valenzuela, J.G., Schuck, P., Ribeiro, J.M., 2003. Inhibition of hemostasis by a high affinity biogenic amine-binding protein from the saliva of a blood-feeding insect. J. Biol. Chem. 278, 4611–4617.

Andersen, J.F., Gudderra, N.P., Francischetti, I.M., Ribeiro, J.M., 2005. The role of salivary lipocalins in blood feeding by Rhodnius prolixus. Arch. Insect Biochem. Physiol. 58, 97–105.

Arakane, Y., Dittmer, N.T., Tomoyasu, Y., Kramer, K.J., Muthukrishnan, S., Beeman, R.W., Kanost, M.R., 2010. Identification, mRNA expression and functional analysis of several yellow family genes in Tribolium castaneum. Insect Biochem. Mol. Biol. 40, 259–266.

Assumpcao, T.C., Alvarenga, P.H., Ribeiro, J.M., Andersen, J.F., , Francischetti, I.M., 2010. Dipetalodipin, a novel multifunctional salivary lipocalin that inhibits platelet aggregation, vasoconstriction, and angiogenesis through unique binding specificity for TXA2, PGF2α, and 15(S)-HETE. J. Biol. Chem. 285, 39001–39012.

Beaufays, J., Adam, B., Menten-Dedoyart, C., Fievez, L., Grosjean, A., Decrem, Y., Prevot, P.P., Santini, S., Brasseur, R., Brossard, M., Vanhaeverbeek, M., Bureau, F., Heinen, E., Lins, L., Vanhamme, L., Godfroid, E., 2008. Ir-LBP, an Ixodes ricinus tick salivary LTB4-binding lipocalin, interferes with host neutrophil function. PLoS One 3, e3987.

Bjorkqvist, J., Nickel, K.F., Stavrou, E., Renne, T., 2014. In vivo activation and functions of the protease factor XII. Thromb. Haemost. 112, 868–875.

Bonvin, P., Power, C.A., Proudfoot, A.E., 2016. Evasins: therapeutic potential of a new family of chemokine-binding proteins from ticks. Front. Immunol. 7, 208.

Calvo, E., Mans, B.J., Andersen, J.F., Ribeiro, J.M., 2006. Function and evolution of a mosquito salivary protein family. J. Biol. Chem. 281, 1935–1942.

Calvo, E., Tokumasu, F., Marinotti, O., Villeval, J.L., Ribeiro, J.M., Francischetti, I.M., 2007. Aegyptin, a novel mosquito salivary gland protein, specifically binds to collagen and prevents its interaction with platelet glycoprotein VI, integrin α2β1, and von Willebrand factor. J. Biol. Chem. 282, 26928–26938.

Calvo, E., Mans, B.J., Ribeiro, J.M., Andersen, J.F., 2009. Multifunctionality and mechanism of ligand binding in a mosquito antiinflammatory protein. Proc. Natl. Acad. Sci. U.S.A. 106, 3728–3733.

Deruaz, M., Frauenschuh, A., Alessandri, A.L., Dias, J.M., Coelho, F.M., Russo, R.C., Ferreira, B.R., Graham, G.J., Shaw, J.P., Wells, T.N., Teixeira, M.M., Power, C.A., Proudfoot, A.E., 2008. Ticks produce highly selective chemokine binding proteins with antiinflammatory activity. J. Exp. Med. 205, 2019–2031.

Drapeau, M.D., Albert, S., Kucharski, R., Prusko, C., Maleszka, R., 2006. Evolution of the Yellow/Major Royal Jelly Protein family and the emergence of social behavior in honey bees. Genome Res. 16, 1385–1394.

Ferguson, L.C., Green, J., Surridge, A., Jiggins, C.D., 2011. Evolution of the insect yellow gene family. Mol. Biol. Evol. 28, 257–272.

Francischetti, I.M., Ribeiro, J.M., Champagne, D., Andersen, J., 2000. Purification, cloning, expression, and mechanism of action of a novel platelet aggregation inhibitor from the salivary gland of the blood-sucking bug, Rhodnius prolixus. J. Biol. Chem. 275, 12639–12650.

Francischetti, I.M., Andersen, J.F., Ribeiro, J.M., 2002. Biochemical and functional characterization of recombinant Rhodnius prolixus platelet aggregation inhibitor 1 as a novel lipocalin with high affinity for adenosine diphosphate and other adenine nucleotides. Biochemistry 41, 3810–3818.

Frauenschuh, A., Power, C.A., Deruaz, M., Ferreira, B.R., Silva, J.S., Teixeira, M.M., Dias, J.M., Martin, T., Wells, T.N., Proudfoot, A.E., 2007. Molecular cloning and characterization of a highly selective chemokine-binding protein from the tick Rhipicephalus sanguineus. J. Biol. Chem. 282, 27250–27258.

Gillespie, R.D., Dolan, M.C., Piesman, J., Titus, R.G., 2001. Identification of an IL-2 binding protein in the saliva of the Lyme disease vector tick, *Ixodes scapularis*. J. Immunol. 166, 4319–4326.

Gomes, R., Oliveira, F., Teixeira, C., Meneses, C., Gilmore, D.C., Elnaiem, D.E., Kamhawi, S., Valenzuela, J.G., 2012. Immunity to sand fly salivary protein LJM11 modulates host response to vector-transmitted leishmania conferring ulcer-free protection. J. Invest. Dermatol. 132, 2735–2743.

Haendler, B., Becker, A., Noeske-Jungblut, C., Kratzschmar, J., Donner, P., Schleuning, W.D., 1995. Expression of active recombinant pallidipin, a novel platelet aggregation inhibitor, in the periplasm of *Escherichia coli*. Biochem. J. 307 (Pt. 2), 465–470.

Han, Q., Fang, J., Ding, H., Johnson, J.K., Christensen, B.M., Li, J., 2002. Identification of *Drosophila melanogaster* yellow-f and yellow-f2 proteins as dopachrome-conversion enzymes. Biochem. J. 368, 333–340.

Hofman, Z., de Maat, S., Hack, C.E., Maas, C., 2016. Bradykinin: inflammatory product of the coagulation system. Clin. Rev. Allergy Immunol. 51, 152–161.

Jablonka, W., Pham, V., Nardone, G., Gittis, A., Silva-Cardoso, L., Atella, G.C., Ribeiro, J.M., Andersen, J.F., 2016. Structure and ligand-binding mechanism of a cysteinyl leukotriene-binding protein from a blood-feeding disease vector. ACS Chem. Biol. 11, 1934–1944.

Jore, M.M., Johnson, S., Sheppard, D., Barber, N.M., Li, Y.I., Nunn, M.A., Elmlund, H., Lea, S.M., 2016. Structural basis for therapeutic inhibition of complement C5. Nat. Struct. Mol. Biol. 23, 378–386.

Keller, P.M., Waxman, L., Arnold, B.A., Schultz, L.D., Condra, C., Connolly, T.M., 1993. Cloning of the cDNA and expression of moubatin, an inhibitor of platelet aggregation. J. Biol. Chem. 268, 5450–5456.

Konik, P., Slavikova, V., Salat, J., Reznickova, J., Dvoroznakova, E., Kopecky, J., 2006. Anti-tumour necrosis factor-alpha activity in *Ixodes ricinus* saliva. Parasite Immunol. 28, 649–656.

Levine, S.J., 2008. Molecular mechanisms of soluble cytokine receptor generation. J. Biol. Chem. 283, 14177–14181.

Li, Z., Zhang, G., Le Breton, G.C., Gao, X., Malik, A.B., Du, X., 2003. Two waves of platelet secretion induced by thromboxane A2 receptor and a critical role for phosphoinositide 3-kinases. J. Biol. Chem. 278, 30725–30731.

Ma, D., Xu, X., An, S., Liu, H., Yang, X., Andersen, J.F., Wang, Y., Tokumasu, F., Ribeiro, J.M., Francischetti, I.M., Lai, R., 2011. A novel family of RGD-containing disintegrins (Tablysin-15) from the salivary gland of the horsefly *Tabanus yao* targets $\alpha_{IIb}\beta_3$ or $\alpha_V\beta_3$ and inhibits platelet aggregation and angiogenesis. Thromb. Haemost. 105, 1032–1045.

Manon-Jensen, T., Kjeld, N.G., Karsdal, M.A., 2016. Collagen-mediated hemostasis. J. Thromb. Haemost. 14, 438–448.

Mans, B.J., Ribeiro, J.M., 2008a. Function, mechanism and evolution of the moubatin-clade of soft tick lipocalins. Insect Biochem. Mol. Biol. 38, 841–852.

Mans, B.J., Ribeiro, J.M., 2008b. A novel clade of cysteinyl leukotriene scavengers in soft ticks. Insect Biochem. Mol. Biol. 38, 862–870.

Mans, B.J., Louw, A.I., Neitz, A.W., 2003. The major tick salivary gland proteins and toxins from the soft tick, *Ornithodoros savignyi*, are part of the tick Lipocalin family: implications for the origins of tick toxicoses. Mol. Biol. Evol. 20, 1158–1167.

Mans, B.J., Calvo, E., Ribeiro, J.M., Andersen, J.F., 2007. The crystal structure of D7r4, a salivary biogenic amine-binding protein from the malaria mosquito *Anopheles gambiae*. J. Biol. Chem. 282, 36626–36633.

Mans, B.J., Andersen, J.F., Schwan, T.G., Ribeiro, J.M., 2008a. Characterization of anti-hemostatic factors in the argasid, *Argas monolakensis*: implications for the evolution of blood-feeding in the soft tick family. Insect Biochem. Mol. Biol. 38, 22–41.

Mans, B.J., Ribeiro, J.M., Andersen, J.F., 2008b. Structure, function, and evolution of biogenic amine-binding proteins in soft ticks. J. Biol. Chem. 283, 18721–18733.

Mans, B.J., Andersen, J.F., Francischetti, I.M., Valenzuela, J.G., Schwan, T.G., Pham, V.M., Garfield, M.K., Hammer, C.H., Ribeiro, J.M., 2008c. Comparative sialomics between hard and soft ticks: implications for the evolution of blood-feeding behavior. Insect Biochem. Mol. Biol. 38, 42–58.

Neubig, R.R., Spedding, M., Kenakin, T., Christopoulos, A., 2003. International Union of Pharmacology Committee on Receptor Nomenclature and Drug Classification. XXXVIII. Update on terms and symbols in quantitative pharmacology. Pharmacol. Rev. 55, 597–606.

Nunn, M.A., Sharma, A., Paesen, G.C., Adamson, S., Lissina, O., Willis, A.C., Nuttall, P.A., 2005. Complement inhibitor of C5 activation from the soft tick *Ornithodoros moubata*. J. Immunol. 174, 2084–2091.

Oschatz, C., Maas, C., Lecher, B., Jansen, T., Bjorkqvist, J., Tradler, T., Sedlmeier, R., Burfeind, P., Cichon, S., Hammerschmidt, S., Muller-Esterl, W., Wuillemin, W.A., Nilsson, G., Renne, T., 2011. Mast cells increase vascular permeability by heparin-initiated bradykinin formation in vivo. Immunity 34, 258–268.

Paesen, G.C., Adams, P.L., Harlos, K., Nuttall, P.A., Stuart, D.I., 1999. Tick histamine binding proteins: isolation, cloning, and three-dimensional structure. Mol. Cell 3, 661–671.

Paesen, G.C., Adams, P.L., Nuttall, P.A., Stuart, D.L., 2000. Tick histamine-binding proteins: lipocalins with a second binding cavity. Biochim. Biophys. Acta 1482, 92–101.

Pelosi, P., Mastrogiacomo, R., Iovinella, I., Tuccori, E., Persaud, K.C., 2014a. Structure and biotechnological applications of odorant-binding proteins. Appl. Microbiol. Biotechnol. 98, 61–70.

Pelosi, P., Iovinella, I., Felicioli, A., Dani, F.R., 2014b. Soluble proteins of chemical communication: an overview across arthropods. Front. Physiol. 5, 320.

Pizzolo, G., Chilosi, M., Semenzato, G., 1987. The soluble interleukin-2 receptor in haematological disorders. Br. J. Haematol. 67, 377–380.

Ribeiro, J.M.C., Arca, B., 2009. From sialomes to the sialoverse: an insight into the salivary potion of blood feeding insects. Adv. Insect Physiol. 37, 59–118.

Ribeiro, J.M.C., Walker, F.A., 1994. High affinity histamine-binding and anti-histaminic activity of the salivary NO-carrying heme protein (Nitrophorin) of *Rhodnius prolixus*. J. Exp. Med. 180, 2251–2257.

Ribeiro, J.M., Hazzard, J.M., Nussenzveig, R.H., Champagne, D.E., Walker, F.A., 1993. Reversible binding of nitric oxide by a salivary heme protein from a bloodsucking insect. Science 260, 539–541.

Ribeiro, J.M., Schneider, M., Guimaraes, J.A., 1995. Purification and characterization of prolixin S (nitrophorin 2), the salivary anticoagulant of the blood-sucking bug *Rhodnius prolixus*. Biochem. J. 308 (Pt. 1), 243–249.

Ribeiro, J.M.C., 1982. The antiserotonin and antihistamine activities of salivary secretion of *Rhodnius prolixus*. J. Insect Physiol. 28, 69–75.

Ribeiro, J.M.C., 1995. Blood-feeding arthropods: live syringes or invertebrate pharmacologists? Infect. Agents Dis. 4, 143–152.

Roversi, P., Lissina, O., Johnson, S., Ahmat, N., Paesen, G.C., Ploss, K., Boland, W., Nunn, M.A., Lea, S.M., 2007. The structure of OMCI, a novel lipocalin inhibitor of the complement system. J. Mol. Biol. 369, 784–793.

Roversi, P., Ryffel, B., Togbe, D., Maillet, I., Teixeira, M., Ahmat, N., Paesen, G.C., Lissina, O., Boland, W., Ploss, K., Caesar, J.J., Leonhartsberger, S., Lea, S.M., Nunn, M.A., 2013. Bifunctional lipocalin ameliorates murine immune complex-induced acute lung injury. J. Biol. Chem. 288, 18789–18802.

Sangamnatdej, S., Paesen, G.C., Slovak, M., Nuttall, P.A., 2002. A high affinity serotonin- and histamine-binding lipocalin from tick saliva. Insect Mol. Biol. 11, 79–86.

Schiefner, A., Skerra, A., 2015. The menagerie of human lipocalins: a natural protein scaffold for molecular recognition of physiological compounds. Acc. Chem. Res. 48, 976–985.

Soter, N.A., Lewis, R.A., Corey, E.J., Austen, K.F., 1983. Local effects of synthetic leukotrienes (LTC4, LTD4, LTE4, and LTB4) in human skin. J. Invest. Dermatol. 80, 115–119.

Sugiyama, K., Iyori, M., Sawaguchi, A., Akashi, S., Tame, J.R., Park, S.Y., Yoshida, S., 2014. The crystal structure of the active domain of Anopheles anti-platelet protein, a powerful anti-coagulant, in complex with an antibody. J. Biol. Chem. 289, 16303–16312.

Swieringa, F., Kuijpers, M.J., Heemskerk, J.W., van der Meijden, P.E., 2014. Targeting platelet receptor function in thrombus formation: the risk of bleeding. Blood Rev. 28, 9–21.

Syme, N.R., Dennis, C., Bronowska, A., Paesen, G.C., Homans, S.W., 2010. Comparison of entropic contributions to binding in a "hydrophilic" versus "hydrophobic" ligand-protein interaction. J. Am. Chem. Soc. 132, 8682–8689.

Vancova, I., Slovak, M., Hajnicka, V., Labuda, M., Simo, L., Peterkova, K., Hails, R.S., Nuttall, P.A., 2007. Differential anti-chemokine activity of *Amblyomma variegatum* adult ticks during blood-feeding. Parasite Immunol. 29, 169–177.

Weichsel, A., Andersen, J.F., Champagne, D.E., Walker, F.A., Montfort, W.R., 1998. Crystal structures of a nitric oxide transport protein from a blood-sucking insect. Nat. Struct. Biol. 5, 304–309.

Weston-Davies, W., Couillin, I., Schnyder, S., Schnyder, B., Moser, R., Lissina, O., Paesen, G.C., Nuttall, P., Ryffel, B., 2005. Arthropod-derived protein EV131 inhibits histamine action and allergic asthma. Ann. N.Y. Acad. Sci. 1056, 189–196.

Xu, X., Oliveira, F., Chang, B.W., Collin, N., Gomes, R., Teixeira, C., Reynoso, D., My Pham, V., Elnaiem, D.E., Kamhawi, S., Ribeiro, J.M., Valenzuela, J.G., Andersen, J.F., 2011. Structure and function of a "yellow" protein from saliva of the sand fly *Lutzomyia longipalpis* that confers protective immunity against *Leishmania major* infection. J. Biol. Chem. 286, 32383–32393.

Xu, X., Francischetti, I.M., Lai, R., Ribeiro, J.M., Andersen, J.F., 2012. Structure of protein having inhibitory disintegrin and leukotriene scavenging functions contained in single domain. J. Biol. Chem. 287, 10967–10976.

Xu, X., Chang, B.W., Mans, B.J., Ribeiro, J.M., Andersen, J.F., 2013. Structure and ligand-binding properties of the biogenic amine-binding protein from the saliva of a blood-feeding insect vector of *Trypanosoma cruzi*. Acta Crystallogr. 69, 105–113.

Yoshida, S., Sudo, T., Niimi, M., Tao, L., Sun, B., Kambayashi, J., Watanabe, H., Luo, E., Matsuoka, H., 2008. Inhibition of collagen-induced platelet aggregation by anopheline antiplatelet protein, a saliva protein from a malaria vector mosquito. Blood 111, 2007–2014.

Basic and Translational Research on Sand Fly Saliva: Pharmacology, Biomarkers, and Vaccines

Waldionê de Castro, Fabiano Oliveira, Iliano V. Coutinho-Abreu, Shaden Kamhawi, Jesus G. Valenzuela

National Institutes of Health, Rockville, MD, United States

BACKGROUND

Blood-feeding arthropods when attempting to get a blood meal from mammalian hosts face a highly efficient and redundant barrier, the hemostatic system. Hemostasis in mammals has three main branches: platelet aggregation, vasoconstriction, and the blood coagulation cascade. These three branches work effectively and in an interconnected manner to avoid or reduce blood loss. Blood-feeding insects, therefore, have managed to counteract the efficiency of the hemostatic system and its redundancy by the presence of powerful bioactive molecules in their saliva that are injected into the host during probing and feeding. Molecules such as vasodilators, inhibitors of platelet aggregation, and anticoagulants have been described in blood-feeding arthropods (Ribeiro, 1987, 1995; Ribeiro and Francischetti, 2003). In sand flies, the presence of a salivary vasodilator has been described since the late 1980s, and since then a multitude

of other pharmacological activities have been characterized (Abdeladhim et al., 2014). The first evidence that sand fly saliva had immunomodulatory activity in addition to potent physiological properties came from the work by Titus and Ribeiro who demonstrated that saliva of *Lutzomyia longipalpis* exacerbated *Leishmania* infection (Titus and Ribeiro, 1988). In this work, inoculation with the equivalent of one pair of salivary gland homogenate (SGH) was sufficient to exacerbate cutaneous leishmaniasis 1000-fold higher than *Leishmania* parasites alone. This finding resulted in a new paradigm in vector biology, particularly for sand fly/*Leishmania*/host interactions, and led to the development of new rodent models of disease that incorporated saliva of sand flies into the inoculum with *Leishmania* parasites (Samuelson et al., 1991; Theodos et al., 1991). These and future studies focused on the role of salivary proteins in modulating *Leishmania* infection, and on understanding the role of sand fly saliva in suppression of immune cells (Gillespie et al., 2000; Kamhawi, 2000; Rohousova and Volf, 2006; Titus et al., 2006; Titus and Ribeiro, 1990). At the turn of the century the sand fly saliva research community addressed a new and unexpected element, the adaptive host immune response to sand fly saliva and its implication for leishmaniasis. Early studies demonstrated that an immune response generated against saliva or uninfected sand fly bites conferred protection against cutaneous leishmaniasis (Belkaid et al., 1998; Kamhawi et al., 2000). Simultaneously, new molecular approaches were providing new information on the molecules present in the saliva of sand flies (Charlab et al., 1999). By the early 2000s, transcriptomics and functional genomic approaches accelerated the characterization of sand fly proteins from various sand fly species and revealed the biological function of some. Importantly, their potential use in biomedical research began to gain attention (Abdeladhim et al., 2014).

PHARMACOLOGICAL ACTIVITIES OF SAND FLY SALIVA

Blood-feeding arthropods have evolved to counteract the hemostatic system to obtain a blood meal. Generally, saliva of blood-feeding insects was shown to contain a number of biologically active molecules to inhibit the hemostatic system. In sand flies, these and other new or unexpected salivary activities have been recently characterized. By the end of the 20th century only one sand fly salivary antihemostatic component, the salivary protein maxadilan, was fully characterized at the molecular level (Lerner et al., 1991). More progress was observed at the turn of the 21st century, where technologies such as PCR subtraction hybridization revealed novel sequences in the saliva of sand flies (Charlab et al., 1999). Later on, a number of sand fly salivary gland transcriptomes provided a more complete repertoire of the proteins present in saliva of sand flies (Abdeladhim

et al., 2014; Valenzuela et al., 2004). These technologies, together with functional genomic approaches such as heterologous expression of sand fly salivary recombinant proteins, provided new insights into the biological activities present in sand fly saliva (Table 5.1). Following are descriptions of some of the potent pharmacological activities identified in sand fly saliva.

Antihemostatic Components Described in Sand Fly Saliva

Vasodilators

Vasodilators are molecules that help blood-feeding insects get a blood meal. They achieve this by relaxing the smooth muscle cells at the bite site to increase blood flow. Maxadilan, a protein of 6.8 kDa, was shown to be the salivary vasodilator from *L. longipalpis* saliva (Ribeiro et al., 1989). Maxadilan is a strong agonist for the pituitary adenylate cyclase–activating polypeptide type I receptor (Moro and Lerner, 1997) and acts by altering the levels of intracellular cAMP in smooth muscle cells (Lerner et al., 2007). Interestingly, in the *L. intermedia* salivary gland transcriptome, only small peptides with low similarity to maxadilan were identified (de Moura et al., 2013). Moreover, no maxadilan transcripts were present in *Bichromomyia olmeca* (Abdeladhim et al., 2016), confirming that maxadilan is not present in all New World sand fly species. Additionally, the peptide maxadilan has not been identified from any *Phlebotomus* species to date (Abdeladhim et al., 2016; Abdeladhim et al., 2012). Instead, the vasodilator for some *Phlebotomus* species was shown to be a purine and not a protein. Adenosine was found in millimolar concentrations in the salivary glands of both *P. papatasi* and *P. argentipes* (Ribeiro et al., 1999; Ribeiro and Modi, 2001). Further, adenosine from *P. papatasi* was shown to inhibit both vasoconstriction and platelet aggregation (Ribeiro et al., 1999). Of note, adenosine was not present in saliva of *P. duboscqi* (Kato et al., 2007), a sister species of *P. papatasi*, suggesting that

TABLE 5.1 Biological Activities of Sand Fly Salivary Proteins

Family of Proteins	Molecular Weight (kDa)	Biological Activity	Sand Fly Species	References
Maxadilan	6	Vasodilator	*Lutzomyia longipalpis*	Ribeiro et al. (1989)
Apyrase	36	Ectoadenosine diphosphatase, inhibitor of platelet aggregation	All sand fly species	Charlab et al. (1999)
Lufaxin	32	Anticoagulant, inhibitor of factor Xa	All sand fly species	Collin et al. (2012)
Small odorant-binding protein-like	15	Inhibitor of contact activation, heparin binding	*Phlebotomus duboscqi* All sand fly species	Alvarenga et al. (2013)
Yellow-related protein	43	Biogenic amine–binding protein	All sand fly species	Xu et al. (2011)
Lundep	44	Endonuclease	*Lutzomyia longipalpis* Most sand fly species	Chagas et al. (2014)
SALO	11	Inhibitor of classical pathway of complement	New world sand flies	Ferreira et al. (2016)
5′ Nucleotidase	61	Nucleotidase	*L. longipalpis*	Charlab et al. (1999)
Adenosine deaminase	56	Hydrolysis of adenosine	*L. longipalpis* *P duboscqi*	Charlab et al. (2000) and Kato et al. (2007)
Hyaluronidase	42	Degradation of hyaluronan	All sand fly species	Cerna et al. (2002) and Charlab et al. (1999)

other molecules or proteins may be the vasodilator for *Phlebotomus* species other than *P. papatasi* and *P. argentipes*.

Biogenic Amine–Binding Proteins

Biogenic amines are mediators that induce responses such as vasoconstriction, platelet activation, itching, pain, and increased vascular permeability (Oliveira et al., 2007; Xanthos et al., 2008). These mediators include serotonin, histamine, norepinephrine, and epinephrine. Salivary proteins that remove these small molecules with high affinity binding were described in saliva of mosquitoes and named "kratagonists" (Ribeiro et al., 2010). In mosquitoes, these kratagonists belong to a family of odorant-binding proteins (Calvo et al., 2009). In sand flies, proteins from a completely

different family, the yellow family of proteins, were characterized as biogenic amine–binding proteins (Xu et al., 2011). Members of the yellow family of proteins are among the most abundant in saliva of sand flies. They are approximately 43 kDa in size and are present in all sand flies studied so far (Abdeladhim et al., 2014). Interestingly, yellow proteins are only present in saliva of sand flies and are not present in saliva of other blood-feeding insects (Valenzuela et al., 2004). The sequence of sand fly salivary yellow proteins is similar to the *drosophila* Dopa Decarboxylase enzyme; however, the sand fly yellow proteins do not have Dopa Decarboxylase enzymatic activity (Xu et al., 2011). In *L. longipalpis*, the yellow family is composed of the salivary proteins LJM11, LJM17, and LJM111. These three yellow proteins were shown to bind biogenic amines, including serotonin, catecholamines, and histamine. The crystal structure of LJM11 and LJM111 was solved, and the amino acids responsible for binding to biogenic amines were identified. Importantly, these amino acid residues are conserved in yellow proteins from other sand fly species (Abdeladhim et al., 2012), suggesting they share the same function. This also implies that binding to biogenic amines to facilitate feeding and avoid reactions induced by these mediators occurs in all sand fly species.

Inhibitors of Platelet Aggregation

Platelets represent one of the first defenses of the hemostatic system against blood loss. They are activated by various agonists including collagen and adenosine diphosphate (ADP) (Hughes, 2013). When activated, platelets change shape and membrane composition. This allows factors of the blood coagulation cascade to bind to the membrane, resulting in platelet aggregation and platelet plug formation (Monroe and Hoffman, 2012). The activity of platelets, therefore, represents an obstacle for blood feeding arthropods that needs to be counteracted. Apyrase (EC 3.6.1.5) is one of the most common inhibitors of platelet aggregation in arthropod saliva, hydrolyzing ATP and ADP but not AMP (Ribeiro, 1987). Hydrolysis of ADP, a potent agonist of platelet aggregation, by apyrases destroys its activity and prevents plug formation.

Mosquito salivary glands contain an apyrase with sequence similarity to 5′ nucleotidases (Champagne et al., 1995). In sand flies, the apyrase belongs to a novel family first discovered in bedbugs and subsequently in humans (Valenzuela et al., 2001b; Valenzuela et al., 1998). Presently, this family of apyrases has been found in all sand fly species researched so far (Abdeladhim et al., 2016; Anderson et al., 2006). The enzyme is about 36kDa and requires calcium for its activity. In addition to sand flies and humans, the apyrase sequence has homologs in *Drosophila melanogaster*, *Caenorhabditis elegans*,

and the protozoan *Cryptosporidium parvum* (Valenzuela et al., 1998).

Another inhibitor of platelet aggregation is a small peptide containing an RGD motif that was identified as the most abundant transcript in the salivary glands of *L. longipalpis* (Valenzuela et al., 2004). However, its biological function remained unknown until recently where its homolog, identified from the salivary glands of *L. ayacuchensis*, was shown to have dual antihemostatic properties (Kato et al., 2015). The salivary peptide, named Ayadualin, inhibits collagen and ADP induced platelet aggregation by interfering with the binding of integrin $\alpha IIb\beta 3$ to fibrinogen. Ayadualin also inhibits the intrinsic pathway of the blood coagulation cascade by inhibiting the activation of FXII to FXIIa. These RGD containing peptides are only present in *Lutzomyia* sand flies and have not been reported from any *Phlebotumus* sand flies to date.

Anticoagulants

The blood coagulation cascade is part of the hemostatic system. It utilizes a number of proenzymes and enzymes to sequentially activate the clotting cascade to prevent blood loss. Coagulation results in the activation of thrombin and in fibrin production from fibrinogen to produce the blood clot. Blood feeding arthropods contain potent anticoagulant molecules in their saliva. Though anticoagulant activity from sand fly saliva has been reported since 1999 (Charlab et al., 1999), the molecule responsible for the anticlotting effect was only recently discovered, and from several species. Saliva of the sand fly *L. longipalpis* contains LJL143, a protein of 32kDa with anticoagulant activity (Collin et al., 2012). The sequence of this protein is unique to sand flies and has no homologies to other proteins or to any known type of anticoagulants. LJL143 was renamed Lufaxin due to its Factor Xa inhibitory activity (Collin et al., 2012). Lufaxin orthologs are present in all sand fly species researched so far indicating that Lufaxin may be the common anticoagulant for all sand fly species (Anderson et al., 2006; Hostomska et al., 2009; Valenzuela et al., 2004). Finding a Lufaxin ortholog from *Bichromomyia olmeca* with Factor Xa inhibitory activity reinforces this supposition (Abdeladhim et al., 2016).

Inhibitor of Contact Pathway Activation

When mast cells and platelets are activated, they release anionic polymers, like polyphosphate and heparin, which are known to stimulate the contact pathway of coagulation. PdSP15 from saliva of the sand fly *P. duboscqi*, member of the SP15-like odorant binding family proteins was recently shown to act as contact activation inhibitor. PdSP15 binds, with high affinity, to these anionic polymers and, inhibits the activation of FXII and FXI, as well as the cleavage of FXI by FXIIa or thrombin, and to inhibit other anionic surface-mediated reactions (Alvarenga et al., 2013). Thus, this protein may act in the sand fly bite site preventing the effects of mast cell activation by inhibiting the coagulation cascade and possibly other process of hemostasis and inflammation.

Anticomplement Activities

The complement system is one of the first lines of defense against pathogens. This system has three main pathways, classical, alternative and lectin pathways (Walport, 2001a,b). Complement activation results in direct killing of target cells and the induction of adaptive cellular and humoral responses. The anaphylotoxins C3a and C5a generated during complement activation are important for phagocyte recruitment and also produce microcirculatory alterations (Walport, 2001a,b). These alterations include vasoconstriction, platelet aggregation, and increased vascular permeability that directly affect the ability of a blood feeder insect to get a blood meal. Importantly, after the insect has acquired a blood meal the host complement system can affect the integrity of the midgut epithelium and consequently the survival of the insect (Barros et al., 2009). Recently, saliva

of the sand fly *L. longipalpis* was shown to have anticomplement activity, inhibiting both the classical as well as the alternative pathways of complement activation (Cavalcante et al., 2003). SALO, an 11 kDa protein present only in *Lutzomyia* sand flies (New World), and with no homology to proteins from other organisms, was characterized as the inhibitor of the classical pathway of complement (Cavalcante et al., 2003; Valenzuela et al., 2004). SALO is a specific inhibitor of the classical pathway of complement, and does not affect the alternative or lectin pathways of complement (Ferreira et al., 2016).

Other Salivary Activities From Sand Flies

Salivary Endonucleases

Transcripts coding for salivary endonucleases were identified and New and Old World sand flies (Abdeladhim et al., 2014). Only one endonuclease was fully characterized from sand fly saliva to date. Lundep, a secreted protein from the salivary glands of *L. longipalpis*, was identified as a potent endonuclease (Valenzuela et al., 2004). Functionally, it is suggested that Lundep may facilitate blood feeding by lowering the local viscosity induced by DNA released from damaged host cells at the site of sand fly bites (Chagas et al., 2014). Also, Lundep may contribute to the antithrombotic and antiinflammatory functions of saliva by hydrolyzing the DNA scaffold of neutrophil extracellular traps (NETs) at the bite site (Chagas et al., 2014).

5′ Nucleotidase/Phosphodiesterase

A sand fly protein with sequence similarity to 5′-nucleotidase (EC 3.1.3.5) and phosphodiesterase was identified in the salivary glands of *L. longipalpis* (Charlab et al., 1999). The 5′ nucleotidase activity was dependent on calcium and magnesium for its activity. This enzyme may increase the levels of adenosine by the hydrolysis of AMP. Adenosine from the saliva of *P. papatasi* was shown to inhibit both vasoconstriction and platelet aggregation (Ribeiro et al., 1999).

Adenosine Deaminase

A protein with high similarity to the enzyme adenosine deaminase (EC 3.5.4.4) was identified from the salivary glands of *L. longipalpis* (Charlab et al., 1999). The recombinant form of *L. longipalpis* adenosine deaminase hydrolyzed adenosine into inosine (Charlab et al., 2000). Interestingly, *P. duboscqi* also has in its salivary glands an active and secreted adenosine deaminase (Kato et al., 2007). Adenosine deaminase in sand flies may help to neutralize pain perception caused by adenosine. Additionally, hydrolysis of adenosine by adenosine deaminase generates inosine, which has been shown to inhibit the production of pro-inflammatory cytokines (Hasko et al., 2000).

Hyaluronidase

Hyaluronidases (EC 3.2.1.35) are enzymes that cleave hyaluronic acid, a glycosaminoglycan found in the extracellular matrix in vertebrates, and have been found in the venom of snakes and spiders (Biner et al., 2015). This enzyme has been referred to as a "spreading factor" because it helps spread toxins by degradation of the extracellular matrix. Hyaluronidase activity and the sequence coding for this enzyme were first identified from the salivary glands of *L. longipalpis* (Charlab et al., 1999). Further studies have shown it to be present in all tested sand fly species (Cerna et al., 2002). This protein is postulated to help blood-feeding arthropods by increasing permeability of host skin to other biologically active salivary components that facilitate blood feeding such as anticoagulants and other antihemostatic components.

IMMUNOMODULATION OF IMMUNE CELLS BY SAND FLY SALIVA

Besides its role in hemostasis, sand fly saliva also affects the host immune system, including innate and acquired immune responses. Most of the studies focused on characterizing immunomodulatory and immunogenic salivary proteins from two important vectors of leishmaniasis: *L. longipalpis*, the principal vector of *L. infantum* in Latin America, and *P. papatasi*, a widely distributed vector of *Leishmania major* in the Old World.

T Cells and Antigen-Presenting Cells

T lymphocytes are important components of the adaptive immune response acting in the protection against infection with many microorganisms. CD4+ T cells are pivotal components of adaptive immune responses that can differentiate into several T helper (T_H) subsets depending on the environmental cues. T_H1 cells are critical for host defense against intracellular pathogens, while T_H2 cells play an important role in eliminating helminthic infections (Kaiko et al., 2008). Monocytes and macrophages are cells with plasticity that differentiate into several cell subsets that are characterized by their cytokine secretion and surface markers. Also, these cells are important antigen presenters being crucial to the development of the adaptive immune response. *L. longipalpis* and *P. papatasi* saliva modulated T cell and macrophage function by inhibiting the expression of T_H1 cytokines and upregulating the expression of Th2 cytokines by activated macrophages (Hall and Titus, 1995; Mbow et al., 1998; Soares et al., 1998; Theodos and Titus, 1993). Saliva from *P. papatasi* also induced positive chemotaxis in macrophages and altered their production of nitric oxide by downregulating the expression of inducible NO synthase. This inhibitory effect on macrophage activation was caused by adenosine in *P. papatasi* saliva. As expected, the same effect on NO synthesis was not observed in *L. longipalpis* since saliva of this species does not contain adenosine (Katz et al., 2000).

Antigen-presenting cells (APCs) are responsible for the initiation of the adaptive immune response. They act directly on naïve T lymphocytes governing their activation and differentiation. Sand fly saliva affects the phenotype and function of several APCs. Reports have demonstrated that *L. longipalpis* saliva alters the expression of costimulatory molecules on dendritic cells (DCs), professional APCs, and modulates their cytokine secretion. When DCs were generated from human monocytes in the presence of *L. longipalpis* SGH, saliva led to a decrease in CD80, CD86, MHC-II, and CD1a expression (Costa et al., 2004). Saliva from *P. papatasi* and *P. duboscqi* also induced the production of prostaglandin E_2 (PGE$_2$) and interleukin 10 (IL-10) by DCs (Carregaro et al., 2008). PGE$_2$ and IL-10 then acted on other DCs in an autocrine manner reducing the expression of MHC-II and CD86 costimulatory molecules on their surface. Inhibition of the expression of costimulatory molecules compromises the ability of these cells to present antigen and, consequently, to activate T-cell responses.

In addition to inhibition of costimulatory molecule expression, *L. longipalpis* saliva decreases the secretion of tumor necrosis factor (TNF) α and IL-10 and increases the levels of IL-6, IL-8, and IL-12p40 in human monocytes (Costa et al., 2004). Furthermore, human monocytes pretreated with *L. intermedia* SGH followed by lipopolysaccharide (LPS) stimulation showed a significant decrease in IL-10 production and an increase in CD86, CD80, and major histocompatibility complex (MHC) II expression (Menezes et al., 2008). Pretreatment with *L. intermedia* SGH followed by LPS stimulation and *Leishmania braziliensis* infection led to a significant increase in TNF-α, IL-6, and IL-8 but no significant alterations in costimulatory molecule expression (Menezes et al., 2008).

Neutrophils

Neutrophils are the most prevalent cells among leukocytes in the blood, and they are the first cells recruited to inflamed tissues. Their roles in immunity to infection include secretion of different cytokines, phagocytosis, degranulation, and oxidative burst to efficiently kill and eliminate microorganisms. Neutrophils contain a large number of granules containing several different proteins (Abi Abdallah and Denkers, 2012; Nathan, 2006). Studies have demonstrated that sand fly salivary molecules modulate the function of neutrophils. *L. longipalpis* SGH triggered neutrophil apoptosis inducing caspase activation and expression of FasL (Prates et al., 2011). Relevantly, apoptotic neutrophils are associated with increased parasite survival. Using two-photon intravital microscopy, Peters et al. (2008) demonstrated that neutrophils are the first cells to infiltrate the skin after infected and uninfected sand fly bites, arriving within 30 min. These authors described an acute inflammatory response to sand fly transmitted infections, observing a rapid, massive, localized, and sustained neutrophilic infiltrate at localized sand fly bite sites that was attributed to tissue damage. Nevertheless, considering the effect of saliva on neutrophils, a role for saliva in attracting neutrophils to the bite site should not be dismissed.

Recently, the sand fly salivary protein LJM111 was shown to have strong antiinflammatory properties (Grespan et al., 2012). Leukocytes obtained from lymph nodes of immunized mice and treated with SGE or recombinant protein-LJM111 and stimulated with mBSA had a significant reduction of IL-17, TNF, and interferon gamma (IFN-γ) production. LJM111 inhibited neutrophil migration into the peritoneal cavity in OVA-challenged immunized mice. In an experimental antigen-induced arthritis model, rLJM111 reduced neutrophil recruitment and pain sensitivity in mice (Grespan et al., 2012).

SAND FLY SALIVARY RECOMBINANT PROTEINS AS MARKERS OF VECTOR EXPOSURE

In general, arthropod salivary proteins are described as immunosuppressive. However, there is a large body of evidence that some of these salivary proteins stimulate production of antibodies in animals. Studies have shown that mice mount an immune response to *P. papatasi* or *L. longipalpis* salivary proteins (Belkaid et al., 1998; Silva et al., 2005) and hamsters produce antibodies to *P. argentipes* saliva (Ghosh and Mukhopadhyay, 1998). Importantly, antibodies against sand fly saliva are also reported for humans (Gomes et al., 2002; Vinhas et al., 2007) and animal reservoirs (Bahia et al., 2007; Gomes et al., 2007; Rohousova et al., 2005).

The fact that experimental animals, animal reservoirs, and humans develop a strong humoral immune response to sand fly salivary proteins led to the hypothesis that sand fly salivary proteins could be used as markers of vector exposure (Barral et al., 2000). There are many aspects that favor this hypothesis: (1) some of these proteins are very immunogenic in animals and humans; (2) some of these proteins are present only in sand flies; (3) some of these proteins may be divergent enough among various sand fly species that they could be used to distinguish a specific sand fly vector; and (4) individual proteins can be reproducibly produced in a recombinant form in large quantities and therefore be more specific in assays than total saliva.

The first successful recombinant proteins to be used as potential markers of exposure were the yellow proteins from *L. longipalpis* saliva, specifically LJM11 and LJM17. These two proteins were expressed in mammalian cells, and they were strongly recognized by animals and humans exposed to *L. longipalpis* bites (Teixeira et al., 2010). Additionally, antibodies to these proteins did not cross-react with saliva from other sand fly species. These proteins were further tested against naturally exposed humans in field conditions (Souza et al., 2010). The authors demonstrated that a combination of LJM11 and LJM17 provided a sensitivity similar to whole saliva concluding that these two proteins represent relevant tools for epidemiological studies of sand fly exposure (Souza et al., 2010). LJM11 and LJM17 were also used successfully in field

conditions to monitor seroconversion to *L. longipalpis* bites in sentinel chickens in visceral leishmaniasis endemic regions in Brazil (Soares et al., 2013). Furthermore, a protein of 32 kDa from *P. papatasi* salivary glands was immunogenic in mice and was one of the most recognized antigens by antibodies of exposed humans from Tunisia where *P. papatasi* is prevalent (Belkaid et al., 1998; Marzouki et al., 2012). This protein is highly expressed in the salivary glands of *P. papatasi* and shares identity with a collagen-like protein (Abdeladhim et al., 2012). Named PpSP32, this antigenic protein was used successfully as a tool to determine vector exposure of humans to *P. papatasi* in Tunisia (Marzouki et al., 2015) and in Saudi Arabia (Mondragon-Shem et al., 2015).

For animal reservoirs, three salivary proteins, SP01B, SP01, and SP03B were shown to be effective markers of vector exposure to *P. perniciosus* in dogs (Drahota et al., 2014). Both rPorSP24 and rPorSP65 from the sand fly *P. orientalis* were also identified as promising markers of vector exposure in dogs (Sima et al., 2016).

The fact that specific sand fly salivary proteins can be used as markers of vector exposure in humans and animal reservoirs opens new avenues for their use for public health purposes. More work is needed to optimize production of these proteins to a diagnostic grade level. However, results from field studies are promising and suggest this could be a worthwhile investment that will provide better tools to measure vector exposure in epidemiological studies and during control interventions.

SAND FLY SALIVA AS A VACCINE AGAINST LEISHMANIASIS

Salivary molecules from arthropod vectors of disease are unconventional antigens for antiparasitic vaccines. This novel concept was put forward based on the observation that vector salivary molecules enhance parasite establishment and counteracting them would be beneficial to the mammal host (Titus and Ribeiro, 1988). These salivary molecules are injected into the upper dermal region of the skin to facilitate blood feeding. However, saliva and *Leishmania* are delivered into the skin by the same tiny sand fly proboscis, through independent channels. As such, they are present in such close contact at the bite site that modulation of the microenvironment by immunity against salivary molecules will affect parasites at their most vulnerable point in the mammalian host.

A proof of concept of the protective potential of sand fly saliva against *Leishmania* was successfully demonstrated after mice sensitized with SGH, or by uninfected bites, from *P. papatasi* mounted an immune response that protected them from a challenge with *L. major* coinjected with saliva or delivered by infected sand fly bites, respectively (Belkaid et al., 1998; Kamhawi et al., 2000). In another experiment, immunization with Maxadilan, a 7 kDa molecule from the saliva of the vector *L. longipalpis*, induced a cellular and humoral immune response capable of protecting mice from a challenge with *L. major* in the presence of whole saliva (Morris et al., 2001). Protection from leishmaniasis, observed either with preexposure to uninfected sand fly bites or injection with SGH of *P. papatasi*, was mediated by rapid recruitment to the site of the bite of predominantly CD4 T lymphocytes producing IFNγ in what can be characterized as a T_H1 delayed-type hypersensitivity response. This T_H1 type immune response has been repeatedly shown to be the correlate of protection against *Leishmania* infection for most of the protective sand fly salivary proteins studied so far (Abdeladhim et al., 2014).

From Salivary Gland Homogenate to Recombinant Salivary Proteins

Based on the observation that sand fly SGH or bites of uninfected sand flies produced an immune response that protected animals against leishmaniasis, the next step revolved around the search for, and identification of, *immunogenic*

salivary proteins that induced a T_H1 immune response. This represented a major step forward from proof of concept to the development of a product that can be plausibly marketed for use in vaccines. The first sand fly salivary molecule identified as a DTH-inducing candidate was PpSP15, an abundant molecule present in saliva of *P. papatasi* (Abdeladhim et al., 2012; Valenzuela et al., 2001a). Relevantly, vaccination with PpSP15, as a protein or DNA vaccine, protected mice from a challenge with needle-inoculated *L. major* in the presence of *P. papatasi* saliva (Oliveira et al., 2008; Valenzuela et al., 2001a). This represented the first sand fly salivary molecule that could protect against cutaneous leishmaniasis. Importantly, antibodies were not necessary for protection, since B cell-deficient mice, that did not produce antibodies against PpSP15 but produced a DTH response to this salivary protein, were equally protected from cutaneous leishmaniasis, suggesting that the cellular immune response was responsible for this protective effect (Valenzuela et al., 2001b).

Sand fly salivary proteins from other sand fly species were also tested against cutaneous leishmaniasis (CL) and other types of leishmaniasis, including visceral leishmaniasis (VL), the fatal form of the disease. Immunization with LJM19 DNA, a protein of 11 kDa from saliva of the sand fly *L. longipalpis* produced a Th1 immune response in hamsters and conferred a strong protection from fatal VL when challenged with a co-inoculum of *L. infantum* and *L. longipalpis* saliva (Gomes et al., 2008). This was the first sand fly salivary molecule to show protection in a model of VL, indicating the potential of salivary proteins as a vaccine against visceral leishmaniasis. Interestingly, LJM19 DNA vaccination also protected hamsters against cutaneous leishmaniasis from a challenge with *L. braziliensis* in the presence of saliva from its vector *L. intermedia* (Tavares et al., 2011). This suggested that a homolog of LJM19 in saliva of *L. intermedia* was able to recall the immune response mounted against LJM19. Homologs of LJM19 are present in all New World sand fly vectors studied so far (Abdeladhim et al., 2016), and as such, this molecule could prove to be a strong vaccine candidate against leishmaniasis throughout the Americas.

Immunization with the salivary protein LJM11, a 43-kDa protein from *L. longipalpis*, as a DNA vaccine (Xu et al., 2011) or as a recombinant protein (Gomes et al., 2012) resulted in a specific and long-term protective immunity against needle-inoculated *L. major* plus saliva, or *L. major*-infected sand fly bites. In both studies, protection was correlated with production of IFNγ from CD4 T lymphocytes. The protective potential of LJM11 was further tested and validated by immunization with a *Listeria monocytogenes*–expressing LJM11 as a vehicle (Abdallah et al., 2014). Immunized mice elicited a specific and robust anti-*Leishmania* T$_H$1 immune response that correlated with strong protection against *L. major*-infected *L. longipalpis* bites. Despite the potential of this protein as a protective vaccine, antibodies to LJM11 were associated with disease, Pemphigus foliaceus (Fogo Selvagem), in leishmaniasis endemic areas (Qian et al., 2015, 2012). Pemphigus foliaceus is an autoimmune disease caused by antibodies targeting the self-protein desmoglein. Desmoglein, mostly present on the surface of keratinocytes, is a component of desmosomes and important for adhesion. Antibodies present in animals exposed to *L. longipalpis* bites or immunized with LJM11 recognized desmoglein despite a lack on sequence similarity (Qian et al., 2012). Therefore, in an *L. longipalpis* abundant area where individuals are constantly bitten, mounting an antibody response to LJM11 could trigger the disease in genetically prone individuals. In light of these findings, and despite its protective potential against leishmaniasis, the pursuit of LJM11 as a vaccine for humans has been sidelined.

LJL143, another molecule from *L. longipalpis* saliva, shows promise as a canine vaccine.

This is significant, since dogs are reservoirs of *L. infantum* parasites, and they are a major source of human infection (Moreno and Alvar, 2002). Canines immunized with LJL143, as a DNA primer followed by a boost with LJL143-recombinant protein and posteriorly with canarypox virus–expressing LJL143, were able to induce IFNγ and IL-12 cytokine production upon challenge with *L. infantum*-infected *L. longipalpis* sand flies. Moreover, macrophages cultured from peripheral blood mononuclear cells of vaccinated canines could be stimulated to kill *L. infantum* upon the addition of autologous T cells and *L. longipalpis* SGH (Collin et al., 2009). Overall, LJL143 is a promising molecule since it is immunogenic in dogs, the main reservoir of VL in the Americas; it shares no similarity with any mammalian molecule; and it is present in all species of sand flies studied so far (Abdeladhim et al., 2014).

A departure from the observation that immunization with whole sand fly saliva leads to protective immunity against leishmaniasis was shown for *L. intermedia* saliva. In this particular case, upon challenge with *L. intermedia* saliva and *L. braziliensis*, Balb/c mice immunized with *L. intermedia* saliva exhibited a delayed onset of disease, but no measureable level of protection or exacerbation (de Moura et al., 2007). Nevertheless, when salivary proteins in *L. intermedia* saliva were tested individually through DNA immunizations, Linb-11, a molecule of 4.5 kDa, induced a predominantly cellular immune response characterized by a T$_H$1–mediated DTH response that protected Balb/c mice upon challenge with *L. braziliensis* and *L. intermedia* saliva (de Moura et al., 2013). Taken together, these data indicate that even when the overriding immunity generated by a whole salivary mixture is not effective, protection can be achieved by selecting a molecule that induces a T$_H$1-DTH response. These results have major implications for endemic areas. Possibly, individuals that fail to produce a protective immune response to whole saliva can still benefit from vaccination with a distinct molecule that induces the appropriate antisaliva protective immunity.

TRANSLATIONAL ASPECTS OF SAND FLY SALIVA

From the above, it is clear that certain salivary molecules are antigenic and confer robust protection against both CL and VL in various animal models of infection (Table 5.2). Additionally, data show that upon challenge, immunity to saliva generates a rapid response that acts during the initial phase of parasite establishment at the bite site. Acknowledging the multipronged complexity of the infectious inoculum from a *Leishmania*-infected sand fly bite that strongly favors parasite survival, the antigenic potential of saliva, a constant major component of the inoculum, could be exploited to overcome it.

From Mouse Models to Nonhuman Primates

The protective effect of sand fly salivary proteins observed in small rodents can be replicated in rhesus monkeys. Prior exposure to uninfected sand fly bites led to robust protection against a challenge with 50 *L. major*-infected *P. dubosqci* sand flies (Oliveira et al., 2015). As predicted from experiments in mice, the protective response in monkeys can again be correlated to development of an antisaliva immune response observed 48h after challenge and characterized by the production of IFNγ. Impressively, antisaliva protection induced a more robust anti-*Leishmania* T_H1 immune response by multifunctional CD4$^+$T cells producing IFNγ and IL-2 that correlated with a low total burden of disease. Interestingly, 30% of the rhesus monkeys failed to develop an immune response against sand fly saliva. These animals developed a weak anti-*Leishmania* specific immune response after challenge, and they presented with a higher burden of disease more like controls. This study alludes to the potential heterogeneity of the human immune response to sand fly saliva that is a pattern not previously observed in inbred mice.

From the Bench to the Field: Studying Sand Fly Immunity in Humans

In nature, the percentage of sand flies carrying a *Leishmania* infection is very low (Ready, 2013). Therefore, there is a high probability that inhabitants of areas where sand flies are abundant are exposed to sand fly saliva through uninfected bites. If this is the case, a major question that eludes us is "how come exposure to uninfected sand fly bites protects mice while not all people living in endemic areas are protected against leishmaniasis?" One answer could be that some individuals are not sufficiently exposed to sand fly saliva to mount a *Leishmania*-protective saliva-based immune response. This assumption may explain why leishmaniasis in endemic areas is usually more prevalent in children (Carvalho et al., 2015; Chappuis et al., 2007), and why the severity of disease is worse in tourists or individuals migrating from leishmaniasis-free areas (de Silva et al., 2015). Another explanation may be that not all individuals respond to uninfected sand fly bites in the same manner and at the same pace. That is reflected in the work done with rhesus monkeys where 30% of the animals did not mount an antisaliva protective immune response after three exposures and succumbed to CL (Oliveira et al., 2015). This observation was further explored in residents of an endemic area of CL in Mali where *P. duboscqi* is abundant (Oliveira et al., 2013). Similar to the rate of responsiveness observed in monkeys, only 64% of the tested residents developed a DTH immune response to saliva following experimental exposure to uninfected bites. It is worthwhile noting, however, that in a subset of DTH positive individuals, the antisaliva immunity observed at the bite site was T_H1-biased, an immune response predictive of protection against *Leishmania*

TABLE 5.2   Protective Sand Fly Salivary Proteins Tested in Animal Models of Leishmaniasis

Salivary Protein	Sand Fly Species	Immunization Protocol	Challenge	Animal Model	References
Maxadilan	*Lutzomyia longipalpis*	Recombinant protein	*Leishmania major* + *L. longipalpis* salivary gland homogenate (SGH)	Mice	Morris et al. (2001)
PpSP15	*Phlebotomus papatasi*	Native protein/DNA	*L. major* + *P. papatasi* SGH	Mice	Valenzuela et al. (2001a,b)
PpSP15	*P. papatasi*	DNA	*L. major* + *P. papatasi* SGH	Mice	Oliveira et al. (2008)
LJM19	*L. longipalpis*	DNA	*Leishmania infantum* + *L. longipalpis* SGH	Hamsters	Gomes et al. (2008)
LJM19	*L. longipalpis*	DNA	*Leishmania braziliensis* + *Leishmania intermedia* SGH	Hamsters	Tavares et al. (2011)
LJM11	*L. longipalpis*	DNA	*L. major* + *L. longipalpis* SGH	Mice	Xu et al. (2011)
LJM19	*L. longipalpis*	DNA + KMP11-DNA	*L. infantum* + *L. longipalpis* SGH	Hamsters	da Silva et al. (2011)
LJM11	*L. longipalpis*	DNA	*Leishmania chagasi* + *L. longipalpis* SGH	Mice	Gomes et al. (2012)
Linb-11	*L. intermedia*	DNA	*L. braziliensis* + *L. intermedia* SGH	Mice	de Moura et al. (2013)
LJM11	*L. longipalpis*	*Listeria monocytogenes*–expressing LJM11	*L. major* + *L. longipalpis* SGH	Mice	Abdallah et al. (2014)
PpSP15	*P. papatasi*	DNA + *Leishmania* CPA/CPB	*L. major* + *P. papatasi* SGH	Mice	Zahedifard et al. (2014)
LJM11	*L. longipalpis*	*L. monocytogenes*-expressing LJM11	*L. major*-infected *L. longipalpis* bites	Mice	Abdallah et al. (2014)
PpSP15	*P. papatasi*	*Leishmania tarentolae*-expressing PpSP15	*L. major* + *P. papatasi* SGH	Mice	Katebi et al. (2015)
PdSP15	*P. duboscqi*	DNA + recombinant protein	*L. major*-infected *P. duboscqi* bites	Monkey	Oliveira et al. (2015)
LJM19	*L. longipalpis*	Recombinant protein	*Leishmania donovani* + *L. longipalpis* SGH	Hamsters	Fiuza et al. (2016)

parasites. When peripheral blood from study subjects was stimulated with sand fly saliva, a mixed cytokine profile was observed where 23% produced a T_H1-dominated immune response and 25% a Th2-dominated pattern of responses (Oliveira et al., 2013). The latter may reflect a pool of unprotected hosts that on an encounter with an infected sand fly would not have the benefit of the T_H1-biased protective antisaliva immunity.

Combining Sand Fly and Leishmania Antigens

There is enough evidence to suggest that a specific sand fly salivary protein can produce a protective immune response to *Leishmania* in rodents, animal reservoirs such as dogs, and in nonhuman primates. However, perhaps an ideal or effective anti-*Leishmania* vaccine should not be composed of either a salivary protein alone or a *Leishmania* antigen alone. We put forward the notion that a combination vaccine using both *Leishmania* and vector salivary antigens may have a better chance at success. Recent studies assessing heterologous combination vaccines suggest that they work more efficiently than using one source of antigen. One study showed the synergic potential of combining PpSP15-DNA immunization with nonpathogenic *L. tarentole*-expressing *Leishmania* cysteine proteinases. The combination vaccine showed greater protection than either PpSP15 or *L. tarentole*-expressing *Leishmania* cysteine proteinases alone (Zahedifard et al., 2014). Interestingly, of the various combinations tested, the best protection was observed by heterologous priming with PpSP15 alone followed by a booster with a combination of PpSP15 DNA and live-recombinant *L. tarentolae*-expressing cysteine proteinases (Zahedifard et al., 2014). In another study, optimal protection against *L. major* in the presence of *P. papatasi* SGH was also observed with a live-recombinant *L. tarentolae* expressing PpSP15 in the presence of CpG ODN adjuvants (Katebi et al., 2015).

Importantly, PpSP15 represents a good choice as a vaccine candidate. Despite sequence variation in the protein among *P. papatasi* populations from different endemic areas in the Middle East, its predicted MHC class II antigenic motifs have remained conserved (Ramalho-Ortigao et al., 2015). This suggests that PpSP15 and its homologs may retain their antigenicity despite some sequence variation. The strong protection conferred by PdSP15, a homolog of PpSP15 in the vector *P. duboscqi*, against CL in large nonhuman primates, despite a large infectious inoculum of 50 infected *L. major*-infected *P. duboscqi* sand flies (Oliveira et al., 2015), validates the predicted conservancy of antigenicity in PpSP15 homologs. Importantly, protected rhesus macaques displayed a T_H1-CD4 expansion and IFNγ production, reflecting the same correlates of protection observed in protected mice. Protective cross-reactivity has also been reported for animals exposed to *P. papatasi* sand fly bites and challenged with *P. duboscqi* plus *L. major* (Lestinova et al., 2015), further demonstrates that a vaccine from either of these sand fly vector species will likely protect populations across the distribution range of both.

Another example of the improved efficacy of combination vaccines was demonstrated when *L. braziliensis* antigens were combined with SGH from *Lutzomyia longipalpis* and used to vaccinate canines in the presence of the adjuvant saponin (Giunchetti et al., 2008). Immunized animals developed both anti–sand fly saliva and anti-*Leishmania* immune responses and displayed a greater reduction in splenic parasitic loads compared to either antigen alone or to mock immunized controls (Aguiar-Soares et al., 2014). In yet another study, hamsters immunized with recombinant LJM19, a protein in *L. longipalpis* saliva, in combination with *L. donovani* centrin 1 knockout parasites controlled the infection more effectively than either antigen alone after intradermal challenge with wild-type parasites and saliva (Fiuza et al., 2016). Overall, the above examples suggest that combining a *Leishmania* antigen(s) and a vector salivary antigen is likely to enhance the effectiveness of anti-*Leishmania* vaccines.

EVOLUTION OF SAND FLY SALIVARY PROTEINS

Sand fly salivary proteins protect a variety of animal models against *Leishmania* infection in immunization trials. In addition, such proteins can be used as markers of exposure to sand fly bites (Coutinho-Abreu et al., 2015), which is a proxy of vector exposure, and in some instances risk for *Leishmania* infection. Understanding the molecular evolution of salivary proteins can inform our choice of vaccines and markers of exposure. Additionally, the rate of molecular divergence of such protein encoding gene families can be used as a good indicator of the extent of sequence polymorphisms in MHC class II binding sites (Anderson et al., 2006; Ramalho-Ortigao et al., 2015). This information can significantly impact the efficacy of a vaccine or a marker of exposure at a population level.

Salivary proteins are among the most divergent protein families in insect disease vectors. In the comparative genome analysis of 16 anopheles mosquito species, salivary protein encoding genes were not only among the fastest evolving protein families at the sequence level, but also among the protein families displaying the fastest rate of gene turnover (Neafsey et al., 2015). Additionally, multiple cases of convergent evolution have been described for salivary proteins (Calvo et al., 2006), pointing to the variety of ways the proteins responsible for specific biological processes have emerged in salivary glands of disease vectors.

It is worth noting some examples of convergent evolution for salivary proteins shared between sand flies and other insect disease vectors (Valenzuela et al., 2001b; Valenzuela et al., 1998). In mosquitoes and sand flies, proteins from different gene families prevent blood coagulation. In the mosquito *Aedes albopictus*, a serine protease inhibitor named Alboserpin has been identified as a factor Xa (FXa) inhibitor (Calvo et al., 2011). Factor Xa is a component of the blood coagulation cascade. On the other hand, in the saliva of the sand fly *L. longipalpis*, FXa inhibition is performed by a novel protein named Lufaxin (Collin et al., 2012). The Lufaxin sequence is novel and does not share any similarities with other serine protease inhibitors (Collin et al., 2012). Lufaxin and Alboserpin perform the same molecular function without sharing a common ancestor, an example of natural selection pressure at work to independently evolve a molecule that inhibits FXa in these blood-feeding vectors. Another case of convergent evolution has been observed in mosquitoes and sand flies for proteins that prevent platelet aggregation. The clumping of platelets at the biting site disrupts blood flow and in turn blood engorgement. As ATP and ADP are the main mediators of platelet aggregation, blood-sucking insects secrete enzymes (apyrases) that catalyze the breakdown of ATP and ADP into AMP and orthophosphate. Different from other disease vectors that mediate ATP breakdown with an apyrase of the 5′-nucleotidase family (Champagne et al., 1995), the apyrase activity in sand flies is performed by a protein from the Cimex protein family (Valenzuela et al., 2001b; Valenzuela et al., 1998).

Within sand flies, salivary protein encoding genes have also been evolving in different directions to cope with specific feeding needs in different habitats. Sand flies are generically classified as belonging to New World and Old World groups that refers to their geographic distribution. To date, sand fly salivary transcriptomes were obtained for nine Old World sand fly species belonging to five subgenera of the genus Phlebotomus (Abdeladhim et al., 2012; Anderson et al., 2006; Hostomska et al., 2009; Kato et al., 2006; Martin-Martin et al., 2013; Oliveira et al., 2006; Rohousova et al., 2012; Valenzuela et al., 2001a; Vlkova et al., 2014). In regards to New World sand flies, four species belonging to three different genera had their salivary gland transcriptomes sequenced (Abdeladhim et al., 2016; de Moura et al., 2013; Kato et al., 2013; Valenzuela et al., 2004). Comparative analysis of the salivary protein families revealed different modes of molecular evolution that led to their diversification in sand flies.

Sand fly saliva encompasses proteins unique to either New World or Old World sand flies as well as protein families common to all sand fly species (Table 5.3). The phylogenetic trees of five out of seven salivary protein-encoding gene families shared by all sand fly species correlate well with the sand fly species phylogeny, constructed based on ITS-2 sequences (Aransay et al., 2000). Such a pattern indicates that those gene families have been evolving under purifying (negative) selection or are selective neutral (Barton and Etheridge, 2004). On the other hand, the Silk and D7 salivary protein gene family trees diverged considerably from the sand fly species phylogeny. Signs of diversifying (positive) selection were noted for the clades belonging to the Phlebotomus and Paraphlebotomus subgenera (Old World sand flies) in the D7 and Silk phylogenies, and for the New World sand fly clade in the Silk phylogeny. Although less striking, signatures of diversifying selection were also observed for a few codons belonging to antigen-5 and apyrase in the clades encompassing Phlebotomus and Paraphlebotomus sand flies. Taken together, these findings indicate that the salivary protein encoding genes shared by all sand flies have been evolving at a faster pace in members of the Old World Phlebotomus and Paraphlebotomus subgenera. Further phylogenetic studies need to be performed in order to shed light on the importance of the fast evolving pace of salivary proteins in such a taxonomic group (Abdeladhim et al., 2016).

Different from silk and D7 salivary protein phylogenies, the yellow protein gene family phylogeny is in accordance with the sand fly species phylogeny. Nonetheless, analysis of the two yellow-related protein paralogs in a single tree reveals that the SP42 and SP44 proteins are more similar to their paralogs in the same species than to their orthologs in different species (Abdeladhim et al., 2016). This suggests that gene conversion has been taking place to maintain the sequence similarity between paralogs in the same species. Therefore, the yellow-related protein family is evolving by concerted evolution (Innan and Kondrashov, 2010). Although gene conversion can restrain the emergence of new functions upon gene duplication, it leads to an increase in the net amount of a protein in the organism, which can also be advantageous (Innan and Kondrashov, 2010).

Molecular evolution of salivary protein families unique to New World sand flies revealed not only different modes of gene emergence but also different paces of gene diversification as well as specific signatures of protein structure. Amongst the salivary gene families unique to New World sand flies, the c-type lectin, the manose binding lectin, and the spider toxin-like gene families share paralogs expressed in other tissues of sand flies and other unrelated arthropods (Abdeladhim et al., 2016). The emergence of such genes in salivary glands may have happened due to the acquisition of a salivary gland specific promoter upon duplication of a gene expressed in another tissue (Hahn, 2009; Innan and Kondrashov, 2010). Hence, such gene families seem to have emerged by a mechanism called neofunctionalization (subfunctionalization) (Hahn, 2009; Innan and Kondrashov, 2010).

The molecular divergence of the salivary protein encoding gene families unique to New World sand flies are overall greater than their counterparts in Old World sand flies. The gene families unique to New World sand flies are under diversifying or relaxed purifying selection, which is rarely seen for the gene families shared among Old World sand flies. By the same token, gene duplication events are more often detected in the gene families of the former than the latter.

Overall, salivary protein gene families unique to New World sand flies diverge at a faster pace than their counterparts in the Old World (Abdeladhim et al., 2016). Another interesting

TABLE 5.3 Sand Fly Salivary Protein Families Across Multiple Sand Fly Species

Family of Proteins	Bichromomyia olmeca	Lutzomyia longipalpis	Lutzomyia ayacuyensis	Nyssomyia intermedia	Phlebotomus papatasi	Phlebotomus duboscqi
PROTEINS UNIQUE TO NEW WORLD SAND FLIES						
Toxin-like	✔	–	–	✔	–	–
Arg-Gly-Asp–containing	✔	✔	✔	✔	–	–
C-type lectin	✔	✔	✔	✔	–	–
Maxadilan	–	✔	–	✔	–	–
14 kDa SP	✔	✔	–	✔	–	–
ML domain	✔	–	–	✔	–	–
5′ Nucleotidase	–	✔	–	–	–	–
9 kDa SP	–	✔	✔	–	–	–
SALO-like (10 kDa)	✔	✔	–	✔	–	–
11.5 kDa SP	✔	✔	✔	–	–	–
71 kDa proteins	✔	✔	✔	–	–	–
5 kDa protein	✔	–	–	✔	–	–
MOLECULES OR PROTEINS UNIQUE TO OLD WORLD SAND FLIES						
Adenosine	–	–	–	–	✔	–
Pyrophosphatase	–	–	–	–	–	✔
Phospholipase A2	–	–	–	–	–	–
SP16-like	–	–	–	–	✔	–
ParSP23-like	–	–	–	–	–	–
SP2.5 kDa-like	–	–	–	–	✔	✔
ParSP25-like	–	–	–	–	–	–
5 kDa-like (ParSP15)	–	–	–	–	–	–
PROTEINS SHARED BETWEEN NEW WORLD AND OLD WORLD SAND FLIES						
Small odorant-binding protein	✔	✔	✔	✔	✔	✔
Yellow-related	✔	✔	✔	✔	✔	✔
Antigen 5-related	✔	✔	✔	✔	✔	✔
Lufaxin	✔	✔	✔	✔	✔	✔
D7-related	✔	✔	✔	✔	✔	✔
Apyrase	✔	✔	✔	✔	✔	✔
Endonuclease	✔	✔	✔	✔	–	–
Hyaluronidase	✔	✔	–	✔	–	–
Silk-related	✔	✔	✔	✔	✔	✔
Adenosine deaminase	✔	✔	–	–	–	✔
56.6 kDa SP	✔	✔	–	–	✔	✔
PPTSP38.8-like	✔	–	–	–	✔	–
16.1 kDa	–	✔	–	–	–	✔

Phlebotomus sergenti	Phlebotomus ariasi	P. orientalis	P. tobbi	P. perniciosus	P. arabicus	P. argentipes
–	–	–	–	–	–	–
–	–	–	–	–	–	–
–	–	–	–	–	–	–
–	–	–	–	–	–	–
–	–	–	–	–	–	–
–	–	–	–	–	–	–
–	–	–	–	–	–	–
–	–	–	–	–	–	–
–	–	–	–	–	–	–
–	–	–	–	–	–	–
–	–	–	–	–	–	–
–	–	–	–	–	–	✔
–	–	✔	–	✔	✔	✔
–	✔	✔	–	✔	✔	–
✔	–	✔	–	–	✔	✔
–	✔	–	–	–	✔	–
–	–	–	–	–	–	–
–	✔	✔	✔	✔	✔	✔
–	✔	✔	✔	✔	–	–
✔	✔	✔	✔	✔	✔	✔
✔	✔	✔	✔	✔	✔	✔
✔	✔	✔	✔	✔	✔	✔
✔	✔	✔	✔	✔	✔	✔
✔	✔	✔	✔	✔	✔	✔
✔	✔	✔	✔	✔	✔	✔
–	✔	✔	–	✔	✔	✔
–	–	✔	✔	✔	✔	–
✔	✔	✔	✔	✔	✔	✔
–	–	–	–	✔	–	–
✔	✔	–	✔	✔	✔	–
✔	✔	–	✔	✔	✔	–
–	✔	–	–	–	✔	–

aspect to the molecular evolution of salivary gland protein families unique to New World sand flies is the presence of conserved cysteine residue signatures. Such amino acid signatures are present in five out of seven gene families unique to New World sand flies whereas only three gene families shared between Old World and New World sand flies bear such cysteine signatures. This suggests that *denovo* emergence of salivary gland gene families may rely on cysteine building blocks (Abdeladhim et al., 2016).

All in all, the phylogenetic analysis of sand fly salivary proteins unveils the most conserved and divergent sets of molecules amongst sand flies from different continents. When such proteins are to be applied in vaccine trials, the proteins unique to New World sand flies should be used with caution, as very high degrees of sequence divergence among species were noticed. The extent of sequence polymorphisms at the population level should be analyzed before vaccination trial so as to assess the impact of the divergence on the MHC class II epitopes (Kato et al., 2006; Ramalho-Ortigao et al., 2015). On the other hand, the use of salivary proteins shared amongst all sand fly species in vaccination trial would likely result in more homogenous protection among individuals, as such proteins display low levels of sequence polymorphism overall.

TRANSLATIONAL OPPORTUNITIES AND FUTURE DIRECTIONS

From basic research on sand fly salivary proteins three main translational opportunities emerge: (1) potential translation of pharmacological activities from sand fly saliva to products in biomedicine; (2) use of salivary proteins as markers of vector exposure for epidemiological studies; and, (3) use of salivary proteins as a component of a human leishmaniasis vaccine. For biomedicine, the knowledge that sand fly saliva has pharmacological activities is not novel;

however, identification of the proteins responsible for these activities make such molecules more suitable candidates to be developed as potential pharmacological products for human use. One of the most recent and likely candidates is the anticomplement salivary protein SALO (Ferreira et al., 2016). This protein specifically inhibits the classical pathway of complement; it is small and it does not produce antibodies in the absence of adjuvants. Currently, monoclonal antibodies against specific proteins of the complement cascade are given to alleviate complement adverse events in patients. SALO is a very small protein that acts directly and specifically to inhibit the initial steps of the classical pathway of complement and could provide a new and less expensive alternative. Importantly, the SALO sequence is unique with no homologs in humans. The other salivary protein with potential is Lufaxin, a potent and specific Factor Xa inhibitor with no sequence similarity in humans. Factor Xa is a component of the blood coagulation cascade and Lufaxin may represent a novel choice of anticoagulant. Lufaxin also had antiinflammatory properties that can be channeled for use in biomedicine.

The use of sand fly salivary recombinant proteins as markers of vector exposure is a newly developing area with great potential. Preliminary work with various recombinant salivary proteins is encouraging and promises to be an area of significant growth in the next few years. There are already a good number of recombinant salivary proteins from sand flies that have shown both specificity and immunogenicity as markers of vector exposure that work well in laboratory as well as field conditions. These include LJM11 and LJM17 from *L. longipalpis* as markers for exposure in humans and dogs (Teixeira et al., 2010); PpSP32, a recombinant protein that can be used as a marker for *P. papatasi* exposure in humans (Marzouki et al., 2012, 2015); and there are SP01B, SP01, SP03B, and rPorSP24 from *P. perniciosus* and *P. orientalis* that can be used in reservoir animals such as dogs (Drahota et al., 2014; Sima et al., 2016). The next step is to make robust and sensitive diagnostic kits with these or any other candidate salivary proteins. A technical cautioning, recombinant proteins produced in bacteria may work well for dogs or other small animals but may not work for humans. The background observed with the use of recombinant salivary proteins produced by bacteria and human sera can be avoided by using other heterologous expression systems such as insect cells or mammalian cells that have shown empirically to be very specific and clean. Another alternative will be to produce synthetic peptides that are immunogenic and that will not have any bacterial byproduct in the preparation. Another important challenge for this approach is to select the best protein or region of the protein that can distinguish the vector at the species level. Sand flies share many salivary proteins and candidate salivary markers will need to be tested for specificity at the genus or species level.

For the potential use of sand fly saliva as vaccines, significant advances were made in the last few years including identification of the correlates of protection from leishmaniasis as a T_H1 type cellular immune response induced by distinct salivary proteins (Gomes et al., 2012; Oliveira et al., 2008, 2015). Another significant advance was validation of observations made in rodent models of infection in humans. A recent study demonstrated that humans also develop a T_H1 immune response to sand fly salivary proteins (Oliveira et al., 2013), suggesting that the protective effect observed in rodents may translate to humans. Importantly, in the last few years, candidate vaccines have been narrowed down to only a few, LJL143, LJM19, and PpSP15. A significant advance for this field and for the likelihood to move this concept forward toward a human vaccine is the vaccine efficacy of PdSP15, a homolog of PpSP15, in nonhuman primates, that induced protection after challenge with *L. major*-infected sand flies,. Among these, two salivary proteins, LJL143 and LJM19,

were shown to be potential vaccines or components of a vaccine against visceral leishmaniasis. LJL143 is immunogenic in dogs and in vitro data show that immunity against this protein can effectively activate infected macrophages to kill *Leishmania* (Collin et al., 2009). The advantage of this salivary vaccine candidate is that it is present in all sand flies tested so far. This protein may therefore cover various regions of the world where sand fly vectors of visceral leishmaniasis are prevalent. LJM19 is a salivary vaccine candidate shown to protect hamsters against the fatal outcome of visceral leishmaniasis. The advantage of this salivary protein is that it is a small molecule of about 11 kDa and it was also shown to work effectively as a combination vaccine in hamsters. Further work needs to be done with these two proteins to determine their immunogenicity in humans and their potential use in combination vaccines with current *Leishmania* antigens. The T_H1-inducing property, in the absence of adjuvant, of these two salivary proteins and their strong protective effect observed in animals suggest that these proteins are suitable candidates to move forward for human studies. The next steps for these vaccine candidates will be to focus on product development to achieve clinical grade material to be used in preclinical studies and importantly in human clinical trials to test their safety and immunogenicity, and ultimately their efficacy.

In the last two decades there have been significant advances in research on sand fly saliva. Novel molecules have been identified and characterized; novel biological activities have been described; a comprehensive analysis of the transcripts of various sand fly species around the world have been completed; novel markers of vector exposure to humans and dogs have been discovered; and, the potential to use a vector salivary protein as a component of a vaccine against a parasite in humans became a credible prospect. Moreover, this will open up not only the potential use of sand fly salivary proteins as vaccines against leishmaniasis, but it may pave the road for similar studies of arthropod saliva for other vector-borne diseases.

The catalog of sand fly salivary proteins with unknown function is still large, and more work needs to be done to discover their biological roles and activities. Current technologies, including CRISPR/Cas9, can help determine the impact of some of these salivary proteins in blood feeding, transmission, and immunity. Nevertheless, the profound impact of sand fly saliva on our physiology and health is abundantly clear.

References

Abdallah, D.S.A., Bitar, A.P., Oliveira, F., Meneses, C., Park, J.J., Mendez, S., Kamhawi, S., Valenzuela, J.G., Marquis, H., 2014. A Listeria monocytogenes-based vaccine that secretes sand fly salivary protein LJM11 confers long-term protection against vector-transmitted leishmania major. Infect. Immun. 82, 2736–2745.

Abdeladhim, M., Coutinho-Abreu, I.V., Townsend, S., Pasos-Pinto, S., Sanchez, L., Rasouli, M., Guimaraes-Costa, A.B., Aslan, H., Franchischetti, I.M., Oliveira, F., Becker, I., Kamhawi, S., Ribeiro, J.M., Jochim, R.C., Valenzuela, J.G., 2016. Molecular diversity between salivary proteins from new world and old world sand flies with emphasis on Bichromomyia olmeca, the sand fly vector of Leishmania mexicana in Mesoamerica. PLoS Negl. Trop. Dis. 10, e0004771.

Abdeladhim, M., Jochim, R.C., Ben Ahmed, M., Zhioua, E., Chelbi, I., Cherni, S., Louzir, H., Ribeiro, J.M., Valenzuela, J.G., 2012. Updating the salivary gland transcriptome of Phlebotomus papatasi (Tunisian strain): the search for sand fly-secreted immunogenic proteins for humans. PLoS One 7, e47347.

Abdeladhim, M., Kamhawi, S., Valenzuela, J.G., 2014. What's behind a sand fly bite? The profound effect of sand fly saliva on host hemostasis, inflammation and immunity. Infect. Genet. Evol. 28, 691–703.

Abi Abdallah, D.S., Denkers, E.Y., 2012. Neutrophils cast extracellular traps in response to protozoan parasites. Front. Immunol. 3, 382.

Aguiar-Soares, R.D., Roatt, B.M., Ker, H.G., Moreira, N., Mathias, F.A., Cardoso, J.M., Gontijo, N.F., Bruna-Romero, O., Teixeira-Carvalho, A., Martins-Filho, O.A., Correa-Oliveira, R., Giunchetti, R.C., Reis, A.B., 2014. LBSapSal-vaccinated dogs exhibit increased circulating T-lymphocyte subsets (CD4(+) and CD8(+)) as well as a reduction of parasitism after challenge with Leishmania infantum plus salivary gland of Lutzomyia longipalpis. Parasit. Vectors 7, 61.

Alvarenga, P.H., Xu, X., Oliveira, F., Chagas, A.C., Nascimento, C.R., Franchischetti, I.M., Juliano, M.A., Juliano, L., Scharfstein, J., Valenzuela, J.G., Ribeiro, J.M., Andersen, J.F., 2013. Novel family of insect salivary inhibitors blocks contact pathway activation by binding to polyphosphate, heparin, and dextran sulfate. Arterioscler. Thromb. Vasc. Biol. 33, 2759–2770.

Anderson, J.M., Oliveira, F., Kamhawi, S., Mans, B.J., Reynoso, D., Seitz, A.E., Lawyer, P., Garfield, M., Pham, M., Valenzuela, J.G., 2006. Comparative salivary gland transcriptomics of sandfly vectors of visceral leishmaniasis. BMC Genomics 7, 52.

Aransay, A.M., Scoulica, E., Tselentis, Y., Ready, P.D., 2000. Phylogenetic relationships of phlebotomine sandflies inferred from small subunit nuclear ribosomal DNA. Insect Mol. Biol. 9, 157–168.

Bahia, D., Gontijo, N.F., Leon, I.R., Perales, J., Pereira, M.H., Oliveira, G., Correa-Oliveira, R., Reis, A.B., 2007. Antibodies from dogs with canine visceral leishmaniasis recognise two proteins from the saliva of Lutzomyia longipalpis. Parasitol. Res. 100, 449–454.

Barral, A., Honda, E., Caldas, A., Costa, J., Vinhas, V., Rowton, E.D., Valenzuela, J.G., Charlab, R., Barral-Netto, M., Ribeiro, J.M., 2000. Human immune response to sand fly salivary gland antigens: a useful epidemiological marker? Am. J. Trop. Med. Hyg. 62, 740–745.

Barros, V.C., Assumpcao, J.G., Cadete, A.M., Santos, V.C., Cavalcante, R.R., Araujo, R.N., Pereira, M.H., Gontijo, N.F., 2009. The role of salivary and intestinal complement system inhibitors in the midgut protection of triatomines and mosquitoes. PLoS One 4, e6047.

Barton, N.H., Etheridge, A.M., 2004. The effect of selection on genealogies. Genetics 166, 1115–1131.

Belkaid, Y., Kamhawi, S., Modi, G., Valenzuela, J., Noben-Trauth, N., Rowton, E., Ribeiro, J., Sacks, D.L., 1998. Development of a natural model of cutaneous leishmaniasis: powerful effects of vector saliva and saliva preexposure on the long-term outcome of Leishmania major infection in the mouse ear dermis. J. Exp. Med. 188, 1941–1953.

Biner, O., Trachsel, C., Moser, A., Kopp, L., Langenegger, N., Kampfer, U., von Ballmoos, C., Nentwig, W., Schurch, S., Schaller, J., Kuhn-Nentwig, L., 2015. Isolation, N-glycosylations and function of a Hyaluronidase-like enzyme from the venom of the Spider Cupiennius salei. PLoS One 10, e0143963.

Calvo, E., Mans, B.J., Andersen, J.F., Ribeiro, J.M., 2006. Function and evolution of a mosquito salivary protein family. J. Biol. Chem. 281, 1935–1942.

Calvo, E., Mans, B.J., Ribeiro, J.M., Andersen, J.F., 2009. Multifunctionality and mechanism of ligand binding in a mosquito antiinflammatory protein. Proc. Natl. Acad. Sci. U.S.A. 106, 3728–3733.

Calvo, E., Mizurini, D.M., Sa-Nunes, A., Ribeiro, J.M., Andersen, J.F., Mans, B.J., Monteiro, R.Q., Kotsyfakis, M., Francischetti, I.M., 2011. Alboserpin, a factor Xa inhibitor from the mosquito vector of yellow fever, binds heparin and membrane phospholipids and exhibits antithrombotic activity. J. Biol. Chem. 286, 27998–28010.

Carregaro, V., Valenzuela, J.G., Cunha, T.M., Verri Jr., W.A., Grespan, R., Matsumura, G., Ribeiro, J.M., Elnaiem, D.E., Silva, J.S., Cunha, F.Q., 2008. Phlebotomine salivas inhibit immune inflammation-induced neutrophil migration via an autocrine DC-derived PGE2/IL-10 sequential pathway. J. Leukoc. Biol. 84, 104–114.

Carvalho, A.M., Amorim, C.F., Barbosa, J.L., Lago, A.S., Carvalho, E.M., 2015. Age modifies the immunologic response and clinical presentation of American tegumentary leishmaniasis. Am. J. Trop. Med. Hyg. 92, 1173–1177.

Cavalcante, R.R., Pereira, M.H., Gontijo, N.F., 2003. Anti-complement activity in the saliva of phlebotomine sand flies and other haematophagous insects. Parasitology 127, 87–93.

Cerna, P., Mikes, L., Volf, P., 2002. Salivary gland hyaluronidase in various species of phlebotomine sand flies (Diptera: psychodidae). Insect Biochem. Mol. Biol. 32, 1691–1697.

Chagas, A.C., Oliveira, F., Debrabant, A., Valenzuela, J.G., Ribeiro, J.M., Calvo, E., 2014. Lundep, a sand fly salivary endonuclease increases Leishmania parasite survival in neutrophils and inhibits XIIa contact activation in human plasma. PLoS Pathog. 10, e1003923.

Champagne, D.E., Smartt, C.T., Ribeiro, J.M., James, A.A., 1995. The salivary gland-specific apyrase of the mosquito Aedes aegypti is a member of the 5'-nucleotidase family. Proc. Natl. Acad. Sci. U.S.A. 92, 694–698.

Chappuis, F., Sundar, S., Hailu, A., Ghalib, H., Rijal, S., Peeling, R.W., Alvar, J., Boelaert, M., 2007. Visceral leishmaniasis: what are the needs for diagnosis, treatment and control? Nat. Rev. Microbiol. 5, 873–882.

Charlab, R., Rowton, E.D., Ribeiro, J.M., 2000. The salivary adenosine deaminase from the sand fly Lutzomyia longipalpis. Exp. Parasitol. 95, 45–53.

Charlab, R., Valenzuela, J.G., Rowton, E.D., Ribeiro, J.M., 1999. Toward an understanding of the biochemical and pharmacological complexity of the saliva of a hematophagous sand fly Lutzomyia longipalpis. Proc. Natl. Acad. Sci. U.S.A. 96, 15155–15160.

Collin, N., Assumpcao, T.C., Mizurini, D.M., Gilmore, D.C., Dutra-Oliveira, A., Kotsyfakis, M., Sa-Nunes, A., Teixeira, C., Ribeiro, J.M., Monteiro, R.Q., Valenzuela, J.G., Francischetti, I.M., 2012. Lufaxin, a novel factor Xa inhibitor from the salivary gland of the sand fly Lutzomyia longipalpis blocks protease-activated receptor 2 activation and inhibits inflammation and

thrombosis in vivo. Arterioscler. Thromb. Vasc. Biol. 32, 2185–2198.

Collin, N., Gomes, R., Teixeira, C., Cheng, L., Laughinghouse, A., Ward, J.M., Elnaiem, D.E., Fischer, L., Valenzuela, J.G., Kamhawi, S., 2009. Sand fly salivary proteins induce strong cellular immunity in a natural reservoir of visceral leishmaniasis with adverse consequences for Leishmania. PLoS Pathog. 5, e1000441.

Costa, D.J., Favali, C., Clarencio, J., Afonso, L., Conceicao, V., Miranda, J.C., Titus, R.G., Valenzuela, J., Barral-Netto, M., Barral, A., Brodskyn, C.I., 2004. Lutzomyia longipalpis salivary gland homogenate impairs cytokine production and costimulatory molecule expression on human monocytes and dendritic cells. Infect. Immun. 72, 1298–1305.

Coutinho-Abreu, I.V., Guimaraes-Costa, A.B., Valenzuela, J.G., 2015. Impact of insect salivary proteins in blood feeding, host immunity, disease, and in the development of biomarkers for vector exposure. Curr. Opin. Insect. Sci. 10, 98–103.

de Moura, T.R., Oliveira, F., Novais, F.O., Miranda, J.C., Clarencio, J., Follador, I., Carvalho, E.M., Valenzuela, J.G., Barral-Netto, M., Barral, A., Brodskyn, C., de Oliveira, C.I., 2007. Enhanced Leishmania braziliensis infection following pre-exposure to sand fly saliva. PLoS Negl. Trop. Dis. 1, e84.

de Moura, T.R., Oliveira, F., Carneiro, M.W., Miranda, J.C., Clarencio, J., Barral-Netto, M., Brodskyn, C., Barral, A., Ribeiro, J.M., Valenzuela, J.G., de Oliveira, C.I., 2013. Functional transcriptomics of wild-caught Lutzomyia intermedia salivary glands: identification of a protective salivary protein against Leishmania braziliensis infection. PLoS Negl. Trop. Dis. 7, e2242.

da Silva, R.A., Tavares, N.M., Costa, D., Pitombo, M., Barbosa, L., Fukutani, K., Miranda, J.C., de Oliveira, C.I., Valenzuela, J.G., Barral, A., Soto, M., Barral-Netto, M., Brodskyn, C., 2011. 'DNA vaccination with KMP11 and Lutzomyia longipalpis salivary protein protects hamsters against visceral leishmaniasis'. Acta Trop. 120, 185–190.

de Silva, A.A., Pacheco e Silva Filho, A., Sesso Rde, C., Esmeraldo Rde, M., de Oliveira, C.M., Fernandes, P.F., de Oliveira, R.A., de Silva, L.S., de Carvalho, V.P., Costa, C.H., Andrade, J.X., da Silva, D.M., Chaves, R.V., 2015. Epidemiologic, clinical, diagnostic and therapeutic aspects of visceral leishmaniasis in renal transplant recipients: experience from thirty cases. BMC Infect. Dis. 15, 96.

Drahota, J., Martin-Martin, I., Sumova, P., Rohousova, I., Jimenez, M., Molina, R., Volf, P., 2014. Recombinant antigens from Phlebotomus perniciosus saliva as markers of canine exposure to visceral leishmaniases vector. PLoS Negl. Trop. Dis. 8, e2597.

Ferreira, V.P., Fazito Vale, V., Pangburn, M.K., Abdeladhim, M., Mendes-Sousa, A.F., Coutinho-Abreu, I.V., Rasouli, M., Brandt, E.A., Meneses, C., Lima, K.F., Nascimento Araujo, R., Pereira, M.H., Kotsyfakis, M., Oliveira, F., Kamhawi, S., Ribeiro, J.M., Gontijo, N.F., Collin, N., Valenzuela, J.G., 2016. SALO, a novel classical pathway complement inhibitor from saliva of the sand fly *Lutzomyia longipalpis*. Sci. Rep. 6, 19300.

Fiuza, J.A., Dey, R., Davenport, D., Abdeladhim, M., Meneses, C., Oliveira, F., Kamhawi, S., Valenzuela, J.G., Gannavaram, S., Nakhasi, H.L., 2016. Intradermal immunization of *Leishmania donovani* centrin knock-out parasites in combination with salivary protein LJM19 from sand fly vector induces a durable protective immune response in hamsters. PLoS Negl. Trop. Dis. 10, c0004322.

Ghosh, K.N., Mukhopadhyay, J., 1998. The effect of anti-sandfly saliva antibodies on *Phlebotomus argentipes* and *Leishmania donovani*. Int. J. Parasitol. 28, 275–281.

Gillespie, R.D., Mbow, M.L., Titus, R.G., 2000. The immunomodulatory factors of bloodfeeding arthropod saliva. Parasite Immunol. 22, 319–331.

Giunchetti, R.C., Correa-Oliveira, R., Martins-Filho, O.A., Teixeira-Carvalho, A., Roatt, B.M., de Oliveira Aguiar-Soares, R.D., Coura-Vital, W., de Abreu, R.T., Malaquias, L.C., Gontijo, N.F., Brodskyn, C., de Oliveira, C.I., Costa, D.J., de Lana, M., Reis, A.B., 2008. A killed *Leishmania* vaccine with sand fly saliva extract and saponin adjuvant displays immunogenicity in dogs. Vaccine 26, 623–638.

Gomes, R., Oliveira, F., Teixeira, C., Meneses, C., Gilmore, D.C., Elnaiem, D.E., Kamhawi, S., Valenzuela, J.G., 2012. Immunity to sand fly salivary protein LJM11 modulates host response to vector-transmitted *Leishmania* conferring ulcer-free protection. J. Invest. Dermatol. 132, 2735–2743.

Gomes, R., Teixeira, C., Teixeira, M.J., Oliveira, F., Menezes, M.J., Silva, C., De Oliveira, C.I., Miranda, J.C., Elnaiem, D.-E., Kamhawi, S., Valenzuela, J.G., Brodskyn, C.I., 2008. Immunity to a salivary protein of a sand fly vector protects against the fatal outcome of visceral leishmaniasis in a hamster model. Proc. Natl. Acad. Sci. U.S.A. 105, 7845–7850.

Gomes, R.B., Brodskyn, C., de Oliveira, C.I., Costa, J., Miranda, J.C., Caldas, A., Valenzuela, J.G., Barral-Netto, M., Barral, A., 2002. Seroconversion against *Lutzomyia longipalpis* saliva concurrent with the development of anti-*Leishmania chagasi* delayed-type hypersensitivity. J. Infect. Dis. 186, 1530–1534.

Gomes, R.B., Mendonca, I.L., Silva, V.C., Ruas, J., Silva, M.B., Cruz, M.S., Barral, A., Costa, C.H., 2007. Antibodies against *Lutzomyia longipalpis* saliva in the fox Cerdocyon thous and the sylvatic cycle of *Leishmania chagasi*. Trans. R Soc. Trop. Med. Hyg. 101, 127–133.

Grespan, R., Lemos, H.P., Carregaro, V., Verri Jr., W.A., Souto, F.O., de Oliveira, C.J., Teixeira, C., Ribeiro, J.M., Valenzuela, J.G., Cunha, F.Q., 2012. The protein LJM 111 from *Lutzomyia longipalpis* salivary gland extract (SGE) accounts for the SGE-inhibitory effects upon inflammatory parameters in experimental arthritis model. Int. Immunopharmacol. 12, 603–610.

Hahn, M.W., 2009. Distinguishing among evolutionary models for the maintenance of gene duplicates. J. Hered. 100, 605–617.

Hall, L.R., Titus, R.G., 1995. Sand fly vector saliva selectively modulates macrophage functions that inhibit killing of *Leishmania* major and nitric oxide production. J. Immunol. 155, 3501–3506.

Hasko, G., Kuhel, D.G., Nemeth, Z.H., Mabley, J.G., Stachlewitz, R.F., Virag, L., Lohinai, Z., Southan, G.J., Salzman, A.L., Szabo, C., 2000. Inosine inhibits inflammatory cytokine production by a posttranscriptional mechanism and protects against endotoxin-induced shock. J. Immunol. 164, 1013–1019.

Hostomska, J., Volfova, V., Mu, J., Garfield, M., Rohousova, I., Volf, P., Valenzuela, J.G., Jochim, R.C., 2009. Analysis of salivary transcripts and antigens of the sand fly *Phlebotomus arabicus*. BMC Genomics 10, 282.

Hughes, A.L., 2013. Evolution of the salivary apyrases of blood-feeding arthropods. Gene 527, 123–130.

Innan, H., Kondrashov, F., 2010. The evolution of gene duplications: classifying and distinguishing between models. Nat. Rev. Genet. 11, 97–108.

Kaiko, G.E., Horvat, J.C., Beagley, K.W., Hansbro, P.M., 2008. Immunological decision-making: how does the immune system decide to mount a helper T-cell response? Immunology 123, 326–338.

Kamhawi, S., 2000. The biological and immunomodulatory properties of sand fly saliva and its role in the establishment of *Leishmania* infections. Microbe. Infect. 2, 1765–1773.

Kamhawi, S., Belkaid, Y., Modi, G., Rowton, E., Sacks, D., 2000. Protection against cutaneous leishmaniasis resulting from bites of uninfected sand flies. Science 290, 1351–1354.

Katebi, A., Gholami, E., Taheri, T., Zahedifard, F., Habibzadeh, S., Taslimi, Y., Shokri, F., Papadopoulou, B., Kamhawi, S., Valenzuela, J.G., Rafati, S., 2015. *Leishmania tarentolae* secreting the sand fly salivary antigen PpSP15 confers protection against *Leishmania* major infection in a susceptible BALB/c mice model. Mol. Immunol. 67, 501–511.

Kato, H., Anderson, J.M., Kamhawi, S., Oliveira, F., Lawyer, P.G., Pham, V.M., Sangare, C.S., Samake, S., Sissoko, I., Garfield, M., Sigutova, L., Volf, P., Doumbia, S., Valenzuela, J.G., 2006. High degree of conservancy among secreted salivary gland proteins from two geographically distant *Phlebotomus duboscqi* sandflies populations (Mali and Kenya). BMC Genomics 7, 226.

Kato, H., Gomez, E.A., Fujita, M., Ishimaru, Y., Uezato, H., Mimori, T., Iwata, H., Hashiguchi, Y., 2015. Ayadualin, a novel RGD peptide with dual antihemostatic activities from the sand fly Lutzomyia ayacuchensis, a vector of Andean-type cutaneous leishmaniasis. Biochimie 112, 49–56.

Kato, H., Jochim, R.C., Gomez, E.A., Uezato, H., Mimori, T., Korenaga, M., Sakurai, T., Katakura, K., Valenzuela, J.G., Hashiguchi, Y., 2013. Analysis of salivary gland transcripts of the sand fly Lutzomyia ayacuchensis, a vector of Andean-type cutaneous leishmaniasis. Infect. Genet. Evol. 13, 56–66.

Kato, H., Jochim, R.C., Lawyer, P.G., Valenzuela, J.G., 2007. Identification and characterization of a salivary adenosine deaminase from the sand fly Phlebotomus duboscqi, the vector of Leishmania major in sub-Saharan Africa. J. Exp. Biol. 210, 733–740.

Katz, O., Waitumbi, J.N., Zer, R., Warburg, A., 2000. Adenosine, AMP, and protein phosphatase activity in sandfly saliva. Am. J. Trop. Med. Hyg. 62, 145–150.

Lerner, E.A., Iuga, A.O., Reddy, V.B., 2007. Maxadilan, a PAC1 receptor agonist from sand flies. Peptides 28, 1651–1654.

Lerner, E.A., Ribeiro, J.M., Nelson, R.J., Lerner, M.R., 1991. Isolation of maxadilan, a potent vasodilatory peptide from the salivary glands of the sand fly Lutzomyia longipalpis. J. Biol. Chem. 266, 11234–11236.

Lestinova, T., Vlkova, M., Votypka, J., Volf, P., Rohousova, I., 2015. Phlebotomus papatasi exposure cross-protects mice against Leishmania major co-inoculated with Phlebotomus duboscqi salivary gland homogenate. Acta Trop. 144, 9–18.

Martin-Martin, I., Molina, R., Jimenez, M., 2013. Identifying salivary antigens of Phlebotomus argentipes by a 2DE approach. Acta Trop. 126, 229–239.

Marzouki, S., Abdeladhim, M., Abdessalem, C.B., Oliveira, F., Ferjani, B., Gilmore, D., Louzir, H., Valenzuela, J.G., Ben Ahmed, M., 2012. Salivary antigen SP32 is the immunodominant target of the antibody response to Phlebotomus papatasi bites in humans. PLoS Negl. Trop. Dis. 6, e1911.

Marzouki, S., Kammoun-Rebai, W., Bettaieb, J., Abdeladhim, M., Hadj Kacem, S., Abdelkader, R., Gritli, S., Chemkhi, J., Aslan, H., Kamhawi, S., Ben Salah, A., Louzir, H., Valenzuela, J.G., Ben Ahmed, M., 2015. Validation of recombinant salivary protein PpSP32 as a suitable marker of human exposure to Phlebotomus papatasi, the vector of Leishmania major in Tunisia. PLoS Negl. Trop. Dis. 9, e0003991.

Mbow, M.L., Bleyenberg, J.A., Hall, L.R., Titus, R.G., 1998. Phlebotomus papatasi sand fly salivary gland lysate downregulates a Th1, but up-regulates a Th2, response in mice infected with Leishmania major. J. Immunol. 161, 5571–5577.

Menezes, M.J., Costa, D.J., Clarencio, J., Miranda, J.C., Barral, A., Barral-Netto, M., Brodskyn, C., de Oliveira, C.I., 2008. Immunomodulation of human monocytes following exposure to Lutzomyia intermedia saliva. BMC Immunol. 9, 12.

Mondragon-Shem, K., Al-Salem, W.S., Kelly-Hope, L., Abdeladhim, M., Al-Zahrani, M.H., Valenzuela, J.G., Acosta-Serrano, A., 2015. Severity of old world cutaneous leishmaniasis is influenced by previous exposure to sand fly bites in Saudi Arabia. PLoS Negl. Trop. Dis. 9, e0003449.

Monroe, D.M., Hoffman, M., 2012. The clotting system—a major player in wound healing. Haemophilia 18 (Suppl. 5), 11–16.

Moreno, J., Alvar, J., 2002. Canine leishmaniasis: epidemiological risk and the experimental model. Trends Parasitol. 18, 399–405.

Moro, O., Lerner, E.A., 1997. Maxadilan, the vasodilator from sand flies, is a specific pituitary adenylate cyclase activating peptide type I receptor agonist. J. Biol. Chem. 272, 966–970.

Morris, R.V., Shoemaker, C.B., David, J.R., Lanzaro, G.C., Titus, R.G., 2001. Sandfly maxadilan exacerbates infection with Leishmania major and vaccinating against it protects against L. major infection. J. Immunol. 167, 5226–5230.

Nathan, C., 2006. Neutrophils and immunity: challenges and opportunities. Nat. Rev. Immunol. 6, 173–182.

Neafsey, D.E., Waterhouse, R.M., Abai, M.R., Aganezov, S.S., Alekseyev, M.A., Allen, J.E., Amon, J., Arca, B., Arensburger, P., Artemov, G., Assour, L.A., Basseri, H., Berlin, A., Birren, B.W., Blandin, S.A., Brockman, A.I., Burkot, T.R., Burt, A., Chan, C.S., Chauve, C., Chiu, J.C., Christensen, M., Costantini, C., Davidson, V.L., Deligianni, E., Dottorini, T., Dritsou, V., Gabriel, S.B., Guelbeogo, W.M., Hall, A.B., Han, M.V., Hlaing, T., Hughes, D.S., Jenkins, A.M., Jiang, X., Jungreis, I., Kakani, E.G., Kamali, M., Kemppainen, P., Kennedy, R.C., Kirmitzoglou, I.K., Koekemoer, L.L., Laban, N., Langridge, N., Lawniczak, M.K., Lirakis, M., Lobo, N.F., Lowy, E., MacCallum, R.M., Mao, C., Maslen, G., Mbogo, C., McCarthy, J., Michel, K., Mitchell, S.N., Moore, W., Murphy, K.A., Naumenko, A.N., Nolan, T., Novoa, E.M., O'Loughlin, S., Oringanje, C., Oshaghi, M.A., Pakpour, N., Papathanos, P.A., Peery, A.N., Povelones, M., Prakash, A., Price, D.P., Rajaraman, A., Reimer, L.J., Rinker, D.C., Rokas, A., Russell, T.L., Sagnon, N., Sharakhova, M.V., Shea, T., Simao, F.A., Simard, F., Slotman, M.A., Somboon, P., Stegniy, V., Struchiner, C.J., Thomas, G.W., Tojo, M., Topalis, P., Tubio, J.M., Unger, M.F., Vontas, J., Walton, C., Wilding, C.S., Willis, J.H., Wu, Y.C., Yan, G., Zdobnov, E.M., Zhou, X., Catteruccia, F., Christophides, G.K., Collins, F.H., Cornman, R.S., Crisanti, A., Donnelly, M.J., Emrich, S.J., Fontaine, M.C., Gelbart, W., Hahn, M.W., Hansen, I.A., Howell, P.I., Kafatos, F.C., Kellis, M., Lawson, D., Louis, C., Luckhart, S., Muskavitch, M.A., Ribeiro, J.M., Riehle, M.A., Sharakhov, I.V., Tu, Z., Zwiebel, L.J., Besansky, N.J., 2015. Mosquito genomics. Highly evolvable malaria vectors: the genomes of 16 Anopheles mosquitoes. Science 347, 1258522.

Oliveira, F., Kamhawi, S., Seitz, A.E., Pham, V.M., Guigal, P.M., Fischer, L., Ward, J., Valenzuela, J.G., 2006. From transcriptome to immunome: identification of DTH inducing proteins from a Phlebotomus ariasi salivary gland cDNA library. Vaccine 24, 374–390.

Oliveira, F., Lawyer, P.G., Kamhawi, S., Valenzuela, J.G., 2008. Immunity to distinct sand fly salivary proteins primes the anti-Leishmania immune response towards protection or exacerbation of disease. PLoS Negl. Trop. Dis. 2, e226.

Oliveira, F., Rowton, E., Aslan, H., Gomes, R., Castrovinci, P.A., Alvarenga, P.H., Abdeladhim, M., Teixeira, C., Meneses, C., Kleeman, L.T., Guimaraes-Costa, A.B., Rowland, T.E., Gilmore, D., Doumbia, S., Reed, S.G., Lawyer, P.G., Andersen, J.F., Kamhawi, S., Valenzuela, J.G., 2015. A sand fly salivary protein vaccine shows efficacy against vector-transmitted cutaneous leishmaniasis in nonhuman primates. Sci. Transl. Med. 7, 290ra90.

Oliveira, F., Traore, B., Gomes, R., Faye, O., Gilmore, D.C., Keita, S., Traore, P., Teixeira, C., Coulibaly, C.A., Samake, S., Meneses, C., Sissoko, I., Fairhurst, R.M., Fay, M.P., Anderson, J.M., Doumbia, S., Kamhawi, S., Valenzuela, J.G., 2013. Delayed-type hypersensitivity to sand fly saliva in humans from a leishmaniasis-endemic area of Mali is Th1-mediated and persists to midlife. J. Invest. Dermatol. 133, 452–459.

Oliveira, M.C., Pelegrini-da-Silva, A., Parada, C.A., Tambeli, C.H., 2007. 5-HT acts on nociceptive primary afferents through an indirect mechanism to induce hyperalgesia in the subcutaneous tissue. Neuroscience 145, 708–714.

Peters, N.C., Egen, J.G., Secundino, N., Debrabant, A., Kimblin, N., Kamhawi, S., Lawyer, P., Fay, M.P., Germain, R.N., Sacks, D., 2008. In vivo imaging reveals an essential role for neutrophils in leishmaniasis transmitted by sand flies. Science 321, 970–974.

Prates, D.B., Araujo-Santos, T., Luz, N.F., Andrade, B.B., Franca-Costa, J., Afonso, L., Clarencio, J., Miranda, J.C., Bozza, P.T., Dosreis, G.A., Brodskyn, C., Barral-Netto, M., Borges, V.M., Barral, A., 2011. Lutzomyia longipalpis saliva drives apoptosis and enhances parasite burden in neutrophils. J. Leukoc. Biol. 90, 575–582.

Qian, Y., Jeong, J.S., Abdeladhim, M., Valenzuela, J.G., Aoki, V., Hans-Filhio, G., Rivitti, E.A., Diaz, L.A., Cooperative Group on Fogo Selvagem Research, 2015. IgE anti-LJM11 sand fly salivary antigen may herald the onset of fogo selvagem in endemic Brazilian regions. J. Invest. Dermatol. 135, 913–915.

Qian, Y., Jeong, J.S., Maldonado, M., Valenzuela, J.G., Gomes, R., Teixeira, C., Evangelista, F., Qaqish, B., Aoki, V., Hans Jr., G., Rivitti, E.A., Eaton, D., Diaz, L.A., 2012. Cutting edge: Brazilian pemphigus foliaceus anti-desmoglein 1 autoantibodies cross-react with sand fly salivary LJM11 antigen. J. Immunol. 189, 1535–1539.

Ramalho-Ortigao, M., Coutinho-Abreu, I.V., Balbino, V.Q., Figueiredo Jr., C.A., Mukbel, R., Dayem, H., Hanafi, H.A., El-Hossary, S.S., Fawaz Eel, D., Abo-Shehada, M., Hoel, D.F., Stayback, G., Wadsworth, M., Shoue, D.A., Abrudan, J., Lobo, N.F., Mahon, A.R., Emrich,

S.J., Kamhawi, S., Collins, F.H., McDowell, M.A., 2015. Phlebotomus papatasi SP15: mRNA expression variability and amino acid sequence polymorphisms of field populations. Parasit. Vectors 8, 298.

Ready, P.D., 2013. Biology of phlebotomine sand flies as vectors of disease agents. Annu. Rev. Entomol. 58, 227–250.

Ribeiro, J.M., 1987. Role of saliva in blood-feeding by arthropods. Annu. Rev. Entomol. 32, 463–478.

Ribeiro, J.M., 1995. Blood-feeding arthropods: live syringes or invertebrate pharmacologists? Infect. Agents Dis. 4, 143–152.

Ribeiro, J.M., Francischetti, I.M., 2003. Role of arthropod saliva in blood feeding: sialome and post-sialome perspectives. Annu. Rev. Entomol. 48, 73–88.

Ribeiro, J.M., Katz, O., Pannell, L.K., Waitumbi, J., Warburg, A., 1999. Salivary glands of the sand fly Phlebotomus papatasi contain pharmacologically active amounts of adenosine and 5'-AMP. J. Exp. Biol. 202, 1551–1559.

Ribeiro, J.M., Mans, B.J., Arca, B., 2010. An insight into the sialome of blood-feeding Nematocera. Insect. Biochem. Mol. Biol. 40, 767–784.

Ribeiro, J.M., Modi, G., 2001. The salivary adenosine/ AMP content of Phlebotomus argentipes Annandale and Brunetti, the main vector of human kala-azar. J. Parasitol. 87, 915–917.

Ribeiro, J.M., Vachereau, A., Modi, G.B., Tesh, R.B., 1989. A novel vasodilatory peptide from the salivary glands of the sand fly Lutzomyia longipalpis. Science 243, 212–214.

Rohousova, I., Ozensoy, S., Ozbel, Y., Volf, P., 2005. Detection of species-specific antibody response of humans and mice bitten by sand flies. Parasitology 130, 493–499.

Rohousova, I., Subrahmanyam, S., Volfova, V., Mu, J., Volf, P., Valenzuela, J.G., Jochim, R.C., 2012. Salivary gland transcriptomes and proteomes of Phlebotomus tobbi and Phlebotomus sergenti, vectors of leishmaniasis. PLoS Negl. Trop. Dis. 6, e1660.

Rohousova, I., Volf, P., 2006. Sand fly saliva: effects on host immune response and Leishmania transmission. Folia Parasitol. (Praha) 53, 161–171.

Samuelson, J., Lerner, E., Tesh, R., Titus, R., 1991. A mouse model of Leishmania braziliensis braziliensis infection produced by coinjection with sand fly saliva. J. Exp. Med. 173, 49–54.

Silva, F., Gomes, R., Prates, D., Miranda, J.C., Andrade, B., Barral-Netto, M., Barral, A., 2005. Inflammatory cell infiltration and high antibody production in BALB/c mice caused by natural exposure to Lutzomyia longipalpis bites. Am. J. Trop. Med. Hyg. 72, 94–98.

Sima, M., Ferencova, B., Warburg, A., Rohousova, I., Volf, P., 2016. Recombinant salivary proteins of Phlebotomus orientalis are suitable antigens to measure exposure of domestic animals to sand fly bites. PLoS Negl. Trop. Dis. 10, e0004553.

Soares, B.R., Souza, A.P., Prates, D.B., de Oliveira, C.I., Barral-Netto, M., Miranda, J.C., Barral, A., 2013. Seroconversion of sentinel chickens as a biomarker for monitoring exposure to visceral leishmaniasis. Sci. Rep. 3, 2352.

Soares, M.B., Titus, R.G., Shoemaker, C.B., David, J.R., Bozza, M., 1998. The vasoactive peptide maxadilan from sand fly saliva inhibits TNF-alpha and induces IL-6 by mouse macrophages through interaction with the pituitary adenylate cyclase-activating polypeptide (PACAP) receptor. J. Immunol. 160, 1811–1816.

Souza, A.P., Andrade, B.B., Aquino, D., Entringer, P., Miranda, J.C., Alcantara, R., Ruiz, D., Soto, M., Teixeira, C.R., Valenzuela, J.G., de Oliveira, C.I., Brodskyn, C.I., Barral-Netto, M., Barral, A., 2010. Using recombinant proteins from Lutzomyia longipalpis saliva to estimate human vector exposure in visceral Leishmaniasis endemic areas. PLoS Negl. Trop. Dis. 4, e649.

Tavares, N.M., Silva, R.A., Costa, D.J., Pitombo, M.A., Fukutani, K.F., Miranda, J.C., Valenzuela, J.G., Barral, A., de Oliveira, C.I., Barral-Netto, M., Brodskyn, C., 2011. Lutzomyia longipalpis saliva or salivary protein LJM19 protects against Leishmania braziliensis and the saliva of its vector, Lutzomyia intermedia. PLoS Negl. Trop. Dis. 5, e1169.

Teixeira, C., Gomes, R., Collin, N., Reynoso, D., Jochim, R., Oliveira, F., Seitz, A., Elnaiem, D.E., Caldas, A., de Souza, A.P., Brodskyn, C.I., de Oliveira, C.I., Mendonca, I., Costa, C.H., Volf, P., Barral, A., Kamhawi, S., Valenzuela, J.G., 2010. Discovery of markers of exposure specific to bites of Lutzomyia longipalpis, the vector of Leishmania infantum chagasi in Latin America. PLoS Negl. Trop. Dis. 4, e638.

Theodos, C.M., Ribeiro, J.M., Titus, R.G., 1991. Analysis of enhancing effect of sand fly saliva on Leishmania infection in mice. Infect. Immun. 59, 1592–1598.

Theodos, C.M., Titus, R.G., 1993. Salivary gland material from the sand fly Lutzomyia longipalpis has an inhibitory effect on macrophage function in vitro. Parasite. Immunol. 15, 481–487.

Titus, R.G., Bishop, J.V., Mejia, J.S., 2006. The immunomodulatory factors of arthropod saliva and the potential for these factors to serve as vaccine targets to prevent pathogen transmission. Parasite. Immunol. 28, 131–141.

Titus, R.G., Ribeiro, J.M., 1988. Salivary gland lysates from the sand fly Lutzomyia longipalpis enhance Leishmania infectivity. Science 239, 1306–1308.

Titus, R.G., Ribeiro, J.M., 1990. The role of vector saliva in transmission of arthropod-borne disease. Parasitol. Today 6, 157–160.

Valenzuela, J.G., Belkaid, Y., Garfield, M.K., Mendez, S., Kamhawi, S., Rowton, E.D., Sacks, D.L., Ribeiro, J.M., 2001a. Toward a defined anti-Leishmania vaccine targeting vector antigens: characterization of a protective salivary protein. J. Exp. Med. 194, 331–342.

Valenzuela, J.G., Belkaid, Y., Rowton, E., Ribeiro, J.M., 2001b. The salivary apyrase of the blood-sucking sand fly Phlebotomus papatasi belongs to the novel Cimex family of apyrases. J. Exp. Biol. 204, 229–237.

Valenzuela, J.G., Charlab, R., Galperin, M.Y., Ribeiro, J.M., 1998. Purification, cloning, and expression of an apyrase from the bed bug Cimex lectularius. A new type of nucleotide-binding enzyme. J. Biol. Chem. 273, 30583–30590.

Valenzuela, J.G., Garfield, M., Rowton, E.D., Pham, V.M., 2004. Identification of the most abundant secreted proteins from the salivary glands of the sand fly Lutzomyia longipalpis, vector of Leishmania chagasi. J. Exp. Biol. 207, 3717–3729.

Vinhas, V., Andrade, B.B., Paes, F., Bomura, A., Clarencio, J., Miranda, J.C., Bafica, A., Barral, A., Barral-Netto, M., 2007. Human anti-saliva immune response following experimental exposure to the visceral leishmaniasis vector, Lutzomyia longipalpis. Eur. J. Immunol. 37, 3111–3121.

Vlkova, M., Sima, M., Rohousova, I., Kostalova, T., Sumova, P., Volfova, V., Jaske, E.L., Barbian, K.D., Gebre-Michael, T., Hailu, A., Warburg, A., Ribeiro, J.M., Valenzuela, J.G., Jochim, R.C., Volf, P., 2014. Comparative analysis of salivary gland transcriptomes of Phlebotomus orientalis sand flies from endemic and non-endemic foci of visceral leishmaniasis. PLoS Negl. Trop. Dis. 8, e2709.

Walport, M.J., 2001a. Complement. First of two parts. N. Engl. J. Med. 344, 1058–1066.

Walport, M.J., 2001b. Complement. Second of two parts. N. Engl. J. Med. 344, 1140–1144.

Xanthos, D.N., Bennett, G.J., Coderre, T.J., 2008. Norepinephrine-induced nociception and vasoconstrictor hypersensitivity in rats with chronic post-ischemia pain. Pain 137, 640–651.

Xu, X., Oliveira, F., Chang, B.W., Collin, N., Gomes, R., Teixeira, C., Reynoso, D., My Pham, V., Elnaiem, D.E., Kamhawi, S., Ribeiro, J.M., Valenzuela, J.G., Andersen, J.F., 2011. Structure and function of a "yellow" protein from saliva of the sand fly Lutzomyia longipalpis that confers protective immunity against Leishmania major infection. J. Biol. Chem. 286, 32383–32393.

Zahedifard, F., Gholami, E., Taheri, T., Taslimi, Y., Doustdari, F., Seyed, N., Torkashvand, F., Meneses, C., Papadopoulou, B., Kamhawi, S., Valenzuela, J.G., Rafati, S., 2014. Enhanced protective efficacy of nonpathogenic recombinant Leishmania tarentolae expressing cysteine proteinases combined with a sand fly salivary antigen. PLoS Negl. Trop. Dis. 8, e2751.

6

Unique Features of Vector-Transmitted Leishmaniasis and Their Relevance to Disease Transmission and Control

Tiago D. Serafim[1,a], Ranadhir Dey[2,a], Hira L. Nakhasi[2], Jesus G. Valenzuela[1], Shaden Kamhawi[1]

[1]National Institutes of Health, Rockville, MD, United States; [2]Center for Biologics Evaluation and Research, FDA, Silver Spring, MD, United States

OVERVIEW

Leishmaniasis is a vector-borne neglected tropical disease transmitted by phlebotomine sand flies. More precisely, differing species of the protozoan parasite *Leishmania* cause a spectrum of diseases that we collectively refer to as leishmaniasis. Leishmaniasis is globally distributed encompassing many tropical and temperate regions of the world, and its range is rapidly expanding with climate change (Carvalho et al., 2015; Dujardin et al., 2008; Fischer et al., 2011). It disproportionately afflicts the poorest people and is only second to malaria in terms of parasite-induced fatalities (Chappuis et al., 2007; WHO, 2016).

To date, stable transmission of *Leishmania* parasites is considered to occur exclusively by the bite of an infected phlebotomine sand fly. In addition

to the *Leishmania* species, genetic background of the host as well as environmental factors can influence the clinical features of leishmaniasis that vary from a cutaneous form, resulting in self-healing or chronic skin ulcers contained at the site of an infected bite, to visceral leishmaniasis (VL), a fatal form involving parasite dissemination, primarily to the liver and spleen (Chappuis et al., 2007; WHO, 2016). Despite the impact of leishmaniasis on human health, there are no human vaccines available against any form of leishmaniasis (Gillespie et al., 2016).

Since it is the parasite that cause leishmaniasis, you may be asking why we care how *Leishmania* is transmitted by a sand fly? In this chapter, we will demonstrate the profound influence of sand fly transmission of *Leishmania* on initiation and establishment of leishmaniasis. We will provide an account of vector-derived

[a] These authors contributed equally to this work.

Arthropod Vector: Controller of Disease Transmission, Volume 2
http://dx.doi.org/10.1016/B978-0-12-805360-7.00006-X

factors currently characterized as part of the infectious inoculum introduced with the parasites into host skin, and how those saliva molecules transform the bite site into a welcoming immune-modified environment that benefits parasite establishment. In addition to how the complex composition of the infectious inoculum influences the host immune response at the bite site, we will address how the infected sand fly alters its feeding behavior as a result of coevolution with the *Leishmania* parasites to further promote successful transmission to the host.

Considering the above-mentioned factors, enhanced parasite virulence following vector transmission is not surprising (Peters et al., 2008, 2009; Peters and Sacks, 2009; Rogers, 2012; Rogers et al., 2004). Unquestionably, this has significant ramifications for disease pathology, vaccine design, and drug efficacies. We continue the journey into vector-transmission of leishmaniasis by exploring the evidence pertaining to immunogenicity of various components of the infectious inoculum, and the rational for why and how we can use these elements to enhance vaccine efficacy. In addition, we will briefly address the importance of models of vector-transmitted *Leishmania* in assessing pathology and drug efficacy.

Finally, in light of the explosion of new research technologies in the last decade, we will conclude our chapter by exploring new tools that have impacted leishmaniasis research, and how they can be used to further our understanding of vector transmission of leishmaniasis.

LIFE CYCLE OF *LEISHMANIA* IN THE SAND FLY VECTOR

Transmission of *Leishmania* parasites by the bite of infected sand flies has been known since the beginning of the 20th century (Shortt et al., 1931). Since then, significant progress has been made in our understanding of vector–parasite

interactions and transmission (Bates, 2007; Bates and Rogers, 2004; Dostalova and Volf, 2012; Sacks and Kamhawi, 2001). These advances highlighted the complexity of vector transmission and revealed a multitude of evolutionary adaptations undertaken by the parasites and their vectors. The more these relationships are investigated, the more it is realized that there are significant gaps in our knowledge as to how *Leishmania* develops and differentiates inside the vector.

The outcome of successful parasite transmission is infection or disease. This relies largely on development of infective parasite forms: the metacyclic promastigote. After a series of changes to cell morphology that occur within the confinement of the sand fly midgut (Fig. 6.1), the infective promastigote arises in the anterior portion and becomes available for transmission to the host (Dostalova and Volf, 2012; Killick-Kendrick, 1990; Lawyer et al., 1990; Sacks and Perkins, 1984). The series of changes that precede the development of infective promastigotes usually happen in a time frame of 8–12 days. These changes are poorly understood despite the fact that the forms themselves were validated early on by independent researchers (Lawyer et al., 1990; Walters et al., 1989a,b).

After an infected blood meal, the female sand fly retains blood-containing amastigotes within a synthesized Type 1 peritrophic matrix (PM) for up to 72 h (Secundino et al., 2005; Volf et al., 1994) where they differentiate into procyclics to start their journey through several well-defined and functionally distinct promastigote forms (Dostalova and Volf, 2012; Kamhawi, 2006). Supported by blood nutrients, procyclics, the earliest form of promastigotes, actively divide within the peritrophic matrix successfully resisting the action of midgut digestive enzymes (Dostalova and Volf, 2012; Gossage et al., 2003; Kamhawi, 2006). Approximately, 48 h post-blood meal (hpb), procyclics begin their differentiation into nondividing nectomonads, large forms whose function is to traverse the PM before the

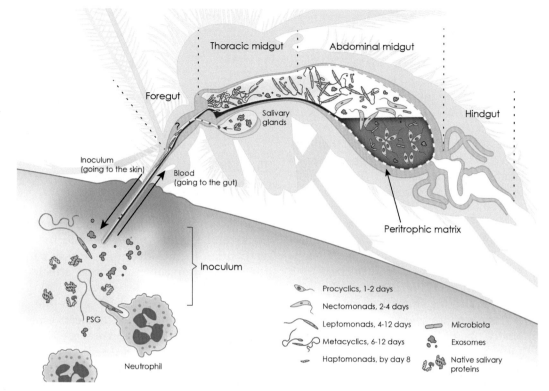

FIGURE 6.1 Development of *Leishmania* species in the digestive tract of a competent sand fly vector. After ingestion of an infected blood meal, amastigotes begin to differentiate into the first promastigote, the procyclic. Procyclic promastigotes actively divide within the confinement of the peritrophic matrix and differentiate into nectomonad promastigotes after 2–4 days. The active nectomonads escape from the peritrophic matrix and attach to the midgut wall avoiding excretion with the digested blood meal. The next round of differentiation culminates with appearance of lepto-monad promastigotes. These forms actively multiply and colonize the anterior portion of the midgut where they secrete a promastigote secretory gel (PSG, in *yellow*). Leptomonads also give rise to metacyclic promastigotes, the infective form, at the anterior portion of the midgut. Haptomonad promastigotes become evident around day 8 and strongly attach to the sto-modeal valve with no clear function or origin. Microbiota and exosomes are present throughout the midgut. Upon taking a subsequent blood meal, an infected sand fly regurgitates into the skin of a mammalian host due to a midgut blocked by PSG and parasites. Saliva is also secreted into the bite site to facilitate blood feeding. Collectively, components of the infectious inoculum act on immune cells arriving at the bite site.

blood meal is excreted (Dostalova and Volf, 2012; Kamhawi, 2006). Anchoring to the midgut wall is a well-described and crucial event in the para-site life cycle inside the vector preventing the excretion and loss of *Leishmania* parasites along with the digested blood. Without attachment, gut peristalsis can also impair establishment of the infection by promoting parasite excretion (Kamhawi, 2006; Kamhawi et al., 2004; Volf and

Myskova, 2007). Midgut attachment is the corner stone of vector competence, at least for restric-tive sand fly vectors that are permissible to only one *Leishmania* species, such as *Phlebotomus papa-tasi* and *Phlebotomus sergenti* (Kamhawi, 2006; Volf and Myskova, 2007). This series of events are not so clear for permissive vectors such as *Lutzomyia longipalpis* that can support mature transmissible infections with most *Leishmania* species, at least under experimental conditions (Kamhawi, 2006; Myskova et al., 2016; Volf and Myskova, 2007). In addition to anchoring parasites, the highly active nectomonads are also suggested as the form responsible for the early stages of anterior midgut colonization (Gossage et al., 2003).

Following successful binding to the midgut epithelium, a third round of differentiation can be observed by day 4 with the appearance of the leptomonad promastigote (Bates, 2007; Gossage et al., 2003; Kamhawi, 2006). This promastigote is the main multiplicative form in the *Leishmania* life cycle within the vector (Gossage et al., 2003). The leptomonad form generates the parasite density needed for development of a mature infection characterized by a strong colonization of both the anterior part of the midgut and the stomodeal valve at the junction between the midgut and foregut. Additionally, one of the main functions attributed to this form is production of the promastigote secretory gel (PSG) (Bates, 2007; Rogers, 2012; Rogers et al., 2004). The PSG is crucial for transmission dynamics, changing vector feeding behavior and modulating the mammalian host immune response. At the valve, another promastigote form, the haptomonad, can be found strongly bound at the cuticular surface by hemidesmosome-like attachment (Walters et al., 1989a,b). The origin and function of the haptomonads are still unclear.

Metacyclics, the promastigote form infective to mammals, is thought to arise from leptomonads and is the main mode of its dissemination, through transmission, to various hosts. The rise of metacyclic promastigotes is known as metacyclogenesis. In the midgut, metacyclics can be found in low numbers by day 6 and accumulate in proportion and quantity with time, reaching as high as 70%–80% of total parasites occupying the midgut around days 10 to 12 postinfection. Apart from a few studies to understand how infectious promastigote arises (Cunningham et al., 2001; Sacks

and Perkins, 1984; Serafim et al., 2012), this crucial event in the parasite life cycle is poorly understood.

THE USUAL SUSPECTS: COMPONENTS OF THE INFECTIOUS INOCULUM

The Metacyclic Parasite

Metacyclic promastigotes were identified more than 30 years ago as the parasite form responsible for infection of the mammalian host (Sacks and Perkins, 1984). Despite the global distribution and health burden of leishmaniasis, a neglected tropical disease that afflicts more than 300 million people yearly (WHO, 2016), fundamental aspects of the parasite life cycle, especially regarding how the infective metacyclic is generated, remain unknown, compromising our understanding of disease pathogenesis and potentially impacting success of control efforts.

Infective promastigotes received the "metacyclic" designation from early studies that associated their appearance to a stationary phase that followed a period of intense growth in vitro (culture) or in vivo (sand fly vector), and this phenomenon preceded the ability of these parasites to infect the mammalian host (Sacks and Perkins, 1984). As described for other trypanosomatids such as *Trypanosoma cruzi* (Duschak et al., 2006), the "meta" prefix refers to the final stages of *Leishmania* parasite development inside the vector that leads to the rise of an infective form adapted to insure the continuation of the life cycle in the definitive host. As such, it is accepted that metacyclics are terminally differentiated nondivisible cell forms (Sadlova et al., 2010).

As mentioned previously in this section, during its life cycle inside the vector (Fig. 6.1), *Leishmania* parasites go through sequential morphological changes that culminate with the appearance of infectious metacyclics through a process known as metacyclogenesis (Serafim et al., 2012; Silva and Sacks, 1987) and that the predecessor of the metacyclic form is most likely the leptomonad promastigote (Gossage et al., 2003). These conclusions, similar to those made for other differentiation steps, are based on microscopic observations of a dominant cell form in the vector midgut at a particular time period postinfection. As such, the prevalence of the leptomonad promastigote just before the rise of the metacyclics pointed to these forms as being the precursors of the metacyclic form. Apart from these temporal and morphological studies, physiological and/or molecular evidence pertaining to parasite differentiation, including metacyclogenesis, in the vector have been poor. A few studies demonstrated that varying the conditions of in vitro cultures modulates the level of metacyclogenesis increasing the percentage of infective forms in the stationary phase (Kamhawi, 2006; Serafim et al., 2012; Zakai et al., 1998). The appearance of metacyclics was associated to a general status of stress induced by triggers such as nutrient depletion of the media, a decrease in pH, and by anaerobic conditions (Kamhawi, 2006; Serafim et al., 2012; Zakai et al., 1998).

Studies investigating the molecular basis of metacyclogenesis are also few. When levels of tetrahydrobiopterin (H_4B) are reduced, by knocking out either the pterin reductase 1 [ptr1] gene or the biopterin transporter 1 [bt1] gene, a significant increase of metacyclic forms can be found in the stationary phase of culture (Cunningham et al., 2001). H_4B is the product of biopterin reduction by the ptr1 enzyme and is an essential cofactor for several hydroxylase enzymes and for the catalytic activity of nitric oxide synthase. The active form of pteridines is used in different cellular biological processes by Leishmania parasites. Opposite to its mammalian and insect hosts, Leishmania spp. are pteridine auxotrophs, meaning they have to acquire such compounds from their hosts by salvage pathways in order to maintain a sustainable cell homeostasis. Despite its effect in vitro, there is still no evidence that H_4B is essential for metacyclic generation in vivo. Similarly, Leishmania are obligatory auxotrophs for purines. The absence of nucleosides, especially adenosine, is a nutrient stress that triggers metacyclogenesis in vitro (Serafim et al., 2012). These authors further demonstrated that the presence of a single molecule, adenosine, is able to prevent metacyclogenesis inside the sand fly vector. Moreover, adenosine sensing by Leishmania amazonensis reverted metacyclics to multiplicative promastigotes, suggesting that purine sensing can control Leishmania differentiation from a proliferative noninfective promastigote form (purine presence) to an infective nonproliferative form (purine absence). Independent of its origin and how metacyclics differentiation is controlled, it is a well-documented fact that metacyclic promastigotes are optimized for infection of the mammalian host.

The Promastigote Secretory Gel

When early studies on sand fly and Leishmania interactions were undertaken, several authors observed the appearance of a mysterious gel-like substance during the development of Leishmania parasites in the vector midgut (Lawyer et al., 1987, 1990; Walters et al., 1989a,b). More recent studies (Rogers, 2012; Rogers et al., 2002) confirmed the presence of this gelatinous structure, termed the PSG. Since PSG permeates the anterior midgut full of parasites, the currently prevailing hypothesis arose as to how transmission happens in the mammalian host, the theory of the blocked fly. Once the infected midgut is filled with parasite-containing PSG, it works as a plug, filling and distending the anterior midgut (Rogers, 2012; Rogers et al., 2002). As a result, the infected sand fly cannot feed properly and is forced to regurgitate the gel together with parasites to successfully obtain a blood meal. Rogers et al. (2004) demonstrated that parasites together with PSG were deposited into the skin of a mouse, confirming the significance of blockage by PSG in parasite transmission.

The PSG is observed simultaneously with the appearance of leptomonad promastigotes, both temporally during the development of the infection and physically where it colocalizes with this parasite form within the midgut (Rogers, 2012; Rogers et al., 2002). In support of the latter, when the PSG is dissected from an infected sand fly and put it in culture, it is possible to observe parasite forms embedded in the gel that recover motility and show morphological features of a leptomonad promastigote (Bates, 2007; Rogers, 2012; Rogers et al., 2002). The simultaneous appearance of PSG and leptomonads in the anterior midgut suggested that these forms are the most probable source of the plug. The other common promastigote form that surrounds the extremities of the plug is the metacyclic form. The location of this form at the poles of the PSG is believed to be important for parasite transmission. Therefore, it is likely that leptomonads give rise to metacyclics that escape the PSG and concentrate in the space between the plug and the stomodeal valve, increasing the probability of their transmission to a mammalian host upon regurgitation by an infected female taking a second blood meal.

Components of the PSG were characterized early on after ultrastructural studies on polymeric and other macromolecular compounds secreted from the flagellar pocket of *Leishmania* parasites (Stierhof et al., 1994). The major component of the plug is a noncovalently associated, high–molecular weight, filamentous proteophosphoglycan (fPPG) (Ilg et al., 1996). The fPPG possesses a distinctive composition with a very high carbohydrate content (>75% w/w) and a large quantity of glycosylated amino acids (Ilg et al., 1996). These characteristics qualify the PPG as a parasite-derived compound similar to mammalian mucins. Other components of the PSG include a filamentous acid phosphatase composed of a 100 kDa phosphoglycoprotein. As the major

component of PSG, fPPG is crucial for transmission and infection establishment (Rogers, 2012; Rogers et al., 2004), but considering the "sticky" gelatinous nature of the PSG, we hypothesize that other compounds (secreted or not) from the parasite and from the sand fly may be present in the plug that could have yet undescribed functions and effects for both the vector and the host.

Exosomes

Small vesicles formed from cells as a consequence of membrane blebbing or shedding were observed in the 1970s (Dumaswala and Greenwalt, 1984; Lutz et al., 1977; Taylor et al., 1988). Subsequent study of the maturation of reticulocytes resulted in the first characterization of these extracellular vesicles as exosomes that form by vesiculation of intracellular endosomes and release by exocytosis, a novel secretory pathway (Johnstone et al., 1987). Since then, exosomes have been reported from most eukaryotic cells as a ubiquitous means of intercellular communication with diverse functions in immune regulation and host–parasite interactions (Coakley et al., 2015; Colombo et al., 2014; Hosseini et al., 2013; Li et al., 2006).

Leishmania exosomes are reported for both amastigotes and promastigotes (Coakley et al., 2015). Silverman et al. (2010a) demonstrated exosome secretion as a general mechanism for protein delivery from the parasites to target cells. Since then, several immunomodulatory roles are attributed to *Leishmania* exosomes including neutrophil recruitment, suppression of macrophage/monocyte, and dendritic cell (DC) activation (Coakley et al., 2015; Silverman and Reiner, 2011). Additionally, exosomes exacerbated *Leishmania* infections in mice pretreated with either *L. major* or *L. donovani* exosomes (Silverman et al., 2010b). As such, *Leishmania* exosomes are considered as packages containing virulence factors that promote parasite survival and result in disease enhancement.

Recently, Atayde et al. (2015) provided evidence of exosome secretion in vivo. This study demonstrated prolific exosome secretion by both *L. major* and *L. infantum* promastigotes within the midgut of the sand fly *L. longipalpis*. Additionally, exosomes are egested during bites of infected sand flies and are therefore a novel component of the infectious inoculum of vector-transmitted *Leishmania* (Atayde et al., 2015).

Saliva

Sand flies possess two globular salivary glands that reside in the thorax and deliver saliva into the proboscis to facilitate feeding. These glands have a low complexity containing an average of 20–40 secreted proteins, and inject only nanograms of each protein to combat hemostasis (Abdeladhim et al., 2014). It is important to appreciate that salivation is an obligatory part of feeding, and the pool feeding behavior of sand flies insures that saliva will be mixed with parasites and other components of the infectious inoculum when each is deposited into the feeding site in the skin of the mammalian host. Therefore, saliva is an integral part of the infectious inoculum that will impact success of transmission and disease establishment.

To overcome host hemostasis and facilitate feeding, sand fly saliva has numerous biological activities including, inhibition of platelet aggregation, blood coagulation, and vasoconstriction (Abdeladhim et al., 2014; Ribeiro and Francischetti, 2003). Apart from its direct physiological effects on host hemostasis, sand fly saliva significantly enhances *Leishmania* infection (Titus and Ribeiro, 1988). Further studies corroborated the leishmaniasis exacerbation property of sand fly saliva and related it to a multitude of immunomodulatory effects that favor parasite survival (Abdeladhim et al., 2014; Gomes and Oliveira, 2012; Kamhawi, 2000). Saliva is a key component of the infectious inoculum.

Midgut Microbiota

Most arthropod vectors of human pathogens harbor a rich microbiota in their midgut (Akhoundi et al., 2012; Azambuja et al., 2005; Boissiere et al., 2012; Sant'Anna et al., 2012; Soumana et al., 2013). Importantly, several studies demonstrated the importance of midgut microbial communities in susceptibility of insect vectors to the pathogens they transmit (Cirimotich et al., 2011; Weiss and Aksoy, 2011).

The diversity of sand fly midgut microbiota was investigated in wild-caught specimens of *Lutzomyia* and *Phlebotomus* species (Akhoundi et al., 2012; Dillon et al., 1996; Monteiro et al., 2016; Sant'Anna et al., 2012). Bacteria are also commonly observed in the midgut of colonized sand flies during specimen manipulation (personal communication). However, the effect of these microbes on *Leishmania* development remains unclear, and there is no evidence to date of their transmission by sand fly bite (Finney et al., 2015). The *Leishmania* life cycle takes place in the midgut of the sand fly that contains a rich and abundant microbial community. Microbiota prevalent throughout the midgut exists in combination with the gelatinous "sticky" nature of the PSG that accumulates at the anterior midgut, suggesting that microbiota transmission is possible, if indeed not likely. This likelihood is increased due to the blocked fly feeding mode of sand flies, where they regurgitate parasites and PSG into the host (Rogers, 2012; Rogers et al., 2002). Therefore, it would not be surprising if material trapped in the gelatinous PSG, such as microbiota, would also be introduced into the host during blood feeding and exert its own effect on the host immune system.

We need to consider that mammals including humans host a vast microbial community belonging to four predominant phyla that participate in metabolic processes, tissue development, and immunity (Cho and Blaser, 2012; Pflughoeft and

Versalovic, 2012). These microbial communities inhabit several of our tissues including the skin where they contribute to our health, enhancing innate immunity and limiting establishment of pathogenic invaders including *Leishmania* parasites (Belkaid and Tamoutounour, 2016; Cho and Blaser, 2012; Naik et al., 2012, 2015; Pflughoeft and Versalovic, 2012).

THE SITE OF BITE

We identified known and potential components of the infectious inoculum egested into host skin during bites of *Leishmania*-infected sand flies (Table 6.1). Clearly, the infectious inoculum is complex, and each of its components will be perceived by the host immune system as

TABLE 6.1 Proven or Projected Components of the Infectious Inoculum After Bites of a *Leishmania*-Infected Sand Fly

Component	Origin	Nature	Egestion Verified	Target Cells	Immunological Consequence	References
Metacyclic promastigote	*Leishmania*	Live eukaryotic cell	Yes	Neutrophils; dendritic cells; macrophages	↑Survival to innate immunity attack; ↓activation of dendritic cells and macrophages; disease exacerbation	Carrera et al. (1996), Rogers et al. (2004), Sacks and Perkins (1984), Warburg and Schlein (1986), and Figueiredo et al. (2012); Recommended review: Ribeiro-Gomes and Sacks (2012)
PSG	*Leishmania*	Proteophosphoglycan	Yes	Neutrophils; macrophages	Neutrophil and macrophage recruitment; induce antiinflammatory macrophage profile; disease exacerbation	Rogers et al. (2009), Rogers et al. (2010), and Rogers et al. (2004); Recommended review: Rogers (2012)
Exosomes	*Leishmania*/ Sand fly	Phospholipid/Protein	Yes	Neutrophils	↑IL-17a; disease exacerbation	Atayde et al. (2015); Recommended review: Atayde et al. (2016)
Saliva	Sand fly	Protein/Nucleosides	Yes	Lymphocytes; neutrophils; Macrophages	Neutrophils and macrophages recruitment; ↑Th2 response profile; and disease exacerbation	Belkaid et al. (1998), Carrera et al. (1996), Kamhawi et al. (2000), Titus and Ribeiro (1988), Katz et al. (2000), and Teixeira et al. (2005); Recommended review: Abdeladhim et al. (2014)
Microbiota	Sand fly	Bacteria	No	?	?	

a foreign entity. Indeed, several studies demonstrated the presence of multiple potent immunomodulators that can alter the nature as well as the activation state of both resident and migratory immune cells at the sites of infected bites (Abdeladhim et al., 2014; Atayde et al., 2015; Rogers, 2012). Knowledge of the events governing the early skin phase of *Leishmania* infections for both cutaneous leishmaniasis (CL) and VL is critical to our understanding of disease initiation and establishment. This early phase is a vulnerable time for *Leishmania* parasites that are trying to establish themselves in a hostile environment, yet it is one of the least explored areas in *Leishmania* immunology. We will next provide a brief introduction to the composition of the normal skin and demonstrate how it is profoundly impacted by a bite from a *Leishmania* infected sand fly.

The Steady State Skin

Skin is the largest and most complex organ in our body, acting as the first line of defense of an animal against any invading pathogens. Anatomically, skin comprises the epidermal and dermal compartments. The epidermis is mainly composed of keratinocytes, while the dermis is a fibroblast-rich network of collagen and elastin fibers that provides elasticity (Malissen et al., 2014; Urmacher, 1990). Hair follicles, sebaceous glands, and sweat glands as well as nerve endings are also found in skin. Numerous heterogeneous populations of immune cells also reside within the skin, while others are recruited upon injury and have the capability to return to their original organ upon recovery of host homeostasis. This immune network is very complex, comprised of skin resident antigen presenting cells, phagocytes, mast cells, innate lymphoid cells, and T lymphocytes. These individual subpopulations perform very specialized functions in a coordinated fashion to help the host mount an adequate immune response to a variety of environmental challenges.

The Epidermis

The major cell types present in the epidermis include keratinocytes, Langerhans cells (LCs), and dendritic epidermal T cells (DETCs) comprised of $\gamma\delta$ T cells (Fig. 6.2A).

KERATINOCYTES

The epidermis consists of an outer layer of cells called the stratum corneum, followed by the tight junction-containing stratum granulosum, the stratum spinosum, and a basement membrane (Koster, 2009). Keratinocytes represent the major innate immune cells of the epidermis and express many pattern recognition receptors, including toll-like receptors (TLRs) that recognize a wide variety of pathogens and nod-like receptors that respond to bacterial peptidoglycans and several other viral and fungal components. Overall, these keratinocytes act as a first line of immune defense forming a barrier against entry of certain pathogens into a host. This involves communication with fibroblasts to insure wound closure (Werner et al., 2007). Importantly, keratinocytes can also produce an array of chemokines including CXCL9, CXCL10, CXCL11, CCL20, and others in response to pathogenic stimuli. These chemokines are critical for T cell and macrophages recruitment to the skin. Keratinocytes are also capable of producing cytokines TNFα, IL-1, IL-1β, IL-6, IL-10, IL-18, and IL-33 that are directly involved with orchestrating the downstream immune response (Heath and Carbone, 2013; Koster, 2009; Meephansan et al., 2012; Werner et al., 2007). Significantly, keratinocyte-derived cytokines determine Th1 immunity during the early phase of cutaneous leishmaniasis (Ehrchen et al., 2010).

LANGERHANS CELLS

The LC is a dominant cell type in the epidermis. LCs are specialized epidermal DCs, characterized by the expression of the lectin, langerin/CD207 molecule (Haniffa et al., 2015; Kaplan, 2010; Malissen et al., 2014; Valladeau et al., 2000). LCs serve as sentinel cells that survey the

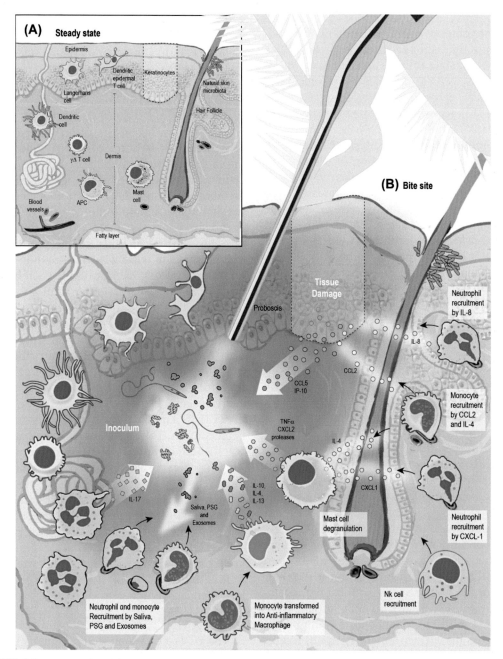

FIGURE 6.2 **Immunological consequences of the infectious inoculum after an infected sand fly bite.** A schematic view of the major cell populations in the skin at steady state (A, inset) and immediately after an infected sand fly bite (B). Mammalian skin is composed of two major compartments, the epidermis and the dermis. The epidermis is mostly composed of keratinocytes, Langerhans cells, and dendritic epidermal T cells. The dermis is highly vascularized and contains mast cells, dermal dendritic cells, macrophages, and γδ T cells. An infected sand fly probes repeatedly to create a blood pool. This results in tissue damage. Tissue damage triggers the wound healing response while components of the infectious inoculum participate by initiating a distinct inflammatory response. Both culminate in recruitment of neutrophils, followed by monocytes, into the bite site in response to various cytokines, chemokines, and proteases. The inflammatory response is initially orchestrated by resident skin cells, such as keratinocytes and degranulating mast cells, and is further amplified by mediators from arriving neutrophils and anti-inflammatory macrophages.

epidermal layer against invading pathogens. Upon microbial infection, LCs become activated, take up pathogen antigens, differentiate, and then migrate to T cell–rich areas of the skin-draining lymph nodes in order to prime naïve T cells (Haniffa et al., 2015; Malissen et al., 2014; Moreno, 2007; Wilson and Villadangos, 2004). LCs are the only cells in the epidermis that express MHC class II molecules.

DENDRITIC EPIDERMAL T CELLS

The γδ T cell population in the epidermis is called DETCs (MacLeod et al., 2013; Malissen et al., 2014). These cells regulate the immune response to wound injury and maintain skin barrier function by activation of epidermal keratinocytes. Activated DETCs express variety of cytokines and chemokines including IFN-γ, IL-2, CCL3, CCL4, and CCL5 (Boismenu et al., 1996; Matsue et al., 1993). The epidermis also supports a few memory αβ T cells (Heath and Carbone, 2013).

It is worth mentioning that the surface of the epidermis is also home to a microbiome consisting of bacteria, fungi, and viruses. Recent studies suggest that this microbiome plays an important role in host physiology including development of immunity against pathogens as well as in wound repair (Belkaid and Tamoutounour, 2016).

The Dermis

The major cell types in the dermis include fibroblasts, mast cells, macrophages, and various DC subsets (Fig. 6.2A). Minor cell populations include innate lymphoid cells such as natural killer (NK) cells and γδ T cells as well as αβ T cells of the adaptive immune response.

FIBROBLASTS

Fibroblasts are a major cell population of the dermis, and they are heterogeneous in terms of both origin and function (Driskell and Watt, 2015). Different subsets of fibroblasts have specific functions that are crucial to maintain skin homeostasis, especially in wound healing.

Fibroblasts produce collagens, glycosaminoglycans, and elastic fibers important for maintaining skin architecture (Driskell and Watt, 2015).

MAST CELLS

Mast cells are an important cell type in skin and are positioned to react with the environmental antigens immediately upon exposure. Mast cells are derived from hematopoietic stem cells, and like monocytes they are also long-lived cells and capable of proliferation upon stimulation. Mast cells act as effector cells by modulating different cells like DCs, monocytes, B cells, T cells, and NKT cells during innate or adaptive immune responses (Galli et al., 2005; Wernersson and Pejler, 2014). During an invasion of the skin, a mast cell is capable of recognizing a pathogen through TLRs as well as a wide range of other receptors that sense pathogens such as Fc receptors that bind pathogen-specific antibodies of the immunoglobulin E (IgE) isotype. Antigen-specific engagement of cell surface bound IgE activates mast cells resulting in degranulation and de novo cytokine synthesis. The degranulation process is very rapid and releases soluble mediators like histamine, leukotrienes, prostaglandins, proteases, and others that participate in the initiation of the immune response. For example, both mast cell–derived chymase and tryptase proteases induce IL-8 from endothelial cells that recruit neutrophils (Huang et al., 1998). Leukotrienes generated by mast cells also promote recruitment of neutrophils (Malaviya and Abraham, 2000). Activated mast cells also secrete chemokines and cytokines such as CCL11 (eotaxin), CXCL8 (IL-8), CXCL10 (IP10), and CCL5 (RANTES) that activate and recruit cells to the site of infection in response to different pathogens (Abraham and St John, 2010). TNFα is an additional proinflammatory cytokine produced by mast cells where it promotes DCs and neutrophil accumulation at the site of infection (McLachlan et al., 2003). Mast cells express both MHC class I and class II molecules and have been reported to process and present antigens (Kambayashi et al., 2008) in a manner normally associated with generation of a Th2 inflammatory responses.

MACROPHAGES

During steady state conditions, macrophages are the most abundant cells in the dermis and unlike LCs and DETCs, skin resident macrophages remain in the dermis (Dupasquier et al., 2004). Monocytes are precursors of macrophages that migrate from the circulation, extravasate from blood vessels, and eventually differentiate into macrophages or DCs (Shi and Pamer, 2011). Macrophages recognize pathogens by TLRs, and once they are activated produce a vast array of cytokines and chemokines depending upon the nature of stimulus. Tissue injury drives rapid recruitment of neutrophils followed by monocytes, and the chemokines CCL2 and CCL5 are actively involved in this process (Serbina and Pamer, 2006; Shi and Pamer, 2011). Skin resident macrophages are highly capable of killing invading pathogens, and they participate in wound repair and resolution of inflammation (Malissen et al., 2014).

DERMAL DENDRITIC CELLS

Dermal dendritic cells (dDCs) do not have exclusive cell markers and comprise many subsets (Haniffa et al., 2015; Malissen et al., 2014). The DC populations are very heterogeneous, and for simplicity can be subdivided on the basis of langerin (C-type lectin) expression (Kaplan, 2010). Langerin positive dDCs are few but are capable of migrating from the lymph nodes to the skin in steady state as well as in response to inflammation. Langerin-negative dDCs are the major DC population in the dermis and they can be further subdivided into two populations based on the expression of CD11b. During inflammation, langerin-negative dDCs mobilize rapidly from the skin to the draining lymph nodes where they contribute to initiating a T cell response.

Skin Disruption at the Bite Site

All blood-feeding arthropod vectors are equipped with piercing or sucking mouth parts. Female sand flies blood feed in order to produce eggs. They probe the skin with a short proboscis approximately 1 mm long that lacerates dermal capillaries to create a pool of blood. In addition to being a pool feeder, infected sand flies with mature *Leishmania* infections have trouble feeding due to blockage of their midgut by parasites and PSG. This causes them to probe frequently, resulting in extensive tissue damage that triggers the wound healing response of the host as they regurgitate the parasite-containing infectious inoculum (Fig. 6.2B). This creates a complex and distinct inflammatory response at the infected bite site.

Wound Repair: A Default Response to Skin Injury

Any time the skin is breached, the host responds with a well-characterized sequential response whose primary objective is to clear damaged cells and invading microbes before beginning tissue repair (Gurtner et al., 2008; Shaw and Martin, 2009). Briefly, upon skin injury the host responds by initiating a cascade of reactions resulting in hemostasis including blood clotting and vasoconstriction. This phase is followed by an inflammatory response dominated by phagocytes, neutrophils followed by monocytes that differentiate to macrophages to phagocytose and clean any damaged cells, debris, or microbes at the site of injury. Macrophages also engulf any expended phagocytes and contribute to angiogenesis and reepithelialization of the wound leading to new tissue formation. Finally, the reepithelialized wound undergoes contraction and remodeling leaving a scar.

Tissue resident mast cells and macrophages initiate the influx of neutrophils immediately after tissue injury (De Filippo et al., 2013; Ng, 2010; Wynn and Vannella, 2016). Upon activation both resident mast cells and macrophages secrete CXCL1 and CXCL2 to attract neutrophils to the injury site (De Filippo et al., 2013). Neutrophils recruitment to the site of tissue damage is a crucial step in controlling tissue injury and infection. Neutrophils are continuously generated in the bone marrow from myeloid precursors and are short-lived cells with a half-life of 8–12h (Kolaczkowska and Kubes, 2013). During sterile inflammation, tissue injury, or pathogen invasion, neutrophils become activated and their longevity increases at the inflamed site and undergo a multistep cascade of events that leads to their recruitment at the wound site (Kolaczkowska and Kubes, 2013). For that reason neutrophils express several chemokines and chemokine receptors to sense and react to tissue damage (Kolaczkowska and Kubes, 2013). Like other phagocytic cells, they either engulf pathogens or kill them extracellularly with the release of lytic molecules crucial for clearance of many invading pathogens (Mocsai, 2013).

At steady state, tissue macrophages are alternatively activated and antiinflammatory in nature, ensuring that homeostasis is maintained in the absence of chronic inflammation (Murray and Wynn, 2011; Wynn and Vannella, 2016). Upon injury, M2-macrophages and newly recruited monocytes promote tissue regeneration through stimulation of epithelial cells and fibroblasts, and inhibit further destruction of the extracellular matrix. Additionally, M2-tissue repairing macrophages promote apoptosis and produce antiinflammatory cytokines and arginase, shown to control the duration of inflammatory responses and to contribute to wound healing (Murray and Wynn, 2011; Wynn and Vannella, 2016). It is important to mention that macrophages are also capable of secreting an array of chemokines, cytokine, and growth factors and are considered major contributors during the wound healing process (Wynn and Vannella, 2016).

Immunological Consequences of the Transmitted Inoculum

The wound repair hemostatic "default" response intrinsic to the immune system can be hijacked by vector-borne pathogens that are transmitted by arthropod bite (Peters et al., 2008; Pingen et al., 2016). This is not surprising since bites trigger the wound healing and tissue repair response, and pathogens deposited in the skin exploit this intrinsic response to promote their own establishment. In the case of vector-transmitted *Leishmania*, parasites delivered by sand fly bite promote their survival by hiding in neutrophils that arrive soon after the skin is pierced by the sand fly proboscis, thus promoting a subsequent better infection of their host cell, the macrophage (Peters et al., 2008). This finding corroborates earlier studies reporting that neutrophils act as a Trojan horse for *Leishmania*, delivering them to macrophages (van Zandbergen et al., 2002, 2004). Interestingly, while in vitro studies show that *Leishmania* parasites delay the apoptosis of neutrophils (Aga et al., 2002; van Zandbergen et al., 2004), a more recent ex vivo analysis of *Leishmania*-infected neutrophils suggests that infective metacyclics enhanced their apoptosis (Ribeiro-Gomes and Sacks, 2012). Other studies demonstrated that parasites release a soluble *Leishmania* chemotactic factor that attracts neutrophils (van Zandbergen et al., 2002) and can resist the killing activity of neutrophil extracellular traps (NETs) as well as escape the antimicrobial machinery of macrophages (Gabriel et al., 2010).

In addition to exploitation of the wound repair mechanism by *Leishmania* parasites and their direct effect on the immune cells encountered in the skin, other components of the infectious inoculum participate in modulation of the sand fly bite site following vector transmission that creates a more hospitable environment for parasite survival. Next, we address known immunological consequences of the infectious inoculum, other than *Leishmania*, after infected sand fly bites.

Promastigote Secretory Gel

PSG is one of the major components of the infectious inoculum regurgitated by infected sand flies and facilitates both CL and VL infections (Rogers et al., 2010). *Leishmania* parasites secrete PSG blocking the anterior part of the midgut. This blockage significantly alters the feeding behavior of infected sand flies and promotes more frequent probing. Importantly, needle injection of *Leishmania* together with PSG exacerbates mice ear lesions as well as potentiates visceralization of parasites (Rogers et al., 2009, 2010). Like saliva, PSG is also capable of recruiting neutrophils and macrophages to the site of injection (Rogers et al., 2009). Moreover, it reduces the efficacy of the *Leishmania*-killing properties of inflammatory monocytes, converting them to an antiinflammatory alternative phenotype of macrophages and inducing arginase metabolism and synthesis of polyamines that promotes parasite survival in hostile macrophages (Rogers et al., 2009). This reinforces the antiinflammatory environment already prevalent at the site of bite to the advantage of *Leishmania* parasites.

Exosomes

A recent study reported that exosomes are secreted by *Leishmania* promastigotes inside the sand fly gut and are part of the infectious inoculum introduced into the host at the bite site (Atayde et al., 2015). Coinjection of in vitro or in vivo exosomes and *Leishmania* parasites into mice footpads exacerbated lesion size and enhanced IL17a production. It is worthwhile mentioning that IL17a is a potent inducer of neutrophil recruitment (Boaventura et al., 2010). The above, together with the immunomodulatory properties attributed to *Leishmania* exosomes in vitro that include neutrophil recruitment as well as suppression of macrophage and dendritic cell activation (Coakley et al., 2015; Silverman and Reiner, 2011), indicate that vector-derived exosomes may have a role in supporting parasite establishment in the early skin phase of infection.

Considering the above, it is not surprising that a distinct property to neutrophil behavior

was observed after sand fly bites. Neutrophils persisted at the bite site for days compared to a transient recruitment following needle-injection of parasites (Peters et al., 2008). This effect was associated to enhanced virulence of vector-transmitted parasites and linked to failure of vaccines effective against needle-challenge when *Leishmania* was transmitted by infected sand fly bites (Peters et al., 2009; Peters and Sacks, 2009). These observations may only be the tip of the iceberg. It is clear that we need to understand how each of these potent immunomodulators act collectively at the bite site and to try to unravel their contribution to the enhanced virulence of vector-transmitted parasites in situ.

Saliva

Infected sand flies deposit *Leishmania* parasites along with other parasite-derived components into the bite site. Similar to other blood feeding arthropods, sand flies need to secrete saliva into the wound to facilitate blood feeding. As described above, sand fly saliva is composed of multiple proteins with diverse functions that evolved mainly to counteract the host's hemostatic system. However, salivary proteins are also capable of inducing a strong inflammatory immune response by activating skin-resident immune cells, including dendritic cells and macrophages (Abdeladhim et al., 2014). Additionally, sand fly saliva has chemotactic activities for both neutrophils and macrophages (de Moura et al., 2010; Prates et al., 2012; Teixeira et al., 2005), and recently a sand fly salivary endonuclease LUNDEP was shown to cleave NETs leading to exacerbation of *Leishmania* infection (Chagas et al., 2014). Sand fly saliva has also been reported to induce apoptosis of neutrophils as well as recruitment of macrophages in an in vitro setting (Abdeladhim et al., 2014, 2016; Oliveira et al., 2013a; Prates et al., 2012). Importantly, a multitude of studies have also demonstrated the capacity of saliva or salivary molecules to modulate immune defenses with an over-all suppressive effect on the host adaptive immune response, biasing it toward a Th2 type that promotes parasite survival and disease establishment (Abdeladhim et al., 2014; Gomes and Oliveira, 2012; Kamhawi, 2000).

BEHAVIORAL MATTERS

Apart from the intrinsic contribution of vector-derived components in the infectious inoculum to *Leishmania* survival, extrinsic factors such as behavior can also influence successful transmission of vector-borne disease. Accumulating evidence suggests that many pathogens alter the behavior of arthropod vectors to promote their transmission. A wide range of pathogens from plant viruses to malaria parasites use elaborate and distinct strategies to manipulate the behavior of their vectors in order to optimize their transmission (Cator et al., 2012; Hurd, 2003; Ingwell et al., 2012; Van Den Abbeele et al., 2010). For example, malaria parasites manipulate the feeding behavior of *Anopheles* mosquitoes at different stages of their development. Moreover, studies report that oocyst-bearing mosquitoes are less persistent at blood feeding compared to those harboring more mature sporozoite-positive infections, insuring better survival of the former and more efficient transmission by the latter (Cator et al., 2012). Another study provided direct evidence that *Plasmodium falciparum*-infected *Anopheles gambiae* show enhanced attraction to the human host compared to their uninfected counterparts, and this behavioral change was attributed to a change in their response to olfactory stimuli within skin odors (Smallegange et al., 2013). Similarly, parasites in *Trypanosoma brucei*-infected tsetse flies suppress salivary gland gene transcription reducing the efficiency of salivation and prolonging feeding time, thus promoting

their transmission to multiple hosts (Van Den Abbeele et al., 2010).

Sand flies also exhibit altered feeding behavior when infected with *Leishmania* (Rogers, 2012). Distinct from trypanosomes and malaria parasites, *Leishmania* do not infect salivary glands and their entire developmental cycle is restricted to the open tube of the midgut (Fig. 6.1). When a sand fly harbors a mature infection, the anterior part of its midgut becomes blocked by a buildup of PSG that contains parasites, PPGs, and other vector-derived components, including exosomes (Atayde et al., 2015; Rogers, 2012). It is easy to imagine how this would impede the ability of an infected sand fly to ingest blood freely, leading to an increase in probing frequency and deficient blood meal repletion. This altered feeding behavior and its association to optimal transmission of *Leishmania* was reported in several studies using different sand fly and parasite species. Those studies demonstrated that in contrast to uninfected or poorly infected sand flies, those with mature infections probed many times more and obtained little or no blood while efficiently transmitting parasites to multiple animals (Beach et al., 1984, 1985; Killick-Kendrick et al., 1977; Maia et al., 2011). *Leishmania* infection of sand flies results in biting persistence when compared to uninfected sand flies with infected females more likely to refeed on a new host after being interrupted, resulting in amplified vectorial capacity of the infected sand fly (Rogers and Bates, 2007). Increased feeding persistence after interruption of engorgement in infected sand flies was finely attuned to the appearance of infective stage metacyclics (Rogers and Bates, 2007). Taken together, these data suggest that an adaptive selective pressure led to the behavioral manipulation of the sand fly vector by *Leishmania* parasites, optimizing the success of each transmission as well as maximizing the number of hosts that become infected by a single sand fly.

CURRENT STATUS OF LEISHMANIASIS CONTROL

Several partnerships that include the governments of Bangladesh, India, and Nepal; the World Health Organization; the Bill and Melinda Gates Foundation; and more recently KalaCORE (the consortium for the control and elimination of VL) strive to eliminate VL as a public health problem from the Indian subcontinent and East Africa (Cameron et al., 2016; Chowdhury et al., 2014). These efforts are making progress and should be applauded, yet we need to be cautious they do not to dampen support for the development of efficacious vaccines. Mathematical models predict that the current focus on vector control and diagnosis and treatment of VL cases would eliminate VL in foci with low to medium endemicity, while areas of high endemicity will require additional interventions (Le Rutte et al., 2016). Moreover, as a vector-borne disease, a constant need for surveillance postcontrol of VL is indicated in the absence of vaccines. Pockets of low transmission may escape control efforts providing the seed for resurgence of transmission and reemergence of disease (Cameron et al., 2016). Furthermore, we still do not know the main source of sand fly infections: is it asymptomatic individuals, treated cases with post-Kala Azar dermal leishmaniasis or clinical VL cases? This leads us to another facet of leishmaniasis, its diverse clinical features. Apart from VL caused by *L. donovani*, the majority of leishmaniasis forms are zoonotic in nature with wild rodents or canids for reservoirs. This again reinforces the need for vaccines.

Components of the Infectious Inoculum and *Leishmania* Vaccines

Despite significant advancement in our understanding of the host immune response against *Leishmania* parasites, there are no licensed human vaccines against any form of leishmaniasis. Early *Leishmania*-based vaccines have failed to protect when tested in clinical trials in endemic regions (Khalil et al., 2000; Sharifi et al., 1998); in contrast, leishmanization worked suggesting that an effective leishmaniasis vaccine is attainable (Gillespie et al., 2016). Moreover, success of leishmanization and failure of killed parasites as effective vaccines suggest that the presence of live parasites, likely as a continuous source of antigen, may be required. This spurred the development of live attenuated and genetically modified vaccines, most showing efficacy in animal models of infection (Saljoughian et al., 2014). Nevertheless, showing efficacy in animal models does not necessarily predict efficacy of the vaccine in humans in an endemic setting. Recent studies have suggested that this may partly be due to their development using a needle-challenge with parasites alone. These studies have demonstrated that needle-initiated infections fail to mimic the virulence of vector-transmitted *Leishmania*, associated to an induction of a robust and rapid pro-parasitic immune response, which might override the antileishmanial response developed by suboptimal candidate vaccines (Peters et al., 2009, 2012).

So let us consider two fundamental facts: (1) the absence of a human vaccine against any form of leishmaniasis and (2) the virulence of an infectious inoculum made up of multiple components. Naturally, you would expect to try to improve the efficacy of a *Leishmania* vaccine by including other immunogenic molecules constantly present at the site of infected bites. Studies addressing the potential of components of the infectious inoculum as *Leishmania* vaccines have mainly focused on PSG and salivary proteins (Abdeladhim et al., 2014; Kamhawi et al., 2014; Rogers et al., 2006). Rogers et al. (2006) demonstrated the protective property of PSG, or a synthetic glycovaccine containing *L. mexicana* glycans, against infection initiated by infected sand fly bites; importantly, neither protected against needle-challenge with parasites demonstrating the relevance of using vector-transmission models when mining for candidate vaccines against *Leishmania*, or other vector-borne infections. As for saliva, after more than a decade since studies demonstrated the capacity of sand fly salivary proteins to protect against *Leishmania* infection (Belkaid et al., 1998; Kamhawi et al., 2000; Valenzuela et al., 2001), researchers and vaccinologists are beginning to take note of their potential value as components of *Leishmania* vaccines (De Luca and Macedo, 2016; Reed et al., 2016). To date, several defined salivary antigens have shown promise as effective *Leishmania* vaccines, some having been challenged by infected vector bites (Abdeladhim et al., 2014; Kamhawi et al., 2014; Oliveira et al., 2015). Additionally, a novel approach aimed at increasing the efficacy of a vaccine combined promising *Leishmania* and salivary vaccine candidates yielding encouraging results for both CL and VL (Fiuza et al., 2016; Zahedifard et al., 2014).

Components of the Infectious Inoculum and Drugs

Since the need for drugs usually occurs at a later stage during the course of infection, the complexity and virulence of the infectious inoculum, and the prevailing immune environment at the bite site, may not be of direct relevance. Nevertheless, compared to needle-initiated infections, the use of natural transmission models by vector bites reported differences in the kinetics of disease establishment and in clinical manifestations of leishmaniasis (Aslan et al., 2016; Peters et al., 2008) that may be pertinent to drug development and application. For instance, as a consequence of the distinct initiation of the immune response at the bite site, using models of vector transmission may reveal new drug targets or risk markers early after the onset of infection. Moreover, since leishmaniasis in all its forms, cutaneous or visceral, initiated in the skin using such models may mimic better the kinetics, dissemination pattern, and pathogenicity of disease.

A BRIGHT FUTURE AWAITS

Basic Research

In the context of *Leishmania* transmission by sand fly vector bites, the skin interface remains a black box, and its relevance to both disease pathogenesis and vaccine design is just being realized. For example, several aspects unique to the bite site are mostly overlooked despite significant advancement in our understanding of *Leishmania* transmission by infected sand flies. Sand flies are pool feeders that lacerate shallow capillaries located in the upper dermis. However, we do not yet know the cell types in the skin that first encounter and are influenced by components of the infectious inoculum. Are epidermal cells such as keratinocytes, and resident dermal cells such as mast cells, involved in the initial immune response to an infected bite? How are they affected by the immunomodulatory components of the infectious inoculum? Addressing these questions would further unravel the secrets behind the virulence and efficacy of vector transmission. Another is a lack of appreciation of the minute size of the bite site and its implication to the success of parasite transmission. A sand fly proboscis is only approximately 10 µm in diameter (Krenn and Aspock, 2012), 100 times smaller than the smallest hypodermic needle, while the volume of blood taken by a fully fed sand fly is only about 0.2–0.3 µL. As a result, probing and feeding creates a highly focalized bite site where contents of the infectious inoculum are delivered into a minuscule area in a concentrated manner enabling them to exert a potent effect on their surrounding microenvironment. Pertinently, the outcome of having several immunomodulatory molecules at the bite site is yet unclear. Do they synergize, antagonize, or independently exert their effect on the cells and parasites around them?

Rapid advances in technology promise to provide the tools to address many of these unanswered questions. Particularly, the development of live tissue imaging technologies with ever more powerful microscopes that penetrate the skin and the availability of dyes, labels, and an array of transgenic mice to visualize cells of interest provide exciting and powerful tools to study the host immune response at the skin interface in real time (Germain, 2015; Germain et al., 2006).

Another underexplored field of research is the use of natural models of transmission to determine the course of disease establishment. This may identify new drug targets or markers of risk of disease considering that infection is initiated in the skin with a low number of parasites (Kimblin et al., 2008; Maia et al., 2011; Secundino et al., 2012) and in the presence of the various immunomodulatory components of the infectious inoculum.

Translational Research

It is important to translate our findings at the bench to the field. This validates observations we make using animal models and standardized methodologies in target species under realistic conditions of genetic diversity and existing confounding factors such as coinfections.

For example, a study undertaken by our group on a naturally exposed population in Mali validated the long-term immunogenicity of sand fly vector saliva, one of the major components of the infectious inoculum, in humans (Oliveira et al., 2013b). This gives credence to laboratory findings demonstrating the protective capacity of distinct immunogenic salivary proteins against *Leishmania* parasites (Abdeladhim et al., 2014; Gomes and Oliveira, 2012) and supports the notion that they would contribute to the betterment of *Leishmania* vaccines. Similarly, field studies have demonstrated the value of antibodies against salivary proteins as markers of vector exposure of use for epidemiological and control studies (Clements et al., 2010; Marzouki et al., 2015; Souza et al., 2010).

CONCLUDING REMARKS

In this chapter, we put forward an argument for the relevance of vector transmission of *Leishmania* parasites by bites of infected vector sand flies to leishmaniasis research. We described components of the infectious inoculum and provided evidence for the profound effect they exert on the early immune response at the bite site, and ultimately on disease evolution. We also referenced reports demonstrating the practical relevance of such components for vaccine development. Despite significant progress in our understanding of vector transmission, the low number of research groups with the capacity to conduct experiments using infected vector sand flies hampers further advances. We need to promote interest in vector biology and the need for groups working with vector-borne pathogens, including leishmaniasis, to invest in infrastructure and personnel to support relevant vector-related research.

Acknowledgments

We are indebted to Ryan Kissinger for patiently working with us to produce the artwork contained in this chapter. This work was funded by the Intramural Research Programs at the NIAID, NIH, and by CBER, FDA, USA.

References

Abdeladhim, M., Kamhawi, S., Valenzuela, J.G., 2014. What's behind a sand fly bite? The profound effect of sand fly saliva on host hemostasis, inflammation and immunity. Infect. Genet. Evol. 28, 691–703.
Abdeladhim, M.I., Coutinho-Abreu, V., Townsend, S., Pasos-Pinto, S., Sanchez, L., Rasouli, M., G.-C., A.B., Aslan, H., Francischetti, I.M., Oliveira, F., Becker, I., Kamhawi, S., Ribeiro, J.M., Jochim, R.C., Valenzuela, J.G., 2016. Molecular diversity between salivary proteins from New World and Old World Sand Flies with Emphasis on *Bichromomyia olmeca*, the Sand Fly Vector of *Leishmania mexicana* in Mesoamerica. PLoS Negl. Trop. Dis. 10, e0004771.
Abraham, S.N., St John, A.L., 2010. Mast cell-orchestrated immunity to pathogens. Nat. Rev. Immunol. 10, 440–452.

Aga, E., Katschinski, D.M., van Zandbergen, G., Laufs, H., Hansen, B., Muller, K., Solbach, W., Laskay, T., 2002. Inhibition of the spontaneous apoptosis of neutrophil granulocytes by the intracellular parasite *Leishmania major*. J. Immunol. 169, 898–905.

Akhoundi, M., Bakhtiari, R., Guillard, T., Baghaei, A., Tolouei, R., Sereno, D., Toubas, D., Depaquit, J., Abyaneh, M.R., 2012. Diversity of the bacterial and fungal microflora from the midgut and cuticle of phlebotomine sand flies collected in North-Western Iran. PLoS One 7, e50259.

Aslan, H., Oliveira, F., Meneses, C., Castrovinci, P., Gomes, R., Teixeira, C., Derenge, C.A., Orandle, M., Gradoni, L., Oliva, G., Fischer, L., Valenzuela, J.G., Kamhawi, S., 2016. New insights into the transmissibility of *Leishmania infantum* from dogs to sand flies: experimental vector-transmission reveals persistent parasite depots at bite sites. J. Infect. Dis. 213, 1752–1761.

Atayde, V.D., Hassani, K., da Silva Lira Filho, A., Borges, A.R., Adhikari, A., Martel, C., Olivier, M., 2016. *Leishmania* exosomes and other virulence factors: impact on innate immune response and macrophage functions. Cell. Immunol 309, 7–18.

Atayde, V.D., Aslan, H., Townsend, S., Hassani, K., Kamhawi, S., Olivier, M., 2015. Exosome secretion by the parasitic Protozoan *Leishmania* within the sand fly midgut. Cell Rep. 13, 957–967.

Azambuja, P., Garcia, E.S., Ratcliffe, N.A., 2005. Gut microbiota and parasite transmission by insect vectors. Trends Parasitol. 21, 568–572.

Bates, P.A., 2007. Transmission of *Leishmania* metacyclic promastigotes by phlebotomine sand flies. Int. J. Parasitol. 37, 1097–1106.

Bates, P.A., Rogers, M.E., 2004. New insights into the developmental biology and transmission mechanisms of *Leishmania*. Curr. Mol. Med. 4, 601–609.

Beach, R., Kiilu, G., Hendricks, L., Oster, C., Leeuwenburg, J., 1984. Cutaneous leishmaniasis in Kenya: transmission of *Leishmania major* to man by the bite of a naturally infected *Phlebotomus duboscqi*. Trans. R. Soc. Trop. Med. Hyg. 78, 747–751.

Beach, R., Kiilu, G., Leeuwenburg, J., 1985. Modification of sand fly biting behavior by *Leishmania* leads to increased parasite transmission. Am. J. Trop. Med. Hyg. 34, 278–282.

Belkaid, Y., Tamoutounour, S., 2016. The influence of skin microorganisms on cutaneous immunity. Nat. Rev. Immunol. 16, 353–366.

Belkaid, Y., Kamhawi, S., Modi, G., Valenzuela, J., Noben-Trauth, N., Rowton, E., Ribeiro, J., Sacks, D.L., 1998. Development of a natural model of cutaneous leishmaniasis: powerful effects of vector saliva and saliva preexposure on the long-term outcome of *Leishmania major* infection in the mouse ear dermis. J. Exp. Med. 188, 1941–1953.

Boaventura, V.S., Santos, C.S., Cardoso, C.R., de Andrade, J., Dos Santos, W.L., Clarencio, J., Silva, J.S., Borges, V.M., Barral-Netto, M., Brodskyn, C.I., Barral, A., 2010. Human mucosal leishmaniasis: neutrophils infiltrate areas of tissue damage that express high levels of Th17-related cytokines. Eur. J. Immunol. 40, 2830–2836.

Boismenu, R., Feng, L., Xia, Y.Y., Chang, J.C., Havran, W.L., 1996. Chemokine expression by intraepithelial gamma delta T cells. Implications for the recruitment of inflammatory cells to damaged epithelia. J. Immunol. 157, 985–992.

Boissiere, A., Tchioffo, M.T., Bachar, D., Abate, L., Marie, A., Nsango, S.E., Shahbazkia, H.R., Awono-Ambene, P.H., Levashina, E.A., Christen, R., Morlais, I., 2012. Midgut microbiota of the malaria mosquito vector *Anopheles gambiae* and interactions with *Plasmodium falciparum* infection. PLoS Pathog. 8, e1002742.

Cameron, M.M., Acosta-Serrano, A., Bern, C., Boelaert, M., den Boer, M., Burza, S., Chapman, L.A., Chaskopoulou, A., Coleman, M., Courtenay, O., Croft, S., Das, P., Dilger, E., Foster, G., Garlapati, R., Haines, L., Harris, A., Hemingway, J., Hollingsworth, T.D., Jervis, S., Medley, G., Miles, M., Paine, M., Picado, A., Poche, R., Ready, P., Rogers, M., Rowland, M., Sundar, S., de Vlas, S.J., Weetman, D., 2016. Understanding the transmission dynamics of *Leishmania donovani* to provide robust evidence for interventions to eliminate visceral leishmaniasis in Bihar, India. Parasit. Vectors 9, 25.

Carrera, L., Gazzinelli, R.T., Badolato, R., Hieny, S., Muller, W., Kuhn, R., Sacks, D.L., 1996. *Leishmania* promastigotes selectively inhibit interleukin 12 induction in bone marrow-derived macrophages from susceptible and resistant mice. J. Exp. Med. 183, 515–526.

Carvalho, B.M., Rangel, E.F., Ready, P.D., Vale, M.M., 2015. Ecological niche modelling predicts southward expansion of *Lutzomyia* (*Nyssomyia*) *flaviscutellata* (Diptera: Psychodidae: Phlebotominae), vector of *Leishmania* (*Leishmania*) *amazonensis* in South America, under climate change. PLoS One 10, e0143282.

Cator, L.J., Lynch, P.A., Read, A.F., Thomas, M.B., 2012. Do malaria parasites manipulate mosquitoes? Trends Parasitol. 28, 466–470.

Chagas, A.C., Oliveira, F., Debrabant, A., Valenzuela, J.G., Ribeiro, J.M., Calvo, E., 2014. Lundep, a sand fly salivary endonuclease increases *Leishmania* parasite survival in neutrophils and inhibits XIIa contact activation in human plasma. PLoS Pathog. 10, e1003923.

Chappuis, F., Sundar, S., Hailu, A., Ghalib, H., Rijal, S., Peeling, R.W., Alvar, J., Boelaert, M., 2007. Visceral leishmaniasis: what are the needs for diagnosis, treatment and control? Nat. Rev. Microbiol. 5, 873–882.

Cho, I., Blaser, M.J., 2012. The human microbiome: at the interface of health and disease. Nat. Rev. Genet. 13, 260–270.

Chowdhury, R., Mondal, D., Chowdhury, V., Faria, S., Alvar, J., Nabi, S.G., Boelaert, M., Dash, A.P., 2014. How far are we from visceral leishmaniasis elimination in Bangladesh? An assessment of epidemiological surveillance data. PLoS Negl. Trop. Dis. 8, e3020.

Cirimotich, C.M., Ramirez, J.L., Dimopoulos, G., 2011. Native microbiota shape insect vector competence for human pathogens. Cell Host Microbe 10, 307–310.

Clements, M.F., Gidwani, K., Kumar, R., Hostomska, J., Dinesh, D.S., Kumar, V., Das, P., Muller, I., Hamilton, G., Volfova, V., Boelaert, M., Das, M., Rijal, S., Picado, A., Volf, P., Sundar, S., Davies, C.R., Rogers, M.E., 2010. Measurement of recent exposure to *Phlebotomus argentipes*, the vector of Indian visceral Leishmaniasis, by using human antibody responses to sand fly saliva. Am. J. Trop. Med. Hyg. 82, 801–807.

Coakley, G., Maizels, R.M., Buck, A.H., 2015. Exosomes and other extracellular vesicles: the new communicators in parasite infections. Trends Parasitol. 31, 477–489.

Colombo, M., Raposo, G., Thery, C., 2014. Biogenesis, secretion, and intercellular interactions of exosomes and other extracellular vesicles. Annu. Rev. Cell Dev. Biol. 30, 255–289.

Cunningham, M.L., Titus, R.G., Turco, S.J., Beverley, S.M., 2001. Regulation of differentiation to the infective stage of the protozoan parasite *Leishmania major* by tetrahydrobiopterin. Science 292, 285–287.

De Filippo, K., Dudeck, A., Hasenberg, M., Nye, E., van Rooijen, N., Hartmann, K., Gunzer, M., Roers, A., Hogg, N., 2013. Mast cell and macrophage chemokines CXCL1/CXCL2 control the early stage of neutrophil recruitment during tissue inflammation. Blood 121, 4930–4937.

De Luca, P.M., Macedo, A.B., 2016. Cutaneous leishmaniasis vaccination: a matter of quality. Front. Immunol. 7, 151.

de Moura, T.R., Oliveira, F., Rodrigues, G.C., Carneiro, M.W., Fukutani, K.F., Novais, F.O., Miranda, J.C., Barral-Netto, M., Brodskyn, C., Barral, A., de Oliveira, C.I., 2010. Immunity to *Lutzomyia intermedia* saliva modulates the inflammatory environment induced by *Leishmania braziliensis*. PLoS Negl. Trop. Dis. 4, e712.

Dillon, R.J., el Kordy, E., Shehata, M., Lane, R.P., 1996. The prevalence of a microbiota in the digestive tract of *Phlebotomus papatasi*. Ann. Trop. Med. Parasitol. 90, 669–673.

Dostalova, A., Volf, P., 2012. *Leishmania* development in sand flies: parasite-vector interactions overview. Parasit. Vectors 5, 276.

Driskell, R.R., Watt, F.M., 2015. Understanding fibroblast heterogeneity in the skin. Trends Cell Biol. 25, 92–99.

Dujardin, J.C., Campino, L., Canavate, C., Dedet, J.P., Gradoni, L., Soteriadou, K., Mazeris, A., Ozbel, Y., Boelaert, M.,

2008. Spread of vector-borne diseases and neglect of Leishmaniasis, Europe. Emerg. Infect. Dis. 14, 1013–1018.

Dumaswala, U.J., Greenwalt, T.J., 1984. Human erythrocytes shed exocytic vesicles in vivo. Transfusion 24, 490–492.

Dupasquier, M., Stoitzner, P., van Oudenaren, A., Romani, N., Leenen, P.J., 2004. Macrophages and dendritic cells constitute a major subpopulation of cells in the mouse dermis. J. Invest. Dermatol. 123, 876–879.

Duschak, V.G., Barboza, M., Garcia, G.A., Lammel, E.M., Couto, A.S., Isola, E.L., 2006. Novel cysteine proteinase in *Trypanosoma cruzi* metacyclogenesis. Parasitology 132, 345–355.

Ehrchen, J.M., Roebrock, K., Foell, D., Nippe, N., von Stebut, E., Weiss, J.M., Munck, N.A., Viemann, D., Varga, G., Muller-Tidow, C., Schuberth, H.J., Roth, J., Sunderkotter, C., 2010. Keratinocytes determine Th1 immunity during early experimental leishmaniasis. PLoS Pathog. 6, e1000871.

Figueiredo, A.B., Serafim, T.D., Marques-da-Silva, E.A., Meyer-Fernandes, J.R., Afonso, L.C., 2012. *Leishmania amazonensis* impairs DC function by inhibiting CD40 expression via A2B adenosine receptor activation. Eur. J. Immunol. 42, 1203–1215.

Finney, C.A., Kamhawi, S., Wasmuth, J.D., 2015. Does the arthropod microbiota impact the establishment of vector-borne diseases in mammalian hosts? PLoS Pathog. 11, e1004646.

Fischer, D., Thomas, S.M., Beierkuhnlein, C., 2011. Modelling climatic suitability and dispersal for disease vectors: the example of a phlebotomine sandfly in Europe. Procedia Environ. Sci. 7, 164–169.

Fiuza, J.A., Dey, R., Davenport, D., Abdeladhim, M., Meneses, C., Oliveira, F., Kamhawi, S., Valenzuela, J.G., Gannavaram, S., Nakhasi, H.L., 2016. Intradermal immunization of *Leishmania donovani* centrin knock-out parasites in combination with salivary protein LJM19 from sand fly vector induces a durable protective immune response in hamsters. PLoS Negl. Trop. Dis. 10, e0004322.

Gabriel, C., McMaster, W.R., Girard, D., Descoteaux, A., 2010. *Leishmania donovani* promastigotes evade the antimicrobial activity of neutrophil extracellular traps. J. Immunol. 185, 4319–4327.

Galli, S.J., Nakae, S., Tsai, M., 2005. Mast cells in the development of adaptive immune responses. Nat. Immunol. 6, 135–142.

Germain, R.N., 2015. Tracking the T cell repertoire. Nat. Rev. Immunol. 15, 730.

Germain, R.N., Miller, M.J., Dustin, M.L., Nussenzweig, M.C., 2006. Dynamic imaging of the immune system: progress, pitfalls and promise. Nat. Rev. Immunol. 6, 497–507.

Gillespie, P.M., Beaumier, C.M., Strych, U., Hayward, T., Hotez, P.J., Bottazzi, M.E., 2016. Status of vaccine research and development of vaccines for leishmaniasis. Vaccine 34, 2992–2995.

Gomes, R., Oliveira, F., 2012. The immune response to sand fly salivary proteins and its influence on *Leishmania* immunity. Front. Immunol. 3, 110.

Gossage, S.M., Rogers, M.E., Bates, P.A., 2003. Two separate growth phases during the development of *Leishmania* in sand flies: implications for understanding the life cycle. Int. J. Parasitol. 33, 1027–1034.

Gurtner, G.C., Werner, S., Barrandon, Y., Longaker, M.T., 2008. Wound repair and regeneration. Nature 453, 314–321.

Haniffa, M., Gunawan, M., Jardine, L., 2015. Human skin dendritic cells in health and disease. J. Dermatol. Sci. 77, 85–92.

Heath, W.R., Carbone, F.R., 2013. The skin-resident and migratory immune system in steady state and memory: innate lymphocytes, dendritic cells and T cells. Nat. Immunol. 14, 978–985.

Hosseini, H.M., Fooladi, A.A., Nourani, M.R., Ghanezadeh, F., 2013. The role of exosomes in infectious diseases. Inflamm. Allergy Drug Targets 12, 29–37.

Huang, C., Friend, D.S., Qiu, W.T., Wong, G.W., Morales, G., Hunt, J., Stevens, R.L., 1998. Induction of a selective and persistent extravasation of neutrophils into the peritoneal cavity by tryptase mouse mast cell protease 6. J. Immunol. 160, 1910–1919.

Hurd, H., 2003. Manipulation of medically important insect vectors by their parasites. Annu. Rev. Entomol. 48, 141–161.

Ilg, T., Stierhof, Y.D., Craik, D., Simpson, R., Handman, E., Bacic, A., 1996. Purification and structural characterization of a filamentous, mucin-like proteophosphoglycan secreted by *Leishmania* parasites. J. Biol. Chem. 271, 21583–21596.

Ingwell, L.L., Eigenbrode, S.D., Bosque-Perez, N.A., 2012. Plant viruses alter insect behavior to enhance their spread. Sci. Rep. 2, 578.

Johnstone, R.M., Adam, M., Hammond, J.R., Orr, L., Turbide, C., 1987. Vesicle formation during reticulocyte maturation. Association of plasma membrane activities with released vesicles (exosomes). J. Biol. Chem. 262, 9412–9420.

Kambayashi, T., Baranski, J.D., Baker, R.G., Zou, T., Allenspach, E.J., Shoag, J.E., Jones, P.L., Koretzky, G.A., 2008. Indirect involvement of allergen-captured mast cells in antigen presentation. Blood 111, 1489–1496.

Kamhawi, S., 2000. The biological and immunomodulatory properties of sand fly saliva and its role in the establishment of *Leishmania* infections. Microbes Infect. 2, 1765–1773.

Kamhawi, S., 2006. Phlebotomine sand flies and *Leishmania* parasites: friends or foes? Trends Parasitol. 22, 439–445.

Kamhawi, S., Belkaid, Y., Modi, G., Rowton, E., Sacks, D., 2000. Protection against cutaneous leishmaniasis resulting from bites of uninfected sand flies. Science 290, 1351–1354.

Kamhawi, S., Ramalho-Ortigao, M., Pham, V.M., Kumar, S., Lawyer, P.G., Turco, S.J., Barillas-Mury, C., Sacks, D.L., Valenzuela, J.G., 2004. A role for insect galectins in parasite survival. Cell 119, 329–341.

Kamhawi, S., Aslan, H., Valenzuela, J.G., 2014. Vector saliva in vaccines for visceral leishmaniasis: a brief encounter of high consequence? Front. Public Health 2, 99.

Kaplan, D.H., 2010. In vivo function of Langerhans cells and dermal dendritic cells. Trends Immunol. 31, 446–451.

Katz, O., Waitumbi, J.N., Zer, R., Warburg, A., 2000. Adenosine, AMP, and protein phosphatase activity in sandfly saliva. Am. J. Trop. Med. Hyg. 62, 145–150.

Khalil, E.A., El Hassan, A.M., Zijlstra, E.E., Mukhtar, M.M., Ghalib, H.W., Musa, B., Ibrahim, M.E., Kamil, A.A., Elsheikh, M., Babiker, A., Modabber, F., 2000. Autoclaved *Leishmania major* vaccine for prevention of visceral leishmaniasis: a randomised, double-blind, BCG-controlled trial in Sudan. Lancet 356, 1565–1569.

Killick-Kendrick, R., 1990. Phlebotomine vectors of the leishmaniases: a review. Med. Vet. Entomol. 4, 1–24.

Killick-Kendrick, R., Leaney, A.J., Ready, P.D., Molyneux, D.H., 1977. *Leishmania* in phlebotomid sandflies. IV. The transmission of *Leishmania* mexicana amazonensis to hamsters by the bite of experimentally infected *Lutzomyia longipalpis*. Proc. R. Soc. Lond. B Biol. Sci. 196, 105–115.

Kimblin, N., Peters, N., Debrabant, A., Secundino, N., Egen, J., Lawyer, P., Fay, M.P., Kamhawi, S., Sacks, D., 2008. Quantification of the infectious dose of *Leishmania major* transmitted to the skin by single sand flies. Proc. Natl. Acad. Sci. U.S.A. 105, 10125–10130.

Kolaczkowska, E., Kubes, P., 2013. Neutrophil recruitment and function in health and inflammation. Nat. Rev. Immunol. 13, 159–175.

Koster, M.I., 2009. Making an epidermis. Ann. N.Y. Acad. Sci. 1170, 7–10.

Krenn, H.W., Aspock, H., 2012. Form, function and evolution of the mouthparts of blood-feeding Arthropoda. Arthropod Struct. Dev. 41, 101–118.

Lawyer, P.G., Young, D.G., Butler, J.F., Akin, D.E., 1987. Development of *Leishmania mexicana* in *Lutzomyia diabolica* and *Lutzomyia shannoni* (Diptera: Psychodidae). J. Med. Entomol. 24, 347–355.

Lawyer, P.G., Ngumbi, P.M., Anjili, C.O., Odongo, S.O., Mebrahtu, Y.B., Githure, J.I., Koech, D.K., Roberts, C.R., 1990. Development of *Leishmania major* in *Phlebotomus duboscqi* and *Sergentomyia schwetzi* (Diptera: Psychodidae). Am. J. Trop. Med. Hyg. 43, 31–43.

Le Rutte, E.A., Coffeng, L.E., Bontje, D.M., Hasker, E.C., Postigo, J.A., Argaw, D., Boelaert, M.C., De Vlas, S.J., 2016. Feasibility of eliminating visceral leishmaniasis from the Indian subcontinent: explorations with a set of deterministic age-structured transmission models. Parasit. Vectors 9, 24.

Li, X.B., Zhang, Z.R., Schluesener, H.J., Xu, S.Q., 2006. Role of exosomes in immune regulation. J. Cell. Mol. Med. 10, 364–375.

Lutz, H.U., Liu, S.C., Palek, J., 1977. Release of spectrin-free vesicles from human erythrocytes during ATP depletion. I. Characterization of spectrin-free vesicles. J. Cell Biol. 73, 548–560.

MacLeod, A.S., Hemmers, S., Garijo, O., Chabod, M., Mowen, K., Witherden, D.A., Havran, W.L., 2013. Dendritic epidermal T cells regulate skin antimicrobial barrier function. J. Clin. Invest. 123, 4364–4374.

Maia, C., Seblova, V., Sadlova, J., Votypka, J., Volf, P., 2011. Experimental transmission of *Leishmania infantum* by two major vectors: a comparison between a viscerotropic and a dermotropic strain. PLoS Negl. Trop. Dis. 5, e1181.

Malaviya, R., Abraham, S.N., 2000. Role of mast cell leukotrienes in neutrophil recruitment and bacterial clearance in infectious peritonitis. J. Leukoc. Biol. 67, 841–846.

Malissen, B., Tamoutounour, S., Henri, S., 2014. The origins and functions of dendritic cells and macrophages in the skin. Nat. Rev. Immunol. 14, 417–428.

Marzouki, S., Kammoun-Rebai, W., Bettaieb, J., Abdeladhim, M., Hadj Kacem, S., Abdelkader, R., Gritli, S., Chemkhi, J., Aslan, H., Kamhawi, S., Ben Salah, A., Louzir, H., Valenzuela, J.G., Ben Ahmed, M., 2015. Validation of recombinant salivary protein PpSP32 as a suitable marker of human exposure to *Phlebotomus papatasi*, the vector of *Leishmania major* in Tunisia. PLoS Negl. Trop. Dis. 9, e0003991.

Matsue, H., Cruz Jr., P.D., Bergstresser, P.R., Takashima, A., 1993. Profiles of cytokine mRNA expressed by dendritic epidermal T cells in mice. J. Invest. Dermatol. 101, 537–542.

McLachlan, J.B., Hart, J.P., Pizzo, S.V., Shelburne, C.P., Staats, H.F., Gunn, M.D., Abraham, S.N., 2003. Mast cell-derived tumor necrosis factor induces hypertrophy of draining lymph nodes during infection. Nat. Immunol. 4, 1199–1205.

Meephansan, J., Tsuda, H., Komine, M., Tominaga, S., Ohtsuki, M., 2012. Regulation of IL-33 expression by IFN-gamma and tumor necrosis factor-alpha in normal human epidermal keratinocytes. J. Invest. Dermatol. 132, 2593–2600.

Mocsai, A., 2013. Diverse novel functions of neutrophils in immunity, inflammation, and beyond. J. Exp. Med. 210, 1283–1299.

Monteiro, C.C., Villegas, L.E., Campolina, T.B., Pires, A.C., Miranda, J.C., Pimenta, P.F., Secundino, N.F., 2016. Bacterial diversity of the American sand fly *Lutzomyia intermedia* using high-throughput metagenomic sequencing. Parasit. Vectors 9, 480.

Moreno, J., 2007. Changing views on Langerhans cell functions in leishmaniasis. Trends Parasitol. 23, 86–88.

Murray, P.J., Wynn, T.A., 2011. Protective and pathogenic functions of macrophage subsets. Nat. Rev. Immunol. 11, 723–737.

Myskova, J., Dostalova, A., Penickova, L., Halada, P., Bates, P.A., Volf, P., 2016. Characterization of a midgut mucin-like glycoconjugate of *Lutzomyia longipalpis* with a potential role in *Leishmania* attachment. Parasit. Vectors 9, 413.

Naik, S., Bouladoux, N., Wilhelm, C., Molloy, M.J., Salcedo, R., Kastenmuller, W., Deming, C., Quinones, M., Koo, L., Conlan, S., Spencer, S., Hall, J.A., Dzutsev, A., Kong, H., Campbell, D.J., Trinchieri, G., Segre, J.A., Belkaid, Y., 2012. Compartmentalized control of skin immunity by resident commensals. Science 337, 1115–1119.

Naik, S., Bouladoux, N., Linehan, J.L., Han, S.J., Harrison, O.J., Wilhelm, C., Conlan, S., Himmelfarb, S., Byrd, A.L., Deming, C., Quinones, M., Brenchley, J.M., Kong, H.H., Tussiwand, R., Murphy, K.M., Merad, M., Segre, J.A., Belkaid, Y., 2015. Commensal-dendritic-cell interaction specifies a unique protective skin immune signature. Nature 520, 104–108.

Ng, M.F., 2010. The role of mast cells in wound healing. Int. Wound J. 7, 55–61.

Oliveira, F., de Carvalho, A.M., de Oliveira, C.I., 2013a. Sand-fly saliva-*Leishmania*-man: the trigger trio. Front. Immunol. 4, 375.

Oliveira, F., Traore, B., Gomes, R., Faye, O., Gilmore, D.C., Keita, S., Traore, P., Teixeira, C., Coulibaly, C.A., Samake, S., Meneses, C., Sissoko, I., Fairhurst, R.M., Fay, M.P., Anderson, J.M., Doumbia, S., Kamhawi, S., Valenzuela, J.G., 2013b. Delayed-type hypersensitivity to sand fly saliva in humans from a leishmaniasis-endemic area of Mali is Th1-mediated and persists to midlife. J. Invest. Dermatol. 133, 452–459.

Oliveira, F., Rowton, E., Aslan, H., Gomes, R., Castrovinci, P.A., Alvarenga, P.H., Abdeladhim, M., Teixeira, C., Meneses, C., Kleeman, L.T., Guimaraes-Costa, A.B., Rowland, T.E., Gilmore, D., Doumbia, S., Reed, S.G., Lawyer, P.G., Andersen, J.F., Kamhawi, S., Valenzuela, J.G., 2015. A sand fly salivary protein vaccine shows efficacy against vector-transmitted cutaneous leishmaniasis in nonhuman primates. Sci. Transl. Med. 7, 290ra90.

Peters, N.C., Sacks, D.L., 2009. The impact of vector-mediated neutrophil recruitment on cutaneous leishmaniasis. Cell. Microbiol. 11, 1290–1296.

Peters, N.C., Egen, J.G., Secundino, N., Debrabant, A., Kimblin, N., Kamhawi, S., Lawyer, P., Fay, M.P., Germain, R.N., Sacks, D., 2008. In vivo imaging reveals an essential role for neutrophils in leishmaniasis transmitted by sand flies. Science 321, 970–974.

Peters, N.C., Kimblin, N., Secundino, N., Kamhawi, S., Lawyer, P., Sacks, D.L., 2009. Vector transmission of *Leishmania* abrogates vaccine-induced protective immunity. PLoS Pathog. 5, e1000484.

Peters, N.C., Bertholet, S., Lawyer, P.G., Charmoy, M., Romano, A., Ribeiro-Gomes, F.L., Stamper, L.W., Sacks, D.L., 2012. Evaluation of recombinant *Leishmania* polyprotein plus glucopyranosyl lipid A stable emulsion vaccines against sand fly-transmitted *Leishmania major* in C57BL/6 mice. J. Immunol. 189, 4832–4841.

Pflughoeft, K.J., Versalovic, J., 2012. Human microbiome in health and disease. Annu. Rev. Pathol. 7, 99–122.

Pingen, M., Bryden, S.R., Pondeville, E., Schnettler, E., Kohl, A., Merits, A., Fazakerley, J.K., Graham, G.J., McKimmie, C.S., 2016. Host inflammatory response to mosquito bites enhances the severity of arbovirus infection. Immunity 44, 1455–1469.

Prates, D.B., Araujo-Santos, T., Brodskyn, C., Barral-Netto, M., Barral, A., Borges, V.M., 2012. New insights on the inflammatory role of Lutzomyia longipalpis saliva in leishmaniasis. J. Parasitol. Res. 2012, 643029.

Reed, S.G., Coler, R.N., Mondal, D., Kamhawi, S., Valenzuela, J.G., 2016. Leishmania vaccine development: exploiting the host-vector-parasite interface. Expert Rev. Vaccines 15, 81–90.

Ribeiro, J.M., Francischetti, I.M., 2003. Role of arthropod saliva in blood feeding: sialome and post-sialome perspectives. Annu. Rev. Entomol. 48, 73–88.

Ribeiro-Gomes, F.L., Sacks, D., 2012. The influence of early neutrophil-Leishmania interactions on the host immune response to infection. Front. Cell. Infect. Microbiol. 2, 59.

Rogers, M.E., 2012. The role of Leishmania proteophosphoglycans in sand fly transmission and infection of the Mammalian host. Front Microbiol. 3, 223.

Rogers, M.E., Bates, P.A., 2007. Leishmania manipulation of sand fly feeding behavior results in enhanced transmission. PLoS Pathog. 3, e91.

Rogers, M.E., Chance, M.L., Bates, P.A., 2002. The role of promastigote secretory gel in the origin and transmission of the infective stage of Leishmania mexicana by the sandfly Lutzomyia longipalpis. Parasitology 124, 495–507.

Rogers, M.E., Ilg, T., Nikolaev, A.V., Ferguson, M.A., Bates, P.A., 2004. Transmission of cutaneous leishmaniasis by sand flies is enhanced by regurgitation of fPPG. Nature 430, 463–467.

Rogers, M.E., Sizova, O.V., Ferguson, M.A., Nikolaev, A.V., Bates, P.A., 2006. Synthetic glycovaccine protects against the bite of Leishmania-infected sand flies. J. Infect Dis. 194, 512–518.

Rogers, M., Kropf, P., Choi, B.S., Dillon, R., Podinovskaia, M., Bates, P., Muller, I., 2009. Proteophosophoglycans regurgitated by Leishmania-infected sand flies target the L-arginine metabolism of host macrophages to promote parasite survival. PLoS Pathog. 5, e1000555.

Rogers, M.E., Corware, K., Muller, I., Bates, P.A., 2010. Leishmania infantum proteophosphoglycans regurgitated by the bite of its natural sand fly vector, Lutzomyia longipalpis, promote parasite establishment in mouse skin and skin-distant tissues. Microbes Infect. 12, 875–879.

Sacks, D., Kamhawi, S., 2001. Molecular aspects of parasite-vector and vector-host interactions in leishmaniasis. Annu. Rev. Microbiol. 55, 453–483.

Sacks, D.L., Perkins, P.V., 1984. Identification of an infective stage of Leishmania promastigotes. Science 223, 1417–1419.

Sadlova, J., Price, H.P., Smith, B.A., Votypka, J., Volf, P., Smith, D.F., 2010. The stage-regulated HASPB and SHERP proteins are essential for differentiation of the protozoan parasite Leishmania major in its sand fly vector, Phlebotomus papatasi. Cell. Microbiol. 12, 1765–1779.

Saljoughian, N., Taheri, T., Rafati, S., 2014. Live vaccination tactics: possible approaches for controlling visceral leishmaniasis. Front. Immunol. 5, 134.

Sant'Anna, M.R., Darby, A.C., Brazil, R.P., Montoya-Lerma, J., Dillon, V.M., Bates, P.A., Dillon, R.J., 2012. Investigation of the bacterial communities associated with females of Lutzomyia sand fly species from South America. PLoS One 7, e42531.

Secundino, N.F., Eger-Mangrich, I., Braga, E.M., Santoro, M.M., Pimenta, P.F., 2005. Lutzomyia longipalpis peritrophic matrix: formation, structure, and chemical composition. J. Med. Entomol. 42, 928–938.

Secundino, N.F., de Freitas, V.C., Monteiro, C.C., Pires, A.C., David, B.A., Pimenta, P.F., 2012. The transmission of Leishmania infantum chagasi by the bite of the Lutzomyia longipalpis to two different vertebrates. Parasit. Vectors 5, 20.

Serafim, T.D., Figueiredo, A.B., Costa, P.A., Marques-da-Silva, E.A., Goncalves, R., de Moura, S.A., Gontijo, N.F., da Silva, S.M., Michalick, M.S., Meyer-Fernandes, J.R., de Carvalho, R.P., Uliana, S.R., Fietto, J.L., Afonso, L.C., 2012. Leishmania metacyclogenesis is promoted in the absence of purines. PLoS Negl. Trop. Dis. 6, e1833.

Serbina, N.V., Pamer, E.G., 2006. Monocyte emigration from bone marrow during bacterial infection requires signals mediated by chemokine receptor CCR2. Nat. Immunol. 7, 311–317.

Sharifi, I., FeKri, A.R., Aflatonian, M.R., Khamesipour, A., Nadim, A., Mousavi, M.R., Momeni, A.Z., Dowlati, Y., Godal, T., Zicker, F., Smith, P.G., Modabber, F., 1998. Randomised vaccine trial of single dose of killed Leishmania major plus BCG against anthroponotic cutaneous leishmaniasis in Bam, Iran. Lancet 351, 1540–1543.

Shaw, T.J., Martin, P., 2009. Wound repair at a glance. J. Cell Sci. 122, 3209–3213.

Shi, C., Pamer, E.G., 2011. Monocyte recruitment during infection and inflammation. Nat. Rev. Immunol. 11, 762–774.

Shortt, H.E., Smith, R., Swaminath, C., Krishnan, K., 1931. Transmission of Indian kala-azar by the bite of Phlebotomus argentipes. Indian J. Med. Res. 18, 1373–1375.

Silva, R., Sacks, D.L., 1987. Metacyclogenesis is a major determinant of Leishmania promastigote virulence and attenuation. Infect. Immun. 55, 2802–2806.

Silverman, J.M., Clos, J., de'Oliveira, C.C., Shirvani, O., Fang, Y., Wang, C., Foster, L.J., Reiner, N.E., 2010a. An exosome-based secretion pathway is responsible for protein export from Leishmania and communication with macrophages. J. Cell Sci. 123, 842–852.

Silverman, J.M., Clos, J., Horakova, E., Wang, A.Y., Wiesgigl, M., Kelly, I., Lynn, M.A., McMaster, W.R., Foster, L.J., Levings, M.K., Reiner, N.E., 2010b. *Leishmania* exosomes modulate innate and adaptive immune responses through effects on monocytes and dendritic cells. J. Immunol. 185, 5011–5022.

Silverman, J.M., Reiner, N.E., 2011. Exosomes and other microvesicles in infection biology: organelles with unanticipated phenotypes. Cell. Microbiol. 13, 1–9.

Smallegange, R.C., van Gemert, G.J., van de Vegte-Bolmer, M., Gezan, S., Takken, W., Sauerwein, R.W., Logan, J.G., 2013. Malaria infected mosquitoes express enhanced attraction to human odor. PLoS One 8, e63602.

Soumana, I.H., Simo, G., Njiokou, F., Tchicaya, B., Abd-Alla, A.M., Cuny, G., Geiger, A., 2013. The bacterial flora of tsetse fly midgut and its effect on trypanosome transmission. J. Invertebr. Pathol. (112 Suppl), S89–S93.

Souza, A.P., Andrade, B.B., Aquino, D., Entringer, P., Miranda, J.C., Alcantara, R., Ruiz, D., Soto, M., Teixeira, C.R., Valenzuela, J.G., de Oliveira, C.I., Brodskyn, C.I., Barral-Netto, M., Barral, A., 2010. Using recombinant proteins from *Lutzomyia longipalpis* saliva to estimate human vector exposure in visceral Leishmaniasis endemic areas. PLoS Negl. Trop. Dis. 4, e649.

Stierhof, Y.D., Ilg, T., Russell, D.G., Hohenberg, H., Overath, P., 1994. Characterization of polymer release from the flagellar pocket of *Leishmania mexicana* promastigotes. J. Cell Biol. 125, 321–331.

Taylor, D.D., Taylor, C.G., Jiang, C.G., Black, P.H., 1988. Characterization of plasma membrane shedding from murine melanoma cells. Int. J. Cancer 41, 629–635.

Teixeira, C.R., Teixeira, M.J., Gomes, R.B., Santos, C.S., Andrade, B.B., Raffaele-Netto, I., Silva, J.S., Guglielmotti, A., Miranda, J.C., Barral, A., Brodskyn, C., Barral-Netto, M., 2005. Saliva from *Lutzomyia longipalpis* induces CC chemokine ligand 2/monocyte chemoattractant protein-1 expression and macrophage recruitment. J. Immunol. 175, 8346–8353.

Titus, R.G., Ribeiro, J.M., 1988. Salivary gland lysates from the sand fly *Lutzomyia longipalpis* enhance *Leishmania* infectivity. Science 239, 1306–1308.

Urmacher, C., 1990. Histology of normal skin. Am. J. Surg. Pathol. 14, 671–686.

Valenzuela, J.G., Belkaid, Y., Garfield, M.K., Mendez, S., Kamhawi, S., Rowton, E.D., Sacks, D.L., Ribeiro, J.M., 2001. Toward a defined anti-*Leishmania* vaccine targeting vector antigens: characterization of a protective salivary protein. J. Exp. Med. 194, 331–342.

Valladeau, J., Ravel, O., Dezutter-Dambuyant, C., Moore, K., Kleijmeer, M., Liu, Y., Duvert-Frances, V., Vincent, C., Schmitt, D., Davoust, J., Caux, C., Lebecque, S., Saeland, S., 2000. Langerin, a novel C-type lectin specific to Langerhans cells, is an endocytic receptor that induces the formation of Birbeck granules. Immunity 12, 71–81.

Van Den Abbeele, J., Caljon, G., de Ridder, K., de Baetselier, P., Coosemans, M., 2010. Trypanosoma brucei modifies the tsetse salivary composition, altering the fly feeding behavior that favors parasite transmission. PLoS Pathog. 6, e1000926.

van Zandbergen, G., Hermann, N., Laufs, H., Solbach, W., Laskay, T., 2002. *Leishmania* promastigotes release a granulocyte chemotactic factor and induce interleukin-8 release but inhibit gamma interferon-inducible protein 10 production by neutrophil granulocytes. Infect. Immun. 70, 4177–4184.

van Zandbergen, G., Klinger, M., Mueller, A., Dannenberg, S., Gebert, A., Solbach, W., Laskay, T., 2004. Cutting edge: neutrophil granulocyte serves as a vector for *Leishmania* entry into macrophages. J. Immunol. 173, 6521–6525.

Volf, P., Myskova, J., 2007. Sand flies and *Leishmania*: specific versus permissive vectors. Trends Parasitol. 23, 91–92.

Volf, P., Killick-Kendrick, R., Bates, P.A., Molyneux, D.H., 1994. Comparison of the haemagglutination activities in gut and head extracts of various species and geographical populations of phlebotomine sandflies. Ann. Trop. Med. Parasitol. 88, 337–340.

Walters, L.L., Chaplin, G.L., Modi, G.B., Tesh, R.B., 1989a. Ultrastructural biology of *Leishmania* (*Viannia*) panamensis (=*Leishmania braziliensis panamensis*) in *Lutzomyia gomezi* (Diptera: Psychodidae): a natural host-parasite association. Am. J. Trop. Med. Hyg. 40, 19–39.

Walters, L.L., Modi, G.B., Chaplin, G.L., Tesh, R.B., 1989b. Ultrastructural development of *Leishmania chagasi* in its vector, *Lutzomyia longipalpis* (Diptera: Psychodidae). Am. J. Trop. Med. Hyg. 41, 295–317.

Warburg, A., Schlein, Y., 1986. The effect of post-bloodmeal nutrition of *Phlebotomus papatasi* on the transmission of *Leishmania major*. Am. J. Trop. Med. Hyg. 35, 926–930.

Weiss, B., Aksoy, S., 2011. Microbiome influences on insect host vector competence. Trends Parasitol. 27, 514–522.

Werner, S., Krieg, T., Smola, H., 2007. Keratinocyte-fibroblast interactions in wound healing. J. Invest. Dermatol. 127, 998–1008.

Wernersson, S., Pejler, G., 2014. Mast cell secretory granules: armed for battle. Nat. Rev. Immunol. 14, 478–494.

WHO, 2016. Leishmaniasis fact sheet, 6 pp. http://www.who.int/mediacentre/factsheets/fs375/en/.

Wilson, N.S., Villadangos, J.A., 2004. Lymphoid organ dendritic cells: beyond the Langerhans cells paradigm. Immunol. Cell Biol. 82, 91–98.

Wynn, T.A., Vannella, K.M., 2016. Macrophages in tissue repair, regeneration, and fibrosis. Immunity 44, 450–462.

Zahedifard, F., Gholami, E., Taheri, T., Taslimi, Y., Doustdari, F., Seyed, N., Torkashvand, F., Meneses, C., Papadopoulou, B., Kamhawi, S., Valenzuela, J.G., Rafati, S., 2014. Enhanced protective efficacy of nonpathogenic recombinant *Leishmania tarentolae* expressing cysteine proteinases combined with a sand fly salivary antigen. PLoS Negl. Trop. Dis. 8, e2751.

Zakai, H.A., Chance, M.L., Bates, P.A., 1998. In vitro stimulation of metacyclogenesis in *Leishmania braziliensis*, *L. donovani*, *L. major* and *L. mexicana*. Parasitology 116, 305–309.

Early Immunological Responses Upon Tsetse Fly–Mediated Trypanosome Inoculation

Guy Caljon[1], Benoît Stijlemans[2,3], Carl De Trez[2,4], Jan Van Den Abbeele[5]

[1]University of Antwerp, Antwerp, Belgium; [2]Vrije Universiteit Brussel (VUB), Brussels, Belgium; [3]VIB Inflammation Research Center, Ghent, Belgium; [4]VIB Structural Biology Research Center (SBRC), Brussels, Belgium; [5]Institute of Tropical Medicine Antwerp (ITM), Antwerp, Belgium

List of Abbreviations

Ad Adipocyte
ADA Adenosine deaminase
Adc Adenylate cyclase
ADP ADP hydrolase
ATP Adenosine-5′-triphosphate
BSF Bloodstream form
B-VSG Bloodstream variant–specific surface glycoprotein
C3–9 Complement factors 3–9
Ca Cartilage
Cb Collagen bundle
CD4, CD8, CD39 Cluster of differentiation 4, 8, 39
Ct Connective tissue
CTL C-type lectin
ECM Extracellular matrix
Ep Epidermis
GIP Glycosylinositolphosphate
GPI Glycosylphosphatidylinositol
HAT Human African trypanosomiasis
IFN-γ Interferon-gamma
IgM, IgG, IgE Immunoglobulin M, G, E
IL-1, -6, -10, -12 Interleukin-1, -6, -10, −12
iNOS Inducible nitric oxide synthase
KHC Kinesin heavy chain

LS Long slender
MCF Metacyclic form
M-VSG Metacyclic variant-specific surface glycoprotein
NO Nitric oxide
p Posterior trypanosome end
PAMP Pathogen-associated molecular pattern
PLC Phospholipase C
PRR Pattern recognition receptor
RBP RNA-binding protein
Rf Reticular fibers
Sc *Stratum corneum*
SIF Stumpy inducing factor
SR-A Type A scavenger receptor
SS Short stumpy
sVSG Soluble VSG
TAg5 Tsetse Antigen 5
TCA Tricarboxylic acid
Th2 T helper cell type 2
TLR Toll-like receptor
TNF Tumor necrosis factor
Tsal Tsetse salivary gland protein
TSIF Trypanosome suppression immunomodulating factor
TTI Tsetse thrombin inhibitor
VSG Variant-specific surface glycoprotein

INTRODUCTION: THE TSETSE FLY-TRYPANOSOME–HOST INTERPHASE

Human African trypanosomiasis (HAT), also known as sleeping sickness, is indigenous for the African continent and is caused by two subspecies of *Trypanosoma brucei*, namely *T. brucei rhodesiense* and *T. brucei gambiense*. Tsetse flies (*Glossina* sp.) are the sole vectors of these human pathogenic parasites that rely either on an anthroponotic (*T. brucei gambiense*) or on a zoonotic transmission cycle (*T. brucei rhodesiense*) (Malvy and Chappuis, 2011). A range of other trypanosome species including *T. congolense*, *T. vivax*, and *T. brucei brucei*, is responsible for the majority of infections in livestock and wild animals and inflict a strong socioeconomic impact. Also here, tsetse flies play an important role in transmission although mechanical carry-over by other hematophagous insects (e.g., Tabanidae and *Stomoxys* sp.), especially of *T. vivax*, is prominent (Cherenet et al., 2004; Desquesnes and Dia, 2003; Sinshaw et al., 2006). More than 30 tsetse species and subspecies exist that have preferences for specific biotopes displaying specific host feeding preferences and differential roles in parasite transmission (Caljon et al., 2014; Leak, 1999). All tsetse flies are obligate blood feeding insects, implying that both male and female flies require a blood meal every 3–4 days during their 3- to 4-month life span (Caljon et al., 2014; Leak, 1999). They rely on a pool feeding (telmophagous) strategy, which involves the laceration of the skin with their proboscis and blood ingestion from a superficial lesion (Lehane, 2005). This feeding event provides an opportunity for parasite transmission to the mammalian host or uptake by the tsetse vector.

Mammalian infections with all these medically and veterinary important trypanosomes occur readily during the probing phase of the tsetse fly blood-feeding process. Hereby, the skin of the vertebrate host is a first anatomical barrier that is breached by the tsetse mouthparts, introducing the parasite and tsetse saliva factors in a microenvironment governed by several physiological and immunological processes directed at preventing pathogen development (Lehane, 2005; Van Den Abbeele et al., 2007). However, the pathogen in concert with vector-derived salivary factors transform the skin barrier into an immune-tolerant organ supporting parasite development (Bernard et al., 2014, 2015; Frischknecht, 2007). For host colonization and maintaining a continuous infection, trypanosomes have evolved various mechanisms to escape elimination by the innate and adaptive immune system and to limit their own proliferation. In the bloodstream, proliferative (long slender, LS) and quiescent (short stumpy, SS) blood stream forms (BSF) of *T. brucei* that vary in cell architecture can be recognized (MacGregor et al., 2012). Differentiation into this quiescent form depends on a quorum sensing mechanism that involves a stumpy inducing factor (Mony et al., 2014; Mony and Matthews, 2015). Upon differentiation into SS morphotypes, parasites exhibit mitochondrial biogenesis and are arrested in the G1/G0 stage (Vassella et al., 1997), preadapting the parasite for transition into procyclic forms if taken up by a tsetse fly (Rico et al., 2013).

Upon ingestion, parasites shed their surface coat and start expressing a new set of surface glycoproteins (procyclins). The shed glycoprotein coat was reported to indirectly reduce the integrity of the peritrophic matrix contributing to successful establishment in the midgut (Aksoy et al., 2016). *T. brucei* parasites first have to overcome this midgut barrier and go through a complex developmental cycle of about 3 weeks to finally reach the tsetse salivary glands and to differentiate into the metacyclic forms (MCF) that can infect a new mammalian host (Rotureau and Van Den Abbeele, 2013; Vickerman, 1985). Life cycles of *T. congolense* and *T. vivax* in tsetse flies are less complex with fewer successive parasite stages and with no establishment in the salivary glands and therefore a shorter extrinsic incubation period of approximately 2 weeks and 1 week, respectively (Rotureau and Van Den Abbeele, 2013). Although the etiology for the various trypanosome species is different, it appears that tsetse-transmitted African trypanosomes in general have evolved to modulate the fly-feeding behavior in favor of enhanced contact with the host and higher chance of parasite transmission (Caljon et al., 2016a). *T. brucei* infection in the salivary gland reduces the expression of the anticoagulant and antiplatelet salivary repertoire, thereby extending the probing time required to locate a suitable blood vessel (Van Den Abbeele et al., 2010). *T. congolense* and *T. vivax* were shown to interact with sensory structures (sensilla) in the mouthparts (proboscis, cibarium) that play a role in the detection of an incoming blood flux (Moloo and Dar, 1985; Molyneux et al., 1979; Vickerman, 1973). The presence of parasites is also thought to reduce the diameter of the proboscis reducing the blood flux (Livesey et al., 1980). These changes in the tsetse vector negatively affect the blood feeding performance, enhance the likelihood of interrupted feeding, and could enhance the vector–host contact intensity and frequency in favor of enhanced parasite transmission.

THE METACYCLIC TRYPANOSOME STAGES: CHARACTERISTICS AND INFECTIVITY

African trypanosomes, amongst which *T. brucei* is most intensively studied, rely on indispensable life cycle changes to adapt to the highly different growth conditions and nutrient availability encountered in the various microenvironments of the tsetse fly vector and the mammalian host (Bringaud et al., 2006; Fenn and Matthews, 2007; Li, 2012; Rodrigues et al., 2014). These include fine-tuning of energy metabolism, organelle reorganization, and biochemical and structural remodeling, which is supported by major changes in gene expression and proliferation status to adapt/survive in the different hosts

(MacGregor et al., 2012). Changes upon transfer to a new host have to occur rapidly and are often regulated at the posttranscriptional level by storing stabilized silent mRNAs in cytoplasmic ribonucleoprotein granules, by extensive regulation by RNA-binding proteins (RBPs) and by autophagy-based protein/organelle-turnover during cellular remodeling (Cassola et al., 2007; Herman et al., 2008; Kolev et al., 2013; Mair et al., 2010; Subota et al., 2011). RBP6 is a master regulator of the *T. brucei* maturation process, triggering procyclic trypanosomes in the tsetse fly to enter into a developmental process that ends in the generation of metacyclic trypanosomes as an essential infective parasite stage (Kolev et al., 2013). Metacyclic trypanosomes arising in the mouthparts or salivary glands are nonproliferative and have a short flagellum and a mitochondrion that loses its cristae, which is accompanied by a reduced tricarboxylic acid cycle enzymatic activity required for energy production (Matthews, 2005; Tetley and Vickerman, 1985). Transcriptomic comparisons between MCF trypanosomes in the tsetse salivary gland and BSF trypanosomes indicated some transcriptional differences between these two different life cycle stages. Although the importance of these transcriptional differences remain to be further explored, several metabolic genes (glycolysis, phosphorus metabolism), surface molecules, transporters (ion, glucose, amino acid, and nucleoside transporters), transcriptional regulators, and translation machinery components are modulated (Telleria et al., 2014). The metabolic features of metacyclic trypanosomes are anticipated to be compatible with preadaptation to survival in the mammalian host given the high availability of glucose in blood as energy source in the glycolytic pathway. It is also known that the expression of variant-specific surface glycoproteins (VSGs) plays a crucial role in the escape from elimination by the host adaptive immune response (see Parasite Escape From Early Immune Elimination section). VSGs are for the first time expressed at the metacyclic stage in the tsetse fly salivary glands (Graham et al., 1990; Rotureau et al., 2012).

Metacyclic variant-specific surface glycoprotein (M-VSG) was shown in *T. brucei* to be expressed by RNA polymerase 1 from one of the five canonical metacyclic genes (Cross et al., 2014; Ramey-Butler et al., 2015), resulting in a homogenous M-VSG expression on the majority of the individual cells but with a diverse antigenic repertoire in the inoculated metacyclic population in order to enhance the success of infection establishment in the host by escaping elimination by circulating host antibodies. Metacyclic *T. brucei* also expresses calflagins, a family of calcium-binding proteins present in lipid raft microdomains in the flagellar membrane that are putatively involved in signaling. These proteins are abundantly present on bloodstream trypanosomes but are absent in most premetacyclic parasite stages (Rotureau et al., 2012). RNAi-based silencing of calflagins indicates that they play a role in infection and resistance to early elimination by antibodies (Emmer et al., 2010).

Experimental mouse infections allowed the determination of the median infectious doses of about seven metacyclic trypanosomes revealing that these forms are very proficient in establishing infections in mammalian hosts (Caljon et al., 2016b). In humans, the subcutaneous infective dose for the metacyclic forms of a *T. brucei rhodesiense* strain is about 300–450 metacyclic parasites (Fairbairn and Burtt, 1946). With a lowest infective dose of <200 and the highest noninfective dose of >1000 parasites, it is clear that beside host differences there are significant interindividual variations (Fairbairn and Burtt, 1946). Upon inoculation in the skin microenvironment, the metacyclic parasites transform into LS blood-stage trypomastigotes, which divide by binary fission in the interstitial spaces at the bite site. Following inoculation, a parasite subpopulation transiently remains and proliferates in proximity of the initial inoculation site following a tsetse-mediated infection. Studies in mice indicate that parasites rapidly progress by gaining access to the lymph, with detectable levels in the draining lymph nodes within 18 h

postinfection (Caljon et al., 2016b). Cannulation of the lymphatics in goats confirmed that parasites can be detected in the lymph within 1–2 days after an infective tsetse fly bite (Barry and Emergy, 1984). In humans, Winterbottom's sign is a marked early feature of sleeping sickness caused by posterior cervical lymphadenopathy (Malvy and Chappuis, 2011). Due to this particular characteristic, microscopic analysis of lymph node aspirates is often incorporated in the parasitological diagnosis of sleeping sickness (Büscher, 2014). Trypanosomes further migrate via the lymphatics into the bloodstream where they are described to switch from the typical M-VSG expression to the expression of a set of bloodstream VSG (B-VSG) genes that allow parasites to continuously escape elimination by host antibodies through a mechanism of antigenic variation (Barry and Emergy, 1984; Barry et al., 1998).

THE TSETSE FLY VECTOR: IMPLICATIONS OF SALIVA AS A VEHICLE

The role of tsetse flies in the trypanosome transmission cycle clearly extends beyond a role as flying syringes. Besides hosting the final differentiation into metacyclic forms, tsetse flies inoculate parasites together with a complex mixture of salivary components at the infection initiation site. Tsetse saliva contains over 250 different proteins of which many are glycosylated, and some are known to interfere with vertebrate host hemostatic responses to enable successful blood feeding via the suppression of vasoconstriction, platelet aggregation, and coagulation (Alves-Silva et al., 2010; Caljon et al., 2010; Cappello et al., 1996, 1998; Li and Aksoy, 2000; Mant and Parker, 1981; Zhao et al., 2015). Information on the interplay between trypanosomes, the host hemostatic system, and the factors in saliva remains very scarce. It can be assumed that salivary protease inhibitors, e.g., involved in the inhibition of thrombin, could also affect enzymes involved in immunological processes such as elastase and cathepsin (Tsujimoto et al., 2012). Also nucleolytic factors could play roles in subverting NETosis (Chagas et al., 2014), a process whereby apoptotic neutrophils release chromatin as extracellular traps to limit the spread of pathogens (Kolaczkowska and Kubes, 2013). Salivary nucleotidases, apyrases, and adenosine deaminases could also play an immunomodulatory role by interfering with purinergic inflammatory responses (Gounaris and Selkirk, 2005). A number of molecules such as ATP, ADP, and adenosine that can be released upon tissue damage at the bite site are known to trigger purinergic receptors involved in regulation of hemostasis and/or inflammation, which could be subverted by the action of these salivary enzymes. ATP represents a danger signal by binding onto the $P2X_7$ receptor on macrophages, dendritic cells, and neutrophils contributing to the processing and secretion of mature interleukin (IL) 1β (Gombault et al., 2012; Karmakar et al., 2016). Salivary nucleotidases and apyrases present in tsetse fly saliva (Alves-Silva et al., 2010; Caljon et al., 2010; Van Den Abbeele et al., 2010, 2007) could potentially inhibit this IL-1β response. These enzymes also use ADP as a substrate, thereby degrading ADP as an important agonist of platelet aggregation (Caljon et al., 2010). Adenosine is known to enhance immunoglobulin E (IgE)–dependent mast cell degranulation. Here, salivary adenosine deaminases present in tsetse saliva (Li and Aksoy, 2000) could convert adenosine to inosine, a nucleoside that is >10-fold less potent in enhancing mast cell degranulation (Tilley et al., 2000) and that moreover displays antiinflammatory properties (Hasko et al., 2000). The saliva also contains a set of pattern recognition molecules such as C-type lectins, galectins, and ficolins (Alves-Silva et al., 2010) that might bind to the parasite surface and shield off pathogen-associated molecular patterns from the host immune system. An immunological implication

of the salivary components is highly relevant, as experimental infections with purified BSF *T. brucei* and *T. congolense* without salivary factors has revealed that the intradermal route of infection is very stringent and can compromise the ability of low/natural numbers of parasites to achieve host colonization (Wei et al., 2011). Using gene deficient mice, tumor necrosis factor α (TNF-α) and inducible nitric oxide synthase (iNOS) were shown to contribute to the innate resistance to intradermal infection with low numbers of BSF *T. congolense* parasites (Wei et al., 2011). Addition of saliva to the inoculation mixture was documented to enhance parasitemia onset following intrapinna BSF *T. brucei* injection, which was linked to a local immunosuppressive effect of saliva in the murine dermis and reducing IL-6 and IL-12 inflammatory gene transcription (Caljon et al., 2006b). A recently identified immunoregulatory peptide in tsetse fly saliva, namely Gloss2, was shown to inhibit MAPK signaling and the secretion of the trypanolytic TNF and the proinflammatory cytokines IFN-γ and IL-6, which could assist parasites to avoid initial elimination (Bai et al., 2015). So far, no nonprotein immunomodulatory components have been identified in tsetse fly saliva. Nevertheless, the salivary gland transcriptome revealed the presence of two transcripts encoding putative prostaglandin E2 synthases (Alves-Silva et al., 2010), but it remains to be shown that PGE2 with inhibitory functions on DC maturation is present in the saliva as described for ticks (Sa-Nunes et al., 2007).

Not only the immunomodulatory potential, but also the intrinsic immunogenicity of salivary factors could play a role in trypanosome transmission. A number of salivary antigens have a strong propensity to induce Th2 cellular responses and to be highly immunogenic/immunodominant in humans and various other animal species, with strong reactivity of antisaliva IgGs against the highly abundant 43–45 kDa Tsal protein family of nucleic acid binding proteins (Caljon et al., 2012, 2006a; Poinsignon et al.,

2007). As observed in experimental coimmunization studies with a heterologous antigen, the immunodominant nature of some salivary constituents could skew the immune response away from other saliva or trypanosomal antigens that could potentially be harmful for the vector or the parasite (Caljon et al., 2006a). Beside IgG responses, also IgEs could be detected, mainly against Tsetse Antigen5 (TAg5) that was found to be a potent allergen with high similarity to allergens present in vespid toxin. TAg5 can trigger hypersensitivity reactions and can also induce anaphylactic reactions (Caljon et al., 2009). It can be assumed that dermal immediate type hypersensitivity reactions could modulate parasite extravasation into the blood circulation of sensitized hosts (Caljon et al., 2009). Also, strong delayed type hypersensitivity responses have been described against salivary components (Ellis et al., 1986), with a potential interplay with the parasite. Experimental immunization of mice against tsetse saliva by repeated bite exposure or by immunization of saliva in adjuvant indeed seemed to have the adverse effect of enhancing parasite infection onset following an infective tsetse bite (Caljon et al., 2006b). It can be assumed that functional genomics studies with the aid of the sequenced tsetse fly genome will result in the identification of more salivary factors that function as immunomodulators or could serve as potential transmission-blocking vaccine antigens in the future (IGGI International *Glossina* Genome Initiative, 2014; Telleria et al., 2014).

HISTOLOGICAL AND ULTRASTRUCTURAL CHANGES IN SKIN FOLLOWING AN INFECTIVE TSETSE FLY BITE

In humans, bites from infected tsetse flies regularly result in the formation of a skin ulceration or chancre (Cochran and Rosen, 1983). This skin ulceration has also been documented during early *T. brucei rhodesiense* infections in vervet monkeys (Thuita et al., 2008) and following *T. congolense* infections in rabbits, goats, calves, and sheep (Akol and Murray, 1982; Dwinger et al., 1987, 1988; Mwangi et al., 1990). The kinetics and size of the chancre correlates with the number of metacyclic parasites that are inoculated into the skin and is due to a local immune response directed against the M-VSG coat (Barry and Emergy, 1984). In goats, the inoculation of a single metacyclic *T. brucei* parasite was reported to be sufficient for the typical ulceration (Dwinger et al., 1987). Also anti-trypanosomal drug treatment was shown to alleviate the dermal response (Akol and Murray, 1985), suggesting that it is primarily parasite induced. Nevertheless, local hypersensitivity reactions induced by saliva might still be responsible for changes in the chancre formation and immunological cell composition. Indeed, dermal biopsies of naive rabbits exposed to noninfective bites reveals a mild edema, a slight distension of dermal vessels, and the accumulation of neutrophils and a few eosinophils. In contrast, hypersensitivity reactions in tsetse preexposed animals result in a much more pronounced edema and cellular infiltration of numerous eosinophils and mast cells (Ellis et al., 1986). Chancres in calves can reach a size of up to 100 mm coinciding with a rapid and heavy infiltration of neutrophils. About 24 h after infection initiation, lymphocytes, plasma cells, and macrophages become more predominant (Akol and Murray, 1982; Roberts et al., 1969). In sheep, elevated levels of B cells and T cells, with a high CD4/CD8 ratio, were observed (Mwangi et al., 1990). CD4[+] T lymphocytes were shown to play a key role in the chancre formation, since in vivo depletion of CD4[+] T cells prior to exposure of calves to infective tsetse fly bites resulted in a significantly reduced skin ulcerative response. No impact of the reduced chancre formation on infectivity and time of first parasite appearance in the blood could be detected (Naessens et al., 2003). Although this needs to be further explored, this would suggest that the local ulcerative response does not strongly impact the infection onset.

Histological and electron microscopy studies have documented the presence of significant numbers of parasites in the local skin reactions induced by various trypanosome species (*T. brucei* and *T. congolense*) (Dwinger et al., 1988; Luckins and Gray, 1978; Mwangi et al., 1995; Wolbach and Binger, 1912). In mice, this locally expanding intradermal trypanosome population was shown to derive directly from the tsetse fly inoculum and not from a reinvasion of the primary inoculation site by parasites in the bloodstream (Caljon et al., 2016b). Transmission electron microscopy studies revealed some host-specific features of the intradermal trypanosome population. In goat and sheep chancres, high percentages (50%–75%) of degenerating trypanosomes were found during the chancre development (Dwinger et al., 1988; Mwangi et al., 1995). In contrast, in New Zealand White rabbits and mice, parasite degeneration was only observed later (Caljon et al., 2016b; Luckins and Gray, 1978). It can therefore be assumed that the degree of ulceration in the various hosts not only depends on the number of inoculated metacyclic trypanosomes but also on the degree of parasite degradation at the primary inoculation site. Interestingly, thermographic imaging revealed significant changes in the skin surface temperature correlating with the intradermal parasite burden (Caljon et al., 2016b). This elevated local temperature could represent a local trigger for tsetse probing and feeding given that tsetse flies can sense elevated local temperatures by thermosensors in the antennae and tarsi (Reinouts van Haga and Mitchell, 1975). As such, parasites that are present in the skin could potentially play a role as early reservoir for uptake by tsetse flies from animals with low to undetectable parasite concentrations in the peripheral blood.

Electron microscopic studies of the dermal trypanosome population revealed intricate interactions of the anterior end of the parasites with adipocytes (Fig. 7.1). Intriguingly, the entanglement left free the flagellar pocket, which is a primary site required for endocytosis (Caljon et al., 2016b). Collagen fibers surrounding the adipocytes (periadipocyte baskets) act synergistically in supporting the interaction of trypanosomes with these skin adipocytes. Dermal parasites in the interstitial spaces were found heavily embedded in collagen fibrous structures and bundles (Caljon et al., 2016b; Luckins and Gray, 1978; Mwangi et al., 1995; Wolbach and Binger, 1912). The dermal trypanosomes were found to be proliferative as observed by the presence of double flagella (Fig. 7.1). Also aberrant proliferative forms of *T. brucei* with multiple nuclei and several axonemes per flagellum have been described (Dwinger et al., 1988). It is probable that *T. brucei* can regulate its interactions with collagen in the extracellular matrix as it was described to secrete an active prolyl oligopeptidase that cleaves collagen (Bastos et al., 2010). The interactions with collagen and adipocytes could be responsible for the retention of parasites in the dermis with yet poorly understood biological implications. Several observations in in vitro *T. brucei* cultures support the beneficial role of collagen and adipocytes in parasite expansion (Balber, 1983; Hirumi et al., 1977). It is possible that these interactions create a physiologically or immunologically privileged site that is beneficial for the inoculated parasites allowing them to proliferate and establish a locally adapted subpopulation. These interactions might provide an advantage to the parasite in terms of metabolic requirements or exposure to cellular or humoral components of the immune system. It also remains to be further established how the dermal parasite population relates or differs from the life cycle stages in the bloodstream and those described to occur in adipose tissue (Trindade et al., 2016). Also the role of the dermal trypanosome population in chronic, low parasitemic models, its contribution as a reservoir for recrudescence, and its susceptibility to drug treatment remains to be further established.

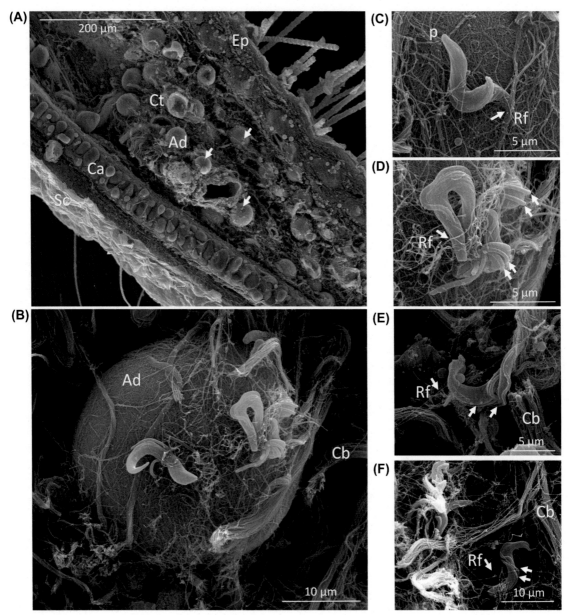

FIGURE 7.1 *Trypanosoma brucei* interaction with dermal adipocytes and collagen fibrous structures following an infective tsetse fly bite. Scanning electron microscopy (SEM) images of dermal ear sections illustrate the presence of parasites in the connective tissue of the ear dermis. Parasites are indicated with *white arrows* (A). A false-colored SEM image (43.5 × 37.6 μm) showing parasites (blue) in intricate interaction with an adipocyte (gray) (B). Intricate interactions of trypanosomes with adipocytes in the connective tissue were observed frequently (C–D). Parasites were found buried with the anterior side in the adipocyte and the posterior side with flagellar pocket exposed to the extracellular environment. Parasites were prominently entangled by reticular fibers (*white arrows*) of the periadipocytic baskets. Parasites in the interstitial spaces were embedded between collagen bundles (E–F). Despite these interactions, trypanosomes were proliferating as many cells had multiple flagella (D–F). *Ad*, adipocyte; *Ca*, cartilage; *Cb*, collagen bundle; *Ct*, connective tissue; *Ep*, epidermis; *p*, posterior trypanosome end; *Rf*, reticular fibers; *Sc*, stratum corneum. *Figure adapted from Caljon, G., Van Reet, N., De Trez, C., Vermeersch, M., Pérez-Morga, D., Van Den Abbeele, J., 2016b. The dermis as a delivery site of* Trypanosoma brucei *for tsetse flies. PLoS Pathog.*

PARASITE ESCAPE FROM EARLY IMMUNE ELIMINATION

Trypanosomes have evolved to successfully establish chronic infections and to escape elimination by the innate and adaptive immune system of a range of vertebrate hosts (Fig. 7.2). Immune recognition of *T. brucei* is mediated by pattern recognition receptors such as scavenger receptor A (Leppert et al., 2007). Indeed, the glycosylinositolphosphate (GIP) moieties in soluble VSG, that is cleaved from the parasite surface coat by an endogenous phospholipase C (Hereld et al., 1986) upon stress situations or released during antigenic variation, is recognized by a Type A scavenger receptor (SR-A) expressed on myeloid cells (macrophages and dendritic cells). This initiates a cascade of subcellular signaling, leading to activation of NF-κB and MAPK pathways and expression of the proinflammatory

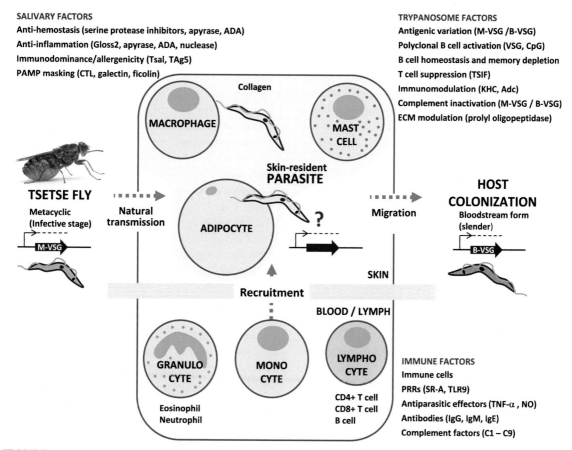

FIGURE 7.2 Model for the early interphase between the host and the tsetse fly transmitted trypanosome. Metacyclic parasites are inoculated by the tsetse fly together with salivary factors into the dermis. The skin is constituted by somatic and homeostatic present immune cells and a number of immune cells and factors that access the dermal site of infection. Salivary factors in concert with parasite factors transform the skin into an environment that is receptive for an expanding parasite population. This dermal population intricately interacts with adipocytes and collagen fibrous structures. From the dermis, the lymph and blood of the host becomes colonized, resulting in a systemic infection. *ADA*, adenosine deaminase; *Adc*, adenylate cyclase; *CTL*, C-type lectin; *ECM*, extracellular matrix; *KHC*, kinesin heavy chain; *M- or B-VSG*, metacyclic or bloodstream variant-specific surface glycoprotein; *PAMP*, pathogen-associated molecular pattern; *PRR*, pattern recognition receptor; *SR-A*, type A scavenger receptor; *TAg5*, tseste antigen five; *TLR*, Toll-like receptor; *Tsal*, tsetse salivary protein; *TSIF*, trypanosome suppression immunomodulating factor.

mediators TNF-α, IL-6, and IL-12p40 (Paulnock et al., 2010). The effect of GIP-sVSG is exacerbated to induce cellular hyperactivation in conditions where macrophages are primed with IFN-γ (Magez et al., 1998), with the produced TNF-α and nitric oxide (NO) playing important roles in early parasite control and reducing the parasite burden (Magez et al., 1997; Wei et al., 2011). *T. brucei* makes use of at least three different mechanisms to suppress those early detrimental inflammatory reactions. First, African trypanosomes exploit tsetse-derived salivary factors to thwart host innate responses and facilitate the host colonization process (Caljon et al., 2006b). Another mechanism relies on an altruistic strategy, where trypanosomes upon engulfment by mononuclear phagocytic cells activate members of a large family of transmembrane receptor-like adenylate cyclases. This activation results in the generation of cyclic adenosine monophosphate (Rolin et al., 1996; Salmon et al., 2012) inside the phagocyte, thereby activating protein kinase A and leading to a reduced synthesis of the trypanolytic cytokine TNF-α (Lucas et al., 1994; Magez et al., 1997). A third strategy relies on secreted factors that are able to modulate the parasite/host interplay. For instance, *T. brucei* was shown to secrete kinesin heavy chain 1 (TbKHC1) through noncanonical pathways (De Muylder et al., 2013). This protein was found to reduce the levels of iNOS activity in favor of elevated arginase activity in a SIGN-R1–/IL-10–dependent fashion. Exposure of mononuclear phagocytes to released VSG prior to an IFN-γ trigger, a situation that could occur at the early infection stage prior to the initial natural killer (NK)–/NKT cell–derived IFN-γ burst (Cnops et al., 2015b), was also shown to downregulate inducible NO synthase and secretion of NO, which was associated with a reduced level of STAT1 phosphorylation (Coller et al., 2003). As a result of the shifted iNOS/arginase balance, lower NO levels and higher availability of L-ornithine, as source for the synthesis of polyamines, contribute to a favorable host environment for parasite outgrowth and establishment of the first peak of parasitemia in the blood (De Muylder et al., 2013). It is likely that other factors secreted through noncanonical pathways play an essential role in virulence (Geiger et al., 2010; Langousis et al., 2016). Recently, it has been shown that *T. brucei rhodesiense* can even transfer virulence factors conferring human serum resistance to *T. brucei brucei* by membranous nanotubes that originate from the flagellar membrane and disassociate into free extracellular vesicles (Szempruch et al., 2016). The highly fusogenic character of these vesicles as well as their content—VSG, flagellar, mitochondrial, and glycosomal proteins and some less abundant constituents such as calflagin and adenylate cyclase—suggest that those may play a significant role in virulence.

Given the strict extracellular nature of African trypanosomes and the fact that trypanosomes induce T-cell independent B cell responses, parasites are very rapidly confronted with the host humoral response (i.e., antibodies). Remodeling of the parasite cell surface (Shimogawa et al., 2015) consisting of a uniform VSG coat is an essential aspect of the humoral escape mechanism. Parasites at the initial stage of infection are already covered by the M-VSGs, which is rapidly replaced by a set of B-VSGs. The *T. brucei* B-VSG coat consists of approximately 5×10^6 homodimers of 50–60 kDa subunits anchored in the plasma membrane by a glycosylphosphatidylinositol anchor (Mehlert et al., 2002). These B-VSG genes are organized in polycistronic transcription units at the telomeres where various mechanisms such as in situ gene activation and genetic rearrangements can yield a nearly inexhaustible repertoire of B-VSG antigenic variants that are structurally related and allow parasites to continuously escape elimination by host antibodies (Hall et al., 2013; Mugnier et al., 2015). Beside its role in antigenic variation, the VSG coat contributes to shielding off buried invariant proteins from recognition by the hosts' immune system and to protecting parasites from complement-mediated lysis (Ferrante and Allison, 1983; Stijlemans et al., 2004). Especially the alternative pathway of complement activation, which occurs in the absence of specific antibodies, may potentially play a crucial role in parasite clearance during the early stage of infection. However, *T. brucei gambiense* parasites were shown to inhibit, through their VSGs, the progression of complement lysis beyond the establishment of the C3 convertase, thereby impairing the generation of the terminal complex (C5–C9), which normally induces trypanolysis (Devine et al., 1986; Ferrante and Allison, 1983). Also the classical pathway, activated by immune complexes with antibodies (Abs), can be overcome by the parasite through the rapid VSG recycling system that removes surface-bound IgGs from the surface (Engstler et al., 2007). Of note, trypanosome infection results in a state of hypocomplementemia probably due to the scavenging of the complement factors by VSG (Balber et al., 1979; Musoke and Barbet, 1977; Rurangirwa et al., 1980), which further creates a less hostile host environment.

Trypanosomes have also developed various mechanisms to suppress or exhaust lymphocyte responses. Severe T-cell suppression has been described as a hallmark of a trypanosome infection, which occurs early after infection. Several immunosuppressive mechanisms are attributed to suppressive macrophages elicited during trypanosomiasis, a feature to which the *T. brucei*–derived trypanosome suppression immunomodulating factor, a parasite membrane protein, contributes (Gomez-Rodriguez et al., 2009). Macrophages, primarily induced during the early infection stage, produce NO and prostaglandins that are responsible for T-cell proliferation impairment (Schleifer and Mansfield, 1993). In addition, these macrophages were shown to have reduced abilities to activate specific T cells given their lower capacity to present antigenic peptides in MHC class II (Namangala et al., 2000). Suppression of T cell proliferative responses is associated with decreased interleukin-2 (IL-2) production and IL-2 receptor (IL-2R) expression (Sileghem et al., 1986, 1987). During the early stage of infection, involvement of IFN-γ and TNF in the inhibition of T cell proliferative responses seems limited to the upregulation of prostaglandins and NO synthesis (Darji et al., 1996). During the late stage of infection, inhibition of T cell proliferation in the lymph nodes occurs through a prostaglandin-/NO-independent pathway whereby IFN-γ, produced by T cells, plays a crucial role (Beschin et al., 1998; Mabbott et al., 1998).

In addition to T cell suppression, a significant degree of B cell dysfunction is observed during trypanosome infections (Assoku et al., 1977). This feature is characterized by a polyclonal B cell activation, resulting in the production of polyspecific and autoreactive antibodies, mainly of the IgM isotype (Buza and Naessens, 1999; Hudson et al., 1976; Radwanska et al., 2000). Hereby, CpG motifs in the trypanosomal genomic DNA trigger TLR9 signaling events and contribute to the polyclonal B cell activation (Drennan et al., 2005; Shoda et al., 2001). Although polyclonal B cell activation is a natural innate immune response induced by many pathogens, it might contribute to parasitic immune evasion by driving unselective differentiation of B cells into short-lived plasmablasts. Furthermore, Fcγ-receptors on phagocytes could become saturated by polyspecific antibodies, thereby reducing the efficiency of opsonization-mediated trypanosome clearance. Besides polyclonal activation, the trypanosome is able to avoid the build-up of a protective humoral response by ablation of B cell lymphopoiesis in primary and secondary lymphoid organs. Already after the first parasitemic peak of a virulent *T. brucei* and *T. congolense* infection, both the bone marrow and splenic B cell compartment are severely compromised (Bockstal et al., 2011; Cnops et al., 2015a; Obishakin et al., 2014; Radwanska et al., 2008). This impairment of different B cell subset homeostasis within the spleen, e.g., immature transitional as well as mature marginal zone and follicular B cells, is mainly due to massive cell death and is likely dependent on IFN-γ signaling (Bockstal et al., 2011; Radwanska et al., 2008). Also, virulent *T. brucei* infection results in the impairment of protective anti-parasite humoral responses and vaccine-induced memory responses against unrelated antigens via its impact on effector B cells, such as plasma cells (De Trez et al., 2015; Radwanska et al., 2008). The degree of follicular B cell ablation seems to relate to parasite virulence and the level of IFN-γ induction, with low parasite burdens following infection of mice with a low virulent *T. brucei gambiense* strain being associated with a normal follicular B cell composition of the spleen (Cnops et al., 2016). Also in *T. brucei gambiense* HAT patients, low parasite levels seem to be associated with limited B cell dysfunction given that historical vaccine-induced antibody levels seemed only moderately reduced (Lejon et al., 2014).

FUTURE DIRECTIONS

Although the annual number of reported HAT cases is steadily declining, history tells us that sleeping sickness can rapidly recrudesce if control efforts are discontinued. Nevertheless, the current trend of reducing numbers of sleeping sickness cases holds promise of moving into the eradication phase for this disease (Aksoy, 2011; Simarro et al., 2015), whereas for animal trypanosomiasis with many problems of reservoir management and drug resistance (Chitanga et al., 2011; Van den Bossche et al., 2011; von Wissmann et al., 2011), this goal is not yet in sight. Despite tremendous efforts in vaccination trials, to date not a single effective vaccine that can protect humans and animals from infection is available. Moreover, the biology of the African trypanosome and the various mechanisms it has developed to exploit and manipulate its vector and host to maintain a successful transmission cycle is fascinating but sheds a pessimistic view on the perspectives toward such protective vaccine. However, a crucial step in natural

trypanosome transmission is the survival of the infective metacyclic forms at the biting site micro-environment and transformation into a stage that is able to continue life in the host lymphatics and bloodstream. Despite the essential nature of this early dermal interphase between the parasite, vector and host, the possibilities of using this as a window of opportunity for intervention are not sufficiently explored. A thorough analysis of the early parasitological and immunological events that shape successful tsetse-mediated trypanosome transmission should provide a solid scientific basis for designing strategies aimed at blocking transmission.

References

Akol, G.W., Murray, M., 1982. Early events following challenge of cattle with tsetse infected with *Trypanosoma congolense*: development of the local skin reaction. Vet. Rec. 110, 295–302.

Akol, G.W., Murray, M., 1985. Induction of protective immunity in cattle by tsetse-transmitted cloned isolates of *Trypanosoma congolense*. Ann. Trop. Med. Parasitol. 79, 617–627.

Aksoy, S., 2011. Sleeping sickness elimination in sight: time to celebrate and reflect, but not relax. PLoS Negl. Trop. Dis. 5, e1008.

Aksoy, E., Vigneron, A., Bing, X., Zhao, X., O'Neill, M., Wu, Y.N., Bangs, J.D., Weiss, B.L., Aksoy, S., 2016. Mammalian African trypanosome VSG coat enhances tsetse's vector competence. Proc. Natl. Acad. Sci. U.S.A. 113, 6961–6966.

Alves-Silva, J., Ribeiro, J.M., Van Den Abbeele, J., Attardo, G., Hao, Z., Haines, L.R., Soares, M.B., Berriman, M., Aksoy, S., Lehane, M.J., 2010. An insight into the sialome of *Glossina morsitans morsitans*. BMC Genom. 11, 213.

Assoku, R.K., Tizard, I.R., Neilsen, K.H., 1977. Free fatty acids, complement activation, and polyclonal B-cell stimulation as factors in the immunopathogenesis of African trypanosomiasis. Lancet 2, 956–959.

Bai, X., Yao, H., Du, C., Chen, Y., Lai, R., Rong, M., 2015. An immunoregulatory peptide from tsetse fly salivary glands of *Glossina morsitans morsitans*. Biochimie 118, 123–128.

Balber, A.E., 1983. Primary murine bone marrow cultures support continuous growth of infectious human trypanosomes. Science 220, 421–423.

Balber, A.E., Bangs, J.D., Jones, S.M., Proia, R.L., 1979. Inactivation or elimination of potentially trypanolytic, complement-activating immune complexes by pathogenic trypanosomes. Infect. Immun. 24, 617–627.

Barry, J.D., Emergy, D.L., 1984. Parasite development and host responses during the establishment of *Trypanosoma brucei* infection transmitted by tsetse fly. Parasitology 88 (Pt. 1), 67–84.

Barry, J.D., Graham, S.V., Fotheringham, M., Graham, V.S., Kobryn, K., Wymer, B., 1998. VSG gene control and infectivity strategy of metacyclic stage *Trypanosoma brucei*. Mol. Biochem. Parasitol. 91, 93–105.

Bastos, I.M., Motta, F.N., Charneau, S., Santana, J.M., Dubost, L., Augustyns, K., Grellier, P., 2010. Prolyl oligopeptidase of *Trypanosoma brucei* hydrolyzes native collagen, peptide hormones and is active in the plasma of infected mice. Microbes Infect. Institut Pasteur 12, 457–466.

Bernard, Q., Jaulhac, B., Boulanger, N., 2014. Smuggling across the border: how arthropod-borne pathogens evade and exploit the host defense system of the skin. J. Investig. Dermatol. 134, 1211–1219.

Bernard, Q., Jaulhac, B., Boulanger, N., 2015. Skin and arthropods: an effective interaction used by pathogens in vector-borne diseases. Eur. J. Dermatol. 25 (Suppl. 1), 18–22.

Beschin, A., Brys, L., Magez, S., Radwanska, M., De Baetselier, P., 1998. *Trypanosoma brucei* infection elicits nitric oxide-dependent and nitric oxide-independent suppressive mechanisms. J. Leucocyte Biol. 63, 429–439.

Bockstal, V., Guirnalda, P., Caljon, G., Goenka, R., Telfer, J.C., Frenkel, D., Radwanska, M., Magez, S., Black, S.J., 2011. T. *brucei* infection reduces B lymphopoiesis in bone marrow and truncates compensatory splenic lymphopoiesis through transitional B-cell apoptosis. PLoS Pathog. 7, e1002089.

Bringaud, F., Riviere, L., Coustou, V., 2006. Energy metabolism of trypanosomatids: adaptation to available carbon sources. Mol. Biochem. Parasitol. 149, 1–9.

Büscher, P., 2014. Diagnosis of African trypanosomiasis. In: Magez, S., Radwanska, M. (Eds.), Trypanosomes and Trypanosomiasis. Springer-Verlag, Wien.

Buza, J., Naessens, J., 1999. Trypanosome non-specific IgM antibodies detected in serum of Trypanosoma congolense-infected cattle are polyreactive. Vet. Immunol. Immunopathol. 69, 1–9.

Caljon, G., Van Den Abbeele, J., Sternberg, J.M., Coosemans, M., De Baetselier, P., Magez, S., 2006a. Tsetse fly saliva biases the immune response to Th2 and induces anti-vector antibodies that are a useful tool for exposure assessment. Int. J. Parasitol. 36, 1025–1035.

Caljon, G., Van Den Abbeele, J., Stijlemans, B., Coosemans, M., De Baetselier, P., Magez, S., 2006b. Tsetse fly saliva accelerates the onset of *Trypanosoma brucei* infection in a mouse model associated with a reduced host inflammatory response. Infect. Immun. 74, 6324–6330.

Caljon, G., Broos, K., De Goeyse, I., De Ridder, K., Sternberg, J.M., Coosemans, M., De Baetselier, P., Guisez, Y., Den Abbeele, J.V., 2009. Identification of a functional Antigen5-related allergen in the saliva of a blood feeding insect, the tsetse fly. Insect Biochem. Mol. Biol. 39, 332–341.

Caljon, G., De Ridder, K., De Baetselier, P., Coosemans, M., Van Den Abbeele, J., 2010. Identification of a tsetse fly salivary protein with dual inhibitory action on human platelet aggregation. PLoS One 5, e9671.

Caljon, G., De Ridder, K., Stijlemans, B., Coosemans, M., Magez, S., De Baetselier, P., Van Den Abbeele, J., 2012. Tsetse salivary gland proteins 1 and 2 are high affinity nucleic acid binding proteins with residual nuclease activity. PLoS One 7, e47233.

Caljon, G., De Vooght, L., Van Den Abbeele, J., 2014. The biology of Tsetse-trypanosome interactions. In: Magez, S., Radwanska, M. (Eds.), Trypanosomes and Trypanosomiasis. Springer-Verlag, Wien.

Caljon, G., De Muylder, G., Durnez, L., Jennes, W., Vanaerschot, M., Dujardin, J.C., 2016a. Alice in microbes' land: adaptations and counter-adaptations of vector-borne parasitic protozoa and their hosts. FEMS Microbiol. Rev.

Caljon, G., Van Reet, N., De Trez, C., Vermeersch, M., Pérez-Morga, D., Van Den Abbeele, J., 2016b. The dermis as a delivery site of *Trypanosoma brucei* for tsetse flies. PLoS Pathog.

Cappello, M., Bergum, P.W., Vlasuk, G.P., Furmidge, B.A., Pritchard, D.I., Aksoy, S., 1996. Isolation and characterization of the tsetse thrombin inhibitor: a potent antithrombotic peptide from the saliva of *Glossina morsitans morsitans*. Am. J. Trop. Med. Hyg. 54, 475–480.

Cappello, M., Li, S., Chen, X., Li, C.B., Harrison, L., Narashimhan, S., Beard, C.B., Aksoy, S., 1998. Tsetse thrombin inhibitor: bloodmeal-induced expression of an anticoagulant in salivary glands and gut tissue of *Glossina morsitans morsitans*. Proc. Natl. Acad. Sci. U.S.A. 95, 14290–14295.

Cassola, A., De Gaudenzi, J.G., Frasch, A.C., 2007. Recruitment of mRNAs to cytoplasmic ribonucleoprotein granules in trypanosomes. Mol. Microbiol. 65, 655–670.

Chagas, A.C., Oliveira, F., Debrabant, A., Valenzuela, J.G., Ribeiro, J.M., Calvo, E., 2014. Lundep, a sand fly salivary endonuclease increases *Leishmania* parasite survival in neutrophils and inhibits XIIa contact activation in human plasma. PLoS Pathog. 10, e1003923.

Cherenet, T., Sani, R.A., Panandam, J.M., Nadzr, S., Speybroeck, N., van den Bossche, P., 2004. Seasonal prevalence of bovine trypanosomosis in a tsetse-infested zone and a tsetse-free zone of the Amhara Region, north-west Ethiopia. Onderstepoort J. Vet. Res. 71, 307–312.

Chitanga, S., Marcotty, T., Namangala, B., Van den Bossche, P., Van Den Abbeele, J., Delespaux, V., 2011. High prevalence of drug resistance in animal trypanosomes without a history of drug exposure. PLoS Negl. Trop. Dis. 5, e1454.

Cnops, J., De Trez, C., Bulte, D., Radwanska, M., Ryffel, B., Magez, S., 2015a. IFN-gamma mediates early B-cell loss in experimental African trypanosomosis. Parasite Immunol. 37, 479–484.

Cnops, J., De Trez, C., Stijlemans, B., Keirsse, J., Kauffmann, F., Barkhuizen, M., Keeton, R., Boon, L., Brombacher, F., Magez, S., 2015b. NK-, NKT- and CD8-derived IFNgamma Drives myeloid cell activation and Erythrophagocytosis, resulting in trypanosomosis-associated Acute anemia. PLoS Pathog. 11, e1004964.

Cnops, J., Kauffmann, F., De Trez, C., Baltz, T., Keirsse, J., Radwanska, M., Muraille, E., Magez, S., 2016. Maintenance of B cells during chronic murine T. brucei gambiense infection. Parasite Immunol.

Cochran, R., Rosen, T., 1983. African trypanosomiasis in the United States. Arch. Dermatol. 119, 670–674.

Coller, S.P., Mansfield, J.M., Paulnock, D.M., 2003. Glycosylinositolphosphate soluble variant surface glycoprotein inhibits IFN-gamma-induced nitric oxide production via reduction in STAT1 phosphorylation in African trypanosomiasis. J. Immunol. 171, 1466–1472.

Cross, G.A., Kim, H.S., Wickstead, B., 2014. Capturing the variant surface glycoprotein repertoire (the VSGnome) of Trypanosoma brucei Lister 427. Mol. Biochem. Parasitol. 195, 59–73.

Darji, A., Beschin, A., Sileghem, M., Heremans, H., Brys, L., De Baetselier, P., 1996. In vitro simulation of immunosuppression caused by Trypanosoma brucei: active involvement of gamma interferon and tumor necrosis factor in the pathway of suppression. Infect. Immun. 64, 1937–1943.

De Muylder, G., Daulouede, S., Lecordier, L., Uzureau, P., Morias, Y., Van Den Abbeele, J., Caljon, G., Herin, M., Holzmuller, P., Semballa, S., Courtois, P., Vanhamme, L., Stijlemans, B., De Baetselier, P., Barrett, M.P., Barlow, J.L., McKenzie, A.N., Barron, L., Wynn, T.A., Beschin, A., Vincendeau, P., Pays, E., 2013. A Trypanosoma brucei kinesin heavy chain promotes parasite growth by triggering host arginase activity. PLoS Pathog. 9, e1003731.

De Trez, C., Katsandegwaza, B., Caljon, G., Magez, S., 2015. Experimental African trypanosome infection by needle passage or natural tsetse fly challenge thwarts the development of collagen-induced arthritis in DBA/1 prone mice via an impairment of antigen specific B cell autoantibody titers. PLoS One 10, e0130431.

Desquesnes, M., Dia, M.L., 2003. Trypanosoma vivax: mechanical transmission in cattle by one of the most common African tabanids, Atylotus agrestis. Exp. Parasitol. 103, 35–43.

Devine, D.V., Falk, R.J., Balber, A.E., 1986. Restriction of the alternative pathway of human complement by intact Trypanosoma brucei subsp. gambiense. Infect. Immun. 52, 223–229.

Drennan, M.B., Stijlemans, B., Van den Abbeele, J., Quesniaux, V.J., Barkhuizen, M., Brombacher, F., De Baetselier, P., Ryffel, B., Magez, S., 2005. The induction of a type 1 immune response following a Trypanosoma brucei infection is MyD88 dependent. J. Immunol. 175, 2501–2509.

Dwinger, R.H., Lamb, G., Murray, M., Hirumi, H., 1987. Dose and stage dependency for the development of local skin reactions caused by Trypanosoma congolense in goats. Acta Tropica 44, 303–314.

Dwinger, R.H., Rudin, W., Moloo, S.K., Murray, M., 1988. Development of Trypanosoma congolense, T vivax and T brucei in the skin reaction induced in goats by infected Glossina morsitans centralis: a light and electron microscopical study. Res. Vet. Sci. 44, 154–163.

Ellis, J.A., Shapiro, S.Z., ole Moi-Yoi, O., Moloo, S.K., 1986. Lesions and saliva-specific antibody responses in rabbits with immediate and delayed hypersensitivity reactions to the bites of Glossina morsitans centralis. Vet. Pathol. 23, 661–667.

Emmer, B.T., Daniels, M.D., Taylor, J.M., Epting, C.L., Engman, D.M., 2010. Calflagin inhibition prolongs host survival and suppresses parasitemia in Trypanosoma brucei infection. Eukaryotic Cell 9, 934–942.

Engstler, M., Pfohl, T., Herminghaus, S., Boshart, M., Wiegertjes, G., Heddergott, N., Overath, P., 2007. Hydrodynamic flow-mediated protein sorting on the cell surface of trypanosomes. Cell 131, 505–515.

Fairbairn, H., Burtt, E., 1946. The infectivity to man of a strain of Trypanosoma rhodesiense transmitted cyclically by Glossina morsitans through sheep and antelope; evidence that man requires a minimum infective dose of metacyclic trypanosomes. Ann. Trop. Med. Parasitol. 40, 270–313.

Fenn, K., Matthews, K.R., 2007. The cell biology of Trypanosoma brucei differentiation. Curr. Opin. Microbiol. 10, 539–546.

Ferrante, A., Allison, A.C., 1983. Alternative pathway activation of complement by African trypanosomes lacking a glycoprotein coat. Parasite Immunol. 5, 491–498.

Frischknecht, F., 2007. The skin as interface in the transmission of arthropod-borne pathogens. Cell. Microbiol. 9, 1630–1640.

Geiger, A., Hirtz, C., Becue, T., Bellard, E., Centeno, D., Gargani, D., Rossignol, M., Cuny, G., Peltier, J.B., 2010. Exocytosis and protein secretion in Trypanosoma. BMC Microbiol. 10, 20.

Gombault, A., Baron, L., Couillin, I., 2012. ATP release and purinergic signaling in NLRP3 inflammasome activation. Front. Immunol. 3, 414.

Gomez-Rodriguez, J., Stijlemans, B., De Muylder, G., Korf, H., Brys, L., Berberof, M., Darji, A., Pays, E., De Baetselier, P., Beschin, A., 2009. Identification of a parasitic immunomodulatory protein triggering the development of suppressive M1 macrophages during African trypanosomiasis. J. Infect. Dis. 200, 1849–1860.

Gounaris, K., Selkirk, M.E., 2005. Parasite nucleotide-metabolizing enzymes and host purinergic signalling. Trends Parasitol. 21, 17–21.

Graham, S.V., Matthews, K.R., Shiels, P.G., Barry, J.D., 1990. Distinct, developmental stage-specific activation mechanisms of trypanosome VSG genes. Parasitology 101 (Pt. 3), 361–367.

Reinouts van Haga, H.A., Mitchell, B.K., 1975. Temperature receptors on tarsi of the tsetse fly Glossina morsitans west. Nature 255, 225–226.

Hall, J.P., Wang, H., Barry, J.D., 2013. Mosaic VSGs and the scale of Trypanosoma brucei antigenic variation. PLoS Pathog. 9, e1003502.

Hasko, G., Kuhel, D.G., Nemeth, Z.H., Mabley, J.G., Stachlewitz, R.F., Virag, L., Lohinai, Z., Southan, G.J., Salzman, A.L., Szabo, C., 2000. Inosine inhibits inflammatory cytokine production by a posttranscriptional mechanism and protects against endotoxin-induced shock. J. Immunol. 164, 1013–1019.

Hereld, D., Krakow, J.L., Bangs, J.D., Hart, G.W., Englund, P.T., 1986. A phospholipase C from Trypanosoma brucei which selectively cleaves the glycolipid on the variant surface glycoprotein. J. Biol. Chem. 261, 13813–13819.

Herman, M., Perez-Morga, D., Schtickzelle, N., Michels, P.A., 2008. Turnover of glycosomes during life-cycle differentiation of Trypanosoma brucei. Autophagy 4, 294–308.

Hirumi, H., Doyle, J.J., Hirumi, K., 1977. African trypanosomes: cultivation of animal-infective Trypanosoma brucei in vitro. Science 196, 992–994.

Hudson, K.M., Byner, C., Freeman, J., Terry, R.J., 1976. Immunodepression, high IgM levels and evasion of the immune response in murine trypanosomiasis. Nature 264, 256–258.

IGGI, International Glossina Genome Initiative Genome sequence of the tsetse fly (Glossina morsitans): vector of African trypanosomiasis. Science 344, 2014, 380–386.

Karmakar, M., Katsnelson, M.A., Dubyak, G.R., Pearlman, E., 2016. Neutrophil P2X7 receptors mediate NLRP3 inflammasome-dependent IL-1beta secretion in response to ATP. Nat. Commun. 7, 10555.

Kolaczkowska, E., Kubes, P., 2013. Neutrophil recruitment and function in health and inflammation. Nat. Rev. Immunol. 13, 159–175.

Kolev, N.G., Ramey-Butler, K., Cross, G.A., Ullu, E., Tschudi, C., 2013. Developmental progression to infectivity in Trypanosoma brucei triggered by an RNA-binding protein. Science 338, 1352–1353.

Langousis, G., Shimogawa, M.M., Saada, E.A., Vashisht, A.A., Spreafico, R., Nager, A.R., Barshop, W.D., Nachury, M.V., Wohlschlegel, J.A., Hill, K.L., 2016. Loss of the BBSome perturbs endocytic trafficking and disrupts virulence of Trypanosoma brucei. Proc. Natl. Acad. Sci. U.S.A. 113, 632–637.

Leak, 1999. Tsetse Biology and Ecology: Their Role in the Epidemiology and Control of Trypanosomosis. Cabi Publishing, Wallingford, UK.

Lehane, M.J., 2005. The Biology of Blood-Sucking in Insects. Cambridge University Press.

Lejon, V., Mumba Ngoyi, D., Kestens, L., Boel, L., Barbe, B., Kande Betu, V., van Griensven, J., Bottieau, E., Muyembe Tamfum, J.J., Jacobs, J., Buscher, P., 2014. Gambiense human african trypanosomiasis and immunological memory: effect on phenotypic lymphocyte profiles and humoral immunity. PLoS Pathog. 10, e1003947.

Leppert, B.J., Mansfield, J.M., Paulnock, D.M., 2007. The soluble variant surface glycoprotein of African trypanosomes is recognized by a macrophage scavenger receptor and induces I kappa B alpha degradation independently of TRAF6-mediated TLR signaling. J. Immunol. 179, 548–556.

Li, Z., 2012. Regulation of the cell division cycle in Trypanosoma brucei. Eukaryotic Cell 11, 1180–1190.

Li, S., Aksoy, S., 2000. A family of genes with growth factor and adenosine deaminase similarity are preferentially expressed in the salivary glands of Glossina m. morsitans. Gene 252, 83–93.

Livesey, J.L., Molyneux, D.H., Jenni, L., 1980. Mechanoreceptor-trypanosome interactions in the labrum of Glossina: fluid mechanics. Acta Trop. 37, 151–161.

Lucas, R., Magez, S., De Leys, R., Fransen, L., Scheerlinck, J.P., Rampelberg, M., Sablon, E., De Baetselier, P., 1994. Mapping the lectin-like activity of tumor necrosis factor. Science 263, 814–817.

Luckins, A.G., Gray, A.R., 1978. An extravascular site of development of Trypanosoma congolense. Nature 272, 613–614.

Mabbott, N.A., Coulson, P.S., Smythies, L.E., Wilson, R.A., Sternberg, J.M., 1998. African trypanosome infections in mice that lack the interferon-gamma receptor gene: nitric oxide-dependent and -independent suppression of T-cell proliferative responses and the development of anaemia. Immunology 94, 476–480.

MacGregor, P., Szoor, B., Savill, N.J., Matthews, K.R., 2012. Trypanosomal immune evasion, chronicity and transmission: an elegant balancing act. Nat. Rev. Microbiol. 10, 431–438.

Magez, S., Geuskens, M., Beschin, A., del Favero, H., Verschueren, H., Lucas, R., Pays, E., de Baetselier, P., 1997. Specific uptake of tumor necrosis factor-alpha is involved in growth control of Trypanosoma brucei. J. Cell Biol. 137, 715–727.

Magez, S., Stijlemans, B., Radwanska, M., Pays, E., Ferguson, M.A., De Baetselier, P., 1998. The glycosyl-inositol-phosphate and dimyristoylglycerol moieties of the glycosylphosphatidylinositol anchor of the trypanosome variant-specific surface glycoprotein are distinct macrophage-activating factors. J. Immunol. 160, 1949–1956.

Mair, G.R., Lasonder, E., Garver, L.S., Franke-Fayard, B.M., Carret, C.K., Wiegant, J.C., Dirks, R.W., Dimopoulos, G., Janse, C.J., Waters, A.P., 2010. Universal features of post-transcriptional gene regulation are critical for *Plasmodium* zygote development. PLoS Pathog. 6, e1000767.

Malvy, D., Chappuis, F., 2011. Sleeping sickness. Clin. Microbiol. Infect. 17, 986–995.

Mant, M.J., Parker, K.R., 1981. Two platelet aggregation inhibitors in tsetse (*Glossina*) saliva with studies of roles of thrombin and citrate in in vitro platelet aggregation. Br. J. Haematol. 48, 601–608.

Matthews, K.R., 2005. The developmental cell biology of *Trypanosoma brucei*. J. Cell Sci. 118, 283–290.

Mehlert, A., Bond, C.S., Ferguson, M.A., 2002. The glycoforms of a *Trypanosoma brucei* variant surface glycoprotein and molecular modeling of a glycosylated surface coat. Glycobiology 12, 607–612.

Moloo, S.K., Dar, F., 1985. Probing by *Glossina morsitans centralis* infected with pathogenic *Trypanosoma* species. Trans. R. Soc. Trop. Med. Hyg. 79, 119.

Molyneux, D.H., Lavin, D.R., Elce, B., 1979. A possible relationship between Salivarian trypanosomes and *Glossina* labrum mechano-receptors. Ann. Trop. Med. Parasitol. 73, 287–290.

Mony, B.M., Matthews, K.R., 2015. Assembling the components of the quorum sensing pathway in African trypanosomes. Mol. Microbiol. 96, 220–232.

Mony, B.M., MacGregor, P., Ivens, A., Rojas, F., Cowton, A., Young, J., Horn, D., Matthews, K., 2014. Genome-wide dissection of the quorum sensing signalling pathway in *Trypanosoma brucei*. Nature 505, 681–685.

Mugnier, M.R., Cross, G.A., Papavasiliou, F.N., 2015. The in vivo dynamics of antigenic variation in *Trypanosoma brucei*. Science 347, 1470–1473.

Musoke, A.J., Barbet, A.F., 1977. Activation of complement by variant-specific surface antigen of *Trypanosoma brucei*. Nature 270, 438–440.

Mwangi, D.M., Hopkins, J., Luckins, A.G., 1990. Cellular phenotypes in *Trypanosoma congolense* infected sheep: the local skin reaction. Parasite Immunol. 12, 647–658.

Mwangi, D.M., Hopkins, J., Luckins, A.G., 1995. *Trypanosoma congolense* infection in sheep: ultrastructural changes in the skin prior to development of local skin reactions. Vet. Parasitol. 60, 45–52.

Naessens, J., Mwangi, D.M., Buza, J., Moloo, S.K., 2003. Local skin reaction (chancre) induced following inoculation of metacyclic trypanosomes in cattle by tsetse flies is dependent on CD4 T lymphocytes. Parasite Immunol. 25, 413–419.

Namangala, B., Brys, L., Magez, S., De Baetselier, P., Beschin, A., 2000. *Trypanosoma brucei brucei* infection impairs MHC class II antigen presentation capacity of macrophages. Parasite Immunol. 22, 361–370.

Obishakin, E., de Trez, C., Magez, S., 2014. Chronic *Trypanosoma congolense* infections in mice cause a sustained disruption of the B-cell homeostasis in the bone marrow and spleen. Parasite Immunol. 36, 187–198.

Paulnock, D.M., Freeman, B.E., Mansfield, J.M., 2010. Modulation of innate immunity by African trypanosomes. Parasitology 137, 2051–2063.

Poinsignon, A., Cornelie, S., Remoue, F., Grebaut, P., Courtin, D., Garcia, A., Simondon, F., 2007. Human/vector relationships during human African trypanosomiasis: initial screening of immunogenic salivary proteins of *Glossina* species. Am. J. Trop. Med. Hyg. 76, 327–333.

Radwanska, M., Magez, S., Michel, A., Stijlemans, B., Geuskens, M., Pays, E., 2000. Comparative analysis of antibody responses against HSP60, invariant surface glycoprotein 70, and variant surface glycoprotein reveals a complex antigen-specific pattern of immunoglobulin isotype switching during infection by *Trypanosoma brucei*. Infect. Immun. 68, 848–860.

Radwanska, M., Guirnalda, P., De Trez, C., Ryffel, B., Black, S., Magez, S., 2008. Trypanosomiasis-induced B cell apoptosis results in loss of protective anti-parasite antibody responses and abolishment of vaccine-induced memory responses. PLoS Pathog. 4, e1000078.

Ramey-Butler, K., Ullu, E., Kolev, N.G., Tschudi, C., 2015. Synchronous expression of individual metacyclic variant surface glycoprotein genes in *Trypanosoma brucei*. Mol. Biochem. Parasitol. 200, 1–4.

Rico, E., Rojas, F., Mony, B.M., Szoor, B., Macgregor, P., Matthews, K.R., 2013. Bloodstream form pre-adaptation to the tsetse fly in *Trypanosoma brucei*. Front. Cell. Infect. Microbiol. 3, 78.

Roberts, C.J., Gray, M.A., Gray, A.R., 1969. Local skin reactions in cattle at the site of infection with *Trypanosoma congolense* by *Glossina morsitans* and *G. tachinoides*. Trans. R. Soc. Trop. Med. Hyg. 63, 620–624.

Rodrigues, J.C., Godinho, J.L., de Souza, W., 2014. Biology of human pathogenic trypanosomatids: epidemiology, lifecycle and ultrastructure. Subcell. Biochem. 74, 1–42.

Rolin, S., Hanocq-Quertier, J., Paturiaux-Hanocq, F., Nolan, D., Salmon, D., Webb, H., Carrington, M., Voorheis, P., Pays, E., 1996. Simultaneous but independent activation of adenylate cyclase and glycosylphosphatidylinositol-phospholipase C under stress conditions in *Trypanosoma brucei*. J. Biol. Chem. 271, 10844–10852.

Rotureau, B., Van Den Abbeele, J., 2013. Through the dark continent: African trypanosome development in the tsetse fly. Front. Cell. Infect. Microbiol. 3, 53.

Rotureau, B., Subota, I., Buisson, J., Bastin, P., 2012. A new asymmetric division contributes to the continuous production of infective trypanosomes in the tsetse fly. Development 139, 1842–1850.

Rurangirwa, F.R., Tabel, H., Losos, G., Tizard, I.R., 1980. Hemolytic complement and serum C3 levels in Zebu cattle infected with *Trypanosoma congolense* and *Trypanosoma vivax* and the effect of trypanocidal treatment. Infect. Immun. 27, 832–836.

Sa-Nunes, A., Bafica, A., Lucas, D.A., Conrads, T.P., Veenstra, T.D., Andersen, J.F., Mather, T.N., Ribeiro, J.M., Francischetti, I.M., 2007. Prostaglandin E2 is a major inhibitor of dendritic cell maturation and function in Ixodes scapularis saliva. J. Immunol. 179, 1497–1505.

Salmon, D., Vanwalleghem, G., Morias, Y., Denoeud, J., Krumbholz, C., Lhomme, F., Bachmaier, S., Kador, M., Gossmann, J., Dias, F.B., De Muylder, G., Uzureau, P., Magez, S., Moser, M., De Baetselier, P., Van Den Abbeele, J., Beschin, A., Boshart, M., Pays, E., 2012. Adenylate cyclases of *Trypanosoma brucei* inhibit the innate immune response of the host. Science 337, 463–466.

Schleifer, K.W., Mansfield, J.M., 1993. Suppressor macrophages in African trypanosomiasis inhibit T cell proliferative responses by nitric oxide and prostaglandins. J Immunol. 151, 5492–5503.

Shimogawa, M.M., Saada, E.A., Vashisht, A.A., Barshop, W.D., Wohlschlegel, J.A., Hill, K.L., 2015. Cell surface proteomics provides insight into stage-specific remodeling of the host-parasite interface in *Trypanosoma brucei*. Mol. Cell. Proteom. 14, 1977–1988.

Shoda, L.K., Kegerreis, K.A., Suarez, C.E., Roditi, I., Corral, R.S., Bertot, G.M., Norimine, J., Brown, W.C., 2001. DNA from protozoan parasites *Babesia bovis, Trypanosoma cruzi*, and *T. brucei* is mitogenic for B lymphocytes and stimulates macrophage expression of interleukin-12, tumor necrosis factor alpha, and nitric oxide. Infect. Immun. 69, 2162–2171.

Sileghem, M., Hamers, R., De Baetselier, P., 1986. Active suppression of interleukin 2 secretion in mice infected with *Trypanosoma brucei* AnTat 1.1.E. Parasite Immunol. 8, 641–649.

Sileghem, M., Hamers, R., De Baetselier, P., 1987. Experimental *Trypanosoma brucei* infections selectively suppress both interleukin 2 production and interleukin 2 receptor expression. Eur. J. Immunol. 17, 1417–1421.

Simarro, P.P., Cecchi, G., Franco, J.R., Paone, M., Diarra, A., Priotto, G., Mattioli, R.C., Jannin, J.G., 2015. Monitoring the progress towards the elimination of gambiense human African trypanosomiasis. PLoS Negl. Trop. Dis. 9, e0003785.

Sinshaw, A., Abebe, G., Desquesnes, M., Yoni, W., 2006. Biting flies and *Trypanosoma vivax* infection in three highland districts bordering lake Tana, Ethiopia. Vet. Parasitol. 142, 35–46.

Stijlemans, B., Conrath, K., Cortez-Retamozo, V., Van Xong, H., Wyns, L., Senter, P., Revets, H., De Baetselier, P., Muyldermans, S., Magez, S., 2004. Efficient targeting of conserved cryptic epitopes of infectious agents by single domain antibodies. African trypanosomes as paradigm. J. Biol. Chem. 279, 1256–1261.

Subota, I., Rotureau, B., Blisnick, T., Ngwabyt, S., Durand-Dubief, M., Engstler, M., Bastin, P., 2011. ALBA proteins are stage regulated during trypanosome development in the tsetse fly and participate in differentiation. Mol. Biol. Cell 22, 4205–4219.

Szempruch, A.J., Sykes, S.E., Kieft, R., Dennison, L., Becker, A.C., Gartrell, A., Martin, W.J., Nakayasu, E.S., Almeida, I.C., Hajduk, S.L., Harrington, J.M., 2016. Extracellular vesicles from *Trypanosoma brucei* mediate virulence factor transfer and cause host anemia. Cell 164, 246–257.

Telleria, E.L., Benoit, J.B., Zhao, X., Savage, A.F., Regmi, S., Alves e Silva, T.L., O'Neill, M., Aksoy, S., 2014. Insights into the trypanosome-host interactions revealed through transcriptomic analysis of parasitized tsetse fly salivary glands. PLoS Negl. Trop. Dis. 8, e2649.

Tetley, L., Vickerman, K., 1985. Differentiation in *Trypanosoma brucei*: host-parasite cell junctions and their persistence during acquisition of the variable antigen coat. J. Cell Sci. 74, 1–19.

Thuita, J.K., Kagira, J.M., Mwangangi, D., Matovu, E., Turner, C.M., Masiga, D., 2008. *Trypanosoma brucei rhodesiense* transmitted by a single tsetse fly bite in vervet monkeys as a model of human African trypanosomiasis. PLoS Negl. Trop. Dis. 2, e238.

Tilley, S.L., Wagoner, V.A., Salvatore, C.A., Jacobson, M.A., Koller, B.H., 2000. Adenosine and inosine increase cutaneous vasopermeability by activating A(3) receptors on mast cells. J. Clin. Invest. 105, 361–367.

Trindade, S., Rijo-Ferreira, F., Carvalho, T., Pinto-Neves, D., Guegan, F., Aresta-Branco, F., Bento, F., Young, S.A., Pinto, A., Van Den Abbeele, J., Ribeiro, R.M., Dias, S., Smith, T.K., Figueiredo, L.M., 2016. *Trypanosoma brucei* parasites occupy and functionally adapt to the adipose tissue in mice. Cell Host Microbe 19, 837–848.

Tsujimoto, H., Kotsyfakis, M., Francischetti, I.M., Eum, J.H., Strand, M.R., Champagne, D.E., 2012. Simukunin from the salivary glands of the black fly *Simulium vittatum* inhibits enzymes that regulate clotting and inflammatory responses. PLoS One 7, e29964.

Van Den Abbeele, J., Caljon, G., Dierick, J.F., Moens, L., De Ridder, K., Coosemans, M., 2007. The *Glossina morsitans* tsetse fly saliva: general characteristics and identification of novel salivary proteins. Insect Biochem. Mol. Biol. 37, 1075–1085.

Van Den Abbeele, J., Caljon, G., De Ridder, K., De Baetselier, P., Coosemans, M., 2010. *Trypanosoma brucei* modifies the tsetse salivary composition, altering the fly feeding behavior that favors parasite transmission. PLoS Pathog. 6, e1000926.

Van den Bossche, P., Chitanga, S., Masumu, J., Marcotty, T., Delespaux, V., 2011. Virulence in *Trypanosoma congolense* Savannah subgroup. A comparison between strains and transmission cycles. Parasite Immunol. 33, 456–460.

Vassella, E., Reuner, B., Yutzy, B., Boshart, M., 1997. Differentiation of African trypanosomes is controlled by a density sensing mechanism which signals cell cycle arrest via the cAMP pathway. J. Cell Sci. 110 (Pt. 21), 2661–2671.

Vickerman, K., 1973. The mode of attachment of *Trypanosoma vivax* in the proboscis of the tsetse fly *Glossina fuscipes*: an ultrastructural study of the epimastigote stage of the trypanosome. J. Protozool. 20, 394–404.

Vickerman, K., 1985. Developmental cycles and biology of pathogenic trypanosomes. Br. Med. Bull. 41, 105–114.

von Wissmann, B., Machila, N., Picozzi, K., Fevre, E.M., deC Bronsvoort, B.M., Handel, I.G., Welburn, S.C., 2011. Factors associated with acquisition of human infective and animal infective trypanosome infections in domestic livestock in Western Kenya. PLoS Negl. Trop. Dis. 5, e941.

Wei, G., Bull, H., Zhou, X., Tabel, H., 2011. Intradermal infections of mice by low numbers of African trypanosomes are controlled by innate resistance but enhance susceptibility to reinfection. J. Infect. Dis. 203, 418–429.

Wolbach, S.B., Binger, C.A., 1912. A contribution to the parasitology of trypanosomiasis. J. Med. Res. 27 83–108.13.

Zhao, X., Alves e Silva, T.L., Cronin, L., Savage, A.F., O'Neill, M., Nerima, B., Okedi, L.M., Aksoy, S., 2015. Immunogenicity and serological cross-reactivity of saliva proteins among different tsetse species. PLoS Negl. Trop. Dis. 9, e0004038.

Mosquito Modulation of Arbovirus–Host Interactions

Stephen Higgs, Yan-Jang S. Huang, Dana L. Vanlandingham

Kansas State University, Manhattan, KS, United States

INTRODUCTION

Blood feeding (hematophagy) has independently evolved in arthropods from at least 21 times in multiple arthropod taxa (Black and Kondratieff, 2005). These hematophagous arthropods obtain blood from a vertebrate in a variety of ways, ranging from simple biting with chewing-type mouthparts, to penetration of the skin with tubular sucking mouthparts that are of a diameter that is small enough to cannulate blood vessels. The latter approach may have arisen from arthropods that had evolved to suck on plant juices (xylem) by penetration of the stem or leaves.

This has perhaps been a consequence of close association with vertebrates, for example in resting places, dens, and nests, that provide protection from adverse conditions.

SALIVA OF HEMATOPHAGOUS ARTHROPODS

Common requirements to enable the imbibement of blood by hematophagous arthropods have resulted in considerable functional overlap of salivary components across the range of

blood-feeding arthropods. Vertebrate hemostasis that has evolved to prevent blood loss involves the capability to restrict blood flow, for example when a vessel is damaged. Underlying mechanisms include physical reduction of vessel diameter (vasoconstriction), rapid repair of the wound by fibrinogen, and clotting as platelets become enmeshed in the fibrous network. In addition to these hemostatic responses, vertebrates also have both cellular and humoral immune responses that help to protect the vertebrate from infection due to foreign bodies, contaminants, and infectious agents that may be introduced into the wound site.

As a result of these vertebrate responses, although the actual constituents may vary widely between different hematophagous arthropods, functionally they all possess salivary substances that enable them to feed successfully, often undetected for hours or even days in the case of some ticks. All secrete vasodilatory substances that can maintain an open vessel with unrestricted blood flow, anticoagulants to prevent clotting and antiplatelet substances. Since the salivary substances themselves may be recognized as foreign, the saliva of most hematophagous arthropods also has immunomodulatory

133

activity. The consequences of this activity with respect to pathogens transmitted by arthropods are discussed in more detail below. Although the focus of this chapter is on arthropod-borne viruses (arboviruses), the effects of saliva on other pathogens is included, since for some of these, our understanding is more advanced than for the viruses. This puts the field in context and perhaps shows researchers useful avenues to pursue in order to improve our understanding of vector–virus–vertebrate interactions.

Due to the potentially painful process of blood acquisition, for instance due to physical cutting of the skin in order to reach capillaries, saliva may also contain substances with anesthetic properties. Many arthropods such as mosquitoes and other biting flies feed very rapidly; however, because of the slow feeding that is typical of hard ticks, the saliva of these arachnids contains substances that cement the mouthparts into the skin. Also, because of the long-feeding process of ticks the saliva of ticks is considerably more complex in comparison with rapid feeding flies. Whereas saliva of mosquitoes may contain 100 different components, that of ticks is far more complex and may contain approximately 700 different proteins (Ribeiro et al., 2006; Chmelař et al., 2016a,b). An explanation for this is that the long engagement period during which the tick's mouthparts are embedded in the vertebrate's skin with periodic release of saliva and processes of fluid exchange likely stimulate the vertebrate's immune responses. These could disrupt the feeding process and potentially harm the feeding tick. However, with a large repertoire of salivary substances and duplication/redundancy of function, the tick can switch between different substances to enable ingestion of relatively large volumes of blood and yet avoid the consequences of the immune responses (Cavassani et al., 2005; Kotsyfakis et al., 2005; Ribeiro and Spielman, 1986; Ribeiro et al., 1985, 1990; Steen et al., 2006; Valenzuela, 2004; Wikel, 1996; Wikel et al., 1996; Brossard and Wikel, 2004).

For rapidly feeding mosquitoes, immune responses to salivary substances would be too slow to have any effect on the feeding individual; however, mosquito saliva also contains pharmacologically active substances (Arca et al., 1999, 2002, 2005; Calvo et al., 2004, 2006a,b; Calvo and Ribeiro, 2006; Champagne, 2004; Champagne et al., 1995; Cross et al., 1994; Edwards et al., 1998; James, 1994; James et al., 1991; Kerlin and Hughes, 1992; Lanfrancotti et al., 2002; Limesand et al., 2000, 2003; Mellinck and van Zeben, 1976; Nascimento et al., 2000; Osorio et al., 1996; Poehling, 1979; Racioppi and Spielman, 1987; Ribeiro, 1987, 1989, 1992, 2000; Ribeiro and Francishetti, 2003; Ribeiro and Nussenzveig, 1993; Ribeiro and Valenzuela, 1999; Ribeiro et al., 1994, 2001, 2004; Schneider et al., 2004, 2006; Stark and James, 1996; Styer et al., 2006; Suwan et al., 2002; Valenzuela, 2002; Valenzuela et al., 2002a,b; Wanasen et al., 2004; Wasserman et al., 2004; Zeidner et al., 1999), including immunosuppressive molecules (Cross et al., 1994; Wanasen et al., 2004; Wasserman et al., 2004; Zeidner et al., 1999; Billingsley et al., 2006), that perhaps evolved to provide an overall effect that is nondeleterious to these rapidly feeding arthropods.

ARTHROPOD SALIVA AND THE VERTEBRATE HOST

Hematophagous arthropods salivate into the vertebrate host. This complex mixture of multiple substances evolved to facilitate efficient blood acquisition whilst minimally impacting the vertebrate host so that the feeding arthropod can be undetected and not affected by host immune responses. Anyone who has been fed upon by arthropods, for example mosquitoes and fleas, will likely have experienced localized inflammatory immediate responses at the feeding site. The intensity of these responses can vary widely between individuals and individuals may become tolerant to the saliva so that with frequent exposure, the responses wane. This tolerance can be species specific, for example an individual may become unreactive to feeding by one mosquito species to which they have been exposed many times, and yet respond to feeding by a different mosquito. Fontaine et al. (2011) provide an excellent review of salivary proteins and host vector interactions.

ARTHROPODS AND ARBOVIRUSES

Many species of hematophagous arthropods are important vectors of infectious agents such as viruses, bacteria, protozoa, helminthes, and nematodes. There are approximately 500 different arthropod-borne viruses (arboviruses) that infect a wide variety of vertebrate hosts, including humans, with potentially fatal consequences. These can include hemorrhagic and encephalitic viruses such as dengue and Japanese encephalitis viruses, respectively. Sylvatic transmission cycles that have evolved over long periods of time between arthropods and wild vertebrates, typically are associated with mild or asymptomatic infections of the vertebrate. In contrast, transmission of these agents to new hosts, including humans in urban cycles, can result in severe symptomatic disease development with occasionally high fatality rates.

Most arboviral infection of arthropod vectors follow a similar pattern that progresses from infection of the digestive system to a disseminated infection of secondary tissues and ultimately infection of the salivary glands so that the virus is transmitted in saliva during feeding on the vertebrate host. The time between when the arthropod takes the infectious viremic meal from an infected vertebrate to being able to transmit the virus during feeding on a subsequent host is known as the extrinsic incubation period. This can vary from a few days to more than a week, being dependent on the species of arthropods, type of virus, infectious dose, and temperature. The relationship between the arthropod and virus can be highly species specific, and whilst some viruses such as West Nile virus (WNV) can infect multiple species of mosquito, others for example dengue, are transmitted by relatively few. The basis for this species specificity is poorly understood.

Some viruses can be transmitted vertically either from one infected stage to another—designated as transstadial transmission in ticks (Benda, 1958; Roseboom and Burgdorfer, 1959; Varma and Smith, 1962) or from an infected arthropod to the offspring, referred to as transovarial transmission (Korenberg and Pchelkina, 1984; Benda, 1958; Chumakov, 1944; Chumakov et al., 1945; Pretzmann et al., 1963; Stockman, 1918; Kondratenko et al., 1970). For some viruses this type of transmission can be a critical component to ensure for survival between seasons or during periods when arthropod–host contact is infrequent.

Additionally, transmission from an infected arthropod to an uninfected arthropod that is feeding in relatively close proximity is well documented. This type of transmission does not depend on the infection of the vertebrate and replication of the virus in the vertebrate and is termed nonviremic or nonsystemic transmission (NVT). Such transmission has been long recognized in tick-borne viruses (Jones et al., 1987; Nuttall and Jones, 1991; Randolph et al., 1996, 2000) and other pathogens (Patrican, 1997; Gern and Rais, 1996; Tonetti et al., 2015; Voorouw, 2015); and more recently has been documented for mosquito-borne viruses (Higgs et al., 2005; Higgs, 2006; McGee et al., 2007; Reisen et al., 2007).

An interesting phenomenon that has been well documented for some tick-borne viruses is so-called salivary activated transmission (SAT). Although SAT has not been specifically reported for mosquito-borne viruses, it is briefly discussed here, because of the common aspects of SAT and NVT. The basis of SAT is an effect of saliva, probably at the vector–host feeding interface, on transmission dynamics. The presence of saliva or salivary gland extract (SGE) can, for example, result in elevated viremic levels and lead to an increase in the rate of tick infection during feeding (Labuda et al., 1993b). Delivery of tick-borne encephalitis virus (TBEV) with saliva or SGE, can enhance infection of feeding ticks in the absence of detectable viremia (i.e., NVT) (Labuda et al., 1993a; Alekseev et al., 1991). This phenomenon can play a critical role in the transmission cycle of TBEV, although the exact mechanism is not fully understood (Jones et al., 1992a,b; Nuttall and Labuda, 2004).

Further information on the development of viral infections in arthropods and transmission cycles is available in many publications and general reviews (Higgs, 2004; Higgs and Vanlandingham, 2016).

The Effects of Arthropod Saliva on Pathogen Development and Disease Development

The secretion of saliva from an infectious arthropod with an arbovirus present in the salivary glands into the feeding site is obviously the critical step for transmission to occur. Whether or not the vertebrate on which the arthropod is feeding becomes infected can depend on multiple factors. The vertebrate may simply not be susceptible to infection with the particular virus that has been delivered. For example, a bird fed upon by a mosquito infected with dengue virus (DENV) or a human fed upon by a tick infected with African swine fever virus (ASFV), will not become infected with DENV or ASFV, respectively. In these examples, their lack of susceptibility may be due to a lack of suitable receptors for these viruses, but there may be other reasons to explain why, despite exposure to a virus, a vertebrate does not become infected.

Immune status is certainly important. Prior exposure, for example, may have resulted in a specific protective immune response so that even though susceptible to infection, antibodies

of the individual host will effectively neutralize the virus. Thus infection is prevented. In contrast, an immunocompromised host may become infected by a pathogen to which it is normally resistant, as for example with opportunistic infections in untreated HIV patients. Infection in itself does not automatically result in symptomatic disease.

In many sylvatic transmission cycles, an amplifying host may have high titers of virus in the blood and yet show no overt signs of infection. Infection of an incidental host with the same virus, however, can result in fatal infection. For chikungunya virus, it has been estimated that 80% of people exposed may develop symptomatic disease, but with Zika virus, 80% may be asymptomatic. In many cases we still do not fully understand the interactions between the vertebrate host and a virus to which it is exposed or the underlying mechanisms that determine the outcome of infection. The relationships between virus and vertebrate are complex and influenced by many factors, and interestingly one of these can be the saliva in which the virus is delivered.

Saliva may be recognized as "foreign" by the host, and exposure to saliva over the course of repeated bites elicits inflammation and a full range of innate and adaptive immune responses. This response is based on migration of immune cells to the feeding site with rapid development of a characteristic local swelling, often manifested as a red spot. With frequent exposure, tolerance can be developed so that a host no longer seems to react to the feeding arthropod. This can be quite specific, for example, one may react to the feeding of one species of mosquito but not to feeding by a different one. Presumably this is because of differences in the salivary components.

Although saliva acts as the medium in which arboviruses are naturally transmitted to the vertebrate host, it should not be viewed as an inert process. There is an increasing body of literature to demonstrate that the effects of saliva on the vertebrate host can have significant impact on the efficiency of the infection process, pathogen establishment and dissemination, and ultimately disease development (Edwards et al., 1998; Limesand et al., 2000, 2003; Osorio et al., 1996; Pingen et al., 2016; Schneider et al., 2004, 2006; Schmid et al., 2016; Styer et al., 2006, 2011). This is perhaps not surprising for tick-borne pathogens given the long feeding process of hard ticks. The infection rate and dissemination of Powassan virus (POWV), for example, is enhanced in the presence of SGE, and disease outcome is subsequently affected (Hermance and Thangamani, 2015). The mechanism of this enhancement involves early recruitment of neutrophil and mononuclear infiltrates into the tick feeding site and subsequent infection of macrophages and fibroblasts by POWV (Hermance et al., 2016). Ultimately, fatal infection can result, with apoptotic infection of macrophages in lymphoid tissues and infection of the spinal cord, brainstem, and neuronal cells of the brain (Santos et al., 2016).

For mosquito-borne viruses, even though mosquito feeding is relatively rapid, the saliva in which the viruses are delivered can have significant impact for viral infections. Enhancement of infection due to saliva has now been documented for viruses in several families. These include the Alphavirus Sindbis virus (Schneider et al., 2004), Bunyaviruses Cache Valley virus (Edwards et al., 1998) and Lacrosse virus (Osorio et al., 1996), the Rhabdovirus vesicular stomatitis virus (Limesand et al., 2000, 2003), and the Flavivirus WNV (Schneider et al., 2006, 2007). For WNV, an interesting effect was that exposure of mice to feeding by uninfected mosquitoes resulted in enhancement of infection, dissemination, and disease severity following challenge of these mice with WNV (Schneider et al., 2007). The greater the exposure, the greater the effect: two exposures resulted in 68% mortality, whilst four exposures resulted in 91% mortality. Underlying mechanisms for the apparently Th2-cell driven

allergic responses in Balb/c mice could include recruitment of target cells to the feeding site due to the presence of salivary proteins, which in effect provides greater opportunity for virus to infect susceptible cells. Additionally, saliva could suppress host antiviral responses, for example by reducing levels of interferons, thereby allowing less restricted viral replication. The specific type of suppression, for example whether it is type I, alpha and beta, or type II, gamma interferon, can vary with the species of arthropod. The overall shifting from a potentially protective Th1 response has been reported for a range of pathogens. Vaccines targeted against salivary proteins to protect animals from arthropod-borne pathogens are discussed below.

A very recent publication (Pingen et al., 2016) demonstrated underlying mechanisms of how host inflammatory responses to mosquito bites can enhance severity of infection with Semliki Forest virus and Bunyamwera virus. Very detailed analysis demonstrated an inflammatory influx of neutrophils into the bite site causing edema and retention of the virus. The neutrophils were associated with a coordinated localized innate immune response that enabled viral entry and infection of susceptible bone marrow–derived myeloid cells. Interestingly, in contrast to suppression of IFN observed in some previous studies, this study revealed enhanced cutaneous IFN-B transcript levels at 24 h postexposure. The authors concluded that saliva delivered to the bite site does not enhance infection due to a specifically evolved function, but rather enhancement results from an inadvertent promotion of neutrophil-dependent inflammation at the site. The IL-1B pathway involved in the localized inflammation was required for effective virus replication and dissemination, which subsequently determined the systemic course of infection and disease outcome.

Of considerable significance, due to the number of cases of dengue that occur every year, is the description that in a mouse model, SGE from *Aedes aegypti* can enhance pathogenesis but only in the presence of enhancing antibodies (Schmid et al., 2016). When inoculated with 10^5 plaque forming units (PFU) of DENV, significant increases in mortality were observed at 5, 6, and 7 days postinoculation. Interestingly, the effect was dose dependent, with no observable enhancement at a dose of 10^6 PFU. Given the association of dengue hemorrhagic fever with antibody-dependent enhancement, the discovery of a compounding influence of mosquito saliva is interesting and adds to the complexity of how the balance of multiple factors can influence arboviral diseases. The mechanism by which saliva influenced dengue virus pathogenesis was determined to be via increased infection in the skin, migration of immune dendritic cells to lymph nodes, and increased permeability of vascular endothelial cells.

An interesting observation reported by Edgar et al. (2016) for influenza and herpes viruses is that the time at which an infection occurs can influence the success and course of infection. The viremia levels in mice infected in the morning were 10-fold higher than those infected in the evening. It was suggested that the mechanism for this significant difference was based on temporal differences in cellular activity and gene expression. Although the research did not include a consideration of vector-borne pathogens, one might wonder if this relationship between severity of acute infections and circadian rhythm could apply since the time at which vectors' host seek and feed and duration of feeding can be variable and species specific. For example, many species of *Aedes* mosquitoes are regarded as daytime or crepuscular feeders whereas most *Culex* species are regarded as dusk and nighttime feeders. If the time of host exposure to a pathogen influences establishment, viremia, and outcome, then based on Edgar et al.'s (2016) observation, transmission of a pathogen by a mosquito that feeds early in the morning may be more consequential than one transmitted by a mosquito feeding at dusk.

Arthropod Saliva as a Target to Protect the Host From Infection

The idea that vaccines directed toward arthropod vectors might impact the pathogens that they transmit, either indirectly by disrupting feeding or directly by influencing the pathogen, has been discussed over a long period of time (e.g., Dubin et al., 1948; Feinsod et al., 1975; Wikel, 1980, 1982). More recent reviews describe how the saliva and associated immunomodulatory factors could be vaccine targets to prevent pathogen transmission and how these might be applied (Leitner et al., 2015; McDowell, 2015; Titus et al., 2006). When vertebrates are fed upon by hematophagous arthropods, immune responses may be induced not only by pathogens that are transmitted in the saliva but also to salivary proteins. The responses to saliva can be demonstrated by, for example, testing exposed vertebrate sera in immunological assays such as Western blotting. It has been suggested that one application for these antisaliva responses could be to evaluate the intensity of exposure to arthropod feeding and by extrapolation the risk of infection. Such markers related to feeding exposure have been discussed for different arthropod vectors including mosquitoes (Ali et al., 2012) and sandflies (Teixeira et al., 2010). Assuming that intensity of the immune response can be correlated with exposure to saliva, a risk index could be developed. In a malaria endemic area for example, if a human has a strong immune response to vector saliva, then it might be interpreted that they are at relatively high risk to infection. Leitner et al. (2015) provide an excellent discussion on immune responses to saliva proteins as biomarkers of exposure.

The first description of salivary enhancement in 1988 (Titus and Ribeiro, 1988) demonstrated that salivary extracts can enhance infectivity of *Leishmania*. Subsequent studies demonstrated that presensitization of mice by exposure to the bites of uninfected sandflies not only abrogated the salivary enhancement, but actually had a controlling effect on infection with reduced subsequent transmission from the infected animals back to sand flies (Kambarri et al., 2000). The mechanism of this protective effect was hypothesized to be via a delayed type hypersensitivity response to the initial sand fly exposure that resulted in a CD4+ cell-dependent infiltration of macrophages and eosinophils into the dermis, with increased numbers of keratinocytes, gamma delta T cells, and Langerhans cells producing IL-12 and IFN-gamma. Studies in nonhuman primates have subsequently demonstrated that a sand fly salivary protein vaccine has efficacy to protect against cutaneous leishmaniasis (Oliveira et al., 2015).

With respect to arboviruses it has been suggested that knowledge of the salivary molecules present in SGE, which react with DENV, might help to identify new targets for anti-dengue strategies (Cao-Lormeau, 2009). The approach would use transmission blocking by generating antibodies to receptors that disrupted the transmission process. Interestingly, research that was designed to demonstrate the utility of the salivary protein D7 for use as a vaccine target to elicit a protective response against WNV unexpectedly found that vaccination enhanced pathogenesis. Furthermore, serum transferred from vaccinated mice increased mortality in mice challenged with WNV (Reagan et al., 2012). The authors concluded that "selection of salivary protein vaccines on the bases of abundance and immunogenicity does not predict efficacy", a lesson for all of those considering use of antisaliva vaccines to protect against arthropod-borne pathogens.

Approaches that are more general could incl ude antisaliva vaccines to target salivary proteins involved in blood acquisition. Such a strategy would likely be most effective against ticks, rather than diptera that feed rapidly. The approach might impact vector fecundity and result in population suppression due to smaller blood meals resulting in fewer eggs being produced. However, a consequence might be that vectors seek more hosts to compensate for the smaller meals, thereby potentially increasing pathogen transmission (Leitner et al., 2015). An excellent overview of how salivary proteins could be targeted to disrupt transmission of a range of pathogens is provided by McDowell (2015), including a discussion on strategies to inhibit feeding, neutralize enhancement, and so-called bystander effects.

CONCLUSIONS

As described with many examples in this chapter, although the saliva of hematophagous arthropods share many common activities related to blood acquisition, the actual substances responsible for specific activities, for example vasodilation, may vary and there is considerable diversity in the saliva of different hematophagous arthropods. Immune responses of the vertebrate host to the saliva of arthropods are complex and not fully understood, especially in the context of how saliva influences the transmission efficiency of different pathogens, establishment of these pathogens in the host, and ultimately disease development and severity. Application of modern genetic and immunological techniques is providing greater insight into the interplay between immune molecules and pathogens, with considerable opportunities to develop new strategies to disrupt pathogen transmission and protect the host from infection. Recent advances in the ability to change vector population structure, for example by releasing *Wolbachia*-infected mosquitoes or with genetically manipulated vectors by releasing insects with dominant lethal genes, have the capacity to impact arthropod-borne pathogens, either directly by altering the capacity of vectors to transmit pathogens or indirectly by suppressing vector populations (Higgs, 2013) The revolutionary technique to introduce gene drive into a wide range of species including mosquitoes is already being considered for pathogen control (National Academies, 2016), and one can imagine that with appropriate knowledge and understanding, this approach could target aspects of the vector–pathogen–host interactions discussed in this chapter.

References

Ali, Z.M., Bakli, M., Fontaine, A., Bakkali, N., Hai, V.V., Audebert, S., Boublik, Y., Pages, F., Remoue, F., Rogier, C., Fraisier, C., Almeras, L., 2012. Assessment of Anopheles salivary antigens as individual exposure biomarkers to species-specific malaria vector bites. Malar. J. 11, 439.

Alekseev, A.N., Chunikhin, S.P., Rukhkyan, M.Y., Stefutkina, L.E., 1991. Possible role of Ixodidae salivary gland substrate as an adjuvant enhancing arbovirus transmission. Med. Parazitol. Parazit. Boleznei 1, 28–31 (in Russian).

Arca, B., Lombardo, F., de Lara Capurro, M., della Torre, A., Dimopoulos, G., James, A.A., Coluzzi, M., 1999. Trapping cDNAs encoding secreted proteins from the salivary glands of the malaria vector Anopheles gambiae. Proc. Natl. Acad. Sci. U.S.A. 96, 1516–1521.

Arca, B., Lombardo, F., Lanfrancotti, A., Spanos, L., Veneri, M., Louis, C., Coluzzi, M., 2002. A cluster of four D7-related genes is expressed in the salivary glands of the African malaria vector Anopheles gambiae. Insect Mol. Biol. 11, 47–55.

Arca, B., Lombardo, F., Valenzuela, J.G., Francischetti, I.M.B., Marinotti, O., Coluzzi, M., Ribeiro, J.M.C., 2005. An updated catalogue of salivary gland transcripts in the adult female mosquito, Anopheles gambiae. J. Exp. Biol. 208, 3971–3986.

Benda, R., 1958. The common tick, Ixodes ricinus, L., as a reservoir and vector of tick borne encephalitis. I. Survival of the virus (strain B3) during the development of the tick under laboratory conditions. J. Hyg. Epidemiol. Microbiol. Immunol. 2, 314–330.

Billingsley, P.F., Baird, J., Mitchell, J.A., Drakeley, C., 2006. Immune interactions between mosquitoes and their hosts. Parasite Immunol. 28, 143–153.

Black, W.C.I.V., Kondratieff, B.C., 2005. Evolution of arthropod disease vectors. In: Marquardt, W.C., Kondratieff, B., Moore, C.G., Freier, J., Hagedorn, H.H., Black III, W., James, A.A., Hemingway, J., Higgs, S. (Eds.), The Biology of Disease Vectors, second ed. Elsevier Academic Press, pp. 9–23. 785 pp. 2005.

Brossard, M., Wikel, S.K., 2004. Tick immunobiology. Parasitology 129, S161–S176.

Calvo, E., Andersen, J., Francischetti, I.M., del Capurro, M., deBianchi, A.G., James, A.A., Ribeiro, J.M.C., Mirinotti, O., 2004. The transcriptome of adult female Anopheles darlingi salivary glands. Insect Mol. Biol. 13, 73–88.

Calvo, E., Mans, B.J., Andersen, J.F., Ribeiro, J.M.C., 2006a. Function and evolution of a mosquito salivary gland protein family. J. Biol. Chem. 281, 1935–1942.

Calvo, E., Pham, V.M., Lombardo, F., Arca, B., Ribeiro, J.M., 2006b. The sialotranscriptome of adult male Anopheles gambiae mosquitoes. Insect Biochem. Mol. Biol. 36, 570–575.

Calvo, E., Ribeiro, J.M.C., 2006. A novel secreted endonuclease from Culex quinquefasciatus salivary glands. J. Exp. Biol. 209, 2651–2659.

Cao-Lormeau, V.-M., 2009. Dengue viruses binding proteins for Aedes aegypti and Aedes polynesiensis salivary glands. Virol. J. 6, 35–38.

Cavassani, K.A., Aliberti, J.C., Dias, A.R., Silva, J.S., Ferreira, B.R., 2005. Tick saliva inhibits differentiation, maturation and function of murine bone-marrow-derived dendritic cells. Immunology 114, 235–245.

Champagne, D.E., 2004. Antihemostatic strategies of blood-feeding arthropods. Curr. Drug Targets Cardiovasc. Haematol. Disord. 4, 375–396.

Champagne, D.E., Smartt, C.T., Ribeiro, J.M.C., James, A.A., 1995. The salivary gland-specific apyrase of the mosquito Aedes aegypti is a member of the 5'-nucleotide family. Proc. Natl. Acad. Sci. U.S.A. 92, 694–698.

Chmelař, J., Kotál, J., Karim, S., Kopacek, P., Francischetti, I.M., Pedra, J.H., Kotsyfakis, M., 2016a. Sialomes and mialomes: a systems-biology view of tick tissues and tick-host interactions. Trends Parasitol. 32, 242–254.

Chmelař, J., Kotál, J., Kopecký, J., Pedra, J.H., Kotsyfakis, M., 2016b. All for one and one for all on the tick-host battlefield. Trends Parasitol. 32, 368–377.

Chumakov, M.P., 1944. Studies on virus encephalitides. 6. Transmission of tick-borne encephalitis to the offspring in Ixodidae ticks and the question of natural reservoirs of this infection. Med. Parazitol. Parazitar. Bolezni 6, 38–40.

Chumakov, M.P., Petrova, S.P., Sondak, V.A., 1945. A study of ultravirus encephalitis. VII. Artificial adaptation of the virus of ticks and Japanese encephalitis to various species of ticks of the family Ixodidae. Med. Parasitol. Moscow 14, 18–24.

Cross, M.L., Cupp, E.W., Enriquez, F.J., 1994. Differential modulation of murine cellular immune responses by salivary gland extract of Aedes aegypti. Am. J. Trop. Med. Hyg. 51, 690–696.

Dubin, Y.N., Reese, J.D., Seamans, L.A., 1948. Attempt to produce protection against mosquitoes by active immunization. J. Immunol. 58, 293–297.

Edgar, R.S., Stangherlin, A., Nagy, A.D., Nicoll, M.P., Efstathiou, S., O'Neill, J.S., Reddy, A.B., 2016. Cell autonomous regulation of herpes and influenza virus infection by the circadian clock. Proc. Natl. Acad. Sci. U.S.A. http://dx.doi.org/10.1073/pnas.1601895113 (e-Published ahead of print).

Edwards, J.F., Higgs, S., Beaty, B.J., 1998. Mosquito feeding-induced enhancement of Cache Valley virus (Bunyaviridae) infection in mice. J. Med. Entomol. 35, 261–265.

Feinsod, F.M., Spielman, A., Waner, J.L., 1975. Neutralization of Sindbis virus by antisera to antigens of vector mosquitoes. Am. J. Trop. Med. Hyg. 24, 533–536.

Fontaine, A., Diouf, I., Bakkali, N., Missé, D., Pagès, F., Fusai, T., Rogier, C., Almeras, L., 2011. Implication of haematophagous arthropod salivary proteins in host-vector interactions. Parasit. Vectors 4, 187.

Gern, L., Rais, O., 1996. Efficient transmission of *Borrelia burgdorferi* between cofeeding *Ixodes ricinus* ticks (Acari: Ixodidae). J. Med. Ent. 33, 189–192.

Hermance, M., Thangamani, S., 2015. Tick saliva enhances Powassan virus transmission to the host influencing its dissemination and the course of disease. J. Virol. 89 (15), 7852–7860.

Hermance, M.E., Santos, R.I., Kelly, B.C., Valbuena, G., Thangamani, S., 2016. Immune cell targets of infection at the tick-skin interface during Powassan virus transmission. PLoS One 11 (5), e0155889.

Higgs, S., 2004. How do mosquito vectors live with their viruses? In: Gillespie, S.H., Smith, G.L., Osbourn, A. (Eds.), Microbe-Vector Interactions in Vector-borne Diseases. Cambridge University Press, Cambridge, pp. 103–137.

Higgs, S., 2006. A novel method of West Nile virus transmission. Contagion 3, 95–97.

Higgs, S., 2013. Alternative approaches to control dengue and chikungunya: transgenic mosquitoes. Public Health 24, 35–42.

Higgs, S., Schneider, B.S., Vanlandingham, D.L., Klingler, K., Gould, E.A., 2005. Nonviremic transmission of West Nile virus. Proc. Natl. Acad. Sci. U.S.A. 102, 8871–8874.

Higgs, S., Vanlandingham, D.L., 2016. Influences of arthropod vectors on encephalitic arboviruses. In: Reiss, C.S. (Ed.), Neurotropic Viral Infections. Neurotropic Retroviruses, DNA Viruses, Immunity and Transmission, vol. 2, second ed. Springer International Publishing, Cham, pp. 371–401. ISBN: 9783319331881.

James, A.A., 1994. Molecular and biochemical analyses of the salivary glands of vector mosquitoes. Bull. Inst. Pasteur 92, 133–150.

James, A.A., Blackmer, K., Marinotti, O., Ghosn, C.R., Racioppi, J.V., 1991. Isolation and characterization of the gene expressing the major salivary gland protein of the female mosquito, *Aedes aegypti*. Mol. Biochem. Parasitol. 44, 245–253.

Jones, L.D., Davies, C.R., Steele, G.M., Nuttall, P.A., 1987. A novel mode of arbovirus transmission involving a nonviremic host. Science 237, 775–777.

Jones, L.D., Hodgson, E., Williams, T., Higgs, S., Nuttall, P.A., 1992a. Saliva activated transmission (SAT) of Thogoto virus: relationship with vector potential of different haematophagous arthropods. Med. Vet. Entomol. 6, 261–265.

Jones, L.D., Matthewson, M., Nuttall, P.A., 1992b. Saliva-activated transmission (SAT) of Thogoto virus: dynamics of SAT factor activity in the salivary glands of *Rhipicephalus appendiculatus*, *Amblyomma variegatum*, and *Boophilus microplus* ticks. Exp. Appl. Acarol. 13, 241–248.

Kambarri, S., Belkaid, Y., Modi, G., Rowton, E., Sacks, D., 2000. Protection against cutaneous leishmaniasis resulting from bites of uninfected sand flies. Science 290, 1351–1354.

Kerlin, R.L., Hughes, S., 1992. Enzymes in saliva from four parasitic arthropods. Med. Vet. Entomol. 6, 121–126.

Kondratenko, V.F., Blagoveschenskaya, N.M., Butenko, A.M., Vyshnivetskaya, L.K., Zarubina, L.V., Milutin, V.N., Kuchin, V.V., Novikova, E.M., Rabinovich, V.D., Shevchenko, S.F., Chumakov, M.P., 1970. Results of virological investigation of ixodid ticks in Crimean hemorrhagic fever focus in Rostov Oblast. Mater, 3. Oblast. Nauchn. Prakt. Konf. (Rostov-on-Don) 29–35.

Korenberg, E.I., Pchelkina, A.A., 1984. Tickborne encephalitis virus titers in engorged adult *Ixodes persulcatus* ticks. Parazitologiia 18, 123–127.

Kotsyfakis, M., Anderson, J.F., Francischetti, I.M., Ribeiro, J.M., 2005. *Ixodes scapularis* can suppress host cysteine protease activity in the sites of blood feeding. In: Abstract 547, 54th annual meeting of the American Society for Tropical Medicine and Hygiene, December 11–15.

Labuda, M., Jones, L.D., Williams, T., Danielova, V., Nuttall, P.A., 1993a. Efficient transmission of tick-borne encephalitis virus between co-feeding ticks. J. Med. Entomol. 30, 295–299.

Labuda, M., Jones, L.D., Williams, T., Nuttall, P.A., 1993b. Enhancement of tick-borne encephalitis virus transmission by salivary gland extracts. Med. Vet. Entomol. 7, 193–196.

Lanfrancotti, A., Lombardo, F., Santolamazza, F., Veneri, M., Castrignano, T., Coluzzi, M., Arca, B., 2002. Novel cDNAs encoding salivary proteins from the malaria vector *Anopheles gambiae*. FEBS Lett. 517, 67–71.

Leitner, W.W., Wali, T., Kincaid, R., Costero-Saint, D.A., 2015. Arthropod vectors and disease transmission: translational aspects. PLoS Negl. Trop. Dis. 9 (11), e0004107.

Limesand, K.H., Higgs, S., Pearson, L.D., Beaty, B.J., 2000. Potentiation of vesicular stomatitis New Jersey virus infection in mice by mosquito saliva. Parasite Immunol. 22, 461–467.

Limesand, K.H., Higgs, S., Pearson, L.D., Beaty, B.J., 2003. The effect of mosquito salivary gland treatment on vesicular stomatitis New Jersey virus replication and interferon α/β expression *in vitro*. J. Med. Entomol. 40, 199–205.

142

8. MOSQUITO-ARBOVIRUS–HOST INTERACTIONS

McDowell, M.A., 2015. Vector-transmitted disease vaccines: targeting salivary proteins in transmission (SPIT). Trends Parasitol. 8, 363–372.

McGee, C.E., Schneider, B.S., Girard, Y.A., Vanlandingham, D.L., Higgs, S., 2007. Nonviremic transmission of West Nile virus: evaluation of the effects of space, time and mosquito species. Am. J. Trop. Med. Hyg. 76, 424–430.

Mellinck, J.J., van Zeben, M.S., 1976. Age related differences of saliva composition in Aedes aegypti. Mosq. News 36, 247–250.

Nascimento, E.P., dos Santos Malafronte, R., Marinotti, O., 2000. Salivary gland proteins of the mosquito Culex quinquefasciatus. Arch. Insect Biochem. Physiol. 43, 9–15.

National Academies of Sciences, Engineering and Medicine, 2016. Gene Drives on the Horizon: Advancing Science, Navigating Uncertainty, and Aligning Research with Public Values. The National Academies Press, Washington, DC. http://dx.doi.org/10.17226/23405.

Nuttall, P.A., Jones, L.D., 1991. Non-viraemic tick-borne virus transmission: mechanism and significance. In: Dusbábek, F., Bukva, V. (Eds.). Modern Acarology. Modern Acarology vol. 2. Academia Prague, The Hague, pp. 3–6.

Nuttall, P.A., Labuda, M., 2004. Tick-host interactions: saliva activated transmission. Parasitology (129) Suppl., S177–S189.

Oliveira, F., Rowton, E., Aslan, H., Gomes, R., Castrovinci, P.A., Alvarenga, P.H., Abdeladhim, M., Teixeira, C., Meneses, C., Kleeman, L.T., Guimarães-Costa, A.B., Rowland, T.E., Gilmore, D., Doumbia, S., Reed, S.G., Lawyer, P.G., Andersen, J.F., Kamhawi, S., Valenzuela, J.G., 2015. A sand fly salivary protein vaccine shows efficacy against vector-transmitted cutaneous leishmaniasis in nonhuman primates. Sci. Transl. Med. 7 (290), 290ra90.

Osorio, J.E., Godsey, M.S., Defoliart, G.R., Yuill, T.M., 1996. La Crosse viremias in white-tailed deer and chipmunks exposed by injection or mosquito bite. Am. J. Trop. Med. Hyg. 54, 338–342.

Patrican, L.A., 1997. Acquisition of Lyme disease spirochaetes by cofeeding Ixodes scapularis ticks. Am. J. Trop. Med. Hyg. 57, 589–593.

Pingen, M., Bryden, S.R., Pondeville, E., Schnettler, E., Kohl, A., Merits, A., Fazakerley, J.K., Graham, G.J., McKimmie, C.S., 2016. Host inflammatory response to mosquito bites enhances the severity of arbovirus infection. Immunity 44 (6), 1455–1469.

Poehling, H.M., 1979. Distribution of specific proteins in the salivary gland lobes of Culicidae and their relation to age and blood sucking. J. Insect Physiol. 25, 3–8.

Pretzmann, G., Loew, J., Radda, A., 1963. Untersuchen in einem Naturherd der Fruhsummer-Meningoencephalitis (FSME) in Niederosterreich 3. Mitteilung: Versuch einer Gesamtdarstellung des Zyklus der FSME in Naturherd Zentr. Bacteriol. Parasitenk Aby Orig. 190, 299–312.

Racioppi, J.V., Spielman, A., 1987. Secretory proteins from the salivary glands of adult Aedes aegypti mosquitoes. Insect Biochem. 17, 503–511.

Randolph, S.E., Gern, L., Nuttall, P.A., 1996. Co-feeding ticks: epidemiological significance for tick-borne pathogen transmission. Parasitol. Today 12, 472–479.

Randolph, S.E., Green, R.M., Peacey, M.F., Rogers, D.J., 2000. Seasonal synchrony: the key to tick-borne encephalitis foci identified by satellite data. Parasitology 121, 15–23

Reagan, K.L., Machain-Williams, C., Wang, T., Blair, C.D., 2012. Immunization of mice with recombinant mosquito salivary protein D7 enhances mortality from subsequent West Nile virus infection via mosquito bite. PLoS Negl. Trop. Dis. 6 (12), e1935.

Reisen, W.K., Fang, Y., Martinez, V., 2007. Is nonviremic transmission of West Nile virus by Culex mosquitoes (Diptera: Culicidae) nonviremic? J. Med. Entomol. 44, 299–302.

Ribeiro, J.M.C., 1987. Role of saliva in blood-feeding by arthropods. Annu. Rev. Entomol. 32, 463–478.

Ribeiro, J.M.C., 1989. Vector saliva and its role in parasite transmission. Exp. Parasitol. 69, 104–106.

Ribeiro, J.M.C., 1992. Characterization of a vasodilator from the salivary glands of the yellow fever mosquito Aedes aegypti. J. Exp. Biol. 165, 61–71.

Ribeiro, J.M.C., 2000. Blood-feeding in mosquitoes: probing time and salivary gland anti-haemostatic activities in representatives of three genera (Aedes, Anopheles, Culex). Med. Vet. Entomol. 14, 142–148.

Ribeiro, J.M., Alarcon-Chaidez, F., Francischetti, I.M., Mans, B.J., Mather, T.N., Valenzuela, J.G., Wikel, S.K., 2006. An annotated catalog of salivary gland transcripts from Ixodes scapularis ticks. Insect Biochem. Mol. Biol. 36, 111–129.

Ribeiro, J.M.C., Charlab, R., Valenzuela, J.G., 2001. The salivary adenosine deaminase activity of the mosquitoes Culex quinquefasciatus and Aedes aegypti. J. Exp. Biol. 204, 2001–2010.

Ribeiro, J.M.C., Charlab, R., Pham, V.M., Garfield, M., Valenzuela, J.G., 2004. An insight into the salivary transcriptome and proteome of the adult mosquito Culex pipiens quinquefasciatus. Insect Biochem. Mol. Biol. 34, 543–563.

Ribeiro, J.M.C., Makoul, G., Levine, J., Robinson, D., Spielman, A., 1985. Antihemostatic, antiinflammatory and immunosuppressive properties of the saliva of a tick, Ixodes dammini. J. Exp. Med. 161, 332–344.

Ribeiro, J.M.C., Nussenzveig, R.H., Tortorella, G., 1994. Salivary vasodilators of Aedes triseriatus and Anopheles gambiae (Diptera: Culicidae). J. Med. Entomol. 31, 747–753.

Ribeiro, J.M.C., Francishetti, I.M.B., 2003. Role of arthropod saliva in blood feeding: sialome and post-sialome perspectives. Annu. Rev. Entomol. 48, 73–88.

Ribeiro, J.M.C., Nussenzveig, R.H., 1993. The salivary catechol oxidase/peroxidase activities of the mosquito *Anopheles albimanus*. J. Exp. Biol. 179, 273–287.

Ribeiro, J.M.C., Spielman, A., 1986. *Ixodes dammini*: salivary anaphylatoxin inactivating activity. Exp. Parasitol. 62, 292–297.

Ribeiro, J.M.C., Valenzuela, J.G., 1999. Purification and cloning of the salivary peroxidase/catechol oxidase of the mosquito *Anopheles albimanus*. J. Exp. Biol. 202, 809–816.

Ribeiro, J.M.C., Weis, J.J., Telford III, S.R., 1990. Saliva of the tick *Ixodes dammini* inhibits neutrophils functions. Exp. Parasitol. 70, 382–388.

Roseboom, L.E., Burgdorfer, W., 1959. Development of Colorado tick fever virus in the Rocky Mountain wood tick, *Dermacentor andersoni*. Am. J. Hyg. 69, 138–145.

Santos, R.I., Hermance, M.E., Gelman, B.B., Thangamani, S., 2016. Spinal cord ventral horns and lymphoid organ involvement in Powassan virus infection in a mouse model. Viruses 8 (8), 220.

Schmid, M.A., Glasner, D.R., Shah, S., Michlmayr, D., Kramer, L.D., Harris, E., 2016. Mosquito saliva increases endothelial permeability in the skin, immune cell migration, and dengue pathogenesis during antibody-dependent enhancement. PLoS Pathog. 12 (6), e1005676.

Schneider, B.S., Soong, L., Zeidner, N.S., Higgs, S., 2004. *Aedes aegypti* salivary gland extracts modulate anti-viral and TH1/TH2 cytokine responses to sindbis virus infection. Viral Immunol. 17, 565–573.

Schneider, B.S., Soong, L., Girard, Y.A., Campbell, G., Mason, P., Higgs, S., 2006. Potentiation of West Nile encephalitis by mosquito feeding. Viral Immunol. 19, 74–82.

Schneider, B.S., McGee, C.E., Jordan, J.M., Stevenson, H.L., Soong, L., Higgs, S., 2007. Prior exposure to uninfected mosquitoes enhances mortality in naturally-transmitted West Nile virus infection. PLoS One 2 (11), e1171.

Stark, K.R., James, A.A., 1996. Salivary gland anticoagulants in culicine and anopheline mosquitoes (Diptera: Culicidae). J. Med. Entomol. 33, 645–650.

Steen, N.A., Barker, S.C., Alewood, P.F., 2006. Proteins in the saliva of the Ixodida (ticks): pharmacological features and biological significance. Toxicon 47, 1–20.

Stockman, S., 1918. Louping-ill. J. Comp. Pathol. 31, 137–193.

Styer, L.M., Bernard, K.A., Kramer, L.D., 2006. Enhanced early West Nile virus infection in young chickens infected by mosquito bite: effect of viral dose. Am. J. Trop. Med. Hyg. 75, 337–345.

Styer, L.M., Lim, P.Y., Louie, K.L., Albright, R.G., Kramer, L.D., Bernard, K.A., 2011. Mosquito saliva causes enhancement of West Nile virus infection in mice. J. Virol. 85, 1517–1527.

Suwan, N., Wilkinson, M.C., Crampton, J.M., Bates, P.A., 2002. Expression of D7 and D7 related proteins in the salivary glands of the human malaria mosquito, *Anopheles stephensi*. Insect Mol. Biol. 11, 223–232.

Teixeira, C., Gomes, R., Collin, N., Reynoso, D., Jochim, R., Oliveira, F., Seitz, A., Elnaiem, D.E., Caldas, A., de Souza, A.P., Brodskyn, C.I., de Oliveira, C.I., Mendonca, I., Costa, C.H., Volf, P., Barral, A., Kamhawi, S., Valenzuela, J.G., 2010. Discovery of markers of exposure specific to bites of *Lutzomyia longipalpis*, the vector of *Leishmania infantum chagasi* in Latin America. PLoS Negl. Trop. Dis. 4, e638.

Titus, R.G., Bishop, J.V., Mejia, J.S., 2006. The immunomodulatory factors of arthropod saliva and the potential for these factors to serve as vaccine targets to prevent pathogen transmission. Parasite Immunol. 28, 131–141.

Titus, R.G., Ribeiro, J.M.C., 1988. Salivary gland lysates from the sand fly *Lutzomyia longipalpis* enhance *Leishmania* infectivity. Science 239, 1306–1308.

Tonetti, N., Voordouw, M.J., Durand, J., Monnier, S., 2015. Genetic variation in transmission success of the Lyme borreliosis pathogen *Borrelia afzelii*. Ticks Tick Borne Dis. 6, 334–343.

Valenzuela, J.G., 2002. High-throughput approaches to study salivary gland proteins and genes from vectors of disease. Insect Biochem. Mol. Biol. 32, 1199–1209.

Valenzuela, J.G., 2004. Blood-feeding arthropod salivary glands and saliva. In: Marquardt, W.C., Kondratieff, B., Moore, C.G., Freier, J., Hagedorn, H.H., Black III, W., James, A.A., Hemingway, J., Higgs, S. (Eds.), The Biology of Disease Vectors. Elsevier Academic, San Diego, pp. 377–386.

Valenzuela, J.G., Charlab, R., Gonzalez, E.C., Miranda-Santos, I.K.F., Marinotti, O., Francischetti, I.M., Ribeiro, J.M.C., 2002a. The D7 family of salivary proteins in blood sucking diptera. Insect Mol. Biol. 11, 149–155.

Valenzuela, J.G., Pham, V.M., Grafield, M.K., Francischetti, I.M.B., Ribeiro, J.M.C., 2002b. Toward a description of the sialome of the adult female mosquito *Aedes aegypti*. Insect Biochem. Mol. Biol. 32, 1101–1122.

Varma, M.G.R., Smith, C.E.G., 1962. Studies of Langat virus (TO 21) in *Haemaphysalis spinigera* Neumann. In: Libikova, H. (Ed.), Biology of Viruses of the Tick-borne Encephalitis Complex. Proceedings of a Symposium on the Biology of Viruses of the Tick-borne Encephalitis Complex, Smolenice, 1960. Academic, New York, pp. 397–400.

Voorouw, M., 2015. Co-feeding transmission in Lyme disease pathogens. Parasitology 142, 290–302.

Wanasen, N., Nussenzveig, R.H., Champagne, D.E., Soong, L., Higgs, S., 2004. Differential modulation of murine host immune response by salivary gland extracts from the mosquitoes *Aedes aegypti* and *Culex quinquefasciatus*. Med. Vet. Entomol. 18, 191–199.

Wasserman, H.A., Singh, S., Champagne, D.E., 2004. Saliva of the yellow fever mosquito, *Aedes aegypti*, modulates murine lymphocyte function. Parasite Immunol. 26, 295–306.

Wikel, S.K., 1980. Host resistance to tick-borne pathogens by virtue of resistance to tick infestation. Ann. Trop. Med. Parasitol. 74, 103–104.

Wikel, S.K., 1982. Immune responses to arthropods and their products. Annu. Rev. Entomol. 27, 21–48.

Wikel, S.W., 1996. Immunology of the tick-host interface. In: Wikel, S.K. (Ed.), The Immunology of Host-ectoparasitic Arthropod Relationships. CAB International, Wallingford, UK, pp. 204–231.

Wikel, S.K., Ramachandra, R.N., Bergman, D.K., 1996. Arthropod modulation of host immune responses. In: Wikel, S.K. (Ed.), Immunology of Host-ectoparasitic Arthropod Relationships. CAB International, Wallingford, UK, pp. 107–130.

Zeidner, N.S., Higgs, S., Happ, C.M., Beaty, B.J., Miller, B.R., 1999. Mosquito feeding modulates Th1 and Th2 cytokines in flavivirus susceptible mice: an effect mimicked by injection of sialokinins, but not demonstrated in flavivirus resistant mice. Parasite Immunol. 21, 35–44.

Tick Saliva: A Modulator of Host Defenses

Stephen Wikel

Quinnipiac University, Hamden, CT, United States

INTRODUCTION

Ticks are effective vectors of numerous bacterial, viral, and protozoan infectious agents of humans and other animal species (Jongejan and Uilenberg, 2004; Dennis and Piesman, 2005; Parola and Raoult, 2001; Parola et al., 2013; Nelder et al., 2016). Ticks and tick-borne diseases have significant medical, veterinary, and economic consequences for developed and developing countries throughout the world (Jongejan and Uilenberg, 2004; Kiss et al., 2012). Factors impacting changing distribution of tick species and infectious agents they transmit include climate change, acaricide resistance, residing in tick habitat, inadequate vector surveillance and control programs, reduced public health infrastructure, international travel, absence of vaccines and therapeutics for many tick-borne diseases, and changes in pathogen virulence. An important consideration for understanding tick–host interactions is the range of animal species that an individual tick species may infest and variations in physiology of those host species. In order to effectively blood feed on multiple host species, tick saliva consists of proteins and nonprotein molecules that are differentially expressed during infestation and display high levels of polymorphisms within gene families (Chmelař et al., 2016a,b).

Research focused on tick interactions with the host immune system advanced at a study pace over the last several decades fostered by the application of increasingly powerful tools to define cellular and molecular underpinnings of tick–host relationships. Significant advances continue to increase both depth and scope of understanding of host innate and adaptive immune responses at the tick–host–pathogen interface, tick countermeasures to immune defenses, and actions of bioactive saliva molecules (Kazimírová and Štibrániová, 2013; Wikel, 2013; Kotál et al., 2015). Contributing to the current expansion of knowledge are transcriptome analyses of tick bite site skin cells during infestation (Heinze et al., 2012, 2014). Next generation sequencing and quantitative proteomics are revealing far more complex salivary gland compositions than previously described (Chmelař et al., 2016a,b).

Tick-induced modulation of host immune defenses evolved to facilitate blood feeding and the actions of saliva immunomodulatory molecules create an environment favorable for tick-borne pathogen transmission and establishment (Wikel, 2013). Saliva immunomodulatory molecules reduce the likelihood that a host will develop immune resistance during an infestation lasting from several days to over a week as well as during subsequent blood feeding by the same tick species. Ability of a tick to modify host innate and adaptive immune defenses may be a factor in evolution of the clade of infectious agents transmitted by a specific tick species. Building further on these relationships, interactions of tick saliva and host defenses must be considered and incorporated in models for studying tick-borne infectious agent interactions with their hosts and development of interventions such as vaccines. To accurately assess pathogenesis and efficacy of vaccines for tick-borne infectious agents, use of tick transmission models is optimal, if not essential.

Objective of this chapter is to overlay what is known about tick stimulation and modulation of innate and adaptive immune responses with the expected sequence of cellular and molecular interactions that occur when immunogens are introduced into a break in the skin. Rationale for this approach is that it provides a picture of the dynamic sequence of tick–host immune interactions and interrelationships. Different tick species stimulate and modulate host immune defenses in potentially very different ways. Those differences and similarities are reflected in conserved and diverse transcripts and proteins in tick salivary gland transcriptomes and proteomes (Chmelař et al., 2016a,b). Tick saliva modulates essentially all major cellular and molecular immune effectors (Kazimírová and Štibrániová, 2013; Wikel, 2013; Kotál et al., 2015). Although gaps exist in our knowledge of tick–host immune interactions, a sufficient body of information exists to establish a framework of how these important arthropod vectors interact with the complex immune networks of cells and molecular mediators of effector and regulatory pathways.

SKIN, TICKS, AND TICK SALIVA

Ticks are telmophages that consume blood from a pool involving dermal blood vessels lacerated by the chelicerae (Lavoipierre, 1965; Bergman, 1996). Tick mouthparts are comprised of a pair of dorsal chelicerae that cut the skin into the dermis to access blood vessels and the underlying hypostome with recurved ventral surface teeth to anchor the tick to the host (Gregson, 1960; Bergman, 1996; Sonenshine and Anderson, 2014). Many tick species produce a protein attachment cement in a pattern related to mouthpart structure to help maintain their hold on host skin (Moorhouse, 1969; Bullard et al., 2016). The pair of palps attached to the basis capituli do not enter the feeding lesion, but contain numerous sensory structures for assessing chemical signals from host skin (Sonenshine and Anderson, 2014).

Duration of blood feeding differs for argasid and ixodid ticks as well as for individual life cycle stages. Argasid ticks consume multiple blood meals that last approximately 2h (Cooley and Kohls, 1944). Ixodid ticks remain attached to the host for periods of approximately 4 days for larvae and nymphs and for potentially longer than a week for female ixodid, three host ticks with the engorged female increasing to approximately 100 times her unfed weight (Kaufman, 1989).

During the process of probing, attachment, and blood feeding, the host is exposed to a complex mixture of proteins, nucleosides, and lipids (Ribeiro et al., 2006; Alarcon-Chaidez et al., 2007; Oliveira et al., 2011; Karim and Ribeiro, 2015; Chmelař et al., 2016a,b) that counteract host defenses by inhibiting host blood coagulation and platelet aggregation; reducing pain and itch responses; modulating inflammation, innate, and adaptive immune responses; and altering the course of wound healing.

SKIN IMMUNE NETWORK

Finding that Langerhans cells of the epidermis trap tick saliva antigens provides a direct link between tick feeding and host innate and adaptive immune response stimulation (Allen et al., 1979b). Skin contains a multitude of cells, soluble mediators, and interactive pathways that are effectors and regulators of inflammation and innate and adaptive immune responses that result in local and systemic resistance to challenges from microbes and parasites at the body surface (Kupper and Fuhlbrigge, 2004; Nestle et al., 2009; Pasparakis et al., 2014; Yazdi et al., 2016). In addition to Langerhans cells with immune function, epidermis consists of keratinocytes, CD8[+] T lymphocytes, and dendritic epidermal T cells that are gamma/delta (γ/δ) receptor positive and occur in mouse but not human epidermis (Pasparakis et al., 2014). Keratinocytes sense pathogen-associated molecular patterns and cellular damage-associated molecular patterns (DAMPs) through their toll-like (TLR) and other pattern recognition receptors (Miyake and Yamasaki, 2012). Recognition of conserved microbial molecular signatures and signaling pathways are well defined for TLRs (West et al., 2006). Stimulation of these keratinocyte pattern recognition receptors initiate and regulate cutaneous inflammation and immunity by producing cytokines and chemokines [tumor necrosis factor (TNF), interleukin (IL)-1, IL-7, IL-8, IL-10, IL-12, and others] that initiate signaling cascades involved in activation and migration of cells of the immune response in epidermis and dermis (Gröne, 2002; Lebre et al., 2007; Nestle et al., 2009; Pasparakis et al., 2014). Keratinocyte-derived chemokines attract neutrophils, monocytes, dendritic cells, natural killer (NK) cells, and T lymphocytes to skin inflammatory foci (Lebre et al., 2007). Keratinocyte-derived cytokines influence activation and maturation of skin dendritic cell subpopulations (Said and Weindl, 2015). In turn, keratinocytes possess numerous cytokine and chemokine receptors that are involved in cross talk within the cutaneous immune system (Tüzün et al., 2007).

Dermis contains myeloid and lymphoid cells that reside within or migrate through the dermal network of stromal fibroblasts, fibrocytes, lymphatic vessels, and blood vessels (Kupper and Fuhlbrigge, 2004; Nestle et al., 2009; Pasparakis et al., 2014; Yazdi et al., 2016). Dermis contains numerous dendritic cells consisting of multiple subsets, macrophages, mast cells, CD4[+] T lymphocytes with alpha/beta (α/β) antigen receptors, and in mice the presence of innate lymphoid cells with γ/δ T cell receptors (Pasparakis et al., 2014). Stimulation of innate and adaptive immune responses in the skin result in an influx of different cell populations, production of multiple cytokines, activation of complement, and movement of dendritic cells and antigens into lymphatics to migrate to draining lymph nodes or to the spleen in the case of blood-borne antigens.

Dendritic cells are orchestrating hubs that interconnect and regulate innate and adaptive immunity. Upon interaction with microbes or other antigens, Langerhans cells take of the antigen, beginning processing of the antigen, and migrate from the epidermis to present antigen to lymphoid cells in lymph nodes (Igyártó and Kaplan, 2013). Langerhans cells promote Th17 helper cell responses with IL-17 production that regulates epithelial inflammatory responses and contributes to influxes of neutrophils and eosinophils (Igyártó et al., 2011; Pappu et al., 2012). Th17 lymphocytes promote neutrophil rich responses to extracellular bacteria and fungi, and they contribute to the continuity of innate and adaptive responses (Stockinger et al., 2007).

Multiple dermal dendritic cell subsets have different functional immune response induction roles in skin and secondary lymphoid tissues (Boltjes and van Wijk, 2014). Dermal CD103[+] dendritic cells can stimulate Th1 lymphocyte polarization (Igyártó et al., 2011). Th2 polarization is linked to antigen presentation by mouse dermal CD301b positive dendritic cells (Gao et al., 2013; Kumamoto et al., 2013). Plasmacytoid dendritic regulate innate and adaptive immune responses by producing large amounts of type I interferons, along with IL-6, IL-12, and chemokines that induce resistance to virus infection, enhance NK cell cytotoxicity, induce NK cell IFN-γ production, and promote both B lymphocyte differentiation and Th1 lymphocyte polarization (Colonna et al., 2004; Gilliet et al., 2008).

Skin macrophages are classified as three distinct subpopulations on the basis of being resident or circulating populations (Mosser and Edwards, 2008; Jakubzick et al., 2013). Macrophages are categorized into proinflammatory, classically activated M1 macrophages, regulatory M2 macrophages, and a third subpopulation identified as wound healing macrophages (Mosser and Edwards, 2008). Macrophages are important presenters of antigen and regulators of immune function that display plasticity in their roles with proinflammatory signals (TNF, IFN-γ, DAMPs), driving Th1 (IL-12) and Th2 (IL-4, IL-13) polarization, and antiinflammatory cytokines (IL-10) promoting M2 activity (Porta et al., 2015). M1 macrophages produce proinflammatory cytokines (TNF, IL-1, IL-6), enhance MHC class II molecule expression, and contribute to Th1 and Th17 polarization (Murray et al., 2014). M2 macrophage downregulation of cutaneous inflammation is mediated by IL-10 and transforming growth factor-β production (Locati et al., 2013; Murray et al., 2014). Intense influx of basophils characterizes skin bite site reactions of hosts expressing acquired resistance to tick infestation (Allen, 1973; Wada et al., 2010). In the presence of IL-4 produced by basophils, monocyte precursors of M1 macrophages can be polarized to the antiinflammatory M2 subtype (Egawa et al., 2013).

Mast cells located around blood vessels and nerves in the dermis have immune effector and regulatory roles in inflammation and innate and adaptive immunity (Benyon, 1989; Otsuka et al., 2016). Mast cell granules contain preformed vasoactive, proinflammatory mediators that can be released rapidly by binding of surface TLRs, complement receptors, complement-derived anaphylatoxins, or antigen binding to cell surface bound immunoglobulin E (IgE) (Morita et al., 2016; Otsuka et al., 2016). Mast cell released cytokines and growth factors have regulatory roles for immune functions of dendritic cell and T lymphocytes, particularly related to Th2 response polarization (Abraham and St John, 2010; Voehringer, 2013).

Basophils are important in tick–host relationships due to their established association with expression of acquired resistance to infestation (Allen, 1973; Wada et al., 2010). Role of basophils in immune regulation and protection against pathogens and parasites expanded with the findings that basophils act as antigen-presenting cells expressing MHC class II, costimulatory molecules, and IL-4 contributing to development of Th2 lymphocyte–mediated immunity (Perrigoue et al., 2009; Sokol and Medzhitov, 2010; Karasuyama and Yamanishi, 2014). Innate lymphoid cells, which occur in the skin, interact with mast cells and basophils to induce Th2-polarized responses (Sonnenberg and Artis, 2015). As a consequence of increased interest in their contributions to protective and pathologic immune responses and development of basophil-deficient mice, the protective roles of basophils in resistance to helminths and ectoparasitic arthropods and their immunoregulatory functions are more fully defined in terms of cytokine, vasoactive mediators, proteases, and activation and recruitment of T lymphocytes and eosinophils (Otsuka and Kabashima, 2015; Eberle and Voehringer, 2016).

Resident and circulating lymphocytes provide protection against infectious agents and contribute to pathological processes in both mouse and human skin (Nestle et al., 2009; Pasparakis et al., 2014). Due to the use of murine models to study tick–host interactions, the focus in this chapter is upon mouse skin lymphocytes. Resident population of T lymphocytes of normal skin is estimated to be approximately twice that of circulating T lymphocytes with Th1 effector memory T lymphocytes being the predominant skin resident phenotype (Clark et al., 2006). Cutaneous immune responses can be initiated and maintained in normal skin by the resident T lymphocyte population alone (Clark et al., 2006). Normal skin contains populations of tissue resident effector memory and memory regulatory (Treg) cells (Rosenblum et al., 2011). Resident and circulating T lymphocytes interact to provide cutaneous immune protection. Virus-specific resident memory CD8 T lymphocytes in the epidermis interact with circulating CD4[+] T lymphocyte that move through the skin and provide helper signals (Gebhardt et al., 2011). Gamma/delta T lymphocytes populate skin early in development where they provide immune surveillance and contribute to inflammatory responses (Sumaria et al., 2011).

Cutaneous injury, inflammation, and innate and adaptive immunity involves the influx, activation, phagocytosis, elaboration of cytokines, and intracellular killing by neutrophils and their interactions with tissue resident cells such as macrophages and endothelium (Nathan, 2006; Kim and Luster, 2015). Neutrophil-rich inflammatory response results in uptake and destruction of extracellular infectious agents often accompanied by tissue damage from released neutrophil granule contents accompanied by subsequent neutrophil participation in tissue repair (Weiss, 1989; Summers et al., 2010; Kolaczkowska and Kubes, 2013). Reciprocal interactions between neutrophil, and other leukocyte, integrins, and additional surface molecules with endothelial cell–expressed adhesion molecules and chemokines are integral for egress of these cells from the vasculature into tissue sites of inflammation and immune interaction (Nourshargh and Alon, 2014). Neutrophils respond to chemotactic gradients in tissues and release cytokines that recruit other leukocytes (Summers

et al., 2010; Kolaczkowska and Kubes, 2013). Neutrophil uptake of microbes is mediated by TLRs, additional pattern recognition receptors, complement receptors, Fc receptors, and by formation of neutrophil extracellular traps (Kolaczkowska and Kubes, 2013). Neutrophils engage in two-way interactions resulting in both induction and regulation of innate and adaptive immune responses including chemokine expression and action on other cells, promoting maturation of dendritic cells and polarization of T lymphocyte responses (Mantovani et al., 2011).

NK cells are historically a component of the innate immune system whose roles include immune surveillance resulting in destruction of transformed and virus-infected cells, production of IFN-γ, and responsiveness to macrophage-derived IL-12 and type I interferons (Vivier et al., 2008; Yokoyama et al., 2004). Nature and complexity of NK cell stimulatory and inhibitory receptors has traditionally been regarded as distinct from the clonally distributed receptors of B and T lymphocytes arising from somatic recombination (Sun and Lanier, 2011). Similarities between NK cells and cytotoxic CD8[+] T lymphocytes are greater than previously realized and confirm NK cells as another cell in the continuum between innate and adaptive immunity. NK cells share similarities with B and T lymphocytes in that they undergo a receptor education process during development; undergo clonal expansion in responding to an infectious agent; and develop memory cells postantigen encounter (Sun and Lanier, 2011; Vivier et al., 2011).

An array of cytokines that are produced and released by epithelial, neural, adipose, myeloid, and lymphoid cells provide communication networks regulating cellular crosstalk, developmental events, inflammation, innate and adaptive immune protective defenses, and pathological responses (Fantuzzi, 2005; Dinarello, 2007; Kothur et al., 2016). Cytokines are categorized as individual gene products and as families such as IL-1 and TNF families (So et al., 2006; Garlanda et al., 2013). Central roles of cytokines in regulating protective, autoimmune, and disease-associated immune responses make cytokines and chemokines prime points for therapeutic interventions to either enhance or downregulate inflammation and innate and adaptive immunity (Astrakhantseva et al., 2014; Siebert et al., 2015). Infectious agents manipulate host cytokine networks during pathogenesis of infection (Smith et al., 2013; Hoffmann et al., 2015).

Complement is a series of highly regulated and sequentially activated serum proteins that interact in three pathways to eliminate microorganisms and remove damaged cells (Zipfel and Skerka, 2009). Alternative complement pathway is always activated in plasma, other biological fluids, cell surfaces, and thus provides a rapidly responding cascade to eliminate microbes and produce bioactive component fragments that include anaphylatoxins, chemotactic molecules, and molecules that coat activating surfaces to promote leukocyte phagocytosis (Pangburn and Müller-Eberhard, 1984; Zipfel et al., 2007).

All of these cells, mediators, and immune effector and regulatory pathways are exposed to saliva of the feeding tick. Cells and molecules of the host inflammatory and immune responses and the composition of tick saliva change during the course of days of infestation. This interplay provides a continuous series of host immune responses to infestation accompanied by tick countermeasures to those defenses. What is known about how ticks stimulate and modulate cells, cellular interactions and soluble mediators of inflammation, and innate and adaptive immunity? Next, tick countermeasures are linked to specific host defense mechanisms to obtain a comprehensive overview of these interactions. Readers are directed for general information about immune cells, mediators, and immune response networks to the superb, comprehensive immunology textbook authored by Abbas et al. (2015).

TICKS: INFLAMMATION AND INNATE IMMUNITY

Initial probing and laceration of the skin cause mechanical damage and introduce saliva into contact with epidermal resident keratinocytes, dendritic epidermal T lymphocytes, Langerhans cells, memory CD8[+] T lymphocytes, and commensal microbes on the epidermal surface (Kazimírová and Štibrániová, 2013; Wikel, 2013; Pasparakis et al., 2014; Kotál et al., 2015). During feeding, tick saliva is trapped in attachment cement, adsorbed into all layers of the epidermis percutaneously, deposited at the dermal–epidermal junction, and introduced directly into the feeding lesion (Allen et al., 1979a). All skin cell types and soluble mediators in the epidermis and dermis are potentially exposed to tick saliva. Immediately upon injury, platelet aggregation, vasoconstriction, and coagulation pathways are initiated.

TICKS AND KERATINOCYTES

What is known about tick interactions with keratinocytes? Keratinocytes are a rich source of a variety of proinflammatory cytokines, chemokines, and antimicrobial peptides that orchestrate migration and activation of other cells in the skin (Gröne, 2002; Lebre et al., 2007; Nestle et al., 2009; Pasparakis et al., 2014). Keratinocytes of mice infested with *I. ricinus* nymphs are positive for the proinflammatory cytokines IL-1 and TNF based on immunostaining of protein and in situ hybridization (Mbow et al., 1994). *I. ricinus* salivary gland extract and the saliva protein Salp15 both inhibit human primary keratinocyte production of the chemokine IL-8, monocyte chemoattractant protein 1 (MCP-1), and antigen-presenting cell activating and mobilizing antimicrobial defensins (Marchal et al., 2011). *I.ricinus* salivary gland extract prepared from three-day–fed adult ticks significantly reduces ability of agonists for TLR3 (poly I:C) and TLR2/TLR1 (*Borrelia burgdorferi* OspC) to induce inflammatory response mediators IL-8, TNF, human beta-defensin antimicrobial peptide 2, and macrophage inflammatory protein 2-alpha (Bernard et al., 2016). Downregulating keratinocyte-derived mediators of inflammation by targeting TLR responsiveness reduces attraction and activation of dendritic cells, neutrophils, and macrophages as well as directly influencing activation of endothelial cells and the recruitment of leukocytes into the bite site.

TICKS AND DENDRITIC CELLS

Langerhans cells are located amongst keratinocytes in the suprabasal region of the epidermis. Immature Langerhans cells are immune sensors that phagocytose process and express antigen peptides on class MHC II molecules; transport those antigens through afferent lymphatics to draining lymph nodes; and present antigen to naïve T lymphocytes in the lymph node paracortex for induction of an immune response (Bancherau and Steinman, 1998; Merad et al., 2008; Clausen and Stoitzner, 2015). Langerhans cells are a logical target for tick-induced modulation due to their pivotal roles at the interface of innate and adaptive immunity (Loré et al., 1998; Carbone et al., 2004; Banchereau et al., 2012).

Examination of ixodid tick attachment sites demonstrated saliva antigens throughout all epidermis layers adjacent to the mouthparts; trapped throughout attachment cement; and complexed with complement and IgG at epidermal–dermal border and within epidermal vesicles of tick resistant hosts (Allen et al., 1979a,b). Argasid tick infestation induced increases in numbers of Langerhans cells during both primary and secondary infestations in the absence of development of acquired resistance to infestation (McLaren et al., 1983). Brief feeding duration of argasid tick late-stage nymphs and adults for approximately 2h (Cooley and Kohls, 1944) was sufficient to induce Langerhans cell population changes. Number of Langerhans cells was significantly reduced at attachment sites of *D. andersoni* larvae by the first day of a primary infestation (Nithiuthai and Allen, 1984a). By fifth day of primary infestation, reduction in numbers of Langerhans cells was 71.5% that of uninfested controls. A second, similar, infestation, at different skin sites than the first infestation, initiated 7 days after completion of the primary exposure to ticks resulted in increased number of Langerhans cells around tick mouthparts (Nithiuthai and Allen, 1984a).

Short wavelength (UVC) and mid-wavelength (UVB) ultraviolet radiation–exposed skin reduced Langerhans cell numbers by 52.5% for 6 days after exposure to UVC (Nithiuthai and Allen, 1984b). Use of UVC irradiation of skin to deplete Langerhans cells to which an initial infestation of *D. andersoni* larvae was applied blocked development of expression of acquired resistance to a subsequent infestation (Nithiuthai and Allen, 1984c). Similar UVC treatment administered prior to a second infestation blocked expression of acquired resistance, indicating the role of Langerhans cells in acquisition and expression of acquired resistance. Additional evidence supporting the role of Langerhans cells in presenting tick saliva antigens was obtained by loading Langerhans cells in vitro with salivary gland antigens from five-day–fed female *D. andersoni* and demonstrating their ability to drive T lymphocyte proliferation in an in vitro blastogenesis assay (Nithiuthai and Allen, 1985). Macrophages can be effectively substituted for Langerhans cells and treatment with anti-MHC class II antibodies abolished the antigen presenting capabilities of both Langerhans cells and macrophages (Nithiuthai and Allen, 1985).

Infestation with *Ixodes scapularis* nymphs induces Th2 polarization with concomitant downregulation of Th1 cytokines (Schoeler et al., 1999). Mice deficient in Langerhans cells exhibit enhanced Th1 responses when infested with *I. scapularis* nymphs compared to wild-type controls establishing a role for these cells in infestation-induced Th2 polarization (Vesely et al., 2009). These tick-induced changes can reduce both inflammation and sensitization to antigen.

Phenotypically distinct dendritic cell populations exist in dermis and functional roles of these subsets of antigen-presenting cells in activation of T lymphocytes are being increasingly well defined in immune activation and tolerance (Nestle et al., 2009; Chu et al., 2011; Clausen and Stoitzner, 2015). Tick saliva targeting of dermal dendritic cell function is an important aspect of modulating host immune defenses at the innate and adaptive immunity interface with significant implications for infectious agent transmission and establishment of infection (Slámová et al., 2011; Wikel, 2013; Kotál et al., 2015).

Rhipicephalus sanguineus saliva inhibits both number and differentiation of bone marrow–derived dendritic cells; increases MHC class II expression; and decreases expression of the costimulatory molecules CD40, B7.1, and B7.2 on dendritic cells stimulated with *Escherichia coli* lipopolysaccharide (Cavassani et al., 2005). *R. sanguineus* saliva also reduces chemotactic responses of bone marrow–derived immature dendritic cells to MIP-1α and MIP-1β and impairs dendritic cell ability to stimulate T lymphocyte cytokines in an antigen-specific manner (Oliveira et al., 2008). *R. sanguineus* saliva suppressed chemotactic responses by reducing expression of the chemokine receptor CCR5. *R. sanguineus* saliva suppresses bone marrow–derived dendritic cell maturation in the presence of stimulation with lipopolysaccharide, resulting in suppression of TNF and IL12p70 production while enhancing production of the antiinflammatory cytokine IL-10

(Oliveira et al., 2010). Saliva-induced suppression of proinflammatory cytokines results from impaired activation of ERK 1/2 and p38 MAP kinases. A 17.7 kDa lipocalin in *Rhipicephalus appendiculatus* saliva modulates human dendritic cell maturation from monocytes and downregulates proinflammatory cytokines (TNF, IL-1, IL-6, IL-12) and inhibits increased expression of the costimulatory molecule CD86 (Preston et al., 2013). In a similar manner, *Amblyomma cajennense* saliva inhibits bone marrow dendritic cell differentiation; reduces expression of CCR5, CCR7, and costimulatory molecules CD40 and B7-2; and downregulates expression of TNF, IL-6, and IL-12p40, while upregulating expression of the antiinflammatory cytokine IL-10 (Carvalho-Costa et al., 2015). These findings suggest conserved molecular strategies for suppression of dendritic cell function among multiple metastriate tick species.

Tick saliva polarization of helper T lymphocytes to a Th2 profile accompanied by downregulation of Th1 cytokines is reported for multiple ixodid species (Wikel, 2013; Kotál et al., 2015). Splenic dendritic cells exposed to *I. ricinus* saliva induce Th2 polarization in vitro and Th2 cells only proliferate in the presence of saliva primed dendritic cells if IL-1 is added to the culture (Mejri and Brossard, 2007). In addition to promoting Th2 polarization, administration of *I. ricinus* saliva fails to stimulate either Th1 or Th17 responses; impairs dendritic cell maturation and migration from sites of cutaneous inflammation to draining lymph nodes; and suppresses lymph node dendritic cell presentation of soluble antigen to T lymphocytes (Skallová et al., 2008).

Dendritic cell modulation by *I. ricinus* saliva impacts tick-borne infectious agents. Treatment of mouse splenic dendritic cells with *I. ricinus* saliva increases the percentage of cells that become infected with tick-borne encephalitis virus, decreases virus-induced apoptosis, and reduces production of TNF and IL-6 by virus-infected dendritic cells (Fialová et al., 2010).

Mouse splenic dendritic cells stimulated with *Borrelia afzelli* spirochetes in the presence of *I. ricinus* saliva display reduced phagocytosis and impaired proliferation and IL-2 production by antigen-specific T lymphocytes (Slámová et al., 2011). *I. ricinus* saliva reduced lipopolysaccharide-stimulated mouse splenic dendritic cell production of IFN-β, but not when similar cells were stimulated with a tick isolated strain of *B. afzelli* (Lieskovská and Kopecký, 2012a). Saliva inhibited nuclear factor kappaB (NF-κB), phosphatidylinositol-3 kinase (P13K/Akt), mitogen-activated protein kinases (MAPK), and extracellular matrix-regulated kinase (Erk1/2) signal transduction pathways (Lieskovská and Kopecký, 2012b).

Bone marrow–derived dendritic cells displayed a dose-dependent inhibition of IL-12 and TNF production when stimulated with TLR-2, TLR-4, or TLR-9 ligands in the presence of *I. scapularis* saliva (Sá-Nunes et al., 2007). Very dilute saliva was inhibitory and suppression was attributed to saliva prostaglandin E2. *I. scapularis* saliva Salp15 protein inhibits human dendritic cell TLR-2 and TLR-4 agonist induced expression of proinflammatory cytokines TNF, IL-6, and IL-12 and blocks T lymphocyte activation assessed in a mixed lymphocyte reaction (Hovius et al., 2008). Dendritic cell functional changes are due to Salp15 binding to the C-type lectin receptor DC-SIGN that activates the serine/threonine kinase Raf-1, which in turns results in MAPK-dependent reduced mRNA stability for TNF and IL-6 and aberrant functioning of the promoter for IL-12 p35 (Hovius et al., 2008).

Saliva of *I. scapularis* contains two cysteine protease inhibitors, sialostatins L and L2 (Kotsyfakis et al., 2007). Sialostatin L present in *I. scapularis* saliva binds cathepsin S and cathepsin L inside dendritic cells, resulting in inhibition of dendritic cell maturation and antigen-specific proliferation of T lymphocytes (Sá-Nunes et al., 2009). *I. scapularis* saliva sialostatins L and L2 both suppressed MIP-1 production by

murine bone marrow–derived dendritic cells infected with *B. burgdorferi* (Lieskovská et al., 2015). Only sialostatin L2 reduced activation of NF-κB and P13K/Akt upon dendritic cell activation with lipoteichoic acid, a TLR-2 ligand. Likewise, sialostatin L2 inhibited activation of Erk1/2 in the presence of *B. burgdorferi* stimulation. Plasmacytoid dendritic cell production of IFN-β is reduced by sialostatin L2 (Lieskovská et al., 2015). Detailed examinations are provided about the impact of tick saliva on TLR and nod-like receptors signaling pathways, inflammasomes and tick-borne pathogens in Chapter 11 of this volume and by Sakhon et al. (2013), Wang et al. (2016), and Shaw et al. (2016).

TICKS AND MONOCYTES/ MACROPHAGES

Macrophages present antigens predominantly to memory phenotype T lymphocytes in the skin and regulate innate and adaptive immune responses through cytokine signaling and their ability to phagocytize and kill microbes (Dinarello, 2007; Gordon and Martinez, 2010; Sica and Mantovani, 2012; Porta et al., 2015). Saliva of numerous tick species modulate macrophage cytokine and chemokine responses and other aspects of macrophage defense functions. *D. andersoni* salivary gland extract prepared at different time points during engorgement suppresses macrophage production of TNF and IL-1 (Ramachandra and Wikel, 1992). Saliva of *R. appendiculatus* inhibits both mRNA transcript and protein synthesis for TNF, IL-1, and the antiinflammatory cytokine IL-10 in addition to reducing production of the important macrophage antimicrobial, nitric oxide (Gwakisa et al., 2001). Macrophage nitric oxide production is greatly suppressed by application of *I. scapularis* saliva to lipopolysaccharide-stimulated murine peritoneal macrophages (Urioste et al., 1994). Suppression of nitric oxide was not due to saliva prostaglandin

E2. *Rhipicephalus microplus* saliva inhibits lipopolysaccharide-stimulated bovine macrophage expression of IL-12 and the costimulatory molecules B7-1 and B7-2, altering important T lymphocyte activation points at the immunologic synapse (Brake and Pérez de León, 2012). *Dermacentor variabilis* saliva prostaglandin E2 suppresses macrophage proinflammatory cytokine production (Poole et al., 2013).

Macrophage and T lymphocyte cytokine responses assessed following infestations with *I. scapularis* or *Ixodes pacificus* nymphs reveal suppression of macrophage cytokines TNF and IL-1 accompanied by down-regulation of the Th1 cytokines IL-2 and IFN-γ accompanied by upregulation of the Th2 cytokine IL-4 (Schoeler et al., 1999, 2000). These macrophage and T lymphocyte cytokine polarizations are consistent with those described for dendritic cells in the presence of saliva derived from other tick species (Wikel, 2013; Kotál et al., 2015).

Saliva of multiple tick species suppress or upregulate specific cytokines in patterns that potentially diminish inflammatory, innate, and adaptive immune effector responses that could negatively impact successful blood feeding. Tick saliva also inhibits chemokines that are important at multiple levels in regulating inflammation. *Amblyomma americanum* and *Haemaphysalis longicornis* saliva contain molecules that mimic the action of macrophage migration inhibitory factor (Jaworski et al., 2001; Umemiya et al., 2007), with biological activities that include stopping and activating macrophages at sites of inflammation (Nishihira, 1998). A potential role for this molecule in tick feeding might be to reduce influx of macrophages within immediate vicinity of mouthparts. Hyalomin-A and B present in *Hyalomma asiaticum asiaticum* saliva inhibit monocyte chemotactic protein-1 in addition to suppressing TNF, upregulating IL-10 and acting as scavangers of free radicals (Wu et al., 2010).

IL-8 is an important chemotactic cytokine for attracting and activating neutrophils at sites of inflammation (Baggiolini et al., 1994). Salivary gland extracts of *Amblyomma variegatum*, *Dermacentor reticulatus*, *Haemaphysalis inermis*, *I. ricinus*, and *R. appendiculatus* reduces levels of detectable IL-8 (Hajnická et al., 2001). Fractions of *D. reticulatus* bind IL-8 and inhibit binding to its receptors on human neutrophils (Hajnická et al., 2001). Antichemokine activity of *A. variegatum* extends beyond IL-8 to include MCP-1, MIP-1, RANTES, and eotaxin (Vancová et al., 2007). Salivary gland extract of *R. appendiculatus* inhibits chemokine activity of IL-8, MCP-1, MIP-1, RANTES, and eotaxin (Vancová et al., 2010). *I. scapularis* saliva inhibits IL-8 secretion by bone marrow–derived macrophages stimulated with TNF and blocks the proinflammatory response of macrophages to *Anaplasma phagocytophilum* (Chen et al., 2012). Chemokines and chemokine receptors encompass multiple structures with diverse functions, including targeting of leukocytes into specific tissue sites (Sallusto and Baggiolini, 2008).

TICKS AND ENDOTHELIAL CELLS

Attraction of leukocytes to specific tissues depends upon complex interactions among endothelial cell and leukocyte-expressed specific combinations of adhesion molecules, integrins, and chemokines that orchestrate inflammatory and immune responses (Ley et al., 2007; Weninger et al., 2014). Tick saliva inhibition of a broad array of chemokines has implications for leukocyte trafficking into tissues at the tick bite site, particularly for the transition from leukocyte rolling to tight adhesion to the endothelium.

In addition to cytokines and chemokines, tick saliva also modulates expression of leukocyte integrins and endothelial cell adhesion molecules. Lymphocytes derived from mice infested with *D. andersoni* nymphs have reduced levels of the integrins leukocyte function associated antigen-1 and very late antigen-4 that interact with endothelial cell adhesion molecules

ICAM-1 and VCAM-1, respectively (Macaluso and Wikel, 2001). Exposure of skin-derived endothelial cells to salivary gland extracts of *D. andersoni* or *I. scapularis* downregulates adhesion molecule expression (Maxwell et al., 2005). *D. andersoni* salivary gland extract consistently suppresses expression of ICAM-1 and variably reduces expression of E-selectin. *I. scapularis* salivary gland extract consistently downregulates both P-selectin and VCAM-1. *I. scapularis* saliva reduces expression of neutrophil surface β2-integrins (CD18) with these molecules being relocated to the rear of a neutrophil moving in the circulation to alter their migration responses (Montgomery et al., 2004). Two disintegrin metalloproteinase-like proteins in *I. scapularis* saliva are linked to β2-integrin suppression (Guo et al., 2009).

Combined modulation of leukocyte integrins and endothelial cell adhesion molecules by both *D. andersoni* and *I. scapularis* salivary gland molecules can reduce the influx of inflammatory and immune cells to the tick bite site. This change in leukocyte migration is supported by the observation that *I. scapularis* bite sites during an initial infestation are characterized by few inflammatory cells immediately surrounding the tick mouthparts (Krause et al., 2009). During a repeated infestation, an intense influx of inflammatory cells occurs at tick mouthparts. A possible explanation for this phenomenon is that repeated infestation induces a host immune response that neutralizes the saliva molecules responsible for reducing the primary infestation accumulation of leukocytes at the site of blood feeding.

Bos taurus indicus cattle breeds are significantly more resistant to tick infestation with *R. microplus* than *Bos taurus taurus* breeds (de Castro and Newson, 1993). This breed genetic difference is attributable, in part, to differences in immune responses to tick infestation (Ramachandra and Wikel, 1995). An ex vivo study examining *R. microplus* bite sites on both genetic lineages reveal increased mRNA for ICAM-1, VCAM-1, and P-selectin in skin biopsies of bite sites on tick susceptible breeds undergoing low infestations (Carvalho et al., 2010). High levels of infestation downregulated these adhesion molecules for both breeds of cattle. High levels of expression of E-selectin occur at bite sites of resistant animals as do influxes of basophils and eosinophils (Carvalho et al., 2010).

TICKS AND NEUTROPHILS

In addition to downregulating surface integrins, tick saliva modulates neutrophil functions to reduce the impact of host inflammatory responses. *I. scapularis* saliva inhibits neutrophil aggregation induced by anaphylatoxin, phagocytosis, granule enzyme release, and phagocytosis of *B. burgdorferi* spirochetes (Ribeiro et al., 1990). A leukotriene B4–binding protein in *I. ricinus* saliva inhibits neutrophil chemotaxis, delays apoptosis, and decreases both accumulation and activation of neutrophils at tick attachment sites (Beaufays et al., 2008). *I. scapularis* saliva contains two proteins homologous to disintegrin metalloproteinases that reduce β2-integrin expression, neutrophil adherence, superoxide production, and *B. burgdorferi* killing (Guo et al., 2009).

Neutrophils infiltrating the bite site of *I. ricinus* transmitting *B. burgdorferi* kill phagocytosed spirochetes by the action of hydrolytic enzymes and reactive oxygen species with the formation of neutrophil extracellular traps (Menten-Dedoyart et al., 2012). *I. ricinus* saliva does not affect extracellular trap formation; however, it does significantly decrease neutrophil hydrogen peroxide production. Two *Ixodes persulcatus* salivary gland Salp 16-like proteins inhibit bovine neutrophil chemoattraction to IL-8 and reduce oxidative burst (Hidano et al., 2014). These findings indicate that suppressed neutrophil function is another component of the overall strategies of ticks to impair inflammation and innate immunity that both provide frontline defense and orchestration of adaptive immune responses.

TICKS, TYPE I INTERFERONS AND NATURAL KILLER CELLS

Type I interferons (α, β) and NK cells are innate immune defenses against viruses as well as regulators of innate and adaptive immunity (Ivashkiv and Donlin, 2014; Sun and Lanier, 2011). Human peripheral blood mononuclear cells stimulated with lipopolysaccharide display increased IFN-α mRNA transcripts; however, addition of *R. appendiculatus* salivary gland extract to that milieu reduces transcript expression, indicating reduced ability to mount an antivirus defense (Fuchsberger et al., 1995). *I. ricinus* saliva reduces IFN-β activation of the STAT-1 transcription factor and inhibits lipopolysaccharide-induced IFN-β production by dendritic cells (Lieskovská and Kopecký, 2012a).

D. reticulatus salivary gland extract decreases normal human NK cell cytotoxicity for an acute myelogenous leukemia cell line target in a short-term chromium release assay (Kubes et al., 1994). Addition of IFN-α to these cultures reverses suppressive effects of tick saliva. Using the same short-term chromium release assay, treatment of effector, but not target, cells with *D. reticulatus* salivary gland extract reduces NK cell binding to target cells (Kubes et al., 2002). NK cell killing activity is also suppressed by *I. ricinus* salivary gland extract (Kopecký and Kuthejlová, 1998). Inhibition of NK cell activity is present in salivary gland extracts of *Amblyomma variegatum* and *Haemaphysalis inermis* (Kubes et al., 2002).

TICKS AND MAST CELLS

Mast cell–derived cytokines, enzymes, lipid mediators, and vasoactive molecules are regulators of immune functions of dendritic cells, T lymphocytes, and immunologic injury (Otsuka et al., 2016; Abraham and St John, 2010; Voehringer, 2013). Histological examination of tick attachment sites reveals increased numbers of mast cells, basophils, and monocytes that correlate with resistance to infestation (Marufu et al., 2014).

Histamine released from skin mast cells at tick attachment sites negatively impacts tick feeding. Bovine resistance to *R. microplus* correlates directly with skin histamine levels (Willadsen et al., 1979). Histamine induces ticks to stop salivating and consuming blood, resulting in detachment from the host (Kemp and Bourne, 1980; Paine et al., 1983). Saliva of numerous tick species contain histamine-binding lipocalins (Sangamnatdej et al., 2002; Tirloni et al., 2014). A saliva histamine releasing factor expressed later during blood feeding is linked to the rapid uptake of blood at the end of engorgement (Dai et al., 2010).

I. scapularis saliva sialostatin L suppresses mast cell IL-9 production and inhibits expression of IL-1 and the transcription factor IRF4 (Klein et al., 2015). This finding is relevant to the observation that tick resistant hosts often have increased numbers of mast cells at tick-feeding sites. A role of IL-9 is to induce mast cell growth (Stassen et al., 2012). An interesting question is to examine host species that develop resistance to ticks infestation and determine if they can neutralize the action of sialostatin L.

TICKS AND BASOPHILS

There exists an increasing awareness of the importance of basophils in protective immune responses against both endoparasites and ectoparasitic arthropods, particularly ticks (Eberle and Voehringer, 2016). Basophils are regulators of dendritic cell antigen presentation in development of Th2-polarized T lymphocyte responses (Perrigoue et al., 2009; Wynn, 2009; Sokol and Medzhitov, 2010; Karasuyama and Yamanishi, 2014).

Basophil-rich cutaneous responses and Th2 polarization are associated with tick infestation.

Saliva of numerous tick species induce Th2 polarization during primary and subsequent infestations (Kazimírová and Štibrániová, 2013; Wikel, 2013; Kotál et al., 2015). Basophil-rich cutaneous infiltrates at tick attachment sites occur in acquired resistance to infestation that is generally expressed during repeated infestations by those host species that develop such resistance (Allen, 1973; Allen et al., 1977; Marufu et al., 2014). Selective and inducible depletion of basophils established the nonredundant role for these cells, and not mast cells, in development and expression of acquired resistance to infestation with *H. longicornis* (Wada et al., 2010).

TICKS AND COMPLEMENT

Blood-feeding endoparasites and ectoparasites evolved multiple strategies for disrupting or dysregulation of alternative, lectin, and classical pathways of complement activation (Schroeder et al., 2009). Ticks are particularly efficient at modulating complement (Wikel, 2013). A functional alternative pathway of complement activation is essential for expression of acquired resistance to infestation with *D. andersoni* (Wikel, 1979). *I. scapularis* saliva inhibits both anaphylatoxin C3a and deposition of C3b from sera of multiple host species (Ribeiro, 1987). *I. scapularis* saliva dissociates alternative pathway C3 convertase into Bb and C3b by the inhibitor Isac (Valenzuela et al., 2000) and multiple Salp20 proteins that are members of the Isac protein family (Tyson et al., 2007). Action of Salp20 includes dissociation of the C3 convertase stabilizing protein properdin from the convertase, resulting in diminished duration and magnitude of C3 cleavage (Tyson et al., 2008). Effective inhibition of the alternative pathway of complement by Salp20 may be attributed, in part, to acting synergistically with the alternative pathway regulatory protein factor H that works in concert with factor I to degrade C3b

and to increase breakdown of C3 convertase (Hourcade et al., 2016).

I. ricinus saliva inhibits C3a formation, factor B cleavage, and C3b deposition (Lawrie et al., 2005). *I. ricinus* salivary gland extract induces cleavage near the C-terminus of the C3 alpha chain that results in a modified C3 molecule that does not create an effective alternative pathway C3 convertase due to altered affinity for factor B (Lawrie et al., 2005). *I. scapularis* and *I. ricinus* suppress alternative pathway C3 convertase activity by divergent methods.

Saliva of the argasid tick *Ornithodoros moubata* contains a lipocalin that binds specifically to C5 and inhibits C5 convertase activities of both alternative and classical pathways of complement activation (Nunn et al., 2005). Although not reported to be tested in this study, this C5-binding lipocalin should also inhibit the mannose-binding complement pathway. *I. scapularis* saliva contains a mannose (lectin)-binding complement pathway inhibitor of the first step of the lectin pathway that blocks mannose-binding lectin from ligand binding (Schuijt et al., 2011). *A. americanum* calreticulin binds C1q, which is the classical pathway recognition component for antigen-reacting IgM and multiple IgG molecules; however, this binding interaction does not block classical pathway of complement activation (Kim et al., 2015).

A MODEL: TICKS MODULATE INFLAMMATION AND INNATE IMMUNITY

This model is a composite of tick countermeasures to host inflammation and innate immune defenses described in the preceding sections and overlaid onto general mechanisms of response to cutaneous injury. Although this model combines findings for different ixodid species, the overall goal is to demonstrate how ticks modulate multiple cells, pathways, and mediators. This model is intended to provide a stimulus for future research that can fill gaps of knowledge about individual cell types, communication among cells, mediators targeted in manipulating these responses, and identifying tick saliva molecules responsible for these phenomena. An initial systems biology approach to defining tick interactions with the host could be potentially highly informative in the context of current information and using the tools of high throughput sequencing, proteomics, and bioinformatics. This information is essential before tick transmission of specific infectious agents is added to this analysis strategy. Specific literature references are not included in this section, since each aspect discussed here was specifically addressed elsewhere in this chapter.

Overarching objective of the feeding tick is to reduce the host inflammatory response at the bite site and to reduce stimulation of adaptive immune responses whose effectors could neutralize biological activities of tick saliva molecules essential for successful blood feeding. Tick-mediated suppression of host defenses cannot be complete, since a balance with fending off opportunistic pathogens and effective immune surveillance needs to be maintained. Objective is to achieve a balance that preserves host survival and yet reduces ability of host inflammatory response and innate and adaptive immune defenses to block tick feeding.

Chelicera digits laceration of the cutaneous surface introduces saliva into epidermal layers and dermis and, depending upon the tick species, secretion of attachment cement around the mouthparts and on the skin surface. Among first cells influenced by tick saliva are epidermal keratinocytes and Langerhans cells. Disrupted skin exposes innate immune receptors of epidermal and dermal host cells to DAMPs and saliva proteins.

Tick saliva inhibits keratinocyte secretion of the proinflammatory cytokines TNF and IL-1, reduces defensins that activate antigen-presenting cells, and reduces expression of the chemokine IL-8. What are the possible outcomes of these changes? Combined downregulation of proinflammatory cytokines and chemokines inhibits activation of dendritic cells, neutrophils, and macrophages that may be attracted to the bite site. Tick saliva reduction of TLR responsiveness impairs innate immune sensing and cell activation. Reducing proinflammatory cytokines and a chemokine such as IL-8 diminishes activation of endothelial cells in the dermis with the potential of downregulating leukocyte transmigration into the bite site.

Among dendritic cells impacted by reduced activating cytokines from keratinocytes are Langerhans cells located in a suprabasal position in the epidermis. Reduced proinflammatory cytokines and chemokines in addition to direct effects of tick saliva could reduce movement of Langerhans cells via the lymphatics into draining lymph nodes as well as their activation and ability to present antigen to specific T lymphocytes. Although tick saliva reduces costimulatory molecule expression on other dendritic cell populations, impact upon Langerhans cell costimulatory molecules remains unclear. Reduced Langerhans cell activation and antigen presentation due to tick saliva diminishes development of antigen-specific effector and memory T lymphocytes that could impact the present and future infestations.

Dermal dendritic cell populations are also reduced by tick saliva in their activation, response to chemokines, expression of costimulatory molecules, and antigen presentation based upon studies utilizing splenic and bone marrow–derived dendritic cells. Downregulation of proinflammatory cytokines and chemokines suppresses activation of other cells, including endothelial cells. Saliva reduction of IL-12 expression reduces Th1 polarization of CD4+ T helper lymphocytes. Reduced IL-12 also suppresses the feedback loop resulting in NK cell production of IFN-γ that activates macrophages during the innate immune response.

Tick saliva modulates macrophage function in a manner essentially similar as that of dendritic cells. Reduction of proinflammatory cytokines and chemokines reduce activation signals for endothelial cells and neutrophils and chemotactic signals for neutrophils, eosinophils, and basophils. Reducing macrophage costimulatory molecules and cytokines would likely have the greatest significance for memory cells that are the predominant T lymphocyte phenotype migrating into the skin.

Migration of leukocytes into the bite site is reduced by tick saliva acting directly to downregulate expression of endothelial cell selectins and other adhesion molecules and neutrophil expression of integrins. Reduced proinflammatory cytokines suppress endothelium activation and diminished chemokine expression on endothelial cells reduces signals for increased affinity of integrins to transform leukocytes from a slow rolling state to tight binding to endothelium. In addition to reducing the number of neutrophils at the bite site, tick saliva impairs oxidative burst, which can impair killing of phagocytosed microbes. Tick saliva also reduces neutrophil chemotactic responses once in the extravascular tissues. Outcome is reduced inflammation.

Mast cells are present around blood vessels in the skin. Mechanical damage as well as complement-derived anaphylatoxins can induce release of mediator as the feeding site is being created. Since tick saliva suppresses dendritic cell IL-12, the IL-4 released from mast cells further promotes Th2 polarization of CD4+ T lymphocytes. Histamine has a direct negative impact on tick feeding. Therefore, tick saliva histamine binding lipocalins are physiologically protective for the feeding tick and reduce impact of histamine on host inflammatory response.

Tick countermeasures to host defenses reveal main goals of reducing inflammation and subverting the adaptive immune response to a potentially less damaging state. Development of acquired resistance to tick feeding and cutaneous hypersensitivity to tick bite demonstrates that for at least some tick–host relationships that repeated exposures to tick saliva induces host adaptive immune responses that neutralize to some degree the biological activity of tick spit components.

TICKS AND ADAPTIVE IMMUNITY

Tick saliva modulation of dendritic cells and macrophages reduces efficiency of antigen presentation and activation of naïve and memory T lymphocytes. Effects of tick saliva on B lymphocyte antigen presentation to helper T lymphocytes remains to be defined. This topic is important due to the role of T lymphocytes in driving B lymphocyte immunoglobulin class switching and development of memory B lymphocyte populations.

Two major effects of tick saliva on T lymphocytes are reduction of cell proliferation and polarization to a Th2 profile with concomitant suppression of Th1 cytokines. Infestation with *D. andersoni* nymphs suppressed the ability of host lymphocytes to respond in vitro to the T lymphocyte mitogen (Wikel, 1982). Subsequently, a 36kDal protein identified in salivary glands of both female and male *D. andersoni* was found to strongly inhibit T lymphocyte proliferation (Bergman et al., 2000). *I. scapularis* saliva contains an IL-2–binding protein that blocks the action of this key cytokine that supports T lymphocyte proliferation (Gillespie et al., 2001). *I. scapularis* saliva Salp15 reduces T lymphocyte proliferation by reducing calcium influxes following engagement of the T cell receptor and reducing IL-2 production (Anguita et al., 2002). Receptor for Salp15 is the helper T lymphocyte surface coreceptor CD4 (Garg et al., 2006) and induced signal transduction pathway inhibition is characterized (Juncadella et al., 2007). *I. scapularis* saliva cystatin sialostatin L inhibits cytotoxic CTLL-2T lymphocyte cell line proliferation (Kotsyfakis et al., 2006).

I. scapularis or *I. pacificus* nymph infestations upregulate Th2 cytokines and downregulate Th1 cytokines confirming that Th2 polarization occurs in vivo (Schoeler et al., 1999, 2000). *I. scapularis* Th2 polarizing ability is attributed, at least in part, to a sphingomyelinase-like saliva enzyme, named IsSMase that programs infested host CD4+ T lymphocytes to produce IL-4 (Alarcon-Chaidez et al., 2009). IsSMase is the first tick saliva molecule that induces Th2 polarization and only a very few identified molecules with this property. *I. ricinus* saliva induces a host response that upregulates both thymus-derived chemotactic agent 3 and macrophage inflammatory protein 2 that are contributing to Th2 polarization (Langhansová et al., 2015).

Tick infestation is reported to suppress both IgM and IgG antibody responses, implying in the case of reduced IgG the possibility of impaired T lymphocyte help for antibody class switching. Infestation with *D. andersoni* larvae reduces the number of splenic B lymphocytes producing IgM to a sheep red blood cell antigen challenge (Wikel, 1985). *R. sanguineus* adult infestation suppresses the IgG response to immunization with serum albumin (Inokuma et al., 1997). Infestation with *I. ricinus* reduces antigen-specific IgM and IgG responses following immunization with albumin with adjuvant (Menten-Dedoyart et al., 2008)

I. ricinus salivary gland extract suppresses B lymphocyte in vitro proliferation in response to lipopolysaccharide (Hannier et al., 2003). An 18 kDal B lymphocyte inhibitory saliva protein was identified (Hannier et al., 2004). A salivary gland factor produced by *Hyalomma asiaticum asiaticum* reduces lipopolysaccharide B lymphocyte proliferation in vitro (Yu et al., 2006). Additional studies are needed to more extensively characterize these B lymphocyte inhibitors and their modes of action.

CONCLUDING REMARKS

Systems biology approaches are a promising way moving forward to build upon current foundations of understanding of salivary gland next generation sequencing and proteomics data, tick–host interface transcriptomics, and an increasing body of knowledge about the cellular and molecular pathway underpinnings of tick-induced immune modulation. Systems biology analyses for prostriate and metastriate tick–host interactions throughout the course of infestation will provide valuable comparisons. After these tick–host relationships are well characterized, an appropriate tick transmitted pathogen can be added to the model for analyses of tick–host–pathogen interactions throughout the course of infestation, pathogen transmission, and for a period of time after the tick detaches.

Addressing several additional questions could provide information that can contribute not only to a more complete picture of interactions at the tick–host interface but to points of potential antitick and tick-borne pathogen control. What we have now are snapshots in time of tick–host interactions. What would be useful is an understanding of the dynamic changes occurring across the continuum of infestation. How does one event impact the next moment of tick feeding and host response? How does salivary gland gene expression, and more importantly secreted protein composition, change throughout the course of infestation? How do members of an individual family of salivary gland genes change throughout infestation? What is the profile of temporal changes in a specific immune function during the entire course of tick feeding and how long do those changes persist postattachment of the feeding tick?

Progress in understanding the many aspects of tick–host interactions is increasing at a significant pace. This topic provides a wealth of opportunities for a cadre of exceptionally talented young investigators and students to build upon those advances and develop strategies that will control disease transmission and positively impact the lives of potentially millions of humans and other animal species.

References

Abbas, A.K., Lichtman, A.H., Pillai, S., 2015. Cellular and Molecular Immunology, eighth ed. Elsevier Saunders, Philadelphia.

Abraham, S.N., St John, A.L., 2010. Mast cell-orchestrated immunity to pathogens. Nat. Rev. Immunol. 10, 440–452.

Alarcon-Chaidez, F.J., Sun, J., Wikel, S.K., 2007. Construction and characterization of a cDNA library from the salivary glands of *Dermacentor andersoni* Stiles (Acari: Ixodidae). Insect Biochem. Mol. Biol. 37, 48–71.

Alarcon-Chaidez, F.J., Boppana, V.D., Hagymasi, A.T., Adler, A.J., Wikel, S.K., 2009. A novel sphingomyelinase-like enzyme in *Ixodes scapularis* tick saliva drives host CD4 T cells to express IL-4. Parasite Immunol. 31, 210–219.

Allen, J.R., 1973. Tick resistance: basophils in skin reactions of resistant guinea pigs. Int. J. Parasitol. 3, 195–200.

Allen, J.R., Doube, B.M., Kemp, D.H., 1977. Histology of bovine skin reactions to *Ixodes holocyclus*, Neuman. Can. J. Comp. Med. 41, 26–35.

Allen, J.R., Khalil, H.M., Graham, J.E., 1979a. The location of tick salivary antigens, complement and immunoglobulin in the skin of guinea-pigs infested with *Dermacentor andersoni*. Immunology 38, 467–472.

Allen, J.R., Khalil, H.M., Wikel, S.K., 1979b. Langerhans cells trap tick salivary gland antigens in tick-resistant guinea pigs. J. Immunol. 122, 563–565.

Anguita, J., Ramamoorthi, N., Hovius, J.W., Das, S., Thomas, V., Persinski, R., Conze, D., Askenase, P.W., Rincón, M., Kantor, F.S., Fikrig, E., 2002. Salp15, an *Ixodes scapularis* salivary protein, inhibits CD4+ T cell activation. Immunity 16, 849–859.

Astrakhantseva, I.V., Efimov, G.A., Drutskaya, M.S., Kruglov, A.A., Nedospasov, S.A., 2014. Modern anti-cytokine therapy of autoimmune diseases. Biochemistry (Mosc) 79, 1308–1321.

Baggiolini, M., Dewald, B., Moser, B., 1994. Interleukin-8 and related chemotactic cytokines–CXC and CC chemokines. Adv. Immunol. 55, 97–179.

Banchereau, J., Steinman, R.M., 1998. Dendritic cells and the control of immunity. Nature 392, 245–252.

Banchereau, J., Thompson-Snipes, L., Zurawski, S., Blanck, J.P., Cao, Y., Clayton, S., Gorvel, J.P., Zurawski, G., Klechevsky, E., 2012. The differential production of cytokines by human Langerhans cells and dermal CD14+ DCs controls CTL priming. Blood 119, 5742–5749.

Beaufays, J., Adam, B., Menten-Dedoyart, C., Fievez, L., Grosjean, A., Decrem, Y., Prévôt, P.P., Santini, S., Brasseur, R., Brossard, M., Vanhaeverbeek, M., Bureau, F., Heinen, E., Lins, L., Vanhamme, L., Godfroid, E., 2008. Ir-LBP, an *Ixodes ricinus* tick salivary LTB4-binding lipocalin, interferes with host neutrophil function. PLoS One 3, e3987.

Benyon, R.C., 1989. The human skin mast cell. Clin. Exp. Allergy 19, 375–387.

Bergman, D.K., 1996. Mouthparts and feeding mechanisms of haematophagous arthropods. In: Wikel, S.K. (Ed.), The Immunology of Host-Ectoparasitic Arthropod Relationships. CAB International, Wallingford, U.K., pp. 30–61.

Bergman, D.K., Palmer, M.J., Caimano, M.J., Radolf, J.D., Wikel, S.K., 2000. Isolation and cloning of a secreted immunosuppressant protein from *Dermacentor andersoni* salivary gland. J. Parasitol. 86, 516–525.

Bernard, Q., Gallo, R.L., Jaulhac, B., Nakatsuji, T., Luft, B., Yang, X., Boulanger, N., 2016. *Ixodes* tick saliva suppresses the keratinocyte cytokine response to TLR2/TLR3 ligands during early exposure to Lyme borreliosis. Exp. Dermatol. 25, 26–31.

Boltjes, A., van Wijk, F., 2014. Human dendritic cell functional specialization in steady-state and inflammation. Front. Immunol. 5, 131.

Brake, D.K., Pérez de León, A.A., 2012. Immunoregulation of bovine macrophages by factors in the salivary glands of *Rhipicephalus microplus*. Parasit. Vectors 5, 38.

Bullard, R., Allen, P., Chao, C.C., Douglas, J., Das, P., Morgan, S.E., Ching, W.M., Karim, S., July 2016. Structural characterization of tick cement cones collected from in vivo and artificial membrane blood-fed Lone Star ticks (*Amblyomma americanum*). Ticks Tick Borne Dis. 7 (5), 880–892.

Carbone, F.R., Belz, G.T., Heath, W.R., 2004. Transfer of antigen between migrating and lymph node-resident DCs in peripheral T-cell tolerance and immunity. Trends Immunol. 25, 655–658.

Carvalho, W.A., Franzin, A.M., Abatepaulo, A.R., de Oliveira, C.J., Moré, D.D., da Silva, J.S., Ferreira, B.R., de Miranda Santos, I.K., 2010. Modulation of cutaneous inflammation induced by ticks in contrasting phenotypes of infestation in bovines. Vet. Parasitol. 167, 260–273.

Carvalho-Costa, T.M., Mendes, M.T., da Silva, M.V., da Costa, T.A., Tiburcio, M.G., Anhê, A.C., Rodrigues Jr., V., Oliveira, C.J., 2015. Immunosuppressive effects of *Amblyomma cajennense* tick saliva on murine bone marrow-derived dendritic cells. Parasit. Vectors 8, 22.

Cavassani, K.A., Aliberti, J.C., Dias, A.R., Silva, J.S., Ferreira, B.R., 2005. Tick saliva inhibits differentiation, maturation and function of murine bone-marrow-derived dendritic cells. Immunology 114, 235–245.

Chen, G., Severo, M.S., Sohail, M., Sakhon, O.S., Wikel, S.K., Kotsyfakis, M., Pedra, J.H., 2012. *Ixodes scapularis* saliva mitigates inflammatory cytokine secretion during *Anaplasma phagocytophilum* stimulation of immune cells. Parasit. Vectors 5, 229.

Chmelař, J., Kotál, J., Karim, S., Kopacek, P., Francischetti, I.M., Pedra, J.H., Kotsyfakis, M., 2016a. Sialomes and Mialomes: a systems-biology view of tick tissues and tick-host interactions. Trends Parasitol. 32, 242–254.

Chmelař, J., Kotál, J., Kopecký, J., Pedra, J.H., Kotsyfakis, M., 2016b. All for one and one for all on the tick-host battle-field. Trends Parasitol. 32, 368–377.

Chu, C.C., Di Meglio, P., Nestle, F.O., 2011. Harnessing dendritic cells in inflammatory skin diseases. Semin. Immunol. 23, 28–41.

Clark, R.A., Chong, B., Mirchandani, N., Brinster, N.K., Yamanaka, K., Dowgiert, R.K., Kupper, T.S., 2006. The vast majority of CLA+ T cells are resident in normal skin. J. Immunol. 176, 4431–4439.

Clausen, B.E., Stoitzner, P., 2015. Functional specialization of skin dendritic cell subsets in regulating T cell responses. Front. Immunol. 6, 534.

Colonna, M., Trinchieri, G., Liu, Y.-J., 2004. Plasmacytoid dendritic cells in immunity. Nat. Immunol. 5, 1219–1226.

Cooley, R.A., Kohls, G.M., 1944. The Argasidae of North America, Central America and Cuba. American Midland Naturalist Monograph 1, pp. 1–152.

Dai, J., Narasimhan, S., Zhang, L., Liu, L., Wang, P., Fikrig, E., 2010. Tick histamine release factor is critical for *Ixodes scapularis* engorgement and transmission of the lyme disease agent. PLoS Pathog. 6, 1001205.

de Castro, J.J., Newson, R.M., 1993. Host resistance in cattle tick control. Parasitol. Today 9, 13–17.

Dennis, D.T., Piesman, J.F., 2005. Overview of tick-borne infections of humans. In: Goodman, J.L., Dennis, D.T., Sonenshine, D.E. (Eds.), Tick-Borne Diseases of Humans. American Society for Microbiology, Washington, D.C., pp. 3–11.

Dinarello, C.A., 2007. Historical insights into cytokines. Eur. J. Immunol. 37 (Suppl. 1), S34–S45.

Eberle, J.U., Voehringer, D., September 2016. Role of baso-phils in protective immunity to parasitic infections. Semin. Immunopathol. 38 (5), 605–613.

Egawa, M., Mukai, K., Yoshikawa, S., Iki, M., Mukaida, N., Kawano, Y., Minegishi, Y., Karasuyama, H., 2013. Inflammatory monocytes recruited to allergic skin acquire an anti-inflammatory M2 phenotype via baso-phil-derived interleukin-4. Immunity 38, 570–580.

Fantuzzi, G., 2005. Adipose tissue, adipokines, and inflam-mation. J. Allergy Clin. Immunol. 115, 911–919.

Fialová, A., Cimburek, Z., Iezzi, G., Kopecký, J., 2010. *Ixodes ricinus* tick saliva modulates tick-borne encephalitis virus infection of dendritic cells. Microbes Infect. 12, 580–585.

Fuchsberger, N., Kita, M., Hajnicka, V., Imanishi, J., Labuda, M., Nuttall, P.A., 1995. Ixodid tick salivary gland extracts inhibit production of lipopolysaccharide-induced mRNA of several different human cytokines. Exp. Appl. Acarol. 19, 671–676.

Gao, Y., Nish, S.A., Jiang, R., Hou, L., Licona-Limón, P., Weinstein, J.S., Zhao, H., Medzhitov, R., 2013. Control of T helper 2 responses by transcription factor IRF4-dependent dendritic cells. Immunity 39, 722–732.

Garg, R., Juncadella, I.J., Ramamoorthi, N., Ashish, Ananthanarayanan, Shobana, K., Thomas, V., Rincón, M., Krueger, J.K., Fikrig, E., Yengo, C.M., Anguita, J., 2006. Cutting edge: CD4 is the receptor for the tick saliva immunosuppressor, Salp15. J. Immunol. 177, 6579–6583.

Garlanda, C., Dinarello, C.A., Mantovani, A., 2013. The interleukin-1 family: back to the future. Immunity 39, 1003–1018.

Gebhardt, T., Whitney, P.G., Zaid, A., Mackay, L.K., Brooks, A.G., Heath, W.R., Carbone, F.R., Mueller, S.N., 2011. Different patterns of peripheral migration by memory CD4+ and CD8+ T cells. Nature 477, 216–219.

Gillespie, R.D., Dolan, M.C., Piesman, J., Titus, R.G., 2001. Identification of an IL-2 binding protein in the saliva of the Lyme disease vector tick, *Ixodes scapularis*. J. Immunol. 166, 4319–4326.

Gilliet, M., Cao, W., Liu, Y.J., 2008. Plasmacytoid dendritic cells: sensing nucleic acids in viral infection and autoim-mune diseases. Nat. Rev. Immunol. 8, 594–606.

Gordon, S., Martinez, F.O., 2010. Alternative activation of mac-rophages: mechanism and functions. Immunity 32, 593–604.

Gregson, J.D., 1960. Morphology and functioning of the mouth-parts of *Dermacentor andersoni* Stiles, Part I. The feeding mechanism in relation to the tick. Acta Trop. 17, 48–72.

Gröne, A., 2002. Keratinocytes and cytokines. Vet. Immunol. Immunopathol. 88, 1–12.

Guo, X., Booth, C.J., Paley, M.A., Wang, X., DePonte, K., Fikrig, E., Narasimhan, S., Montgomery, R.R., 2009. Inhibition of neutrophil function by two tick salivary proteins. Infect. Immun. 77, 2320–2329.

Gwakisa, P., Yoshihara, K., Long To, T., Gotoh, H., Amano, F., Momotani, E., 2001. Salivary gland extract of *Rhipicephalus appendiculatus* ticks inhibits in vitro transcription and secretion of cytokines and production of nitric oxide by LPS-stimulated JA-4 cells. Vet. Parasitol. 99, 53–61.

Hajnická, V., Kocáková, P., Sláviková, M., Slovák, M., Gasperík, J., Fuchsberger, N., Nuttall, P.A., 2001. Anti-interleukin-8 activity of tick salivary gland extracts. Parasite Immunol. 23, 483–489.

Hannier, S., Liversidge, J., Sternberg, J.M., Bowman, A.S., 2003. *Ixodes ricinus* tick salivary gland extract inhibits IL-10 secretion and CD69 expression by mitogen-stimu-lated murine splenocytes and induces hyporesponsive-ness in B lymphocytes. Parasite Immunol. 25, 27–37.

Hannier, S., Liversidge, J., Sternberg, J.M., Bowman, A.S., 2004. Characterization of the B-cell inhibitory protein factor in *Ixodes ricinus* tick saliva: a potential role in enhanced *Borrelia burgdoferi* transmission. Immunology 113, 401–408.

Heinze, D.M., Wikel, S.K., Thangamani, S., Alarcon-Chaidez, F.J., 2012. Transcriptional profiling of the murine cutaneous response during initial and subsequent infestations with *Ixodes scapularis* nymphs. Parasit. Vectors 5, 26.

Heinze, D.M., Carmical, J.R., Aronson, J.F., Alarcon-Chaidez, F., Wikel, S., Thangamani, S., 2014. Murine cutaneous responses to the rocky mountain spotted fever vector, *Dermacentor andersoni*, feeding. Front. Microbiol. 5, 198.

Hidano, A., Konnai, S., Yamada, S., Githaka, N., Isezaki, M., Higuchi, H., Nagahata, H., Ito, T., Takano, A., Ando, S., Kawabata, H., Murata, S., Ohahsi, K., 2014. Suppressive effects of neutrophil by Salp16-like salivary gland proteins from *Ixodes persulcatus* Schulze tick. Insect Mol. Biol. 23, 466–474.

Hoffmann, H.H., Schneider, W.M., Rice, C.M., 2015. Interferons and viruses: an evolutionary arms race of molecular interactions. Trends Immunol. 36, 124–138.

Hourcade, D.E., Akk, A.M., Mitchell, L.M., Zhou, H.F., Hauhart, R., Pham, C.T., 2016. Anti-complement activity of the *Ixodes scapularis* salivary protein Salp20. Mol. Immunol. 69, 62–69.

Hovius, J.W., de Jong, M.A., den Dunnen, J., Litjens, M., Fikrig, E., van der Poll, T., Gringhuis, S.I., Geijtenbeek, T.B., 2008. Salp15 binding to DC-SIGN inhibits cytokine expression by impairing both nucleosome remodeling and mRNA stabilization. PLoS Pathog. 4, e31.

Igyártó, B.Z., Kaplan, D.H., 2013. Antigen presentation by Langerhans cells. Curr. Opin. Immunol. 25, 115–119.

Igyártó, B.Z., Haley, K., Ortner, D., Bobr, A., Gerami-Nejad, M., Edelson, B.T., Zurawski, S.M., Malissen, B., Zurawski, G., Berman, J., Kaplan, D.H., 2011. Skin-resident murine dendritic cell subsets promote distinct and opposing antigen-specific T helper cell responses. Immunity 35, 260–272.

Inokuma, H., Aita, T., Tamura, K., Onishi, T., 1997. Effect of infestation with *Rhipicephalus sanguineus* on the antibody productivity in dogs. Med. Vet. Entomol. 11, 201–202.

Ivashkiv, L.B., Donlin, L.T., 2014. Regulation of type I interferon responses. Nat. Rev. Immunol. 14, 36–50.

Jakubzick, C., Gautier, E.L., Gibbings, S.L., Sojka, D.K., Schlitzer, A., Johnson, T.E., Ivanov, S., Duan, Q., Bala, S., Condon, T., van Rooijen, N., Grainger, J.R., Belkaid, Y., Ma'ayan, A., Riches, D.W., Yokoyama, W.M., Ginhoux, F., Henson, P.M., Randolph, G.J., 2013. Minimal differentiation of classical monocytes as they survey steady-state tissues and transport antigen to lymph nodes. Immunity 39, 599–610.

Jaworski, D.C., Jasinskas, A., Metz, C.N., Bucala, R., Barbour, A.G., 2001. Identification and characterization of a homologue of the pro-inflammatory cytokine macrophage migration inhibitory factor in the tick, *Amblyomma americanum*. Insect Mol. Biol. 10, 323–331.

Jongejan, F., Uilenberg, G., 2004. The global importance of ticks. Parasitology 129, S3–S14.

Juncadella, I.J., Garg, R., Ananthnarayanan, S.K., Yengo, C.M., Anguita, J., 2007. T-cell signaling pathways inhibited by the tick saliva immunosuppressor, Salp15. FEMS Immunol. Med. Microbiol. 49, 433–438.

Karasuyama, H., Yamanishi, Y., 2014. Basophils have emerged as a key player in immunity. Curr. Opin. Immunol. 31, 1–7.

Karim, S., Ribeiro, J.M., 2015. An Insight into the sialome of the lone star tick, *Amblyomma americanum*, with a glimpse on its time dependent gene expression. PLoS One 10, e0131292.

Kaufman, W.R., 1989. Tick-host interaction: a synthesis of current concepts. Parasitol. Today 5, 47–56.

Kazimírová, M., Štibrániová, I., 2013. Tick salivary compounds: their role in modulation of host defences and pathogen transmission. Front. Cell. Infect. Microbiol. 3, 43.

Kemp, D.H., Bourne, A., 1980. *Boophilus microplus*: the effect of histamine on the attachment of cattle-tick larvae-studies in vivo and in vitro. Parasitology 80, 487–496.

Kim, N.D., Luster, A.D., 2015. The role of tissue resident cells in neutrophil recruitment. Trends Immunol. 36, 547–555.

Kim, T.K., Ibelli, A.M., Mulenga, A., 2015. *Amblyomma americanum* tick calreticulin binds C1q but does not inhibit activation of the classical complement cascade. Ticks Tick Borne Dis. 6, 91–101.

Kiss, T., Cadar, D., Spînu, M., 2012. Tick prevention at a crossroad: new and renewed solutions. Vet. Parasitol. 187, 357–366.

Klein, M., Brühl, T.J., Staudt, V., Reuter, S., Grebe, N., Gerlitzki, B., Hoffmann, M., Bohn, T., Ulges, A., Stergiou, N., de Graaf, J., Löwer, M., Taube, C., Becker, M., Hain, T., Dietzen, S., Stassen, M., Huber, M., Lohoff, M., Campos Chagas, A., Andersen, J., Kotál, J., Langhansová, H., Kopecký, J., Schild, H., Kotsyfakis, M., Schmitt, E., Bopp, T., 2015. Tick salivary sialostatin L represses the initiation of immune responses by targeting IRF4-dependent transcription in murine mast cells. J. Immunol. 195, 621–631.

Kolaczkowska, E., Kubes, P., 2013. Neutrophil recruitment and function in health and inflammation. Nat. Rev. Immunol. 13, 159–175.

Kopecký, J., Kuthejlová, M., 1998. Suppressive effect of *Ixodes ricinus* salivary gland extract on mechanisms of natural immunity in vitro. Parasite Immunol. 20, 169–174.

Kotál, J., Langhansová, H., Lieskovská, J., Andersen, J.F., Francischetti, I.M., Chavakis, T., Kopecký, J., Pedra, J.H., Kotsyfakis, M., Chmelař, J., 2015. Modulation of host immunity by tick saliva. J. Proteomics 128, 58–68.

Kothur, K., Wienholt, L., Brilot, F., Dale, R.C., 2016. CSF cytokines/chemokines as biomarkers in neuroinflammatory CNS disorders: a systematic review. Cytokine 77, 227–237.

Kotsyfakis, M., Sá-Nunes, A., Francischetti, I.M., Mather, T.N., Andersen, J.F., Ribeiro, J.M., 2006. Antiinflammatory and immunosuppressive activity of sialostatin L, a salivary cystatin from the tick Ixodes scapularis. J. Biol. Chem. 281, 26298–26307.

Kotsyfakis, M., Karim, S., Andersen, J.F., Mather, T.N., Ribeiro, J.M., 2007. Selective cysteine protease inhibition contributes to blood-feeding success of the tick Ixodes scapularis. J. Biol. Chem. 282, 29256–29263.

Krause, P.J., Grant-Kels, J.M., Tahan, S.R., Dardick, K.R., Alarcon-Chaidez, F., Bouchard, K., Visini, C., Deriso, C., Foppa, I.M., Wikel, S., 2009. Dermatologic changes induced by repeated Ixodes scapularis bites and implications for prevention of tick-borne infection. Vector Borne Zoonotic Dis. 9, 603–610.

Kubes, M., Fuchsberger, N., Labuda, M., Zuffová, E., Nuttall, P.A., 1994. Salivary gland extracts of partially fed Dermacentor reticulatus ticks decrease natural killer cell activity in vitro. Immunology 82, 113–116.

Kubes, M., Kocáková, P., Slovák, M., Sláviková, M., Fuchsberger, N., Nuttall, P.A., 2002. Heterogeneity in the effect of different ixodid tick species on human natural killer cell activity. Parasite Immunol. 24, 23–28.

Kumamoto, Y., Linehan, M., Weinstein, J.S., Laidlaw, B.J., Craft, J.E., Iwasaki, A., 2013. CD301b+ dermal dendritic cells drive T helper 2 cell-mediated immunity. Immunity 39, 733–743.

Kupper, T.S., Fuhlbrigge, R.C., 2004. Immune surveillance in the skin: mechanisms and clinical consequences. Nat. Rev. Immunol. 4, 211–222.

Langhansová, H., Bopp, T., Schmitt, E., Kopecký, J., 2015. Tick saliva increases production of three chemokines including monocyte chemoattractant protein-1, a histamine-releasing cytokine. Parasite Immunol. 37, 92–96.

Lavoipierre, M.M.J., 1965. Feeding mechanisms of blood-sucking arthropods. Nature 208, 302–303.

Lawrie, C.H., Sim, R.B., Nuttall, P.A., 2005. Investigation of the mechanisms of anti-complement activity in Ixodes ricinus ticks. Mol. Immunol. 42, 31–38.

Lebre, M.C., van der Aar, A.M., van Baarsen, L., van Capel, T.M., Schuitemaker, J.H., Kapsenberg, M.L., de Jong, E.C., 2007. Human keratinocytes express functional Toll-like receptor 3, 4, 5, and 9. J. Invest. Dermatol. 127, 331–341.

Ley, K., Laudanna, C., Cybulsky, M.I., Nourshargh, S., 2007. Getting to the site of inflammation: the leukocyte adhesion cascade updated. Nat. Rev. Immunol. 7, 678–689.

Lieskovská, J., Kopecký, J., 2012a. Tick saliva suppresses IFN signalling in dendritic cells upon Borrelia afzelii infection. Parasite Immunol. 34, 32–39.

Lieskovská, J., Kopecký, J., 2012b. Effect of tick saliva on signalling pathways activated by TLR-2 ligand and Borrelia afzelii in dendritic cells. Parasite Immunol. 34, 421–429.

Lieskovská, J., Páleníková, J., Langhansová, H., Campos Chagas, A., Calvo, E., Kotsyfakis, M., Kopecký, J., 2015. Tick sialostatins L and L2 differentially influence dendritic cell responses to Borrelia spirochetes. Parasit. Vectors 8, 275.

Locati, M., Mantovani, A., Sica, A., 2013. Macrophage activation and polarization as an adaptive component of innate immunity. Adv. Immunol. 120, 163–184.

Loré, K., Sönnerborg, A., Spetz, A.L., Andersson, U., Andersson, J., 1998. Immunocytochemical detection of cytokines and chemokines in Langerhans cells and in vitro derived dendritic cells. J. Immunol. Methods 214, 97–111.

Macaluso, K.R., Wikel, S.K., 2001. Dermacentor andersoni: effects of repeated infestations on lymphocyte proliferation, cytokine production, and adhesion-molecule expression by BALB/c mice. Ann. Trop. Med. Parasitol. 95, 413–427.

Mantovani, A., Cassatella, M.A., Costantini, C., Jaillon, S., 2011. Neutrophils in the activation and regulation of innate and adaptive immunity. Nat. Rev. Immunol. 11, 519–531.

Marchal, C., Schramm, F., Kern, A., Luft, B.J., Yang, X., Schuijt, T.J., Hovius, J.W., Jaulhac, B., Boulanger, N., 2011. Antialarmin effect of tick saliva during the transmission of Lyme disease. Infect. Immun. 79, 774–785.

Marufu, M.C., Dzama, K., Chimonyo, M., 2014. Cellular responses to Rhipicephalus microplus infestations in pre-sensitised cattle with differing phenotypes of infestation. Exp. Appl. Acarol. 62, 241–252.

Maxwell, S.S., Stoklasek, T.A., Dash, Y., Macaluso, K.R., Wikel, S.K., 2005. Tick modulation of the in-vitro expression of adhesion molecules by skin-derived endothelial cells. Ann. Trop. Med. Parasitol. 99, 661–672.

Mbow, M.L., Rutti, B., Brossard, M., 1994. Infiltration of CD4+ CD8+ T cells, and expression of ICAM-1, Ia antigens, IL-1 alpha and TNF-alpha in the skin lesion of BALB/c mice undergoing repeated infestations with nymphal Ixodes ricinus ticks. Immunology 82, 596–602.

McLaren, D.J., Worms, M.J., Brown, S.J., Askenase, P.W., 1983. Ornithodorus tartakovskyi: quantitation and ultrastructure of cutaneous basophil responses in the guinea pig. Exp. Parasitol. 56, 153–168.

Mejri, N., Brossard, M., 2007. Splenic dendritic cells pulsed with Ixodes ricinus tick saliva prime naive CD4+T to induce Th2 cell differentiation in vitro and in vivo. Int. Immunol. 19, 535–543.

Menten-Dedoyart, C., Couvreur, B., Thellin, O., Drion, P.V., Herry, M., Jolois, O., Heinen, E., 2008. Influence of the Ixodes ricinus tick blood-feeding on the antigen-specific antibody response in vivo. Vaccine 26, 6956–6964.

Menten-Dedoyart, C., Faccinetto, C., Golovchenko, M., Dupiereux, I., Van Lerberghe, P.B., Dubois, S., Desmet, C., Elmoualij, B., Baron, F., Rudenko, N., Oury, C., Heinen, E., 2012. Neutrophil extracellular traps entrap and kill Borrelia burgdorferi sensu stricto spirochetes and are not affected by Ixodes ricinus tick saliva. J. Immunol. 189, 5393–5401.

Merad, M., Ginhoux, F., Collin, M., 2008. Origin, homeostasis and function of Langerhans cells and other langerin-expressing dendritic cells. Nat. Rev. Immunol. 8, 935–947.

Miyake, Y., Yamasaki, S., 2012. Sensing necrotic cells. Adv. Exp. Med. Biol. 738, 144–152.

Montgomery, R.R., Lusitani, D., De Boisfleury Chevance, A., Malawista, S.E., 2004. Tick saliva reduces adherence and area of human neutrophils. Infect. Immun. 72, 2989–2994.

Moorhouse, D.E., 1969. The attachment of some ixodid ticks to their natural hosts. In: Proceedings of the Second International Congress of Acarology. Hungarian Academy of Sciences, Budapest, Hungary, pp. 319–327.

Morita, H., Saito, H., Matsumoto, K., Nakae, S., September 2016. Regulatory roles of mast cells in immune responses. Semin. Immunopathol. 38 (5), 623–629.

Mosser, D.M., Edwards, J.P., 2008. Exploring the full spectrum of macrophage activation. Nat. Rev. Immunol. 8, 958–969.

Murray, P.J., Allen, J.E., Biswas, S.K., Fisher, E.A., Gilroy, D.W., Goerdt, S., Gordon, S., Hamilton, J.A., Ivashkiv, L.B., Lawrence, T., Locati, M., Mantovani, A., Martinez, F.O., Mege, J.L., Mosser, D.M., Natoli, G., Saeij, J.P., Schultze, J.L., Shirey, K.A., Sica, A., Suttles, J., Udalova, I., van Ginderachter, J.A., Vogel, S.N., Wynn, T.A., 2014. Macrophage activation and polarization: nomenclature and experimental guidelines. Immunity 41, 14–20.

Nathan, C., 2006. Neutrophils and immunity: challenges and opportunities. Nat. Rev. Immunol. 6, 173–182.

Nelder, M.P., Russell, C.B., Sheehan, N.J., Sander, B., Moore, S., Li, Y., Johnson, S., Patel, S.N., Sider, D., 2016. Human pathogens associated with the blacklegged tick Ixodes scapularis: a systematic review. Parasit. Vectors 9, 265.

Nestle, F.O., Di Meglio, P., Qin, J.Z., Nickoloff, B.J., 2009. Skin immune sentinels in health and disease. Nat. Rev. Immunol. 9, 679–691.

Nishihira, J., 1998. Novel pathophysiological aspects of macrophage migration inhibitory factor (review). Int. J. Mol. Med. 2, 17–28.

Nithiuthai, S., Allen, J.R., 1984a. Significant changes in epidermal Langerhans cells of guinea-pigs infested with ticks (Dermacentor andersoni). Immunology 51, 133–141.

Nithiuthai, S., Allen, J.R., 1984b. Effects of ultraviolet irradiation on epidermal Langerhans cells in guinea pigs. Immunology 51, 143–151.

Nithiuthai, S., Allen, J.R., 1984c. Effects of ultraviolet irradiation on the acquisition and expression of tick resistance in guinea-pigs. Immunology 51, 153–159.

Nithiuthai, S., Allen, J.R., 1985. Langerhans cells present tick antigens to lymph node cells from tick-sensitized guinea-pigs. 55, 157–163.

Nunn, M.A., Sharma, A., Paesen, G.C., Adamson, S., Lissina, O., Willis, A.C., Nuttall, P.A., 2005. Complement inhibitor of C5 activation from the soft tick Ornithodoros moubata. J. Immunol. 174, 2084–2091.

Nourshargh Ornithodoros moubata, S., Alon, R., 2014. Leukocyte migration into inflamed tissues. Immunity 41, 694–707.

Oliveira, C.J., Cavassani, K.A., Moré, D.D., Garlet, G.P., Aliberti, J.C., Silva, J.S., Ferreira, B.R., 2008. Tick saliva inhibits the chemotactic function of MIP-1 alpha and selectively impairs chemotaxis of immature dendritic cells by down-regulating cell-surface CCR5. Int. J. Parasitol. 38, 705–716.

Oliveira, C.J., Carvalho, W.A., Garcia, G.R., Gutierrez, F.R., de Miranda Santos, I.K., Silva, J.S., Ferreira, B.R., 2010. Tick saliva induces regulatory dendritic cells: MAP-kinases and Toll-like receptor-2 expression as potential targets. Vet. Parasitol. 167, 288–297.

Oliveira, C.J., Sá-Nunes, A., Francischetti, I.M., Carregaro, V., Anatriello, E., Silva, J.S., Santos, I.K., Ribeiro, J.M., Ferreira, B.R., 2011. Deconstructing tick saliva: non-protein molecules with potent immunomodulatory properties. J. Biol. Chem. 286, 10960–10969.

Otsuka, A., Kabashima, K., 2015. Contribution of basophils to cutaneous immune reactions and Th2-mediated allergic responses. Front. Immunol. 6, 393.

Otsuka, A., Nonomura, Y., Kabashima, K., September 2016. Roles of basophils and mast cells in cutaneous inflammation. Semin. Immunopathol. 38 (5), 563–570.

Paine, S.H., Kemp, D.H., Allen, J.R., 1983. In vitro feeding of Dermacentor andersoni (Stiles): effects of histamine and other mediators. Parasitology 86, 419–428.

Pangburn, M.K., Müller-Eberhard, H.J., 1984. The alternative pathway of complement. Springer Semin. Immunopathol. 7, 163–192.

Pappu, R., Rutz, S., Ouyang, W., 2012. Regulation of epithelial immunity by IL-17 family cytokines. Trends Immunol. 33, 343–349.

Parola, P., Raoult, D., 2001. Ticks and tickborne bacterial diseases in humans: an emerging infectious threat. Clin. Infect. Dis. 32, 897–928.

Parola, P., Paddock, C.D., Socolovschi, C., Labruna, M.B., Mediannikov, O., Kernif, T., Abdad, M.Y., Stenos, J., 2013. Update on tick-borne rickettsioses around the world: a geographic approach. Clin. Microbiol. Rev. 26, 657–702.

Pasparakis, M., Haase, I., Nestle, F.O., 2014. Mechanisms regulating skin immunity and inflammation. Nat. Rev. Immunol. 14, 289–301.

Perrigoue, J.G., Saenz, S.A., Siracusa, M.C., Allenspach, E.J., Taylor, B.C., Giacomin, P.R., Nair, M.G., Du, Y., Zaph, C., van Rooijen, N., Comeau, M.R., Pearce, E.J., Laufer, T.M., Artis, D., 2009. MHC class II-dependent basophil-CD4+ T cell interactions promote Th2 cytokine-dependent immunity. Nat. Immunol. 10, 697–705.

Poole, N.M., Mamidanna, G., Smith, R.A., Coons, L.B., Cole, J.A., 2013. Prostaglandin E2 in tick saliva regulates macrophage cell migration and cytokine profile. Parasit. Vectors 6, 261.

Porta, C., Riboldi, E., Ippolito, A., Sica, A., 2015. Molecular and epigenetic basis of macrophage polarized activation. Semin. Immunol. 27, 237–248.

Preston, S.G., Majtán, J., Kouremenou, C., Rysnik, O., Burger, L.F., Cabezas Cruz, A., Chiong Guzman, M., Nunn, M.A., Paesen, G.C., Nuttall, P.A., Austyn, J.M., 2013. Novel immunomodulators from hard ticks selectively reprogramme human dendritic cell responses. PLoS Pathog. 9, 1003450.

Ramachandra, R.N., Wikel, S.K., 1992. Modulation of host-immune responses by ticks (Acari: Ixodidae): effect of salivary gland extracts on host macrophages and lymphocyte cytokine production. J. Med. Entomol. 29, 818–826.

Ramachandra, R.N., Wikel, S.K., 1995. Effects of *Dermacentor andersoni* (Acari: Ixodidae) salivary gland extracts on *Bos indicus* and *B. taurus* lymphocytes and macrophages: in vitro cytokine elaboration and lymphocyte blastogenesis. J. Med. Entomol. 32, 338–345.

Ribeiro, J.M., 1987. *Ixodes dammini*: salivary anti-complement activity. Exp. Parasitol. 64, 347–353.

Ribeiro, J.M., Weis, J.J., Telford 3rd, S.R., 1990. Saliva of the tick *Ixodes dammini* inhibits neutrophil function. Exp. Parasitol. 70, 382–388.

Ribeiro, J., Alarcon-Chaidez, F., Franciscetti, I.M.B., Mans, B., Mather, T.N., Valenzuela, J.G., Wikel, S.K., 2006. An annotated catalog of salivary gland transcripts from *Ixodes scapularis* ticks. Insect Biochem. Mol. Biol. 36, 111–129.

Rosenblum, M.D., Gratz, I.K., Paw, J.S., Lee, K., Marshak-Rothstein, A., Abbas, A.K., 2011. Response to self antigen imprints regulatory memory in tissues. Nature 480, 538–542.

Sá-Nunes, A., Bafica, A., Lucas, D.A., Conrads, T.P., Veenstra, T.D., Andersen, J.F., Mather, T.N., Ribeiro, J.M., Franciscetti, I.M., 2007. Prostaglandin E2 is a major inhibitor of dendritic cell maturation and function in *Ixodes scapularis* saliva. J. Immunol. 179, 1497–1505.

Sá-Nunes, A., Bafica, A., Antonelli, L.R., Choi, E.Y., Franciscetti, I.M., Andersen, J.F., Shi, G.P., Chavakis, T., Ribeiro, J.M., Kotsyfakis, M., 2009. The immunomodulatory action of sialostatin L on dendritic cells reveals its potential to interfere with autoimmunity. J. Immunol. 182, 7422–7429.

Said, A., Weindl, G., 2015. Regulation of dendritic cell function in inflammation. J. Immunol. Res. 2015, 743169.

Sakhon, O.S., Severo, M.S., Kotsyfakis, M., Pedra, J.H., 2013. A Nod to disease vectors: mitigation of pathogen sensing by arthropod saliva. Front. Microbiol. 4, 308.

Sallusto, F., Baggiolini, M., 2008. Chemokines and leukocyte traffic. Nat. Immunol. 9, 949–952.

Sangamnatdej, S., Paesen, G.C., Slovak, M., Nuttall, P.A., 2002. A high affinity serotonin- and histamine-binding lipocalin from tick saliva. Insect Mol. Biol. 11, 79–86.

Schoeler, G.B., Manweiler, S.A., Wikel, S.K., 1999. *Ixodes scapularis*: effects of repeated infestations with pathogen-free nymphs on macrophage and T lymphocyte cytokine responses of BALB/c and C3H/HeN mice. Exp. Parasitol. 92, 239–248.

Schoeler, G.B., Manweiler, S.A., Wikel, S.K., 2000. Cytokine responses of C3H/HeN mice infested with *Ixodes scapularis* or *Ixodes pacificus* nymphs. Parasite Immunol. 22, 31–40.

Schroeder, H., Skelly, P.J., Zipfel, P.F., Losson, B., Vanderplasschen, A., 2009. Subversion of complement by hematophagous parasites. Dev. Comp. Immunol. 33, 5–13.

Schuijt, T.J., Coumou, J., Narasimhan, S., Dai, J., Deponte, K., Wouters, D., Brouwer, M., Oei, A., Roelofs, J.J., van Dam, A.P., van der Poll, T., Van't Veer, C., Hovius, J.W., Fikrig, E., 2011. A tick mannose-binding lectin inhibitor interferes with the vertebrate complement cascade to enhance transmission of the lyme disease agent. Cell Host Microbe 10, 136–146.

Shaw, D.K., Kotsyfakis, M., Pedra, J.H.F., 2016. For whom the bell Tolls (and Nods): spit-acular saliva. Curr. Trop. Med. Rep. 3, 40–50.

Sica, A., Mantovani, A., 2012. Macrophage plasticity and polarization: in vivo veritas. J. Clin. Invest. 122, 787–795.

Siebert, S., Tsoukas, A., Robertson, J., McInnes, I., 2015. Cytokines as therapeutic targets in rheumatoid arthritis and other inflammatory diseases. Pharmacol. Rev. 67, 280–309.

Skallová, A., Iezzi, G., Ampenberger, F., Kopf, M., Kopecký, J., 2008. Tick saliva inhibits dendritic cell migration, maturation, and function while promoting development of Th2 responses. J. Immunol. 180, 6186–6192.

Slámová, M., Skallová, A., Páleníková, J., Kopecký, J., 2011. Effect of tick saliva on immune interactions between *Borrelia afzelii* and murine dendritic cells. Parasite Immunol. 33, 654–660.

Smith, G.L., Benfield, C.T., Maluquer de Motes, C., Mazzon, M., Ember, S.W., Ferguson, B.J., Sumner, R.P., 2013. Vaccinia virus immune evasion: mechanisms, virulence and immunogenicity. J. Gen. Virol. 94, 2367–2392.

So, T., Lee, S.W., Croft, M., 2006. Tumor necrosis factor/tumor necrosis factor receptor family members that positively regulate immunity. Int. J. Hematol. 83, 1–11.

Sokol, C.L., Medzhitov, R., 2010. Emerging functions of basophils in protective and allergic immune responses. Mucosal Immunol. 3, 129–137.

Sonenshine, D.E., Anderson, J.M., 2014. Mouthparts and digestive system. In: Sonenshine, D.E., Roe, R.M. (Eds.). Sonenshine, D.E., Roe, R.M. (Eds.), Biology of Ticks, vol. 1. Oxford University Press, New York, pp. 122–162.

Sonnenberg, G.F., Artis, D., 2015. Innate lymphoid cells in the initiation, regulation and resolution of inflammation. Nat. Med. 21, 698–708.

Stassen, M., Schmitt, E., Bopp, T., 2012. From interleukin-9 to T helper 9 cells. Ann. N. Y. Acad. Sci. 1247, 56–68.

Stockinger, B., Veldhoen, M., Martin, B., 2007. Th17 T cells: linking innate and adaptive immunity. Semin. Immunol. 19, 353–361.

Sumaria, N., Roediger, B., Ng, L.G., Qin, J., Pinto, R., Cavanagh, L.L., Shklovskaya, E., Fazekas de St Groth, B., Triccas, J.A., Weninger, W., 2011. Cutaneous immunosurveillance by self-renewing dermal gamma/delta T cells. J. Exp. Med. 208, 505–518.

Summers, C., Rankin, S.M., Condliffe, A.M., Singh, N., Peters, A.M., Chilvers, E.R., 2010. Neutrophil kinetics in health and disease. Trends Immunol. 31, 318–324.

Sun, J.C., Lanier, L.L., 2011. NK cell development, homeostasis and function: parallels with CD8$^+$T cells. Nat. Rev. Immunol. 11, 645–657.

Tirloni, L., Reck, J., Terra, R.M., Martins, J.R., Mulenga, A., Sherman, N.E., Fox, J.W., Yates 3rd, J.R., Termignoni, C., Pinto, A.F., Vaz Ida Jr., S., 2014. Proteomic analysis of cattle tick *Rhipicephalus (Boophilus) microplus* saliva: a comparison between partially and fully engorged females. PLoS One 9, e94831.

Tüzün, Y., Antonov, M., Dolar, N., Wolf, R., 2007. Keratinocyte cytokine and chemokine receptors. Dermatol. Clin. 25, 467–476.

Tyson, K., Elkins, C., Patterson, H., Fikrig, E., de Silva, A., 2007. Biochemical and functional characterization of Salp20, an *Ixodes scapularis* tick salivary protein that inhibits the complement pathway. Insect Mol. Biol. 16, 469–479.

Tyson, K.R., Elkins, C., de Silva, A.M., 2008. A novel mechanism of complement inhibition unmasked by a tick salivary protein that binds to properdin. J. Immunol. 180, 3964–3968.

Umemiya, R., Hatta, T., Liao, M., Tanaka, M., Zhou, J., Inoue, N., Fujisaki, K., 2007. *Haemaphysalis longicornis*: molecular characterization of a homologue of the macrophage migration inhibitory factor from the partially fed ticks. Exp. Parasitol. 115, 135–142.

Urioste, S., Hall, L.R., Telford 3rd, S.R., Titus, R.G., 1994. Saliva of the Lyme disease vector, *Ixodes dammini*, blocks cell activation by a nonprostaglandin E2-dependent mechanism. J. Exp. Med. 180, 1077–1085.

Valenzuela, J.G., Charlab, R., Mather, T.N., Ribeiro, J.M., 2000. Purification, cloning, and expression of a novel salivary anticomplement protein from the tick, *Ixodes scapularis*. J. Biol. Chem. 275 (25), 18717–18723.

Vancová, I., Slovák, M., Hajnická, V., Labuda, M., Simo, L., Peterková, K., Hails, R.S., Nuttall, P.A., 2007. Differential anti-chemokine activity of *Amblyomma variegatum* adult ticks during blood-feeding. Parasite Immunol. 29, 169–177.

Vancová, I., Hajnická, V., Slovák, M., Nuttall, P.A., 2010. Anti-chemokine activities of ixodid ticks depend on tick species, developmental stage, and duration of feeding. Vet. Parasitol. 167, 274–278.

Vesely, D.L., Fish, D., Shlomchik, M.J., Kaplan, D.H., Bockenstedt, L.K., 2009. Langerhans cell deficiency impairs *Ixodes scapularis* suppression of Th1 responses in mice. Infect. Immun. 77, 1881–1887.

Vivier, E., Tomasello, E., Baratin, M., Walzer, T., Ugolini, S., 2008. Functions of natural killer cells. Nat. Immunol. 9, 503–510.

Vivier, E., Raulet, D.H., Moretta, A., Caligiuri, M.A., Zitvogel, L., Lanier, L.L., Yokoyama, W.M., Ugolini, S., 2011. Innate or adaptive immunity? The example of natural killer cells. Science 331, 44–49.

Voehringer, D., 2013. Protective and pathological roles of mast cells and basophils. Nat. Rev. Immunol. 13, 362–375.

Wada, T., Ishiwata, K., Koseki, H., Ishikura, T., Ugajin, T., Ohnuma, N., Obata, K., Ishikawa, R., Yoshikawa, S., Mukai, K., Kawano, Y., Minegishi, Y., Yokozeki, H., Watanabe, N., Karasuyama, H., 2010. Selective ablation of basophils in mice reveals their nonredundant role in acquired immunity against ticks. J. Clin. Invest. 120, 2867–2875.

Wang, X., Shaw, D.K., Sakhon, O.S., Snyder, G.A., Sundberg, E.J., Santambrogio, L., Sutterwala, F.S., Dumler, J.S., Shirey, K.A., Perkins, D.J., Richard, K., Chagas, A.C., Calvo, E., Kopecký, J., Kotsyfakis, M., Pedra, J.H., 2016. The tick rotein sialostatin L2 binds to annexin A2 and inhibits NLRC4-mediated inflammasome activation. Infect. Immun. 84, 1796–1805.

Weiss, S.J., 1989. Tissue destruction by neutrophils. N. Engl. J. Med. 320, 365–376.

Weninger, W., Biro, M., Jain, R., 2014. Leukocyte migration in the interstitial space of non-lymphoid organs. Nat. Rev. Immunol. 14, 232–246.

West, A.P., Koblansky, A.A., Ghosh, S., 2006. Recognition and signaling by toll-like receptors. Annu. Rev. Cell Dev. Biol. 22, 409–437.

Wikel, S.K., 1979. Acquired resistance to ticks. Expression of resistance by C4-deficient guinea pigs. Am. J. Trop. Med. Hyg. 28, 586–590.

Wikel, S.K., 1982. Influence of *Dermacentor andersoni* infestation on lymphocyte responsiveness to mitogens. Ann. Trop. Med. Parasitol. 76, 627–632.

Wikel, S.K., 1985. Effects of tick infestation on the plaque-forming cell response to a thymic dependant antigen. Ann. Trop. Med. Parasitol. 79, 195–198.

Wikel, S., 2013. Ticks and tick-borne pathogens at the cutaneous interface: host defenses, tick countermeasures, and a suitable environment for pathogen establishment. Front. Microbiol. 4, 337.

Willadsen, P., Wood, G.M., Riding, G.A., 1979. The relation between skin histamine concentration, histamine sensitivity, and the resistance of cattle to the tick, *Boophilus microplus*. Z. Parasitenkd. 59, 87–93.

Wu, J., Wang, Y., Liu, H., Yang, H., Ma, D., Li, J., Li, D., Lai, R., Yu, H., 2010. Two immunoregulatory peptides with antioxidant activity from tick salivary glands. J. Biol. Chem. 285, 16606–16613.

Wynn, T.A., 2009. Basophils trump dendritic cells as APCs for Th2 responses. Nat. Immunol. 10, 679–681.

Yazdi, A.S., Röcken, M., Ghoreschi, K., 2016. Cutaneous immunology: basics and new concepts. Semin. Immunopathol. 38, 3–10.

Yokoyama, W.M., Kim, S., French, A.R., 2004. The dynamic life of natural killer cells. Annu. Rev. Immunol. 22, 405–429.

Yu, D., Liang, J., Yu, H., Wu, H., Xu, C., Liu, J., Lai, R., 2006. A tick B-cell inhibitory protein from salivary glands of the hard tick, *Hyalomma asiaticum asiaticum*. Biochem. Biophys. Res. Commun. 343, 585–590.

Zipfel, P.F., Skerka, C., 2009. Complement regulators and inhibitory proteins. Nat. Rev. Immunol. 9, 729–740.

Zipfel, P.F., Mihlan, M., Skerka, C., 2007. The alternative pathway of complement: a pattern recognition system. Adv. Exp. Med. Biol. 598, 80–92.

Tick Saliva and Microbial Effector Molecules: Two Sides of the Same Coin

Dana K. Shaw[1], Erin E. McClure[1], Michail Kotsyfakis[2], Joao H.F. Pedra[1]

[1]University of Maryland School of Medicine, Baltimore, MD, United States; [2]Czech Academy of Sciences, Budweis, Czech Republic

List of Abbreviations

AIM2 Absent in melanoma 2
AMP Adenosine monophosphate
ASC Apoptosis-associated speck-like protein containing a carboxy terminal CARD domain
BIR Baculovirus inhibitor-of-apoptosis repeats
CARD Caspase activation and recruitment domain
cGAS/STING Cyclic GMP-AMP synthase/stimulator of interferon genes
CLR C-type lectin receptor
DAMP Damage-associated molecular pattern
GMP Guanosine monophosphate
iE-DAP γ-D-glutamyl-meso-diaminopimelic acid
IFN Interferon
IL Interleukin
LPS Lipopolysaccharide
LRR Leucine-rich repeat
MAPK Mitogen-activated protein kinase
MCP Monocyte chemoattractant protein-1
MDP Muramyl dipeptide
MIP-2 Macrophage inflammatory protein-2
NAIP NLR family apoptosis inhibitory proteins
NBD Nucleotide-binding domain
NF-κB Nucleation factor-κB
NLR Nod-like receptor
NLRC NLR family CARD domain–containing protein
NLRP NLR family PYD domain–containing protein
Nod Nucleotide-binding oligomerization domain

PAMP Pathogen-associated molecular pattern
PRR Pattern recognition receptor
PYD Pyrin domain
RIG I Retinoid acid–inducible gene-I
RIPK2 Receptor-interacting protein kinase 2
ROS Reactive oxygen species
SL2 Sialostatin L2
TLR Toll-like receptor
TNF Tumor necrosis factor

INTRODUCTION

Ticks are blood-feeding ectoparasites that, unlike mosquitoes and some other blood-sucking arthropods, are exclusively hematophagous at all life stages (Chmelař et al., 2016b; Gulia-Nuss et al., 2016; Sonenshine and Roe, 2014a). These arachnids transmit a greater diversity of pathogens than any other blood-feeding arthropods and are the most important disease vectors in the northern hemisphere (Bockenstedt and Wormser, 2014; Koedel et al., 2015; Nelder et al., 2016; Nelson et al., 2015; Piesman and Eisen, 2008; Stanek et al., 2012). Ticks are of significant agricultural and public health relevance and are capable of transmitting many pathogens including bacteria (*Borrelia* spp., *Anaplasma* spp., *Rickettsia* spp., *Ehrlichia* spp., *Francisella tularensis*), protozoa (*Babesia* spp., *Theileria* spp.), and viruses (over 50 species, which constitute 160 named strains and types) (Caulfield and Pritt, 2015; Diuk-Wasser et al., 2016; Dugat et al., 2015; Lina et al., 2016; Mead, 2015; Nelder et al., 2016; Nene et al., 2016; Růžek et al., 2010; Sonenshine and Roe, 2014b; Wagemakers et al., 2015). *Ixodes scapularis* is remarkably capable of transmitting as many as six different microbes at one time including extra- and intracellular bacteria (*Borrelia burgdorferi*, *Borrelia miyamotoi*, *Anaplasma phagocytophilum*, and an *Ehrlichia muris*–type bacterium), the parasite *Babesia microti*, and Powassan virus (Sonenshine and Roe, 2014b); none of the diseases caused by these pathogens are currently preventable by vaccines, making them a significant public health concern.

Ticks are unique when compared with other blood-sucking arthropods for a number of reasons, but are noted especially for their prolonged feeding process, which varies depending on life stage and species of tick (Kaufman, 1989). Ticks of the Argasidae family have a flexible and highly folded, leathery cuticle, enabling them to expand quickly and acquire a blood meal within minutes to hours (Hackman, 1982). Ixodid ticks have a hardened cuticle over parts of their body (the scutum). New endocuticle is synthesized during the approximate 3–7 days required for feeding to accommodate their expansion (ranging from a 10- to 100-fold increase), which prolongs the amount of time required for engorgement (Flynn and Kaufman, 2011; Sonenshine and Roe, 2014a).

A feeding lesion is created by cutting the host's skin with retractable saw-edged, blade-like structures, termed chelicerae, that push the barbed hypostome into the bite-wound using a ratchetlike mechanism (Richter et al., 2013).

This serves to anchor the tick mouthparts in the dermis and severs capillaries, creating a blood pool beneath the skin from which to siphon fluids. Hard ticks secrete cement (a component of saliva) around the wound, which varies in amount depending on the species, and glues the tick in place (Francischetti et al., 2009). This process takes approximately 1–2 days. During this time, the lesion expands due to tissue erosion likely caused by migrating neutrophils, which promotes the flow of blood and other fluids to the bite site (Tatchell and Moorhouse, 1970). The tick will then spend several additional days imbibing blood and other fluids. Engorgement of the tick is facilitated by saliva secreted into the bite site. The saliva contains a plethora of bioactive molecules that inhibit pain, itch responses, hemostasis, innate and adaptive immunity, and wound healing (Gillespie et al., 2001; Guo et al., 2009; Paesen et al., 1999; Poole et al., 2013). The feeding lesion is painless to the host, owing, in part, to the immunomodulatory properties of the saliva, which allows the tick to go unnoticed (Sonenshine and Roe, 2014a). A large body of work has been dedicated to understanding cellular and humoral immune modulation by tick saliva, which has contributed significantly to our understanding of the tick life strategy (Chmelař et al., 2016b; Kotál et al., 2015); however, much less molecular detail is known about salivary molecule manipulation of innate signaling networks (Shaw et al., 2016).

Innate signaling is the first line of defense following physical injury or an immune assault (Brubaker et al., 2015; Janeway and Medzhitov, 2002; Mogensen, 2009; Palm and Medzhitov, 2009). Pattern recognition receptors (PRRs) such as Toll-like (TLRs), nucleotide-binding oligomerization domain (Nod)-like (NLRs), absent in melanoma 2 (AIM2), C-type lectin (CLRs), retinoid acid–inducible gene I-like (RIG I-like), and cyclic GMP-AMP synthase (cGAS)/STING (stimulator of interferon genes) recognize pathogen-associated molecular patterns (PAMPs) and/or self-derived damage-associated molecular patterns (DAMPs) and initiate immune responses (Bortoluci and Medzhitov, 2010; Brubaker et al., 2015; Caruso et al., 2014; Iwasaki and Medzhitov, 2015). Research focusing on NLRs is rapidly advancing, making it one of the best studied PRRs, second only to TLRs. TLRs were the first identified family of PRRs, of which there are currently 10 known types in humans and 12 in mice. All TLRs are transmembrane structures that are found on the surface and within endosomal compartments of host cells, which sense stimuli at the cell membrane interface (Brubaker et al., 2015; Iwasaki and Medzhitov, 2015; Kawai and Akira, 2011; Lemaitre et al., 1996). NLRs are cytosolically localized PRRs that also sense PAMPs, but which can lead to different types of inflammatory responses than TLRs owing to the differences in cellular compartmentalization (Caruso et al., 2014; Motta et al., 2015; Shaw et al., 2016). There are currently 22 characterized NLRs in humans and 34 in mice (Motta et al., 2015). These PRRs can serve as scaffolding proteins, which lead to the formation of a large, multiprotein complex termed the "inflammasome" when activated (Brubaker et al., 2015). For the purpose of this chapter, we will focus on inflammasome signaling, although readers are referred to excellent reviews written in the last few years for more details regarding TLR signaling (Bortoluci and Medzhitov, 2010; Brubaker et al., 2015; Iwasaki and Medzhitov, 2015).

The immune system is commonly likened to a cellular and molecular army that functions in a highly regulated manner to monitor and defend the host against damaging and/ or infectious assaults on the body. Organisms seeking to infect or parasitize a host need to employ evasion tactics to subvert this army and survive. One hallmark of pathogenic organisms is the ability to avoid immune killing; this topic has been extensively explored in microbiology (Guven-Maiorov et al., 2016;

Holmgren et al., 2016; Maldonado et al., 2016; Schuren et al., 2016). Microbial pathogens commonly execute evasion tactics by using effector molecules that directly disarm the immune response. Interestingly, tick saliva molecules impede some of the same immune processes as microbial-derived effector molecules. Both ticks and microbes have a common goal: to evade the host immune response and promote survival. In this sense, saliva and microbial molecules can appropriately be described as immune evaders (Brossard and Wikel, 2004; Kotál et al., 2015; Ribeiro, 1989; Schoeler and Wikel, 2001; Wikel, 1999). Indeed, if genes that encode microbial effectors are mutated or lost, it can lead to pathogen recognition followed by immune-mediated killing. Similarly, if a host is immunized against tick salivary proteins, ticks can no longer evade recognition, which leads to immune-mediated rejection and premature termination of a blood meal (Bell et al., 1979; Kotsyfakis et al., 2008; Narasimhan et al., 2007; Trimnell et al., 2002; Wikel et al., 1997; Wikel and Allen, 1976). Herein, we will describe what is known regarding tick saliva and microbial immune effectors and will discuss the convergent evasion strategies employed by these molecules, despite having radically different origins.

NOD-LIKE RECEPTOR SIGNALING

NLRs are a broad superfamily of scaffolding proteins with 22 characterized in humans and at least 34 present in mice. More recently discovered is the presence of over 200 NLR-coding genes in invertebrates, such as the sea urchin, indicating that the evolutionary origin of NLRs is much more ancient than previously believed (Motta et al., 2015). NLRs generally contain both a nucleotide-binding domain (NBD) and leucine-rich repeats (LRRs) (Caruso et al., 2014). The LRR portion of NLRs is believed to have two functions: (1) autoinhibition, keeping the NLR in an inactive

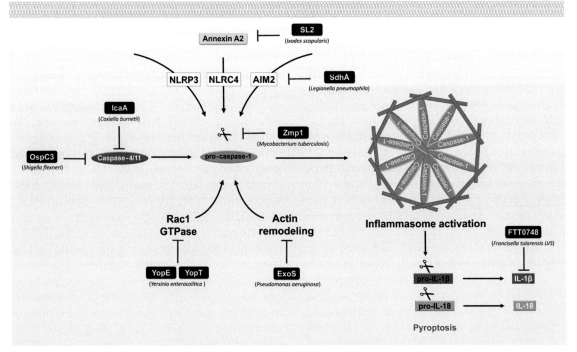

FIGURE 10.1 **Inflammasome targets of bacterial and tick saliva immunomodulators.** Nod-like receptors lead to the formation of a large, multiprotein scaffolding platform termed the inflammasome. NLRP inflammasomes require the ASC signaling adapter molecule to recruit caspase-1 (*blue bars*), whereas NLRC proteins have a CARD domain and can recruit caspase-1 independently of ASC. The AIM2 inflammasome is also ASC-independent, but is not mediated by NLR proteins. Activation of the inflammasome leads to processing of pro-caspase-1 into mature caspase-1, which then cleaves pro-IL-1β (purple) and pro-IL-18 (green) into their mature forms, causing an inflammatory response and pyroptosis. Alternatively, the noncanonical inflammasome is mediated by caspase-11 (mice) or the human homologue, caspase-4 (dark orange), which bypasses NLR proteins and leads to the direct activation of caspase-1. Microbial effectors and tick saliva molecules (black) target and block various steps leading up to inflammasome formation to promote survival of the invading pathogen or the tick. *AIM2*, absent in melanoma 2; *ASC*, apoptosis-associated speck-like protein containing a carboxy terminal CARD domain; *CARD*, Caspase activation and recruitment domain; *NLR*, Nod-like receptor; *NLRC*, NLR family CARD domain–containing protein; *NLRP*, NLR family PYD domain–containing protein. *Adapted from Shaw, D.K., Kotsyfakis, M., Pedra, J.H.F., 2016. For whom the bell tolls (and Nods): spit-acular saliva. Curr. Trop. Med. Rep. 3, 40–50.*

state in the absence of stimuli, and (2) antigen recognition (von Moltke et al., 2013; Rathinam and Fitzgerald, 2016). The NBD domain facilitates oligomerization of the NLRs, which leads to the initiation of a proinflammatory response through nucleation factor-κB (NF-κB) activation, mitogen-activated protein kinase (MAPK) signaling, or the formation of a large multimeric complex, termed the inflammasome, which initiates downstream signaling through caspase activation (Fig. 10.1) (Caruso et al., 2014; Inohara et al., 2001;

von Moltke et al., 2013; Rathinam and Fitzgerald, 2016).

The original members of the NLR superfamily, Nod1 and Nod2, trigger a proinflammatory immune reaction by activating NF-κB in response to peptidoglycan (Caruso et al., 2014; Inohara et al., 2001; Motta et al., 2015). Nod1 recognizes γ-D-glutamyl-meso-diaminopimelic acid, a component found in the peptidoglycan of most Gram-negative bacteria (Chamaillard et al., 2003; Girardin et al., 2003a; Hasegawa et al., 2006). Nod2 recognizes muramyl dipeptide, which is found in the cell wall of both Gram-negative and Gram-positive bacteria (Girardin et al., 2003b; Inohara et al., 2001). Both Nod1 and Nod2 contain NBD, LRR, and caspase activation and recruitment domains (CARD; one on Nod1 and two on Nod2) (Girardin et al., 2003b; Inohara et al., 2001). Activation of NF-κB is initiated after either Nod1 or Nod2 oligomerize via the NBD domain, which recruits receptor-interacting protein kinase 2 (RIPK2) through CARD–CARD interactions (Kobayashi et al., 2002). RIPK2 then undergoes lysine (K)-63 polyubiquitylation, which leads to the activation of NF-κB or may alternatively induce an antimicrobial response through MAPK signaling and the transcription factor, AP1 (Caruso et al., 2014; Inohara et al., 2001; Park et al., 2007; Windheim et al., 2007).

The oligomerization of other NLRs can result in the formation of a large, multiprotein complex that serves as a scaffold for downstream signaling events (von Moltke et al., 2013; Rathinam and Fitzgerald, 2016). Some of these NLRs contain a CARD domain (NLRCs) and others (NLRPs) have a Pyrin domain (PYD), which activates caspase-1 by recruiting the PYD-CARD containing adapter protein, ASC (apoptosis-associated speck-like protein containing a carboxy terminal CARD domain) (von Moltke et al., 2013; Rathinam and Fitzgerald, 2016). Generally speaking, the NLRC4 protein pairs with NLR family apoptosis inhibitory proteins (NAIPs), which have an NBD, an LRR and baculovirus inhibitor-of-apoptosis repeats (von Moltke et al., 2013). Once NLRs oligomerize, caspase-1 is activated by autoproteolysis, which then processes the cytokines interleukin (IL)-1β and IL-18 into their mature forms, potentiating a proinflammatory response (Chavarría-Smith and Vance, 2015; Guo et al., 2015; Lage et al., 2014; Vance, 2015). This is often accompanied by a rapid, lytic form of cell suicide termed pyroptosis,

which can effectively eliminate the niche that an intracellular pathogen may occupy (Fig. 10.1) (Guo et al., 2015; von Moltke et al., 2013; Rathinam and Fitzgerald, 2016).

Inflammasomes are responsive to a variety of PAMPs and assembly of these scaffolding complexes plays a key role in mounting immune responses against infectious agents. Inflammasomes are also activated by some self-derived DAMPs (Rathinam and Fitzgerald, 2016; Zanoni et al., 2016). However, excessive activation can be deleterious to the host and dysregulation is associated with several sterile, inflammatory diseases (i.e., those that are not caused by infectious pathogens) such as atherosclerosis, diabetes, and some forms of cancer. Inflammasome activation is therefore tightly regulated in physiological conditions of the host (von Moltke et al., 2013; Rathinam and Fitzgerald, 2016). For an in-depth overview of inflammasomes, please refer to these excellent reviews (Guo et al., 2015; von Moltke et al., 2013; Rathinam and Fitzgerald, 2016).

TICK SALIVA

Although tick saliva has a crucial physiological function in maintaining water balance and preventing desiccation between molts (Bowman and Sauer, 2004; Gregson, 1967; Sauer, 1977; Waladde et al., 1979), this chapter will focus on the immunomodulatory role that tick saliva has during the process of obtaining a blood meal. Tick saliva has been studied for almost half of a century and contains a myriad of bioactive molecules that facilitate attachment to the host, disarming the localized immune response, and preventing vasoconstriction, coagulation, and wound healing (Chmelar et al., 2011; Chmelař et al., 2016a,b; Fuchsberger et al., 1995; Hovius, 2009; Kotál et al., 2015; Mans, 2011; Oliveira et al., 2011; Ramachandra and Wikel, 1992; Ribeiro, 1989, 1987a; Ribeiro et al., 1985; Ribeiro and Francischetti, 2003; Steen et al., 2006; Stibrániová et al., 2013; Wikel, 2013, 1996; Wikel and Bergman, 1997). Transcriptomic analyses have validated what has long been described: that tick saliva is exceptionally complex (Chmelar et al., 2012; Chmelař et al., 2016b). This is likely a reflection of their unique feeding behavior. The tick genome contains a number of expanded repertoires of proteins with putative immunoregulatory functions, which are differentially expressed throughout the course of a blood meal (Ribeiro et al., 2006). These include factors that inhibit blood clotting such as Salp14 (Narasimhan et al., 2002; Ribeiro et al., 2006), inhibitors of tissue factor pathways such as Ixolaris and Penthalaris (Francischetti et al., 2004, 2002), angiogenesis inhibitors (Francischetti et al., 2005, 2003), factors that induce vasodilation such as prostaglandin E2 and prostacyclin (Gulia-Nuss et al., 2016; Ribeiro et al., 1988; Sá-Nunes et al., 2007), antiinflammatory mediators such as lipocalin (Sangamnatdej et al., 2002), metalloproteases which are known to degrade fibrin (Francischetti et al., 2003) and many more. Due to space constraints and the focus of this chapter, we will not comprehensively cover this topic and instead refer readers to reviews that do (Chmelař et al., 2016b; Francischetti et al., 2009; Kazimírová and Štibrániová, 2013).

There is a large body of research dedicated to investigating inhibitory effects of tick saliva on immune cells. Tick saliva is capable of blocking both phagocytosis and the production of superoxide and nitric oxide from macrophages (Kuthejlová et al., 2001; Kýcková and Kopecký, 2006) as well as neutrophilic functions such as reactive oxygen species production, degranulation and cellular influx to inflammatory sites (Menten-Dedoyart et al., 2012; Ribeiro et al., 1990; Turni et al., 2002). Cellular cross talk is known to be impeded by tick saliva by suppressing signaling molecules produced by macrophages such as TNF, monocyte chemoattractant protein-1 (MCP-1) and macrophage inflammatory protein-2 (MIP-2) (Kýcková and Kopecký, 2006; Langhansová et al., 2015). The effects of tick

saliva can also be cell-type specific, skewing the immune response in favor of the tick. Examples include blocking secretion of proinflammatory cytokines and promoting the production of cytokines that are antiinflammatory or which can skew immune responses, such as the alternative-activating potential of IL-10 on macrophages (Hannier et al., 2003; Kotál et al., 2015). Tick saliva can also polarize inflammation toward a Th2 response rather than Th1 or Th17, block proliferation of specific cell types, such as T and B cells, and manipulate antigen presenting cells such as dendritic cells, which affects the downstream adaptive immune response (Chmelař et al., 2016b; Kotál et al., 2015; Langhansová et al., 2015; Lieskovská et al., 2015a,b; Sá-Nunes et al., 2007). The important role that tick saliva has in manipulating immune cells is clear and has been well studied over the last several decades. However, due to the scope of this chapter, which will focus on interactions with innate immune signaling, we will not comprehensively discuss this topic and instead refer readers to the following reviews (Chmelař et al., 2016b; Francischetti et al., 2009; Kotál et al., 2015).

Much less is known regarding the effect of tick saliva on inflammatory signaling cascades. To date, there is only one characterized molecule, Sialostatin L2, that is known to specifically inhibit inflammasome formation and which has been molecularly characterized (Table 10.1; Fig. 10.1) (Chen et al., 2014; Wang et al., 2016b). As previously mentioned, innate inflammatory signaling cascades are initiated by PRRs, such as (1) NLR, (2) TLR, (3) CLR, (4) AIM2, (5) RIG I-like, and (5) cGAS/STING (Brubaker et al., 2015). Comparatively speaking, tick saliva in the context of TLR signaling is the best studied among PRR-mediated immune cascades, which is discussed in the following review (Shaw et al., 2016). While TLR signaling occurs at the cell membrane interface, either at the surface or within endosomal compartments, NLR signaling senses both PAMPs and DAMPs within the cytosol (Brubaker et al., 2015). The interference of tick

TABLE 10.1 Microbial and Tick Saliva Innate Immunomodulators

Effector Name	Species	Target	Short Description	References
Ats-1	*Anaplasma phagocytophilum*	Mitochondria	Delays apoptosis in neutrophils. Inhibits cytochrome C release when cells undergo either Bax-induced or chemically induced apoptosis. Binds Beclin-1 to hijack host autophagy pathways.	Niu et al. (2010, 2012)
ECH0825	*Ehrlichia chaffeensis*	Mitochondria	Type IV secretion system effector that inhibits Bax-induced apoptosis and upregulates mitochondrial manganese superoxide dismutase (MnSOD) to dampen ROS release and ROS-induced apoptosis. Ultimately prolongs host cell survival.	Liu et al. (2012)
SidF	*Legionella pneumophila*	BNIP3 and Bcl-rambo	PI3 phosphatase that inhibits Bcl-2 protein family members BNIP3 and Bcl-rambo.	Banga et al. (2007)
AnkG	*Coxiella burnetii*	p32	Dot/Icm Type IV secretion system effector that interacts with mitochondrial p32.	Eckart et al. (2014)
CaeA/B	*C. burnetii*	Unknown	Dot/Icm Type IV secretion system effectors. CaeB is an inhibitor of apoptosis and acts upstream of caspase-3 but downstream of Bax activation.	Klingenbeck et al. (2013)
FTL_0325	*Francisella tularensis* LVS	Unknown	Dampens IL-1β secretion by interfering with NF-κB signaling, which prevents the transcription of pro-IL-1β. Delays AIM2 and NLRP3 inflammasome activation through an unknown mechanism. Deletion did not increase intramacrophage bacterial lysis, unlike other *F. tularensis* mutants (see Peng et al., 2011). Termed FTT0831c in the virulent strain, SchuS4.	Dotson et al. (2013)
GogB	*Salmonella enterica* spp. Typhimurium	SCF E3 ubiquitin ligase	Type III secretion system substrate that binds SCF E3 ubiquitin ligase to prevent IκB degradation and therefore also NF-κB activation. Suppresses NF-κB-dependent cytokine secretion and thereby limits tissue damage.	Pilar et al. (2012)

Continued

TABLE 10.1   Microbial and Tick Saliva Innate Immunomodulators—cont'd

Effector Name	Species	Target	Short Description	References
OspG	*Shigella flexneri*	Ubiquitin E2 conjugating enzymes	Protein kinase that inhibits IκB degradation to prevent NF-κB activation and inflammatory cytokine production.	Kim et al. (2005)
IpaH0722	*S. flexneri*	TRAF2	E3 ubiquitin ligase and a Type III secretion system effector that ubiquitylates and causes proteasomal degradation of TRAF2, thereby preventing NF-κB activation through protein kinase C.	Ashida et al. (2013)
LnaB	*L. pneumophila*	NF-κB	Dot/Icm Type IV secretion system effector that activates NF-κB through an unknown mechanism, triggering a transcriptional program that favors *L. pneumophila* replication and survival inside the host cell.	Losick et al. (2010)
LegK1	*L. pneumophila*	IκBα, p100	Dot/Icm Type IV secretion system effector that phosphorylates IκBα and p100, leading to their degradation and liberation of NF-κB, which transcribes a gene set responsible for enhancing *L. pneumophila* infection.	Ge et al. (2012)
SdhA	*L. pneumophila*	AIM2 inflammasome	Dot/Icm Type IV secretion system effector that prevents AIM2 inflammasome activation and subsequent IL-1β secretion and pyroptosis.	Ge et al. (2012)
IcaA	*C. burnetii*	Caspase-11	Secreted by the Dot/Icm Type IV secretion system and inhibits the noncanonical inflammasome in macrophages.	Cunha et al. (2015)
OspC3	*S. flexneri*	Caspase-4	Inhibits caspase-4 (mouse homologue of human capsase-11) to prevent inflammatory cell death in infected epithelial cells.	Kobayashi et al. (2013, p. 3)
ExoS	*Pseudomonas aeruginosa*	IL-1β maturation	Negatively regulates the maturation of pro-IL-1β in a manner that is dependent cytoskeleton rearrangement and ADP ribosyltransferase activity.	Galle et al. (2008)

TABLE 10.1 Microbial and Tick Saliva Innate Immunomodulators—cont'd

Effector Name	Species	Target	Short Description	References
YopE and YopT	*Yersinia enterocolitica*	Caspase-1	Blocks activation of caspase-1 by inhibiting the Rho GTPase, Rac1, in a manner dependent on LIM kinase-1-mediated modifications of the actin cytoskeleton.	Schotte et al. (2004)
YopK	*Yersinia spp.*	Caspase-1	Prevents caspase-1 activation by masking T3SS translocator proteins, YopD and YopB.	Brodsky et al. (2010)
RipA	*F. tularensis* LVS	IL-1β, IL-18	Inhibits the maturation and secretion of IL-1β and IL-18 in a manner that is dependent on caspase-1. Implicated in cell wall integrity.	Huang et al. (2010) and Peng et al. (2011)
FTT0584	*F. tularensis* LVS	IL-1β, IL-18	Suppress production of IL-1β and IL-18 and pyroptosis in macrophages in a caspase-1 and ASC-dependent manner. Implicated in cell wall integrity.	Peng et al. (2011) and Weiss et al. (2007)
FTT0748	*F. tularensis* LVS	IL-1β, IL-18	Suppress production of IL-1β, IL-18 and pyroptosis in macrophages in a caspase-1- and ASC-dependent manner. Predicted transcriptional regulator.	Weiss et al. (2007)
Zmp1	*Mycobacterium tuberculosis*	Caspase-1	Zinc metalloprotease that inhibits the activation of caspase-1 and blocks the maturation of IL-1β.	Master et al. (2008)
SL2	*Ixodes scapularis*	Annexin A2	Binds Annexin A2 and prevents NLRC4 oligomerization, caspase-1 activation and subsequent IL-1β and IL-18 maturation	Wang et al. (2016b)
Ank200	*Ehrlichia chaffeensis*	TNF-α, Stat1, CD48	Type 1 secretion system effector that is translocated across the nucleus to regulate the transcription of several targets, including TNF-α, Stat1, and CD48 to promote bacterial survival.	Zhu et al. (2009)
SodB and SodC	*F. tularensis* LVS	Host ROS	Detoxifies host ROS to prevent MAPK and NF-κB signaling in response to accumulating ROS.	Rabadi et al. (2016)

Continued

TABLE 10.1   Microbial and Tick Saliva Innate Immunomodulators—cont'd

Effector Name	Species	Target	Short Description	References
SdhA	*L. pneumophila*	RIG-I/MDA-5 pathway	Suppresses the Type I interferon response by suppressing the RIG-I/MDA-5 pathway induced after *L. pneumophila* infection.	Monroe et al. (2009)
SpvC	*Salmonella enterica* subsp. Typhimurium	MAPKs	Type III secretion system effector with phosphothreonine lyase activity that suppress MAPK signaling.	Haneda et al. (2012)
OspI	*S. flexneri*	UBC13	Type III secretion system effector that deamidates a key residue in the E2 conjugating enzyme UBC13, preventing TRAF6 activation and acute inflammatory cytokine production.	Sanada et al. (2012)
IpaH4.5	*S. flexneri*	TBK1	E3 ubiquitin ligase and Type III secretion system effector. Polyubiquitylates TBK1, targeting it for proteasomal degradation, to inhibit anti–*S. flexneri* IRF3 activation.	Zheng et al. (2016)
p66	*Borrelia burgdorferi*	β3 integrins	Integrin receptor ligand that leads to the downregulation of MAPK, JAK-STAT, and VEGF signaling pathways after *B. burgdorferi* attaches to host cells.	LaFrance et al. (2011)
NS5	Tick-borne flaviviruses	Prolidase, JAKs	Several roles in suppressing the type I interferon response to viral infection. Binds prolidase to prevent the expression of IFNAR1 on the cell surface, which inhibits further IFN responses. Prevents JAK phosphorylation to inhibit JAK-STAT signaling.	Best et al. (2005) and Lubick et al. (2015)
AnkA	*A. phagocytophilum*	CYBB promoter	Type IV secretion system effector, binds AT-rich DNA sequences and recruits HDAC1 to repress target genes through deacetylation. Leads to neutrophil survival.	Garcia-Garcia et al. (2009) and Rennoll-Bankert et al. (2015)

TABLE 10.1 Microbial and Tick Saliva Innate Immunomodulators—cont'd

Effector Name	Species	Target	Short Description	References
AptA	*A. phagocytophilum*	Vimentin A	Induces Erk1/2 signaling through interaction with vimentin A, which leads to vimentin A reorganization and to prolonged *A. phagocytophilum* survival within the host.	Sukumaran et al. (2011)
TRP47	*Ehrlichia chaffeensis*	PCGF5, FYN, PTPN2, CAP1	Type 1 secretion system effector that interacts with several targets to remodel the host and allow bacterial survival.	Wakeel et al. (2009)
TRP120	*Ehrlichia chaffeensis*	PCGF5	Type 1 secretion system that is SUMOylated, which may enhance its interaction with PCGF5 to promote bacterial survival in the host.	Dunphy et al. (2014)
TRP32	*Ehrlichia chaffeensis*	EF1A1, DAZAP2, FTL, CD63, CD14, PSMB1, RC3H1, TP53I11	Regulates basic eukaryotic cellular processes to promote bacterial survival. Type 1 secretion system substrate, contains tandem repeats, and crosstalks with TRP120	Luo and McBride (2012)
RavZ	*L. pneumophila*	LC3/Atg8	Cysteine protease and a Dot/Icm Type IV secretion system substrate that removes phosphatidylethanolamine from the autophagy proteins LC3/Atg8.	Choy et al. (2012)
BBK32	*B. burgdorferi*	C1	Surface lipoprotein that binds C1 to prevent the activation of the classical complement cascade.	Garcia et al. (2016)
OspC	*B. burgdorferi*	Unknown	Surface lipoprotein that prevents phagocytosis by macrophages.	Carrasco et al. (2015)
CspA	*B. burgdorferi*	Factor H, C7, C9	Inhibits the three branches of the complement cascade—classical, MBL, and alternative cascades—by binding Factor H and Factor H-like proteins and by preventing formation of the terminal membrane attack complex.	Hallström et al. (2013)
BBA70	*B. burgdorferi*	Plasminogen	Binds plasminogen, which, when activated, cleaves C3b and C5 to inhibit the complement cascade.	Koenigs et al. (2013)

AIM2, absent in melanoma 2; *ASC*, apoptosis-associated speck-like protein containing a carboxy terminal CARD domain; *LVS*, live vaccine strain; *NLR*, Nod-like receptor; *NLRC*, NLR family CARD domain–containing protein; *NLRP*, NLR family PYD domain–containing protein.

saliva with NLR immune cascades is of particular interest given the fact that inflammasomes are activated by cellular damage, which takes place at the tick bite site, and the known ability of some tick-borne pathogens to activate NLR signaling (Chattoraj et al., 2013; Shaw et al., 2016; Wang et al., 2016a,b; Yang et al., 2015).

Only a handful of studies have examined if and how tick-borne pathogens can trigger NLR-mediated immune activation. *E. chaffeensis*, an obligate intracellular bacterium of the order Rickettsiales, is known to induce *nod2* and other genes associated with inflammasome formation in various ways, although differential regulation of *tlr* and *nod2* in this experimental set up resulted in varying pathological effects (Chattoraj et al., 2013). While TLR2 induced a protective response against *E. chaffeensis*, Nod2 signaling exacerbated pathogenesis and prompted *Ehrlichia*-induced toxic shock (Chattoraj et al., 2013). In agreement with this, Yang et al. (2015) demonstrated that fatal ehrlichiosis was induced by caspase-11 dependent noncanonical inflammasome activation, which is ultimately governed by type I interferon. Mice deficient in *nlrp3* demonstrated greater bacteria clearance, although they succumbed to infection at the same rate as wild type mice. A deficiency for the interferon-α/β receptor, *ifnar*, conferred resistance to fatal ehrlichiosis, with less liver damage, prolonged survival and lower bacterial burdens. Further research is required to elucidate the mechanistic details of how IFN type I activates the noncanonical inflammasome, as the assays from these studies were limited to cytokine production profiles. Moreover, it will be interesting to evaluate whether the addition of tick saliva or salivary gland extracts will impact noncanonical inflammasome activation in this model. Given that tick saliva and some specific salivary gland molecules are known to inhibit the production of IFN type I, it may very well impact the inflammatory phenotype induced by *E. chaffeensis* infection (Lieskovská et al., 2015b; Lieskovská and Kopecký, 2012).

Another tick transmitted, obligate intracellular bacterium of the order Rickettsiales, *A. phagocytophilum*, has been shown to also activate NLR signaling cascades. A 2012 study published by Sukumaran et al. demonstrated that Nod1 and Nod2 signaling were important for controlling *A. phagocytophilum* bacterial burden in mice (Sukumaran et al., 2012). Transcripts for the adapter signaling molecule for Nod1 and Nod2, receptor-interacting protein kinase-2 (*ripk2*), were significantly up regulated in mice infected with *A. phagocytophilum*. Conversely, mice deficient in *ripk2* showed increased bacterial burden, significant reductions in proinflammatory cytokines IFN-γ and IL-18, and, overall, took longer to clear the infection (Sukumaran et al., 2012). This is peculiar, given that the known agonists of Nod1 and Nod2 are varying forms of peptidoglycan, which *A. phagocytophilum* does not possess (Dunning Hotopp et al., 2006).

Back-to-back studies published by Chen et al. showed that NLR signaling induced by *A. phagocytophilum* stimulation was inhibited by the addition of tick saliva (Chen et al., 2012) and that, specifically, the salivary protein Sialostatin L2 (SL2) blocked NLR-mediated inflammation (Chen et al., 2014). This study demonstrated that SL2 prevented the maturation of proinflammatory cytokines IL-1β and IL-18 by inhibiting NADPH oxidase-mediated production of reactive oxygen species that ultimately blocked caspase-1 activation (Chen et al., 2014). A subsequent study by the same group demonstrated that this process was initiated by the direct interaction and binding of SL2 to the host phospholipid-binding protein, Annexin A2 (Wang et al., 2016b). SL2 binding to Annexin A2 caused inhibition of NLRC4 inflammasome formation, explaining the previously observed phenotype that prevented caspase-1 activation and maturation of IL-1β and IL-18 (Wang et al., 2016b). The NLRC4 inflammasome is typically initiated in response to bacterial type III secretion system components and flagellin (Hu et al., 2015;

Lage et al., 2014; Vance, 2015; Zhang et al., 2015), neither of which *A. phagocytophilum* possess (Dunning Hotopp et al., 2006), demonstrating that there are as-yet unknown inflammasome agonists.

IMMUNE SUBVERSION MEDIATED BY MICROBIAL EFFECTORS

There is a constant evolutionary arms race between hosts and pathogens, each trying to outcompete the other. Immune evasion by pathogens is a common survival strategy for all infectious agents and there are a plethora of subversion strategies (Beachboard and Horner, 2016; Chatterjee et al., 2016; Christensen and Paludan, 2016; Guven-Maiorov et al., 2016; Holmgren et al., 2016; Maldonado et al., 2016; Schuren et al., 2016). As such, the NLR-mediated host immune response, among other immune processes, is targeted by effector molecules to promote survival of the pathogen (Table 10.1; Fig. 10.1). Extensive research has been dedicated to uncovering how pathogens subvert host immune responses. Some strategies can be categorized as subtle evasion tactics, such as varying surface antigen display, while others are outright assaults on the host immune system. An example is the direct injection of microbial effector proteins via a secretion system into the cytosol of the host cell. Other strategies can be considered more indirect such as regulating a physiological process of a host, which in turn would compete with the activation of an immune pathway. An example of this is the *A. phagocytophilum*–mediated induction of autophagy, which directly antagonizes inflammasome formation (Niu et al., 2008; Rathinam and Fitzgerald, 2016; Shi et al., 2012). Outlined below are specific strategies employed by both vector-borne and nonvector-borne pathogens to evade immune recognition and/or clearance, with a specific focus on NLR-mediated signaling events (Table 10.1; Fig. 10.1).

Evasion can be a subtle method for avoiding recognition by the immune system, such as antigenic switching (Julien et al., 2012; McCulloch et al., 2015; Norris, 2014; Palmer et al., 2016; Rudenko, 2011). This is the case with the *B. burgdorferi* variable surface antigen, VlsE, which is essential for the bacteria to establish a persistent infection. On the lp28-1 plasmid of *B. burgdorferi* there are 15 silent cassettes adjacent to the *vslE* expression locus, which are swapped out through segmental gene conversion events in a manner that is dependent on subunits of the Holliday junction branch migrase, *ruvA* and *ruvB* (Dresser et al., 2009; Labandeira-Rey et al., 2003; Labandeira-Rey and Skare, 2001; Lin et al., 2009; Purser and Norris, 2000; Zhang et al., 1997; Zhang and Norris, 1998). *A. phagocytophilum* undergoes antigen switching of a major surface protein by using a Rec-F-dependent mechanism to swap out portions of a hypervariable region of "functional pseudogenes" of the pfam01617 family (Barbet et al., 2003; Lin et al., 2006). *E. chaffeensis* varies p28, one of the most prominent outer surface proteins, by antigen switching as well. This gene family is a series of 21 tandem homologues with varying degrees of sequence similarity (Yu et al., 2000), which are thought to contribute to immune evasion, although this has not yet been directly proven (Reddy et al., 1998; Reddy and Streck, 1999).

Microbial surface antigen variation suggests that complete loss of these proteins would confer a survival disadvantage for the bacteria. However, the expression of some antigenic structures can be completely turned off at different phases of infection or within particular niches where expression of these proteins would be disadvantageous for the microbe (Lawley et al., 2006; Miao et al., 2006; Samuels, 2011; Schwan, 2003). This is the case for *Salmonella* spp., which turns off flagella protein expression after entering host cells. Flagellin is highly antigenic and specifically activates the NLRC4 inflammasome. By shutting off the production of flagellin, *Salmonella* spp. are able to evade the immune response and replicate within the cytosol of macrophages (Lawley et al., 2006; Miao et al., 2006). Other pathogens also suppress the expression of flagella under host conditions, such as *Yersinia enterocolitica* and *Yersinia pseudotuberculosis*. Alternatively, some pathogens have lost the ability to express flagella altogether, perhaps because of the selective pressure imposed by the host immune response. This is the case for *Yersinia pestis*, which contains a frameshift mutation in the flagellar master control operon *flhD*, rendering it a pseudogene (Chain et al., 2004; Minnich and Rohde, 2007; Thomson et al., 2006).

Classical inflammasome activation is a two-step process that requires a priming signal for the production of pro-IL-1β and pro-IL-18, which can be provided by TLR4-dependent recognition of lipopolysaccharide (LPS) (Bauernfeind et al., 2009; Chen and Nuñez, 2010; Franchi et al., 2009; Latz et al., 2013; Stewart et al., 2010). In addition to the lack of flagella, *Y. pestis* also expresses an atypical form of LPS that is less acetylated than the normal hexaacetylated form. The tetraacetylated version of LPS is expressed at higher temperatures (37°C) and is not efficiently recognized by TLR4 (Telepnev et al., 2009), which prevents the production of the proinflammatory precursors needed for inflammasome activation. A number of other pathogens also express varied forms of LPS such as *Francisella* spp., *Brucella* spp., and *Ochrobactrum* spp. or lack LPS synthesis genes altogether such as *Borrelia* spp., *Anaplasma* spp., and *Ehrlichia* spp. (Amano et al., 1987; Diacovich and Gorvel, 2010; Gunn and Ernst, 2007; Lin and Rikihisa, 2003).

Lastly, evasion mechanisms can also include masking an antigenic protein to prevent recognition by the host inflammasome. Such is the case with the *Y. pestis* type 3 secretion system (T3SS) effector protein, YopK (Brodsky et al., 2010). The T3SS is highly immunogenic and components of it elicit the NLRC4 inflammasome. In this scenario, *Y. pestis* secretes YopK through its T3SS into the cytosol of the host cell, but it then remains closely associated with the T3SS

apparatus to prevent recognition by the NLRC4 inflammasome. In an immunoprecipitation assay, YopK specifically associated with both T3SS translocator proteins, YopD and YopB, and when *yopK* is deleted from *Y. pestis*, caspase-1 is robustly activated. These observations support a model where YopK masks the immunogenic T3SS of *Y. pestis* from the inflammasome, demonstrating another form of evasion to prevent host immune recognition (Brodsky et al., 2010).

The most direct immunomodulatory approach is shutting down an immune response altogether through the action of microbial effectors that are secreted into the cytosol of a host cell. Several intracellular pathogens employ this approach to establish an immunoprivileged niche by suppressing the expression of host-derived antibacterial factors, altering/inhibiting signaling intermediates or manipulating autophagy (Hachani et al., 2016; Sansonetti and Di Santo, 2007; Van Avondt et al., 2015; Winchell et al., 2016). By doing so, this ultimately promotes persistence and replication of the pathogen. Although many immunological processes are targeted by such effectors, inflammasome-specific effectors will be focused on in the following section (Fig. 10.1).

Microbial strategies that directly shut down NLR-mediated signaling often target either caspase activation or the subsequent secretion of proinflammatory cytokines IL-1β and IL-18. The *Y. enterocolitica* effectors YopE and, to a lesser extent, YopT were found to inhibit the activation of caspase-1 (Schotte et al., 2004). Schotte et al. determined that the mechanism used by *Yersinia* to block caspase-1 maturation was executed by inhibiting the Rho GTPase, Rac1. Rac1 mediates caspase-1 activation through modifications of the actin cytoskeleton that are dependent on the LIM kinase-1 (Schotte et al., 2004). The authors postulated that the Rac-1-dependent cytoskeleton rearrangements were necessary to provide a molecular scaffold for the subsequent oligomerization and activation of caspase-1 (Schotte et al., 2004). *Mycobacterium* spp. also inhibits the activation of caspase-1 and blocks the maturation of IL-1β. Master et al. (2008) discovered that a bacterial Zn^{2+} metalloprotease, Zmp1, had a crucial role in this process and that the corresponding loss of this gene conferred a survival disadvantage for *Mycobacterium* spp. by resulting in the inability of the pathogen to arrest phagolysosomal fusion.

Other microbial effectors target the noncanonical inflammasome that is mediated by caspase-11 in mice and caspase-4 in humans. The *Shigella flexneri* effector, OspC3, and *Coxiella burnetii* effector, IcaA, both target this mode of inflammasome activation. Kobayashi et al. discovered that the *S. flexneri* effector, OspC3, prevented acute inflammatory cell death mediated by caspase-4 activation. OspC3 specifically inhibited heterodimerization of human caspase-4-p19 and caspase-4-p10 by blocking the predicted catalytic pocket of caspase-4 with the conserved X_1-Y-X_2-D-X_3 OspC3 motif (Kobayashi et al., 2013). The *C. burnetii* effector IcaA also inhibits noncanonical activation of the NLRP3 inflammasome by targeting caspase-11 in mice. Cunha et al. (2015) used the surrogate bacterium, *Legionella pneumophila*, due to the limited genetic tools available for *C. burnetii*. *L. pneumophila* and *C. burnetii* are both under the Legionellales order and are evolutionarily close. Importantly, *L. pneumophila* is well studied and genetically tractable, making it an appealing surrogate model to study *C. burnetii* effectors. By transforming and expressing the *C. burnetii* effector IcaA in *L. pneumophila*, caspase-11 activation was suppressed in host macrophages (Cunha et al., 2015). Moreover, an engineered *icaA C. burnetii* mutant lost the ability to suppress caspase-11 in macrophages, leading to inflammasome activation; however, this was only observed when macrophages were coinfected with *icaA C. burnetii* mutants and *L. pneumophila*. Interestingly, single infections with only *C. burnetii icaA* mutants retained the ability to block inflammasome activation. This suggests that *C. burnetii* has redundant mechanisms, aside from

icaA, to suppress caspase-11-dependent inflammasome activation induced by *C. burnetii*, but not by other bacteria such as *L. pneumophila* during coinfections (Cunha et al., 2015).

Another mechanism of inflammasome inhibition is directly blocking the maturation of pro-IL-1β and pro-IL-18, which are strategies used by *Pseudomonas aeruginosa* and *F. tularensis* (Galle et al., 2008; Huang et al., 2010; Jones et al., 2011; Peng et al., 2011; Weiss et al., 2007). The *P. aeruginosa* effector, ExoS, negatively regulates the maturation of pro-IL-1β in a manner that is dependent on its ADP ribosyltransferase activity (Galle et al., 2008). Interestingly, the authors note that the ADP ribosyltransferase activity is a mechanism that ExoS uses to modulate cytoskeleton dynamics. This is reminiscent of the *Yersinia* effector YopE, which also exerts inhibitory effects by modifying the actin cytoskeleton (Schotte et al., 2004). *F. tularensis* inhibits the maturation of IL-1β and IL-18 in a manner that is dependent on caspase-1 and ASC, as demonstrated by Weiss et al. (2007). In this study, a microarray-based negative selection technique was used to screen *F. tularensis* for virulence factors. The screen determined that mutants in *FTT0584* and *FTT0748* induced higher secreted levels of IL-1β and IL-18 and exhibited more pyroptosis, which was abrogated when the infection was performed in macrophages deficient in capase-1 and ASC (Weiss et al., 2007). Huang et al. (2010) demonstrated that deletion of the *F. tularensis* gene, *ripA*, also caused significant levels of IL-1β, IL-18, and TNF-α to be released from resting macrophages in a caspase-1-dependent manner. Although complementation restored the wild-type phenotype, suggesting specificity, another study demonstrated that lacking *ripA* and *FTT0584* caused the bacteria to lyse more readily, which led to more AIM2 inflammasome activation (Peng et al., 2011). These genes are predicted to be involved in bacterial cell wall integrity and therefore indirectly prevent inflammasome formation by preventing cell lysis. Several other genes that have been previously implicated in inflammasome inhibition also fell into this category such as MviN, FopA, FTN1217, WbtA, and LpxH (Peng et al., 2011). Interestingly, FTT0748 was not grouped with the other effectors in this study and is instead predicted to be a transcriptional regulator (Jones et al., 2011; Peng et al., 2011). The precise mechanism that FTT0748 exerts on inflammasome suppression has not yet been elucidated.

OUTLOOK

Comparisons can be drawn between the immune evasion mechanisms mediated by tick saliva and pathogenic effectors derived from microbes. Both impede host immune signaling pathways to promote evasion and survival of the parasitizing organism. One classic example is complement evasion. *B. burgdorferi*, for example, avoids complement-mediated killing by expressing CspA, CspZ, and proteins from the OspE/F family (also collectively known as complement regulator-acquiring surfaces proteins) (Alitalo et al., 2002; Brooks et al., 2005; Bykowski et al., 2008; de Taeye et al., 2013; Garcia et al., 2016; Hallström et al., 2013; Hellwage et al., 2001; Kenedy et al., 2009; Kenedy and Akins, 2011; Kraiczy et al., 2000; McDowell et al., 2003; Rogers et al., 2009; Rogers and Marconi, 2007). Similarly, tick saliva and salivary gland extracts are also known to inhibit complement pathways (Couvreur et al., 2008; Daix et al., 2007; Lawrie et al., 1999; Nunn et al., 2005; Schuijt et al., 2008; Silva et al., 2016; Tyson et al., 2007; Valenzuela et al., 2000; Wikel, 2013). For example, *I. scapularis* saliva prevents the release of anaphylatoxin C3a by inhibiting C3b deposition (Ribeiro, 1987b; Ribeiro and Spielman, 1986) and *I. ricinus* saliva binds factor P, which inhibits the C3 converatase, and therefore, the alternative complement pathway (Couvreur et al., 2008; Daix et al., 2007).

The mechanisms used to evade host immune responses vary between subtle tactics, such as antigenic variation and atypical variations of PAMPs, to more aggressive approaches such as the direct blockade of host immune pathways. With this perspective in mind, it is clear that the immunomodulatory function of microbial effectors and tick saliva molecules is an example of convergent evolution. However, when compared to bacterial effectors, considerably less is known about the molecular actions of saliva molecules on host defenses. Historically, research has focused on how tick saliva inhibits cell-mediated immunity, coagulation, and wound healing, which have contributed tremendously toward our understanding of the tick's life strategy and how this aids in pathogen transmission. Considering the advancements that have been made in molecular techniques, we offer the opinion that the focus going forward can now be shifted towards a molecular understanding of tick saliva and suggest that the vast body of research amassed on microbial effector mechanisms can be used as a road map for discovery.

It is clear that common immune evasion strategies have emerged more than once throughout the evolutionary history of parasitism. We can exploit and build on what is known about immunomodulatory mechanisms of microbial effectors to expand our understanding of how tick saliva manipulates host immune pathways. For example, one can envision a scenario where competition assays with a microbial effector known to target a particular pathway could be used to study the suspected function of a saliva molecule. A common immune target between a saliva molecule and a microbial effector could be insightful for guiding molecular characterization, such as what types of protein–protein interactions are occurring, which host domains are being targeted, if and what posttranslational modifications are taking place, and how these processes are regulated. While it is unlikely that tick saliva molecules and microbial effector proteins will be identical, given their divergent origins, the overall function, subcellular target, cell surface receptor, or mechanism of inhibition may be conserved.

Conversely, both bacterial effectors and saliva molecules can be used as tools to expand on our knowledge of host cell biology itself. Remarkable progress has been made toward understanding host cell processes and how immune responses are mounted, but there is still a significant amount that is unknown. This is illustrated, in particular, by the rapidly expanding field of NLR signaling and inflammasome biology. Because microbial and tick saliva effectors have been elegantly tailored over millions of years through coevolution alongside hosts to evade immune responses, there is potential to exploit these molecules as tools to shed insight on immune processes we have yet to discover.

One question remaining is how tick saliva molecules enter host cells to modify inflammatory processes that originate in the cytosol, such as NLR signaling pathways. It is unclear if these proteins can be taken up by host cells through pinocytosis, delivered via exosomes, taken up through a channel or if a combination of these mechanisms is used. One possible strategy for examining the effects of a soluble saliva molecule, which is known to function in the cytosol of host cells, is to use a microbial surrogate for delivery of the protein. This strategy could also be used with multiple saliva molecules to address potential synergistic interactions. A cloning strategy could be employed that attaches a secretion signal to the N-terminal coding region of a gene(s) and then transforming the gene(s) into a surrogate microbe, such as *L. pneumophila*. Subsequent studies could be performed with various inflammatory stimuli to evaluate the effects that the saliva molecule(s) has on cytosolic inflammation. This method could also be used with a labeled protein(s) and then used to monitor where the

molecule is trafficked, before and after inflammatory stimuli. Immunoprecipitation strategies may be particularly advantageous for characterizing specific interactions by targeting labeled saliva proteins to identify interacting host molecules.

Finally, if we can identify the immune pathways that are commonly targeted by microbes and tick saliva, there is potential for the development of new pharmaceuticals. For example, if a tick saliva molecule could directly compete for a binding site exploited by a pathogen, it could be developed as an infectious disease therapeutic. This is especially relevant given the rapid rise in antibiotic-resistant pathogens. Similarly, the antiinflammatory properties of microbial effectors and/or tick saliva molecules can be used to treat pathology from sterile inflammatory diseases such as asthma, autoimmune diseases, or detrimental inflammation caused by trauma, such as brain injury (Chalouhi et al., 2012; Chen and Nuñez, 2010; Frosali et al., 2015; Horka et al., 2012; Klein et al., 2015; Kobayashi et al., 2013; Master et al., 2008; Sá-Nunes et al., 2007; Sousa et al., 2015). The damage resulting from these illnesses causes a significant amount of morbidity and in some cases can result in death. From this perspective, the possibility of novel therapeutics derived from either microbes or tick saliva would be particularly powerful.

In summary, there is tremendous potential to exploit the knowledge that has been amassed on pathogen effector mechanisms to investigate similar immunomodulatory processes of tick saliva. This information can be used experimentally as a tool to dissect molecular details of host cell biology and immunity as well as translationally to develop therapeutics for inflammatory diseases. With new molecular techniques readily available and the growing trend of "big data sets," there are increasing opportunities for advancements to be made toward understanding immune evasion mechanisms of tick saliva.

References

Alitalo, A., Meri, T., Lankinen, H., Seppälä, I., Lahdenne, P., Hefty, P.S., Akins, D., Meri, S., 2002. Complement inhibitor factor H binding to Lyme disease spirochetes is mediated by inducible expression of multiple plasmid-encoded outer surface protein E paralogs. J. Immunol. 169, 3847–3853.

Amano, K., Tamura, A., Ohashi, N., Urakami, H., Kaya, S., Fukushi, K., 1987. Deficiency of peptidoglycan and lipopolysaccharide components in *Rickettsia tsutsugamushi*. Infect. Immun. 55, 2290–2292.

Ashida, H., Nakano, H., Sasakawa, C., 2013. *Shigella* IpaH0722 E3 ubiquitin ligase effector targets TRAF2 to inhibit PKC-NF-κB activity in invaded epithelial cells. PLoS Pathog. 9, e1003409.

Banga, S., Gao, P., Shen, X., Fiscus, V., Zong, W.-X., Chen, L., Luo, Z.-Q., 2007. *Legionella pneumophila* inhibits macrophage apoptosis by targeting pro-death members of the Bcl2 protein family. Proc. Natl. Acad. Sci. U.S.A. 104, 5121–5126.

Barbet, A.F., Meeus, P.F.M., Bélanger, M., Bowie, M.V., Yi, J., Lundgren, A.M., Alleman, A.R., Wong, S.J., Chu, F.K., Munderloh, U.G., Jauron, S.D., 2003. Expression of multiple outer membrane protein sequence variants from a single genomic locus of *Anaplasma phagocytophilum*. Infect. Immun. 71, 1706–1718.

Bauernfeind, F.G., Horvath, G., Stutz, A., Alnemri, E.S., MacDonald, K., Speert, D., Fernandes-Alnemri, T., Wu, J., Monks, B.G., Fitzgerald, K.A., Hornung, V., Latz, E., 2009. Cutting edge: NF-κB activating pattern recognition and cytokine receptors license NLRP3 inflammasome activation by regulating NLRP3 expression. J. Immunol. 183, 787–791.

Beachboard, D.C., Horner, S.M., 2016. Innate immune evasion strategies of DNA and RNA viruses. Curr. Opin. Microbiol. 32, 113–119.

Bell, J.F., Stewart, S.J., Wikel, S.K., 1979. Resistance to tick-borne *Francisella tularensis* by tick-sensitized rabbits: allergic klendusity. Am. J. Trop. Med. Hyg. 28, 876–880.

Best, S.M., Morris, K.L., Shannon, J.G., Robertson, S.J., Mitzel, D.N., Park, G.S., Boer, E., Wolfinbarger, J.B., Bloom, M.E., 2005. Inhibition of interferon-stimulated JAK-STAT signaling by a tick-borne flavivirus and identification of NS5 as an interferon antagonist. J. Virol. 79, 12828–12839.

Bockenstedt, L.K., Wormser, G.P., 2014. Review: unraveling Lyme disease. Arthritis. Rheumatol. 66, 2313–2323.

Bortoluci, K.R., Medzhitov, R., 2010. Control of infection by pyroptosis and autophagy: role of TLR and NLR. Cell. Mol. Life. Sci. 67, 1643–1651.

Bowman, A.S., Sauer, J.R., 2004. Tick salivary glands: function, physiology and future. Parasitology (129 Suppl.), S67–S81.

Brodsky, I.E., Palm, N.W., Sadanand, S., Ryndak, M.B., Sutterwala, F.S., Flavell, R.A., Bliska, J.B., Medzhitov, R., 2010. A *Yersinia* secreted effector protein promotes virulence by preventing inflammasome recognition of the type III secretion system. Cell Host Microbe 7, 376–387.

Brooks, C.S., Vuppala, S.R., Jett, A.M., Alitalo, A., Meri, S., Akins, D.R., 2005. Complement regulator-acquiring surface protein 1 imparts resistance to human serum in *Borrelia burgdorferi*. J. Immunol. 175, 3299–3308.

Brossard, M., Wikel, S.K., 2004. Tick immunobiology. Parasitology (129 Suppl.), S161–S176.

Brubaker, S.W., Bonham, K.S., Zanoni, I., Kagan, J.C., 2015. Innate immune pattern recognition: a cell biological perspective. Annu. Rev. Microbiol. 33, 257–290.

Bykowski, T., Woodman, M.E., Cooley, A.E., Brissette, C.A., Wallich, R., Brade, V., Kraiczy, P., Stevenson, B., 2008. *Borrelia burgdorferi* complement regulator-acquiring surface proteins (BbCRASPs): expression patterns during the mammal-tick infection cycle. Int. J. Med. Microbiol. 1 (298 Suppl.), 249–256.

Carrasco, S.E., Troxell, B., Yang, Y., Brandt, S.L., Li, H., Sandusky, G.E., Condon, K.W., Serezani, C.H., Yang, X.F., 2015. Outer surface protein OspC is an antiphagocytic factor that protects *Borrelia burgdorferi* from phagocytosis by macrophages. Infect. Immun. 83, 4848–4860.

Caruso, R., Warner, N., Inohara, N., Núñez, G., 2014. NOD1 and NOD2: signaling, host defense, and inflammatory disease. Immunity 41, 898–908.

Caulfield, A.J., Pritt, B.S., 2015. Lyme disease coinfections in the United States. Clin. Lab. Med. 35, 827–846.

Chain, P.S.G., Carniel, E., Larimer, F.W., Lamerdin, J., Stoutland, P.O., Regala, W.M., Georgescu, A.M., Vergez, L.M., Land, M.L., Motin, V.L., Brubaker, R.R., Fowler, J., Hinnebusch, J., Marceau, M., Medigue, C., Simonet, M., Chenal-Francisque, V., Souza, B., Dacheux, D., Elliott, J.M., Derbise, A., Hauser, L.J., Garcia, E., 2004. Insights into the evolution of *Yersinia pestis* through whole-genome comparison with *Yersinia pseudotuberculosis*. Proc. Natl. Acad. Sci. U.S.A. 101, 13826–13831.

Chalouhi, N., Ali, M.S., Jabbour, P.M., Tjoumakaris, S.I., Gonzalez, L.F., Rosenwasser, R.H., Koch, W.J., Dumont, A.S., September 2012. Biology of intracranial aneurysms: role of inflammation. J. Cereb. Blood Flow Metab. 32 (9), 1659–1676. http://dx.doi.org/10.1038/jcbfm.2012.84.

Chamaillard, M., Hashimoto, M., Horie, Y., Masumoto, J., Qiu, S., Saab, L., Ogura, Y., Kawasaki, A., Fukase, K., Kusumoto, S., Valvano, M.A., Foster, S.J., Mak, T.W., Nuñez, G., Inohara, N., 2003. An essential role for NOD1 in host recognition of bacterial peptidoglycan containing diaminopimelic acid. Nat. Immunol. 4, 702–707.

Chatterjee, S., Basler, C.F., Amarasinghe, G.K., Leung, D.W., 2016. Molecular mechanisms of innate immune inhibition by non-segmented negative-sense RNA viruses. J. Mol. Biol. 428 (17). http://dx.doi.org/10.1016/j.jmb.2016.07.017.

Chattoraj, P., Yang, Q., Khandai, A., Al-Hendy, O., Ismail, N., 2013. TLR2 and Nod2 mediate resistance or susceptibility to fatal intracellular *Ehrlichia* infection in murine models of ehrlichiosis. PLoS One 8 (3). http://dx.doi.org/10.1371/journal.pone.0058514.

Chavarría-Smith, J., Vance, R.E., 2015. The NLRP1 inflammasomes. Immunol. Rev. 265, 22–34.

Chen, G.Y., Nuñez, G., 2010. Sterile inflammation: sensing and reacting to damage. Nat. Rev. Immunol. 10, 826–837.

Chen, G., Severo, M.S., Sohail, M., Sakhon, O.S., Wikel, S.K., Kotsyfakis, M., Pedra, J.H., 2012. *Ixodes scapularis* saliva mitigates inflammatory cytokine secretion during *Anaplasma phagocytophilum* stimulation of immune cells. Parasites Vector 5, 229.

Chen, G., Wang, X., Severo, M.S., Sakhon, O.S., Sohail, M., Brown, L.J., Sircar, M., Snyder, G.A., Sundberg, E.J., Ulland, T.K., Olivier, A.K., Andersen, J.F., Zhou, Y., Shi, G.-P., Sutterwala, F.S., Kotsyfakis, M., Pedra, J.H.F., 2014. The tick salivary protein sialostatin L2 inhibits caspase-1-mediated inflammation during *Anaplasma phagocytophilum* infection. Infect. Immun. 82, 2553–2564.

Chmelar, J., Oliveira, C.J., Rezacova, P., Francischetti, I.M.B., Kovarova, Z., Pejler, G., Kopacek, P., Ribeiro, J.M.C., Mares, M., Kopecky, J., Kotsyfakis, M., 2011. A tick salivary protein targets cathepsin G and chymase and inhibits host inflammation and platelet aggregation. Blood 117, 736–744.

Chmelar, J., Calvo, E., Pedra, J.H.F., Francischetti, I.M.B., Kotsyfakis, M., 2012. Tick salivary secretion as a source of antihemostatics. J. Proteomics 75, 3842–3854.

Chmelař, J., Kotál, J., Karim, S., Kopacek, P., Francischetti, I.M.B., Pedra, J.H.F., Kotsyfakis, M., 2016a. Sialomes and Mialomes: a systems-biology view of tick tissues and tick-host interactions. Trends. Parasitol. 32, 242–254.

Chmelař, J., Kotál, J., Kopecký, J., Pedra, J.H.F., Kotsyfakis, M., 2016b. All for one and one for all on the tick–host battlefield. Trends. Parasitol. 32, 368–377.

Choy, A., Dancourt, J., Mugo, B., O'Connor, T.J., Isberg, R.R., Melia, T.J., Roy, C.R., 2012. The *Legionella* effector RavZ inhibits host autophagy through irreversible Atg8 deconjugation. Science 338, 1072–1076.

Christensen, M.H., Paludan, S.R., 2016. Viral evasion of DNA-stimulated innate immune responses. Cell. Mol. Immunol. http://dx.doi.org/10.1038/cmi.2016.06.

Couvreur, B., Beaufays, J., Charon, C., Lahaye, K., Gensale, F., Denis, V., Charloteaux, B., Decrem, Y., Prévôt, P.-P., Brossard, M., Vanhamme, L., Godfroid, E., 2008. Variability and action mechanism of a family of anticomplement proteins in *Ixodes* ricinus. PloS One 3, e1400.

Cunha, L.D., Ribeiro, J.M., Fernandes, T.D., Massis, L.M., Khoo, C.A., Moffatt, J.H., Newton, H.J., Roy, C.R., Zamboni, D.S., 2015. Inhibition of inflammasome activation by *Coxiella burnetii* type IV secretion system effector IcaA. Nat. Commun. 6, 10205.

Daix, V., Schroeder, H., Praet, N., Georgin, J.-P., Chiappino, I., Gillet, L., de Fays, K., Decrem, Y., Leboulle, G., Godfroid, E., Bollen, A., Pastoret, P.-P., Gern, L., Sharp, P.M., Vanderplasschen, A., 2007. *Ixodes* ticks belonging to the *Ixodes ricinus* complex encode a family of anticomplement proteins. Insect. Mol. Biol. 16, 155–166.

de Taeye, S.W., Kreuk, L., van Dam, A.P., Hovius, J.W., Schuijt, T.J., 2013. Complement evasion by *Borrelia burgdorferi*: it takes three to tango. Trends. Parasitol. 29, 119–128.

Diacovich, L., Gorvel, J.-P., 2010. Bacterial manipulation of innate immunity to promote infection. Nat. Rev. Microbiol. 8, 117–128.

Diuk-Wasser, M.A., Vannier, E., Krause, P.J., 2016. Coinfection by *Ixodes* tick-borne pathogens: ecological, epidemiological, and clinical consequences. Trends. Parasitol. 32, 30–42.

Dotson, R.J., Rabadi, S.M., Westcott, E.L., Bradley, S., Catlett, S.V., Banik, S., Harton, J.A., Bakshi, C.S., Malik, M., 2013. Repression of inflammasome by *Francisella tularensis* during early stages of infection. J. Biol. Chem. 288, 23844–23857.

Dresser, A.R., Hardy, P.-O., Chaconas, G., 2009. Investigation of the genes involved in antigenic switching at the *vlsE* locus in *Borrelia burgdorferi*: an essential role for the RuvAB branch migrase. PLoS Pathog. 5. http://dx.doi.org/10.1371/journal.ppat.1000680.

Dugat, T., Lagrée, A.-C., Maillard, R., Boulouis, H.-J., Haddad, N., 2015. Opening the black box of *Anaplasma phagocytophilum* diversity: current situation and future perspectives. Front. Cell. Infect. Microbiol. 5. http://dx.doi.org/10.3389/fcimb.2015.00061.

Dunning Hotopp, J.C., Lin, M., Madupu, R., Crabtree, J., Angiuoli, S.V., Eisen, J.A., Eisen, J., Seshadri, R., Ren, Q., Wu, M., Utterback, T.R., Smith, S., Lewis, M., Khouri, H., Zhang, C., Niu, H., Lin, Q., Ohashi, N., Zhi, N., Nelson, W., Brinkac, L.M., Dodson, R.J., Rosovitz, M.J., Sundaram, J., Daugherty, S.C., Davidsen, T., Durkin, A.S., Gwinn, M., Haft, D.H., Selengut, J.D., Sullivan, S.A., Zafar, N., Zhou, L., Benahmed, F., Forberger, H., Halpin, R., Mulligan, S., Robinson, J., White, O., Rikihisa, Y., Tettelin, H., 2006. Comparative genomics of emerging human ehrlichiosis agents. PLoS Genet. 2, e21.

Dunphy, P.S., Luo, T., McBride, J.W., 2014. *Ehrlichia chaffeensis* exploits host SUMOylation pathways to mediate effector-host interactions and promote intracellular survival. Infect. Immun. 82, 4154–4168.

Eckart, R.A., Bisle, S., Schulze-Luehrmann, J., Wittmann, I., Jantsch, J., Schmid, B., Berens, C., Lührmann, A., 2014. Antiapoptotic activity of *Coxiella burnetii* effector protein AnkG is controlled by p32-dependent trafficking. Infect. Immun. 82, 2763–2771.

Flynn, P.C., Kaufman, W.R., 2011. Female ixodid ticks grow endocuticle during the rapid phase of engorgement. Exp. Appl. Acarol. 53, 167–178.

Franchi, L., Eigenbrod, T., Núñez, G., 2009. Cutting edge: TNF-α mediates sensitization to ATP and silica via the NLRP3 inflammasome in the absence of microbial stimulation. J. Immunol. 183, 792–796.

Francischetti, I.M.B., Valenzuela, J.G., Andersen, J.F., Mather, T.N., Ribeiro, J.M.C., 2002. Ixolaris, a novel recombinant tissue factor pathway inhibitor (TFPI) from the salivary gland of the tick, *Ixodes scapularis*: identification of factor X and factor Xa as scaffolds for the inhibition of factor VIIa/tissue factor complex. Blood 99, 3602–3612.

Francischetti, I.M.B., Mather, T.N., Ribeiro, J.M.C., 2003. Cloning of a salivary gland metalloprotease and characterization of gelatinase and fibrin(ogen)lytic activities in the saliva of the Lyme disease tick vector *Ixodes scapularis*. Biochem. Biophys. Res. Commun. 305, 869–875.

Francischetti, I.M.B., Mather, T.N., Ribeiro, J.M.C., 2004. Penthalaris, a novel recombinant five-Kunitz tissue factor pathway inhibitor (TFPI) from the salivary gland of the tick vector of Lyme disease, *Ixodes scapularis*. Thromb. Haemost. 91, 886–898.

Francischetti, I.M.B., Mather, T.N., Ribeiro, J.M.C., 2005. Tick saliva is a potent inhibitor of endothelial cell proliferation and angiogenesis. Thromb. Haemost. 94, 167–174.

Francischetti, I.M., Sá-Nunes, A., Mans, B.J., Santos, I.M., Ribeiro, J.M.C., 2009. The role of saliva in tick feeding. Front. Biosci. 14, 2051–2088.

Frosali, S., Pagliari, D., Gambassi, G., Landolfi, R., Pandolfi, F., Cianci, R., 2015. How the intricate interaction among toll-like receptors, microbiota, and intestinal immunity can influence gastrointestinal pathology. J. Immunol. Res. 2015, 489821.

Fuchsberger, N., Kita, M., Hajnicka, V., Imanishi, J., Labuda, M., Nuttall, P.A., 1995. Ixodid tick salivary gland extracts inhibit production of lipopolysaccharide-induced mRNA of several different human cytokines. Exp. Appl. Acarol. 19, 671–676.

Galle, M., Schotte, P., Haegman, M., Wullaert, A., Yang, H.J., Jin, S., Beyaert, R., 2008. The *Pseudomonas aeruginosa* Type III secretion system plays a dual role in the regulation of caspase-1 mediated IL-1β maturation. J. Cell. Mol. Med. 12, 1767–1776.

Garcia, B.L., Zhi, H., Wager, B., Höök, M., Skare, J.T., 2016. *Borrelia burgdorferi* BBK32 inhibits the classical pathway by blocking activation of the C1 complement complex. PLoS Pathog. 12, e1005404.

Garcia-Garcia, J.C., Rennoll-Bankert, K.E., Pelly, S., Milstone, A.M., Dumler, J.S., 2009. Silencing of host cell CYBB gene expression by the nuclear effector AnkA of the intracellular pathogen *Anaplasma phagocytophilum*. Infect. Immun. 77, 2385–2391.

Ge, J., Gong, Y.-N., Xu, Y., Shao, F., 2012. Preventing bacterial DNA release and absent in melanoma 2 inflammasome activation by a *Legionella* effector functioning in membrane trafficking. Proc. Natl. Acad. Sci. U.S.A. 109, 6193–6198.

Gillespie, R.D., Dolan, M.C., Piesman, J., Titus, R.G., 2001. Identification of an IL-2 binding protein in the saliva of the Lyme disease vector tick, *Ixodes scapularis*. J. Immunol. 166, 4319–4326.

Girardin, S.E., Boneca, I.G., Carneiro, L.A.M., Antignac, A., Jéhanno, M., Viala, J., Tedin, K., Taha, M.-K., Labigne, A., Zäthringer, U., Coyle, A.J., DiStefano, P.S., Bertin, J., Sansonetti, P.J., Philpott, D.J., 2003a. Nod1 detects a unique muropeptide from Gram-negative bacterial peptidoglycan. Science 300, 1584–1587.

Girardin, S.E., Boneca, I.G., Viala, J., Chamaillard, M., Labigne, A., Thomas, G., Philpott, D.J., Sansonetti, P.J., 2003b. Nod2 is a general sensor of peptidoglycan through muramyl dipeptide (MDP) detection. J. Biol. Chem. 278, 8869–8872.

Gregson, J.D., 1967. Observations on the movement of fluids in the vicinity of the mouthparts of naturally feeding *Dermacentor andersoni* Stiles. Parasitology 57, 1–8.

Gulia-Nuss, M., Nuss, A.B., Meyer, J.M., Sonenshine, D.E., Roe, R.M., Waterhouse, R.M., Sattelle, D.B., de la Fuente, J., Ribeiro, J.M., Megy, K., Thimmapuram, J., Miller, J.R., Walenz, B.P., Koren, S., Hostetler, J.B., Thiagarajan, M., Joardar, V.S., Hannick, L.I., Bidwell, S., Hammond, M.P., Young, S., Zeng, Q., Abrudan, J.L., Almeida, F.C., Ayllón, N., Bhide, K., Bissinger, B.W., Bonzon-Kulichenko, E., Buckingham, S.D., Caffrey, D.R., Caimano, M.J., Croset, V., Driscoll, T., Gilbert, D., Gillespie, J.J., Giraldo-Calderón, G.I., Grabowski, J.M., Jiang, D., Khalil, S.M.S., Kim, D., Kocan, K.M., Koči, J., Kuhn, R.J., Kurtti, T.J., Lees, K., Lang, E.G., Kennedy, R.C., Kwon, H., Perera, R., Qi, Y., Radolf, J.D., Sakamoto, J.M., Sánchez-Gracia, A., Severo, M.S., Silverman, N., Šimo, L., Tojo, M., Tornador, C., Van Zee, J.P., Vázquez, J., Vieira, F.G., Villar, M., Wespiser, A.R., Yang, Y., Zhu, J., Arensburger, P., Pietrantonio, P.V., Barker, S.C., Shao, R., Zdobnov, E.M., Hauser, F., Grimmelikhuijzen, C.J.P., Park, Y., Rozas, J., Benton, R., Pedra, J.H.F., Nelson, D.R., Unger, M.F., Tubio, J.M.C., Tu, Z., Robertson, H.M., Shumway, M., Sutton, G., Wortman, J.R., Lawson, D., Wikel, S.K., Nene, V.M., Fraser, C.M., Collins, F.H., Birren, B., Nelson, K.E., Caler, E., Hill, C.A., 2016. Genomic insights into the *Ixodes scapularis* tick vector of Lyme disease. Nat. Commun. 7, 10507.

Gunn, J.S., Ernst, R.K., 2007. The structure and function of *Francisella* lipopolysaccharide. Ann. N. Y. Acad. Sci. 1105, 202–218.

Guo, X., Booth, C.J., Paley, M.A., Wang, X., DePonte, K., Fikrig, E., Narasimhan, S., Montgomery, R.R., 2009. Inhibition of neutrophil function by two tick salivary proteins. Infect. Immun. 77, 2320–2329.

Guo, H., Callaway, J.B., Ting, J.P.-Y., 2015. Inflammasomes: mechanism of action, role in disease, and therapeutics. Nat. Med. 21, 677–687.

Guven-Maiorov, E., Tsai, C.-J., Nussinov, R., 2016. Pathogen mimicry of host protein-protein interfaces modulates immunity. Semin. Cell Dev. Biol. http://dx.doi.org/10.1016/j.semcdb.2016.06.004.

Hachani, A., Wood, T.E., Filloux, A., 2016. Type VI secretion and anti-host effectors. Curr. Opin. Microbiol. 29, 81–93.

Hackman, R.H., 1982. Structure and function in tick cuticle. Annu. Rev. Entomol. 27, 75–95.

Hallström, T., Siegel, C., Mörgelin, M., Kraiczy, P., Skerka, C., Zipfel, P.F., 2013. CspA from *Borrelia burgdorferi* inhibits the terminal complement pathway. MBio 4 (4). http://dx.doi.org/10.1128/mBio.00481-13.

Haneda, T., Ishii, Y., Shimizu, H., Ohshima, K., Iida, N., Danbara, H., Okada, N., 2012. *Salmonella* type III effector SpvC, a phosphothreonine lyase, contributes to reduction in inflammatory response during intestinal phase of infection. Cell. Microbiol. 14, 485–499.

Hannier, S., Liversidge, J., Sternberg, J.M., Bowman, A.S., 2003. *Ixodes ricinus* tick salivary gland extract inhibits IL-10 secretion and CD69 expression by mitogen-stimulated murine splenocytes and induces hyporesponsiveness in B lymphocytes. Parasite. Immunol. 25, 27–37.

Hasegawa, M., Yang, K., Hashimoto, M., Park, J.-H., Kim, Y.-G., Fujimoto, Y., Nuñez, G., Fukase, K., Inohara, N., 2006. Differential release and distribution of Nod1 and Nod2 immunostimulatory molecules among bacterial species and environments. J. Biol. Chem. 281, 29054–29063.

Hellwage, J., Meri, T., Heikkilä, T., Alitalo, A., Panelius, J., Lahdenne, P., Seppälä, I.J., Meri, S., 2001. The complement regulator factor H binds to the surface protein OspE of *Borrelia burgdorferi*. J. Biol. Chem. 276, 8427–8435.

Holmgren, A.M., McConkey, C.A., Shin, S., 2016. Outrunning the Red Queen: bystander activation as a means of outpacing innate immune subversion by intracellular pathogens. Cell. Mol. Immunol. http://dx.doi.org/10.1038/cmi.2016.36.

Horka, H., Staudt, V., Klein, M., Taube, C., Reuter, S., Dehzad, N., Andersen, J.F., et al., March 15, 2012. The tick salivary protein sialostatin L inhibits the Th9-derived production of the asthma-promoting cytokine IL-9 and is effective in the prevention of experimental asthma. J. Immunol. 188 (6), 2669–2676.

Hovius, J.W.R., 2009. Spitting image: tick saliva assists the causative agent of Lyme disease in evading host skin's innate immune response. J. Invest. Dermatol. 129, 2337–2339.

Hu, Z., Zhou, Q., Zhang, C., Fan, S., Cheng, W., Zhao, Y., Shao, F., Wang, H.-W., Sui, S.-F., Chai, J., 2015. Structural and biochemical basis for induced self-propagation of NLRC4. Science 350, 399–404.

Huang, M.T.-H., Mortensen, B.L., Taxman, D.J., Craven, R.R., Taft-Benz, S., Kijek, T.M., Fuller, J.R., Davis, B.K., Allen, I.C., Brickey, W.J., Gris, D., Wen, H., Kawula, T.H., Ting, J.P.-Y., 2010. Deletion of *ripA* Alleviates suppression of the inflammasome and MAPK by *Francisella tularensis*. J. Immunol. 185, 5476–5485.

Inohara, N., Ogura, Y., Chen, F.F., Muto, A., Nuñez, G., 2001. Human Nod1 confers responsiveness to bacterial lipopolysaccharides. J. Biol. Chem. 276, 2551–2554.

Iwasaki, A., Medzhitov, R., 2015. Control of adaptive immunity by the innate immune system. Nat. Immunol. 16, 343–353.

Janeway, C.A., Medzhitov, R., 2002. Innate immune recognition. Annu. Rev. Immunol. 20, 197–216. http://dx.doi.org/10.1146/annurev.immunol.20.083001.084359.

Jones, J.W., Broz, P., Monack, D.M., 2011. Innate immune recognition of *Francisella tularensis*: activation of type-I interferons and the inflammasome. Front. Microbiol. 2, 16.

Julien, J.-P., Lee, P.S., Wilson, I.A., 2012. Structural insights into key sites of vulnerability on HIV-1 Env and influenza HA. Immunol. Rev. 250, 180–198.

Kaufman, W.R., 1989. Tick-host interaction: a synthesis of current concepts. Parasitol. Today. 5, 47–56.

Kawai, T., Akira, S., 2011. Toll-like receptors and their crosstalk with other innate receptors in infection and immunity. Immunity 34, 637–650.

Kazimírová, M., Štibrániová, I., 2013. Tick salivary compounds: their role in modulation of host defenses and pathogen transmission. Front. Cell. Infect. Microbiol. 3. http://dx.doi.org/10.3389/fcimb.2013.00043.

Kenedy, M.R., Akins, D.R., 2011. The OspE-related proteins inhibit complement deposition and enhance serum resistance of *Borrelia burgdorferi*, the Lyme disease spirochete. Infect. Immun. 79, 1451–1457.

Kenedy, M.R., Vuppala, S.R., Siegel, C., Kraiczy, P., Akins, D.R., 2009. CspA-mediated binding of human factor H inhibits complement deposition and confers serum resistance in *Borrelia burgdorferi*. Infect. Immun. 77, 2773–2782.

Kim, D.W., Lenzen, G., Page, A.-L., Legrain, P., Sansonetti, P.J., Parsot, C., 2005. The *Shigella flexneri* effector OspG interferes with innate immune responses by targeting ubiquitin-conjugating enzymes. Proc. Natl. Acad. Sci. U.S.A. 102, 14046–14051.

Klein, M., Brühl, T.J., Staudt, V., Reuter, S., Grebe, N., Gerlitzki, B., Hoffmann, M., et al., July 15, 2015. Tick salivary sialostatin L represses the initiation of immune responses by targeting IRF4-dependent transcription in murine mast cells. J. Immunol. 195 (2), 621–631.

Klingenbeck, L., Eckart, R.A., Berens, C., Lührmann, A., 2013. The *Coxiella burnetii* type IV secretion system substrate CaeB inhibits intrinsic apoptosis at the mitochondrial level. Cell. Microbiol. 15, 675–687.

Kobayashi, K., Inohara, N., Hernandez, L.D., Galán, J.E., Núñez, G., Janeway, C.A., Medzhitov, R., Flavell, R.A., 2002. RICK/Rip2/CARDIAK mediates signalling for receptors of the innate and adaptive immune systems. Nature 416, 194–199.

Kobayashi, T., Ogawa, M., Sanada, T., Mimuro, H., Kim, M., Ashida, H., Akakura, R., Yoshida, M., Kawalec, M., Reichhart, J.-M., Mizushima, T., Sasakawa, C., 2013. The *Shigella* OspC3 effector inhibits caspase-4, antagonizes inflammatory cell death, and promotes epithelial infection. Cell Host Microbe 13, 570–583.

Koedel, U., Fingerle, V., Pfister, H.-W., 2015. Lyme neuroborreliosis-epidemiology, diagnosis and management. Nat. Rev. Neurol. 11, 446–456.

Koenigs, A., Hammerschmidt, C., Jutras, B.L., Pogoryelov, D., Barthel, D., Skerka, C., Kugelstadt, D., Wallich, R., Stevenson, B., Zipfel, P.F., Kraiczy, P., 2013. BBA70 of *Borrelia burgdorferi* is a novel plasminogen-binding protein. J. Biol. Chem. 288, 25229–25243.

Kotál, J., Langhansová, H., Lieskovská, J., Andersen, J.F., Francischetti, I.M.B., Chavakis, T., Kopecký, J., Pedra, J.H.F., Kotsyfakis, M., Chmelař, J., 2015. Modulation of host immunity by tick saliva. J. Proteomics 128, 58–68.

Kotsyfakis, M., Anderson, J.M., Andersen, J.F., Calvo, E., Francischetti, I.M.B., Mather, T.N., Valenzuela, J.G., Ribeiro, J.M.C., 2008. Cutting edge: immunity against a "silent" salivary antigen of the Lyme vector *Ixodes scapularis* impairs its ability to feed. J. Immunol. 181, 5209–5212.

Kraiczy, P., Hunfeld, K.P., Breitner-Ruddock, S., Würzner, R., Acker, G., Brade, V., 2000. Comparison of two laboratory methods for the determination of serum resistance in *Borrelia burgdorferi* isolates. Immunobiology 201, 406–419.

Kuthejlová, M., Kopecký, J., Štěpánová, G., Macela, A., 2001. Tick salivary gland extract inhibits killing of *Borrelia afzelii* spirochetes by mouse macrophages. Infect. Immun. 69, 575–578.

Kýcková, K., Kopecký, J., 2006. Effect of tick saliva on mechanisms of innate immune response against *Borrelia afzelii*. J. Med. Entomol. 43, 1208–1214.

Labandeira-Rey, M., Skare, J.T., 2001. Decreased infectivity in *Borrelia burgdorferi* strain B31 is associated with loss of linear plasmid 25 or 28-1. Infect. Immun. 69, 446–455.

Labandeira-Rey, M., Seshu, J., Skare, J.T., 2003. The absence of linear plasmid 25 or 28-1 of *Borrelia burgdorferi* dramatically alters the kinetics of experimental infection via distinct mechanisms. Infect. Immun. 71, 4608–4613.

LaFrance, M.E., Pierce, J.V., Antonara, S., Coburn, J., 2011. The *Borrelia burgdorferi* integrin ligand P66 affects gene expression by human cells in culture. Infect. Immun. 79, 3249–3261.

Lage, S.L., Longo, C., Branco, L.M., da Costa, T.B., Buzzo Cde, L., Bortoluci, K.R., 2014. Emerging concepts about NAIP/NLRC4 inflammasomes. Front. Immunol. 5, 309.

Laguna, R.K., Creasey, E.A., Li, Z., Valtz, N., Isberg, R.R., 2006. A *Legionella pneumophila*-translocated substrate that is required for growth within macrophages and protection from host cell death. Proc. Natl. Acad. Sci. U.S.A. 103, 18745–18750.

Langhansová, H., Bopp, T., Schmitt, E., Kopecký, J., 2015. Tick saliva increases production of three chemokines including monocyte chemoattractant protein-1, a histamine-releasing cytokine. Parasite. Immunol. 37, 92–96.

Latz, E., Xiao, T.S., Stutz, A., 2013. Activation and regulation of the inflammasomes. Nat. Rev. Immunol. 13, 397–411.

Lawley, T.D., Chan, K., Thompson, L.J., Kim, C.C., Govoni, G.R., Monack, D.M., 2006. Genome-wide screen for *Salmonella* genes required for long-term systemic infection of the mouse. PLoS Pathog. 2, e11. http://dx.doi.org/10.1371/journal.ppat.0020011.

Lawrie, C.H., Randolph, S.E., Nuttall, P.A., 1999. *Ixodes* ticks: serum species sensitivity of anticomplement activity. Exp. Parasitol. 93, 207–214.

Lemaitre, B., Nicolas, E., Michaut, L., Reichhart, J.-M., Hoffmann, J.A., 1996. The dorsoventral regulatory gene cassette spätzle/Toll/cactus controls the potent antifungal response in *Drosophila* adults. Cell 86, 973–983.

Lieskovská, J., Kopecký, J., 2012. Tick saliva suppresses IFN signalling in dendritic cells upon *Borrelia afzelii* infection. Parasite. Immunol. 34, 32–39.

Lieskovská, J., Páleníková, J., Langhansová, H., Chagas, A.C., Calvo, E., Kotsyfakis, M., Kopecký, J., 2015a. Tick sialostatins L and L2 differentially influence dendritic cell responses to *Borrelia* spirochetes. Parasites Vectors 15 (8), 275.

Lieskovská, J., Páleníková, J., Širmarová, J., Elsterová, J., Kotsyfakis, M., Campos Chagas, A., Calvo, E., Růžek, D., Kopecký, J., 2015b. Tick salivary cystatin sialostatin L2 suppresses IFN responses in mouse dendritic cells. Parasite. Immunol. 37, 70–78.

Lin, M., Rikihisa, Y., 2003. *Ehrlichia chaffeensis* and *Anaplasma phagocytophilum* lack genes for lipid A biosynthesis and incorporate cholesterol for their survival. Infect. Immun. 71, 5324–5331.

Lin, Q., Zhang, C., Rikihisa, Y., 2006. Analysis of involvement of the RecF pathway in p44 recombination in *Anaplasma phagocytophilum* and in *Escherichia coli* by using a plasmid carrying the p44 expression and p44 donor Loci. Infect. Immun. 74, 2052–2062.

Lin, T., Gao, L., Edmondson, D.G., Jacobs, M.B., Philipp, M.T., Norris, S.J., 2009. Central role of the Holliday junction helicase RuvAB in *vlsE* recombination and infectivity of *Borrelia burgdorferi*. PLoS Pathog. 5, e1000679. http://dx.doi.org/10.1371/journal.ppat.1000679.

Lina, T.T., Farris, T., Luo, T., Mitra, S., Zhu, B., McBride, J.W., 2016. Hacker within! *Ehrlichia chaffeensis* effector driven phagocyte reprogramming strategy. Front. Cell. Infect. Microbiol. 6, 58.

Liu, H., Bao, W., Lin, M., Niu, H., Rikihisa, Y., 2012. *Ehrlichia* type IV secretion effector ECH0825 is translocated to mitochondria and curbs ROS and apoptosis by upregulating host MnSOD. Cell. Microbiol. 14, 1037–1050.

Losick, V.P., Haenssler, E., Moy, M.-Y., Isberg, R.R., 2010. LnaB: a *Legionella pneumophila* activator of NF-κB. Cell. Microbiol. 12, 1083–1097.

Lubick, K.J., Robertson, S.J., McNally, K.L., Freedman, B.A., Rasmussen, A.L., Taylor, R.T., Walts, A.D., Tsuruda, S., Sakai, M., Ishizuka, M., Boer, E.F., Foster, E.C., Chiramel, A.I., Addison, C.B., Green, R., Kastner, D.L., Katze, M.G., Holland, S.M., Forlino, A., Freeman, A.F., Boehm, M., Yoshii, K., Best, S.M., 2015. Flavivirus antagonism of type I interferon signaling reveals prolidase as a regulator of IFNAR1 surface expression. Cell Host Microbe. 18, 61–74.

Luo, T., McBride, J.W., 2012. *Ehrlichia chaffeensis* TRP32 interacts with host cell targets that influence intracellular survival. Infect. Immun. 80, 2297–2306.

Maldonado, R.F., Sá-Correia, I., Valvano, M.A., 2016. Lipopolysaccharide modification in Gram-negative bacteria during chronic infection. FEMS. Microbiol. Rev. 40, 480–493.

Mans, B.J., 2011. Evolution of vertebrate hemostatic and inflammatory control mechanisms in blood-feeding arthropods. J. Innate Immun. 3, 41–51.

Master, S.S., Davis, A.S., Rampini, S.K., Keller, C., Ehlers, S., Springer, B., Sander, P., Deretic, V., 2008. *Mycobacterium tuberculosis* prevents inflammasome activation. Cell Host Microbe 3, 224–232.

McCulloch, R., Morrison, L.J., Hall, J.P.J., 2015. DNA recombination strategies during antigenic variation in the African trypanosome. Microbiol. Spectr. 3 MDNA3-0016-2014.

McDowell, J.V., Wolfgang, J., Tran, E., Metts, M.S., Hamilton, D., Marconi, R.T., 2003. Comprehensive analysis of the factor H binding capabilities of *Borrelia* species associated with Lyme disease: delineation of two distinct classes of factor H binding proteins. Infect. Immun. 71, 3597–3602.

Mead, P.S., 2015. Epidemiology of Lyme disease. Infect. Dis. Clin. North. Am. 29, 187–210.

Menten-Dedoyart, C., Faccinetto, C., Golovchenko, M., Dupiereux, I., Van Lerberghe, P.-B., Dubois, S., Desmet, C., Elmoualij, B., Baron, F., Rudenko, N., Oury, C., Heinen, E., Couvreur, B., 2012. Neutrophil extracellular traps entrap and kill *Borrelia burgdorferi sensu stricto* spirochetes and are not affected by *Ixodes ricinus* tick saliva. J. Immunol. 189, 5393–5401.

Miao, E.A., Alpuche-Aranda, C.M., Dors, M., Clark, A.E., Bader, M.W., Miller, S.I., Aderem, A., 2006. Cytoplasmic flagellin activates caspase-1 and secretion of interleukin 1β via Ipaf. Nat. Immunol. 7, 569–575.

Minnich, S.A., Rohde, H.N., 2007. A rationale for repression and/or loss of motility by pathogenic *Yersinia* in the mammalian host. Adv. Exp. Med. Biol. 603, 298–310.

von Moltke, J., Ayres, J.S., Kofoed, E.M., Chavarría-Smith, J., Vance, R.E., 2013. Recognition of bacteria by inflammasomes. Annu. Rev. Immunol. 31, 73–106.

Mogensen, T.H., 2009. Pathogen recognition and inflammatory signaling in innate immune defenses. Clin. Microbiol. Rev. 22, 240–273.

Monroe, K.M., McWhirter, S.M., Vance, R.E., 2009. Identification of host cytosolic sensors and bacterial factors regulating the type I interferon response to *Legionella pneumophila*. PLoS Pathog. 5, e1000665.

Motta, V., Soares, F., Sun, T., Philpott, D.J., 2015. NOD-like receptors: versatile cytosolic sentinels. Physiol. Rev. 95, 149–178.

Narasimhan, S., Koski, R.A., Beaulieu, B., Anderson, J.F., Ramamoorthi, N., Kantor, F., Cappello, M., Fikrig, E., 2002. A novel family of anticoagulants from the saliva of *Ixodes scapularis*. Insect. Mol. Biol. 11, 641–650.

Narasimhan, S., DePonte, K., Marcantonio, N., Liang, X., Royce, T.E., Nelson, K.F., Booth, C.J., Koski, B., Anderson, J.F., Kantor, F., Fikrig, E., 2007. Immunity against *Ixodes scapularis* salivary proteins expressed within 24 hours of attachment thwarts tick feeding and impairs *Borrelia* transmission. PLoS One 2, e451.

Nelder, M.P., Russell, C.B., Sheehan, N.J., Sander, B., Moore, S., Li, Y., Johnson, S., Patel, S.N., Sider, D., 2016. Human pathogens associated with the blacklegged tick *Ixodes scapularis*: a systematic review. Parasites Vectors 9, 265.

Nelson, C.A., Saha, S., Kugeler, K.J., Delorey, M.J., Shankar, M.B., Hinckley, A.F., Mead, P.S., 2015. Incidence of clinician-diagnosed Lyme disease, United States, 2005–2010. Emerg. Infect. Dis. 21, 1625–1631.

Nene, V., Kiara, H., Lacasta, A., Pelle, R., Svitek, N., Steinaa, L., 2016. The biology of *Theileria parva* and control of East Coast fever – current status and future trends. Ticks Tick Borne Dis. 7, 549–564 TTP8-STVM Special Issue.

Niu, H., Yamaguchi, M., Rikihisa, Y., 2008. Subversion of cellular autophagy by *Anaplasma phagocytophilum*. Cell. Microbiol. 10, 593–605.

Niu, H., Kozjak-Pavlovic, V., Rudel, T., Rikihisa, Y., 2010. *Anaplasma phagocytophilum* Ats-1 is imported into host cell mitochondria and interferes with apoptosis induction. PLoS Pathog. 6, e1000774.

Niu, H., Xiong, Q., Yamamoto, A., Hayashi-Nishino, M., Rikihisa, Y., 2012. Autophagosomes induced by a bacterial Beclin 1 binding protein facilitate obligatory intracellular infection. Proc. Natl. Acad. Sci. U.S.A. 109, 20800–20807.

Norris, S.J., 2014. *Vls* Antigenic variation systems of Lyme disease *Borrelia*: eluding host immunity through both random, segmental gene conversion and framework heterogeneity. Microbiol. Spectr. 2 (6). http://dx.doi.org/10.1128/microbiolspec.MDNA3-0038-2014.

Nunn, M.A., Sharma, A., Paesen, G.C., Adamson, S., Lissina, O., Willis, A.C., Nuttall, P.A., 2005. Complement inhibitor of C5 activation from the soft tick *Ornithodoros moubata*. J. Immunol. 174, 2084–2091.

Oliveira, C.J.F., Sá-Nunes, A., Francischetti, I.M.B., Carregaro, V., Anatriello, E., Silva, J.S., de Miranda Santos, I.K.F., Ribeiro, J.M.C., Ferreira, B.R., 2011. Deconstructing tick saliva. J. Biol. Chem. 286, 10960–10969.

Paesen, G.C., Adams, P.L., Harlos, K., Nuttall, P.A., Stuart, D.I., 1999. Tick histamine-binding proteins: isolation, cloning, and three-dimensional structure. Mol. Cell. 3, 661–671.

Palm, N.W., Medzhitov, R., 2009. Pattern recognition receptors and control of adaptive immunity. Immunol. Rev. 227, 221–233.

Palmer, G.H., Bankhead, T., Seifert, H.S., 2016. Antigenic variation in bacterial pathogens. Microbiol. Spectr. 4. http://dx.doi.org/10.1128/microbiolspec.VMBF-0005-2015.

Park, J.-H., Kim, Y.-G., Shaw, M., Kanneganti, T.-D., Fujimoto, Y., Fukase, K., Inohara, N., Núñez, G., 2007. Nod1/RICK and TLR signaling regulate chemokine and antimicrobial innate immune responses in mesothelial cells. J. Immunol. 179, 514–521.

Peng, K., Broz, P., Jones, J., Joubert, L.-M., Monack, D., 2011. Elevated AIM2-mediated pyroptosis triggered by hypercytotoxic Francisella mutant strains is attributed to increased intracellular bacteriolysis. Cell. Microbiol. 13, 1586–1600.

Piesman, J., Eisen, L., 2008. Prevention of tick-borne diseases. Annu. Rev. Entomol. 53, 323–343.

Pilar, A.V.C., Reid-Yu, S.A., Cooper, C.A., Mulder, D.T., Coombes, B.K., 2012. GogB is an anti-inflammatory effector that limits tissue damage during Salmonella infection through interaction with human FBXO22 and Skp1. PLoS Pathog. 8, e1002773.

Poole, N.M., Mamidanna, G., Smith, R.A., Coons, L.B., Cole, J.A., 2013. Prostaglandin E2 in tick saliva regulates macrophage cell migration and cytokine profile. Parasites Vectors 6, 261.

Purser, J.E., Norris, S.J., 2000. Correlation between plasmid content and infectivity in Borrelia burgdorferi. Proc. Natl. Acad. Sci. U.S.A. 97, 13865–13870.

Rabadi, S.M., Sanchez, B.C., Varanat, M., Ma, Z., Catlett, S.V., Melendez, J.A., Malik, M., Bakshi, C.S., 2016. Antioxidant defenses of Francisella tularensis modulate macrophage function and production of proinflammatory cytokines. J Biol. Chem. 291, 5009–5021.

Ramachandra, R.N., Wikel, S.K., 1992. Modulation of host-immune responses by ticks (Acari: Ixodidae): effect of salivary gland extracts on host macrophages and lymphocyte cytokine production. J. Med. Entomol. 29, 818–826.

Rathinam, V.A.K., Fitzgerald, K.A., 2016. Inflammasome complexes: emerging mechanisms and effector functions. Cell 165, 792–800.

Reddy, G.R., Streck, C.P., 1999. Variability in the 28-kDa surface antigen protein multigene locus of isolates of the emerging disease agent Ehrlichia chaffeensis suggests that it plays a role in immune evasion. Mol. Cell Biol. Res. Commun. 1, 167–175.

Reddy, G.R., Sulsona, C.R., Barbet, A.F., Mahan, S.M., Burridge, M.J., Alleman, A.R., 1998. Molecular characterization of a 28 kDa surface antigen gene family of the tribe Ehrlichiae. Biochem. Biophys. Res. Commun. 247, 636–643.

Rennoll-Bankert, K.E., Garcia-Garcia, J.C., Sinclair, S.H., Dumler, J.S., 2015. Chromatin-bound bacterial effector ankyrin A recruits histone deacetylase 1 and modifies host gene expression. Cell. Microbiol. 17, 1640–1652.

Ribeiro, J.M., 1987a. Role of saliva in blood-feeding by arthropods. Annu. Rev. Entomol. 32, 463–478.

Ribeiro, J.M., 1987b. Ixodes dammini: salivary anti-complement activity. Exp. Parasitol. 64, 347–353.

Ribeiro, J.M., 1989. Role of saliva in tick/host interactions. Exp. Appl. Acarol. 7, 15–20.

Ribeiro, J.M.C., Francischetti, I.M.B., 2003. ROLE OF ARTHROPOD SALIVA IN BLOOD FEEDING: role of arthropod saliva in blood feeding: sialome and post-sialome perspectives. Annu. Rev. Entomol. 48, 73–88.

Ribeiro, J.M., Spielman, A., 1986. Ixodes dammini: salivary anaphylatoxin inactivating activity. Exp. Parasitol. 62, 292–297.

Ribeiro, J.M., Makoul, G.T., Levine, J., Robinson, D.R., Spielman, A., 1985. Antihemostatic, antiinflammatory, and immunosuppressive properties of the saliva of a tick, Ixodes dammini. J. Exp. Med. 161, 332–344.

Ribeiro, J.M., Makoul, G.T., Robinson, D.R., 1988. Ixodes dammini: evidence for salivary prostacyclin secretion. J. Parasitol. 74, 1068–1069.

Ribeiro, J.M., Weis, J.J., Telford, S.R., 1990. Saliva of the tick Ixodes dammini inhibits neutrophil function. Exp. Parasitol. 70, 382–388.

Ribeiro, J.M.C., Alarcon-Chaidez, F., Francischetti, I.M.B., Mans, B.J., Mather, T.N., Valenzuela, J.G., Wikel, S.K., 2006. An annotated catalog of salivary gland transcripts from Ixodes scapularis ticks. Insect. Biochem. Mol. Biol. 36, 111–129.

Richter, D., Matuschka, F.-R., Spielman, A., Mahadevan, L., 2013. How ticks get under your skin: insertion mechanics of the feeding apparatus of Ixodes ricinus ticks. Proc.: Biol. Sci. 280, 20131758.

Rogers, E.A., Marconi, R.T., 2007. Delineation of species-specific binding properties of the CspZ protein (BBH06) of Lyme disease spirochetes: evidence for new contributions to the pathogenesis of Borrelia spp. Infect. Immun. 75, 5272–5281.

Rogers, E.A., Abdunnur, S.V., McDowell, J.V., Marconi, R.T., 2009. Comparative analysis of the properties and ligand binding characteristics of CspZ, a factor H binding protein, derived from Borrelia burgdorferi isolates of human origin. Infect. Immun. 77, 4396–4405.

Rudenko, G., 2011. African trypanosomes: the genome and adaptations for immune evasion. Essays. Biochem. 51, 47–62.

Růžek, D., Dobler, G., Donoso Mantke, O., 2010. Tick-borne encephalitis: pathogenesis and clinical implications. Travel Med. Infect. Dis. 8, 223–232.

Sá-Nunes, A., Bafica, A., Lucas, D.A., Conrads, T.P., Veenstra, T.D., Andersen, J.F., Mather, T.N., Ribeiro, J.M.C., Francischetti, I.M.B., 2007. Prostaglandin E2 is a major inhibitor of dendritic cell maturation and function in Ixodes scapularis saliva. J. Immunol. 179, 1497–1505.

Samuels, D.S., 2011. Gene regulation in *Borrelia burgdorferi*. Annu. Rev. Microbiol. 65, 479–499.

Sanada, T., Kim, M., Mimuro, H., Suzuki, M., Ogawa, M., Oyama, A., Ashida, H., Kobayashi, T., Koyama, T., Nagai, S., Shibata, Y., Gohda, J., Inoue, J., Mizushima, T., Sasakawa, C., 2012. The *Shigella flexneri* effector OspI deamidates UBC13 to dampen the inflammatory response. Nature 483, 623–626.

Sangamnatdej, S., Paesen, G.C., Slovak, M., Nuttall, P.A., 2002. A high affinity serotonin- and histamine-binding lipocalin from tick saliva. Insect. Mol. Biol. 11, 79–86.

Sansonetti, P.J., Di Santo, J.P., 2007. Debugging how bacteria manipulate the immune response. Immunity 26, 149–161.

Sauer, J.R., 1977. Acarine salivary glands–physiological relationships. J. Med. Entomol. 14, 1–9.

Schoeler, G.B., Wikel, S.K., 2001. Modulation of host immunity by haematophagous arthropods. Ann. Trop. Med. Parasitol. 95, 755–771.

Schotte, P., Denecker, G., Van Den Broeke, A., Vandenabeele, P., Cornelis, G.R., Beyaert, R., 2004. Targeting Rac1 by the *Yersinia* effector protein YopE inhibits caspase-1-mediated maturation and release of interleukin-1β. J. Biol. Chem. 279, 25134–25142.

Schuijt, T.J., Hovius, J.W.R., van Burgel, N.D., Ramamoorthi, N., Fikrig, E., van Dam, A.P., 2008. The tick salivary protein Salp15 inhibits the killing of serum-sensitive *Borrelia burgdorferi* sensu lato isolates. Infect. Immun. 76, 2888–2894.

Schuren, A.B., Costa, A.I., Wiertz, E.J., 2016. Recent advances in viral evasion of the MHC Class I processing pathway. Curr. Opin. Immunol. 40, 43–50.

Schwan, T.G., 2003. Temporal regulation of outer surface proteins of the Lyme-disease spirochaete Borrelia burgdorferi. Biochem. Soc. Trans. 31, 108–112.

Shaw, D.K., Kotsyfakis, M., Pedra, J.H.F., 2016. For whom the bell tolls (and Nods): spit-acular saliva. Curr. Trop. Med. Rep. 3, 40–50.

Shi, C.-S., Shenderov, K., Huang, N.-N., Kabat, J., Abu-Asab, M., Fitzgerald, K.A., Sher, A., Kehrl, J.H., 2012. Activation of autophagy by inflammatory signals limits IL-1β production by targeting ubiquitinated inflammasomes for destruction. Nat. Immunol. 13, 255–263.

Silva, N.C.S., Vale, V.F., Franco, P.F., Gontijo, N.F., Valenzuela, J.G., Pereira, M.H., Sant'Anna, M.R.V., Rodrigues, D.S., Lima, W.S., Fux, B., Araujo, R.N., 2016. Saliva of Rhipicephalus (Boophilus) microplus (Acari: Ixodidae) inhibits classical and alternative complement pathways. Parasites Vectors 9, 445.

Sonenshine, D.E., Roe, R.M. (Eds.), 2014a. Biology of Ticks Volume 1, second ed. Oxford University Press, New York.

Sonenshine, D.E., Roe, R.M. (Eds.), 2014b. Biology of Ticks Volume 2, second ed. Oxford University Press, New York.

Sousa, A.C., Szabó, M.P., Oliveira, C.J., Silva, M.J., August 2015. Exploring the anti-tumoral effects of tick saliva and derived components. Toxicon 102, 69–73.

Stanek, G., Wormser, G.P., Gray, J., Strle, F., 2012. Lyme borreliosis. Lancet 379, 461–473.

Steen, N.A., Barker, S.C., Alewood, P.F., 2006. Proteins in the saliva of the Ixodida (ticks): pharmacological features and biological significance. Toxicon 47, 1–20.

Stewart, C.R., Stuart, L.M., Wilkinson, K., van Gils, J.M., Deng, J., Halle, A., Rayner, K.J., Boyer, L., Zhong, R., Frazier, W.A., Lacy-Hulbert, A., Khoury, J.E., Golenbock, D.T., Moore, K.J., 2010. CD36 ligands promote sterile inflammation through assembly of a Toll-like receptor 4 and 6 heterodimer. Nat. Immunol. 11, 155–161.

Stibrániová, I., Lahová, M., Bartíková, P., 2013. Immunomodulators in tick saliva and their benefits. Acta. Virol. 57, 200–216.

Sukumaran, B., Mastronunzio, J.E., Narasimhan, S., Fankhauser, S., Uchil, P.D., Levy, R., Graham, M., Colpitts, T.M., Lesser, C.F., Fikrig, E., 2011. *Anaplasma phagocytophilum* AptA modulates Erk1/2 signalling. Cell. Microbiol. 13, 47–61.

Sukumaran, B., Ogura, Y., Pedra, J.H.F., Kobayashi, K.S., Flavell, R.A., Fikrig, E., 2012. Receptor interacting protein-2 contributes to host defense against *Anaplasma phagocytophilum* infection. FEMS Immunol. Med. Microbiol. 66, 211–219.

Tatchell, R.J., Moorhouse, D.E., 1970. Neutrophils: their role in the formation of a tick feeding lesion. Science 167, 1002–1003.

Telepnev, M.V., Klimpel, G.R., Haithcoat, J., Knirel, Y.A., Anisimov, A.P., Motin, V.L., 2009. Tetraacylated lipopolysaccharide of *Yersinia pestis* can inhibit multiple Toll-like receptor-mediated signaling pathways in human dendritic cells. J. Infect. Dis. 200, 1694–1702.

Thomson, N.R., Howard, S., Wren, B.W., Holden, M.T.G., Crossman, L., Challis, G.L., Churcher, C., Mungall, K., Brooks, K., Chillingworth, T., Feltwell, T., Abdellah, Z., Hauser, H., Jagels, K., Maddison, M., Moule, S., Sanders, M., Whitehead, S., Quail, M.A., Dougan, G., Parkhill, J., Prentice, M.B., 2006. The complete genome sequence and comparative genome analysis of the high pathogenicity *Yersinia enterocolitica* strain 8081. PLoS Genet. 2, e206.

Trimnell, A.R., Hails, R.S., Nuttall, P.A., 2002. Dual action ectoparasite vaccine targeting "exposed" and "concealed" antigens. Vaccine 20, 3560–3568.

Turni, C., Lee, R.P., Jackson, L.A., 2002. Effect of salivary gland extracts from the tick, *Boophilus microplus*, on leucocytes from Brahman and Hereford cattle. Parasite. Immunol. 24, 355–361.

Tyson, K., Elkins, C., Patterson, H., Fikrig, E., de Silva, A., 2007. Biochemical and functional characterization of Salp20, an *Ixodes scapularis* tick salivary protein that inhibits the complement pathway. Insect. Mol. Biol. 16, 469–479.

Valenzuela, J.G., Charlab, R., Mather, T.N., Ribeiro, J.M., 2000. Purification, cloning, and expression of a novel salivary anticomplement protein from the tick, *Ixodes scapularis*. J. Biol. Chem. 275, 18717–18723. http://dx.doi.org/10.1074/jbc.M001486200.

Van Avondt, K., van Sorge, N.M., Meyaard, L., 2015. Bacterial immune evasion through manipulation of host inhibitory immune signaling. PLoS Pathog. 11.

Vance, R.E., 2015. The NAIP/NLRC4 inflammasomes. Curr. Opin. Immunol. 32, 84–89.

Wagemakers, A., Staarink, P.J., Sprong, H., Hovius, J.W.R., 2015. *Borrelia miyamotoi*: a widespread tick-borne relapsing fever spirochete. Trends. Parasitol. 31, 260–269.

Wakeel, A., Kuriakose, J.A., McBride, J.W., 2009. An *Ehrlichia chaffeensis* tandem repeat protein interacts with multiple host targets involved in cell signaling, transcriptional regulation, and vesicle trafficking. Infect. Immun. 77, 1734–1745.

Waladde, S.M., Kemp, D.H., Rice, M.J., 1979. Feeding electrograms and fluid uptake measurements of cattle tick *Boophilus microplus* attached on artificial membranes. Int. J. Parasitol. 9, 89–95.

Wang, X., Shaw, D.K., Hammond, H.L., Sutterwala, F.S., Rayamajhi, M., Shirey, K.A., Perkins, D.J., Bonventre, J.V., Velayutham, T.S., Evans, S.M., Rodino, K.G., VieBrock, L., Scanlon, K.M., Carbonetti, N.H., Carlyon, J.A., Miao, E.A., McBride, J.W., Kotsyfakis, M., Pedra, J.H.F., 2016a. The prostaglandin E2-EP3 receptor axis regulates *Anaplasma phagocytophilum*-mediated NLRC4 inflammasome activation. PLoS Pathog. 12, e1005803.

Wang, X., Shaw, D.K., Sakhon, O.S., Snyder, G.A., Sundberg, E.J., Santambrogio, L., Sutterwala, F.S., Dumler, J.S., Shirey, K.A., Perkins, D.J., Richard, K., Chagas, A.C., Calvo, E., Kopecký, J., Kotsyfakis, M., Pedra, J.H.F., 2016b. The tick protein sialostatin L2 binds to annexin A2 and inhibits NLRC4-mediated inflammasome activation. Infect. Immun. 84 (6), 1796–1805. http://dx.doi.org/10.1128/IAI.01526-15.

Weiss, D.S., Brotcke, A., Henry, T., Margolis, J.J., Chan, K., Monack, D.M., 2007. *In vivo* negative selection screen identifies genes required for *Francisella* virulence. Proc. Natl. Acad. Sci. U.S.A. 104, 6037–6042. http://dx.doi.org/10.1073/pnas.0609675104.

Wikel, S.K., 1996. Host immunity to ticks. Annu. Rev. Entomol. 41, 1–22.

Wikel, S.K., 1999. Tick modulation of host immunity: an important factor in pathogen transmission. Int. J. Parasitol. 29, 851–859.

Wikel, S., 2013. Ticks and tick-borne pathogens at the cutaneous interface: host defenses, tick countermeasures, and a suitable environment for pathogen establishment. Front. Microbiol. 4. http://dx.doi.org/10.3389/fmicb.2013.00337.

Wikel, S.K., Allen, J.R., 1976. Acquired resistance to ticks. I. Passive transfer of resistance. Immunology 30, 311–316.

Wikel, S.K., Bergman, D., 1997. Tick-host immunology: significant advances and challenging opportunities. Parasitol. Today. 13, 383–389.

Wikel, S.K., Ramachandra, R.N., Bergman, D.K., Burkot, T.R., Piesman, J., 1997. Infestation with pathogen-free nymphs of the tick *Ixodes scapularis* induces host resistance to transmission of *Borrelia burgdorferi* by ticks. Infect. Immun. 65, 335–338.

Winchell, C.G., Steele, S., Kawula, T., Voth, D.E., 2016. Dining in: intracellular bacterial pathogen interplay with autophagy. Curr. Opin. Microbiol. 29, 9–14.

Windheim, M., Lang, C., Peggie, M., Plater, L.A., Cohen, P., 2007. Molecular mechanisms involved in the regulation of cytokine production by muramyl dipeptide. Biochem. J. 404, 179–190.

Yang, Q., Stevenson, H.L., Scott, M.J., Ismail, N., 2015. Type I interferon contributes to noncanonical inflammasome activation, mediates immunopathology, and impairs protective immunity during fatal infection with lipopolysaccharide-negative Ehrlichiae. Am. J. Pathol. 185, 446–461.

Yu, X., McBride, J.W., Zhang, X., Walker, D.H., 2000. Characterization of the complete transcriptionally active *Ehrlichia chaffeensis* 28 kDa outer membrane protein multigene family. Gene 248, 59–68.

Zanoni, I., Tan, Y., Gioia, M.D., Broggi, A., Ruan, J., Shi, J., Donado, C.A., Shao, F., Wu, H., Springstead, J.R., Kagan, J.C., 2016. An endogenous caspase-11 ligand elicits interleukin-1 release from living dendritic cells. Science 352, 1232–1236.

Zhang, J.-R., Norris, S.J., 1998. Genetic Variation of the *Borrelia burgdorferi* Gene vlsE involves cassette-specific, segmental gene conversion. Infect. Immun. 66, 3698–3704.

Zhang, J.R., Hardham, J.M., Barbour, A.G., Norris, S.J., 1997. Antigenic variation in Lyme disease borreliae by promiscuous recombination of VMP-like sequence cassettes. Cell 89, 275–285.

Zhang, L., Chen, S., Ruan, J., Wu, J., Tong, A.B., Yin, Q., Li, Y., David, L., Lu, A., Wang, W.L., Marks, C., Ouyang, Q., Zhang, X., Mao, Y., Wu, H., 2015. Cryo-EM structure of the activated NAIP2-NLRC4 inflammasome reveals nucleated polymerization. Science 350, 404–409.

Zheng, Z., Wei, C., Guan, K., Yuan, Y., Zhang, Y., Ma, S., Cao, Y., Wang, F., Zhong, H., He, X., 2016. Bacterial E3 ubiquitin ligase IpaH4.5 of *Shigella flexneri* targets TBK1 to dampen the host antibacterial response. J. Immunol. 196, 1199–1208.

Zhu, B., Nethery, K.A., Kuriakose, J.A., Wakeel, A., Zhang, X., McBride, J.W., 2009. Nuclear translocated *Ehrlichia chaffeensis* ankyrin protein interacts with a specific adenine-rich motif of host promoter and intronic Alu elements. Infect. Immun. 77, 4243–4255.

CHAPTER

11

Tsetse Fly Saliva Proteins as Biomarkers of Vector Exposure

Jan Van Den Abbeele[1], Guy Caljon[2]

[1]Institute of Tropical Medicine Antwerp (ITM), Antwerp, Belgium; [2]University of Antwerp, Antwerp, Belgium

THE TSETSE FLY AS VECTOR OF AFRICAN TRYPANOSOMES; AFRICAN TRYPANOSOMIASIS AND VECTOR CONTROL

The blood-feeding tsetse flies (Diptera: Glossinidae) are the only cyclical vectors of the protozoan parasite African trypanosome (*Trypanosoma* sp.) that cause a series of devastating diseases—African trypanosomiasis—in humans and domesticated animals. Human African trypanosomiasis (HAT) or sleeping sickness, caused by *Trypanosoma brucei gambiense* and *T. brucei rhodesiense*, is endemic to 36 countries in sub-Saharan Africa, with about 70 million inhabitants of this region at risk (Simarro et al., 2012). In 2012, WHO targeted gambiense-HAT (g-HAT) as a public health problem to be eliminated by 2020 (Holmes, 2014). The final goal will be the sustainable disease elimination by 2030, defined as the interruption of the transmission of g-HAT. This elimination is considered feasible because of the epidemiological vulnerability of the disease, the current state of control, the availability of strategies and tools, and international commitment and political will (Courtin et al., 2015; Franco et al., 2014; Lehane et al., 2016).

While g-HAT has reached the point where eradication is being put forward as an achievable goal, animal African trypanosomiasis (AAT) caused by *Trypanosoma congolense* and *Trypanosoma vivax* remains one of the most significant infectious disease threats to sub-Saharan livestock particularly affecting cattle (Morrison et al., 2016). Occurrence of the animal disease is reported in 37 countries covering some 8.7 million km², about a third of Africa's land area, and some 46 million cattle in this region are estimated to be at risk of contracting AAT (Alsan, 2015; PAAT Technical and Scientific Series, 2010). In sub-Saharan Africa, the availability of productive livestock is essential to significantly improve agriculture and to alleviate hunger, food insecurity, and poverty. Therefore, the current presence of tsetse and AAT in this large area can rightfully be considered as one of the major causes of hunger and poverty (Feldmann et al., 2005).

Both sexes of adult tsetse feed exclusively on vertebrate blood, every 2–5 days (Leak, 1999) and contribute to disease transmission. There are 31 species and subspecies of tsetse flies (*Glossina*) confined to sub-Saharan Africa (Gooding and

Krafsur, 2005; Krafsur, 2009). These tsetse species are further subdivided into three groups (subgenera) according to different criteria such as external (genital) morphology and their habitat preference, among others. Species of the fusca group (subgenus *Austenina*) typically occur in lowland rainforests of West and Central Africa or the border areas of the forest and isolated relic forests and are of little or no economic importance (Vreysen et al., 2013). Species of the palpalis group (subgenus *Nemorhina*) are more associated with coastal habitats, degraded forests of West-Africa and riverine vegetation, but some also extend into savannah regions along river systems (gallery forest). The species belonging to this group are important vectors of AAT in West-Africa and g-HAT in West- and Central-Africa (*Glossina palpalis palpalis*, *Glossina palpalis gambiensis*, *Glossina fuscipes fuscipes* and *G. f. quanzensis*) (Van den Bossche et al., 2010; Vreysen et al., 2013). All species belonging to the morsitans group (subgenus *Glossina* sensu stricto) are restricted to dryer regions such as savannah woodlands and their distribution and abundance often correspond with that of wild animals (Jordan, 1986). *Glossina morsitans* spp. and *Glossina pallidipes* are the most important species of this group and are major vectors of AAT and HAT in Eastern and Southern Africa (Vreysen et al., 2013). Morsitans group tsetse flies are thus mainly responsible for the transmission of the two important animal-disease causing trypanosomes, *T. congolense* and *T. vivax*. For the latter parasite, other non-tsetse blood-feeding insects like tabanids are also playing an important role in transmission (Desquesnes and Dia, 2003; Sinshaw et al., 2006).

Tsetse-transmitted trypanosomiasis is an intersectoral problem lying at the heart of African rural development affecting human health, livestock health, and rural development. Therefore, the benefits arising from African trypanosomiasis control overlaps among these three sectors (PAAT Technical and Scientific Series, 2010). Several complementary methods are being used

to fight against tsetse-transmitted trypanosomiasis, targeting the infected host as well as the tsetse vector. For the human diseases, the two forms of HAT, g-HAT in West and Central Africa and Rhodesian HAT (r-HAT) found in East and Southern Africa, strongly differ in epidemiology and control. For g-HAT, there are no significant nonhuman hosts and mass screening and treatment of affected humans with trypanocidal drugs have resulted in good progress in its control (Auty et al., 2016). Moreover, a recent study demonstrated that vector control, in combination with the medical control, was decisive in reducing sleeping sickness transmission and could speed up the HAT elimination process (Courtin et al., 2015). The aim in HAT foci is not to eradicate tsetse but to stop transmission by reducing tsetse–human contact, and modeling suggests that this does not require complete removal of tsetse flies (Lehane et al., 2016). To be successful, it is estimated that 4–5 years of tsetse control continuation is required depending on the distribution of the parasite in the human population and/or the existence of reservoir hosts (Lehane et al., 2016). Long-term elimination of r-HAT is considered to be unfeasible due to the existence of animal reservoirs, mainly in wildlife. Here, control of r-HAT has been achieved through various methods of vector control; for example, a combination of aerial spraying and odor-baited targets was used to eliminate tsetse and trypanosomiasis from the Okavango Delta, Botswana (Auty et al., 2016; Kgori et al., 2006). Animal African trypanosomiasis remains one of the most significant livestock diseases in sub-Saharan Africa, particularly affecting cattle (Yaro et al., 2016). Targeting the parasite with specific drugs is an important way to control the disease but have not produced the expected results due to the development and wide spread of trypanocidal resistance (Delespaux and de Koning, 2007). Control then relies on implementation of local, integrated control strategies by communities or farmers that must take into account the eco-epidemiological context and the cattle rearing system to be sustainable. In most situations, farmers and communities will need to conduct tsetse control (local integrated pest management) in order to reduce their density enough to prevent or reduce transmission of AAT and lower the probability that trypanosome strains resistant to trypanocides establish and spread within cattle populations (Bouyer et al., 2013).

The control of the tsetse vector is based upon the use of a variety of complementary techniques and combinations depending on fly species, agro-ecology, and the capacity of local and national tsetse control agencies: insecticide-treated targets and/or animals, area-wide insecticide applications, and in some areas the Sterile Insect technique (Bouyer et al., 2013, 2014; Courtin et al., 2015; Schofield and Kabayo, 2008; Solano et al., 2010; Sow et al., 2012; Torr et al., 2005; Vreysen et al., 2000, 2013). During the last decade, the African Union (AU) launched the Pan African Tsetse and Trypanosomiasis Eradication Campaign (PATTEC), a continent wide initiative that focuses on the progressive elimination of discrete tsetse-infested areas (Kabayo, 2002) using several combinations of the above mentioned complementary methods.

Regular evaluation in time of the efficacy of the vector control efforts and of the intervention success is essential and remains an important challenge. Currently, entomological evaluation is performed by the use of sentinel traps in order to estimate the apparent tsetse fly density per trap and per day. These entomological surveys are associated with important constraints. Most of the time, the deployment and regular monitoring of the traps in very large areas with poor accessibility, is costly and demanding in terms of human resources and logistics. Moreover, these traps have a low fly capturing efficiency and, when set up in fixed sites, only provide a rough estimate of host exposure to tsetse bites especially in areas where herds are very mobile (Dama et al., 2013a; Somda et al., 2016). In addition, it has been suggested that traps become less efficient when tsetse densities decrease (Gouteux et al., 2001). Given the

above described constraints of the conventional entomological surveys, an easy-to-use and rapid test that allows a semiquantitative estimate on a regular basis of tsetse fly exposure in the target host population would be a highly valuable tool in the follow-up of the efficacy of the applied and/or ongoing tsetse fly control activities in a specific area. For this, it has been demonstrated for a number of blood-feeding arthropods that salivary proteins inoculated during the feeding process induce humoral immune responses in the mammalian host that could be used as measurable markers for exposure to bites (reviewed in Fontaine et al., 2011).

THE *GLOSSINA* SIALOME: CHARACTERISTICS AND DIVERSITY OF SALIVARY COMPONENTS

Both male and female tsetse flies are obligatory blood feeders on a variety of vertebrate hosts. They rely on a pool-feeding strategy, which involves the laceration of the skin with their proboscis and blood ingestion from a superficial lesion. Once the skin is pierced, the proboscis is often partially withdrawn before being thrust again at a slightly different angle to probe for suitable blood vessels and to enhance the blood pool formation (Caljon et al., 2014a; Lehane, 2005). During this blood feeding event a tsetse fly will inject a few micrograms of saliva into the mammalian host (Van Den Abbeele et al., 2007). As it is the case for all blood-feeding arthropods, the saliva of adult tsetse flies contains a large repertoire of physiologically active saliva components that are essential for efficient blood feeding and digestion because they counteract the complex physiological responses of the host that impede blood feeding, including coagulation, blood platelet aggregation, and vasoconstriction. The sialome of *G. morsitans morsitans* identified by the analysis of a salivary gland EST-library and high throughput next-generation sequencing was demonstrated to contain over 250 proteins that are possibly associated with blood feeding (Alves-Silva et al., 2010; IGGI, 2014). So far, no nonprotein immunomodulatory components have been identified in tsetse fly saliva although the presence of two prostaglandin E2 synthase transcripts has been described (Alves-Silva et al., 2010), suggesting a possibility for the presence of PGE2 in tsetse saliva. Since both male and female tsetse flies are obligate blood feeders it can be assumed that the composition of their sialome is similar although no experimental data are currently available to confirm this.

Based on different experimental and analytical approaches, the major secreted tsetse salivary gland proteins are known to include the anticoagulant thrombin inhibitor (tsetse thrombin inhibitor, TTI), two putative adenosine deaminases (tsetse salivary growth factors 1 and 2, TSGF-1 and TSGF-2), salivary apyrases (two 5′ nucleotidase-related proteins, 5-Nuc and Sgp3), an antigen5-related allergen (Tsetse Antigen 5, TAg5), and the abundant tsetse salivary proteins 1 and 2 (Tsal1 and Tsal2), that were demonstrated to be high-affinity nucleic acid–binding proteins with residual nuclease activity (Caljon et al., 2009, 2010, 2012; Cappello et al., 1996, 1998; Li and Aksoy, 2000; Li et al., 2001; Van Den Abbeele et al., 2007; Zhao et al., 2015). It was shown that this tsetse sialome and its associated biological activity is modified in flies with *T. brucei*–infected salivary glands due to a parasite-associated downregulation of specific genes resulting in a strong reduction of the major secreted proteins and related biological activities in the saliva (Van Den Abbeele et al., 2010). In a recent study, the sialome protein profiles from four tsetse species in two subgenera (*G. morsitans morsitans*, *G. pallidipes*, *G. palpalis gambiensis* and *G. fuscipes fuscipes*) were compared by SDS-PAGE analysis and were found to show a high degree of similarity. However, a more detailed study of the adenosine deaminase growth factor (ADGF) family at the genomic and transcriptional level clearly revealed species-specific differences indicating that the relative abundance and the variety of the different major saliva proteins differ according to the tsetse fly species (Zhao et al., 2015).

HOST ANTIBODIES AGAINST TSETSE SALIVA PROTEINS

Saliva of blood-feeding arthropods is a complex potion of immunogenic proteins that induce various immunological responses in the exposed mammalian host resulting in the production of different immunoglobulin classes such as IgE (allergic reaction), IgM, and IgG.

IgM is produced as the first antibody isotype in the primary antibody response and has a short persistence. Detection of IgM is described to be useful as a highly sensitive indicator of recent exposure to vector salivary proteins (e.g., Orlandi-Pradines et al., 2007; Schwarz et al., 2010). A recent study on *Triatoma infestans* revealed that this IgM response is indeed highly sensitive to the specific detection of recent exposure to the bug but no association was observed between level of exposure and IgM antibody levels (Schwarz et al., 2010). So far, the IgM-response in the mammalian host to tsetse fly exposure has not been documented nor explored.

The second and major immunoglobulins that are produced following IgM are the IgGs. Here, the observed link between antisaliva IgG responses and blood-feeding vector exposure, as well as the decrease of the antibody titers after a period of nonexposure, has favored the exploration of the potential use of immunogenic saliva as an immunological marker of vector exposure (Fontaine et al., 2011).

Studies using various tsetse fly species have shown that salivary components are immunogenic in a variety of mammalian hosts (i.e., humans, cattle, pigs) with the induction of a variety of antisaliva IgGs. Experimental data indicated that these antibodies do not hamper the fly survival nor its feeding process and blood meal uptake (Caljon et al., 2006).

These immunoglobulin responses to tsetse fly saliva have been detected in humans living in HAT or tsetse-endemic regions in Uganda (Caljon et al., 2006), Democratic Republic of Congo (Poinsignon et al., 2007, 2008b), and Guinea (Dama et al., 2013a). Cattle from tsetse-infested areas in Burkina Faso and Southeastern Uganda and experimentally exposed to tsetse fly bites displayed elevated levels of antisaliva antibodies (Somda et al., 2013; Zhao et al., 2015). Experimentally exposed pigs to *G. morsitans morsitans* bites also displayed a specific IgG response toward tsetse saliva proteins (Caljon et al., 2014b). In the above mentioned studies it has been demonstrated that salivary proteins of several tsetse fly species are recognized by the circulating antisaliva antibodies indicating a significant degree of serological cross-reactivity of saliva proteins among different tsetse species. In contrast, low cross-reactivity was observed toward exposure to other biting flies such as *Stomoxys* and Tabanids. The study of Zhao et al. (2015) showed that the protein composition and immunogenicity of sialome components varied between tsetse fly species displaying higher similarity and cross reactivity in the same subgenus. Sera obtained from cattle from disease endemic areas of Africa displayed an immunogenicity profile reflective of tsetse species distribution.

The amount of antisaliva IgG's that can be detected in the different hosts relates to some extent to the tsetse bite exposure frequency. In experimental mice, a single fly bite was sufficient to induce detectable antisaliva IgG level response with an estimated half-life of 36–40 days. Specific antibody responses could be detected for more than a year after initial exposure, and a single bite was sufficient to boost antisaliva immunity in different hosts (mice, pigs). In pigs, the apparent antibody clearance rate was faster with an estimation of the antisaliva IgG half-life of 15 days (Caljon et al., 2014b). This observed clearance rate is consistent with the fast decline of anti-*G. morsitans submorsitans* saliva IgG levels observed in cattle within 10 weeks after cessation of tsetse

exposure (Somda et al., 2013). Various exposure frequencies to tsetse bites (in experimental mice and pigs) clearly resulted in different antisaliva IgG levels with an average 3- to 4-fold difference in specific antibody concentration between a low exposure and a high exposure regimen. This is in line with observations made for cattle that were experimentally exposed to different biting intensities (Somda et al., 2013). Moreover, comparative studies of cattle from tsetse epidemic and free areas illustrate that antisaliva antibody titers vary in rainy and dry seasons where tsetse densities fluctuate (Somda et al., 2013). Analysis of sera from cattle aged 8–15 years showed a stronger immunological response to specific tsetse sialome antigens than cattle less than 8 months old, confirming that antisaliva antibody titers increase over time related to frequency of fly challenge (Zhao et al., 2015). Nevertheless, the magnitude of these responses was quite low even in older cattle that likely received many tsetse bites, suggesting that cattle repeatedly exposed to bites may eventually gain tolerance to the bites of those species (Zhao et al., 2015). Also humans living in HAT endemic areas display significant differences in anti-tsetse saliva IgGs depending on the study site (Poinsignon et al., 2008a,b; Dama et al., 2013a). In the active HAT-foci and areas where tsetse flies were present at high densities, the specific IgG responses in humans were significantly higher than in areas where tsetse flies are absent or only present at low densities.

1D- and 2D-electrophoresis and immunoblotting of salivary proteins provided evidence for the immunogenicity of several of the major protein families including the Tsal proteins, the ADGF proteins, 5'nucleotidase (5'Nuc), and Antigen 5 (TAg5)-related proteins (Caljon et al., 2006; Dama et al., 2013a). In addition, immune screening of a phage salivary gland cDNA expression library resulted in the identification of the immunogenic *G. morsitans morsitans* salivary gland proteins Sgp1, Sgp2, and Sgp3 (Van Den Abbeele et al., 2007). Recently, a study of Zhao et al. (2015) described the analysis of different sialome fractions of *G. morsitans morsitans* by LC-MS/MS and confirmed TAg5, Tsal1/2, and Sgp3 as major immunogenic proteins whereas the 5′nucleotidase-related protein and four members of the ADGF family were identified as the major nonimmunogenic saliva proteins. Mice experimentally subjected to tsetse bites display varying immunological responses against the abundant saliva proteins (Caljon et al., 2006; Zhao et al., 2015) with an observed variable immunogenicity of similar proteins in different tsetse fly species (Zhao et al., 2015). Here, one member of the ADGF family (TSGF-2) of *G. morsitans morsitans* was found to be nonimmunogenic in mice whereas the same saliva protein of *G. fuscipes* induces a strong immunogenic response suggesting that the host immune response toward saliva is tsetse species dependent.

TOOLS TO DETECT ANTISALIVA ANTIBODIES IN THE MAMMALIAN HOST SERUM: QUALITATIVE AND SEMIQUANTITATIVE DETERMINATION OF BITE EXPOSURE

Immunogenic components of saliva have been exploited as biomarkers for exposure to different arthropod bites, including ticks, sandflies, mosquitoes, and triatomes (e.g., Schwarz et al., 1990; Vinhas et al., 2007; Marzouki et al., 2012; Souza et al., 2010; Billingsley et al., 2006; Poinsignon et al., 2008a; Rizzo et al., 2011; Schwarz et al., 2009; Fontaine et al., 2011). Here, the correlation between arthropod bite exposure and the level of antisaliva IgG antibodies is the basis of the development of serology-based tools. Antisaliva IgM response (explored for *Triatoma infestans* saliva; Schwarz et al., 2009, 2010) was demonstrated to be highly sensitive to the detection of bug exposure but no association could be found between level of exposure and IgM antibody level. This suggested that the short-lived IgM response could be used as an indicator for a very recent exposure of the host, in complementation with the measurement of antisaliva IgG levels.

The potential use of tsetse saliva as a biomarker of exposure has been explored in an experimental setting using mice, cattle, and pigs and in the natural situation using sera from humans and cattle in tsetse-infested versus tsetse free areas (Fig. 11.1). An overview of the limited amount of published studies is presented in Table 11.1. Serology-based tools in different formats (indirect ELISA, competitive ELISA; see Fig. 11.2) were developed in order to compare semiquantitatively the antisaliva IgG levels in low- and high-exposed host groups versus nonexposed controls.

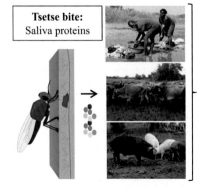

FIGURE 11.1 Schematic overview of the current methodologies used to estimate the antisaliva IgG response in a mammalian host after tsetse bite exposure.

TABLE 11.1 Overview of the Different Assays That Were Developed and Applied to Estimate the Antitsetse Saliva IgG Response in a Variety of the Mammalian Hosts

Host	Sample Origin	Assay Format	Antigen Used	Antisaliva IgG Response	References
Human	Active Human African trypanosomiasis (HAT) foci Democratic Republic Congo	Immunoblot	WSE-Gpg/Gff/Gmm/Gt	Immunogenic profiles specific to *Glossina* species	Poinsignon et al. (2007)
Human	Active HAT foci – Democratic Republic Congo	Indirect ELISA	WSE-Gff	Significant higher levels of IgG response in exposed individuals from HAT/tsetse-endemic area	Poinsignon et al. (2008b)
Human	Active HAT foci – Uganda	Indirect ELISA	WSE-Gmm; Gmm-recTsal1/2	Significant higher levels of IgG response in exposed individuals from HAT/tsetse-endemic area	Caljon et al. (2006)
Human	Active HAT foci – Guinea Historical HAT focus/tsetse endemic – Burkina Faso	Immunoblot Indirect ELISA	WSE-Gpg	Significant higher levels of IgG response in exposed individuals from HAT/tsetse-endemic area	Dama et al. (2013a)
Human	Active HAT foci – Guinea Historical HAT focus/tsetse endemic – Burkina Faso	Indirect ELISA	$Tsgf_{18-43}$ Synthetic peptide	IgG response toward the synthetic peptide were significantly higher in the human samples from the two tsetse-exposed areas	Dama et al. (2013b)
Human	Active HAT foci Democratic Republic Congo	Indirect ELISA	$Tsgf_{18-43}$ Synthetic peptide	Significant decrease of Tsgf18-43 ELISA titers in sentinel villages from vector control area.	Courtin et al. (2015)
Cattle	Tsetse endemic sites in Burkina Faso Experimentally exposed animals	Indirect ELISA	WSE-Gpg, Gt, Gms	Significant increase of cattle antisaliva IgG response in dry season (>tsetse exposure) and in AAT-infected animals Quick rise of antisaliva IgG after tsetse exposure Distinction between cows according to exposure level Gms WSE allowed to detect exposure to all tsetse species/low cross-reactivity to other blood-sucking arthropods	Somda et al. (2013)
Cattle (+mice)	Tsetse-endemic area in South-eastern Uganda Bovines used to maintain Gff colony	Immunoblot	Gmm, Gpd, Gpg, Gff	Saliva from different fly species generate varying immune response signatures; cattle displayed an antisaliva IgG profile reflective of the tsetse species distribution	Zhao et al. (2015)

Continued

TABLE 11.1 Overview of the Different Assays That Were Developed and Applied to Estimate the Antitsetse Saliva IgG Response in a Variety of the Mammalian Hosts—cont'd

Host	Sample Origin	Assay Format	Antigen Used	Antisaliva IgG Response	References
Cattle	Tsetse endemic sites in Burkina Faso Experimentally exposed animals	Immunoblot Indirect ELISA	WSE-*Gms* Tsal1$_{52-75}$ synthetic peptide	IgG responses to the Tsal1$_{52-75}$ peptide was specific of tsetse exposure in naturally and experimentally exposed animals. Anti-Tsal1$_{52-75}$ response was absent in tsetse highly exposed animals	Somda et al. (2016)
Pigs (+Mice)	Experimentally exposed animals	Indirect ELISA	WSE-*Gmm, Gpd, Gpg, Gff Gmm*-recTsal1	Anti-GmmRecTsal1 response is a sensitive indicator that can differentiate various degree of tsetse exposure. Antisaliva Abs persisted relatively long and were efficiently boosted upon reexposure. Low cross-reactivity toward exposure to other blood sucking insects	Caljon et al. (2014b)
Pigs (+Mice)	Experimentally exposed animals	Nanobody-based competitive ELISA	WSE-*Gmm, Gpd, Gpg, Gff Gmm*-recTsal1	Increased sensitivity and accuracy compared to previously described indirect ELISA using *Gmm*-recTsal1 Suitable to be applied to a range of host species without test adaptation. No cross-reactivity toward exposure to other blood sucking insects	Caljon et al. (2015)

Gff, Glossinafuscipes fuscipes; Gmm, Glossina morsitans morsitans; Gms, Glossina morsitans submorsitans; Gpd, Glossina pallidipes; Gpg, Glossina palpalis gambiensis; Gt, Glossina tachinoides; Tsal1, tsetse salivary proteins 1; Tsgf, tsetse salivary growth factor; WSE, whole saliva extract.

INDIRECT ELISA **COMPETITIVE ELISA**

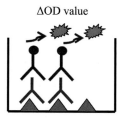   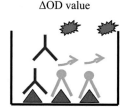

▲ Tsetse saliva Ag

⅄ 1°Ab (host plasma/anti-saliva IgG)

⅄ 2° Ab (anti-host IgG – HRP conjugated)

⅄ Ab-HRP specific to a defined immunogenic

tsetse antigen (e.g anti-Tsal Nanobody-HRP)

FIGURE 11.2 Two serological antitsetse saliva IgG detection assays: indirect ELISA versus competitive ELISA.

An ideal saliva-based biomarker assay for tsetse bite exposure should be (1) specific for tsetse saliva proteins and suitable for different tsetse fly species i.e., pan-tsetse cross reactivity, (2) semiquantitative allowing to estimate the tsetse bite exposure in a host population or in a selection of individual animals, also at low exposure (high sensitivity), (3) compatible to be used in a variety of different mammalian host species, and (4) easy to produce in standardized way compliant with appropriate quality standards. To be reliable and useful, this biomarker assay should have a high sensitivity (≥ 0.9) as well as high specificity (≥ 0.9). The sensitivity measures the proportion of positives that are correctly identified (true tsetse fly bite exposure; possibility of low exposure detection), whereas the specificity measures the proportion of negatives (nonexposed to tsetse bite) that are correctly identified as such.

To detect specific antisaliva IgGs in a variety of hosts, different tsetse saliva antigens were explored ranging from whole saliva extract up to recombinant proteins and synthetic peptides. Here, the use of whole saliva extract as antigen for detecting specific IgG responses presents constraints such as problems of standardized

mass production and reproducibility. Also, cross-reactivity with antibodies directed against similar saliva antigens of other blood-feeding arthropods could compromise the specificity of the tsetse exposure biomarker when using whole saliva extract in an indirect ELISA format. However, several studies reported only a limited cross-reactivity of tsetse salivary antigens and salivary extracts of other blood-feeding flies as tabanids and *Stomoxys* (Caljon et al., 2014a,b; Somda et al., 2013).

A first study on the evaluation of human antibody response specific to tsetse exposure showed that *G. morsitans morsitans* whole saliva extract as well a recombinant Tsal protein allowed to detect anti-tsetse response in Ugandan human serum samples from a *G. fuscipes fuscipes* HAT-endemic region (Caljon et al., 2006). This was the first report that tsetse saliva proteins could be used as biomarker for tsetse fly exposure and that a strong cross-reactivity exists between saliva of different tsetse fly species. Similarly, a strong IgG response against *G. fuscipes fuscipes* whole saliva extract was detected in exposed individuals inhabiting a HAT-endemic region in the Democratic Republic of Congo, whether infected or not with the *T. brucei gambiense* (Poinsignon et al., 2008b). Moreover, the evaluation of the antisaliva IgG response in different villages allowed to discriminate the difference of *Glossina* exposure according to the studied sites suggesting that this biomarker could be a suitable indicator of human tsetse exposure and its spatial variations. Immune reactivity of human plasma collected within active HAT foci in coastal Guinea, historical foci where tsetse flies are still present (South-West Burkina Faso), and a tsetse free area (Bobo-Dioulasso, Burkina Faso) was measured by an indirect ELISA against whole saliva extracts of *G. palpalis gambiensis*. In the active HAT foci and areas where tsetse flies were present in high densities, specific IgG responses were significantly higher than in those areas where tsetse flies were either absent or only present at low densities (Dama et al., 2013a). The above reported studies clearly indicate that the IgG response against tsetse whole saliva extract or a recombinant Tsal protein could provide a suitable epidemiological biomarker of human exposure to tsetse bites. Recently, a synthetic peptide derived from the *G. morsitans morsitans* ADGF-related TSGF-1 protein (peptide derived from the N-terminal region; amino acids 18–43) was evaluated using the above described panel of human West African plasmas and appeared a promising candidate to assess human exposure to tsetse flies as antibody responses toward the synthetic peptide were low in the three control groups and were significantly higher in the human samples from the two tsetse-exposed areas. Moreover, significantly higher anti-Tsgf$_{18-43}$ were detected in sleeping sickness patients compared to healthy controls (Dama et al., 2013b). Recently, this anti-Tsgf$_{18-43}$ was successfully used to assess the evolution of human tsetse contacts during a vector control intervention in Guinea. In agreement with the entomological data, a highly significant decrease of Tsgf1$_{18-43}$ ELISA titers was observed in sentinel villages from the vector control area in 2013. Analysis of the distribution of anti-Tsgf1$_{18-43}$ IgG levels in the vector control area in 2012 and 2013 suggested that a major impact of target deployment was to reduce the number of high responders that are likely to be the most exposed to tsetse bites (Courtin et al., 2015). Indeed, the population of the vector control area (where the tsetse densities were reduced with 80%) demonstrated a significant decrease of anti-Tsgf antibody titers just one year after the initial start of the tsetse control activities. Although an important proportion of inhabitants still exhibited intermediate antibody titers indicating that they were probably still exposed to tsetse flies, the proportion of high responders dropped sharply from 25% to less than 4% suggesting that intense exposure to tsetse bites had been strongly reduced in the population (Courtin et al., 2015).

So far, only a few studies have explored the use of antisaliva IgG response to estimate tsetse exposure in cattle. Serum immune reactivity to whole saliva extracts of *G. palpalis gambiensis*, *G. tachinoides*, and *G. morsitans submorsitans* was evaluated in experimentally exposed cows and in cattle from both tsetse free and tsetse-infested regions in Burkina Faso (Somda et al., 2013). In the tsetse infested area, cattle IgG responses to the whole saliva extracts were significantly higher during the dry season when exposure of the herds to tsetse is more intense. The whole saliva extract of *G. morsitans submorsitans* enabled to detect exposure to all tsetse species with low cross-reactivity to other blood-feeding arthropods (ticks, stable flies, tabanids). Moreover, it allowed to distinguish between groups of cows that were exposed to tsetse fly bites at a different intensity. In a recent study, it was confirmed that the abundant Tsal1 tsetse saliva protein is the best salivary antigen candidate to develop a highly specific biomarker of cattle exposure (Somda et al., 2016). The other member of this abundant saliva protein, Tsal2, was found not suitable as cross-reactions were observed in cattle that were exposed to stable flies. A synthetic peptide derived from the *G. morsitans morsitans* Tsal1 protein (spanning the Tsal1 amino acid sequence aa52-75) was demonstrated in an indirect ELISA format to detect a specific IgG-response after low tsetse exposure in both naturally and experimentally exposed hosts. Surprisingly, high tsetse bite exposure resulted in an absence of anti-Tsal1$_{52-75}$ IgG response in these cows despite the presence of a strong Ab response to whole saliva extract. The authors concluded that this indicates that the immunogenicity of single epitopes of immunogenic saliva proteins can differ according to the exposure conditions (Somda et al., 2016).

Recently, an improved and simplified assay to detect the tsetse fly–induced antibody responses against the Tsal protein family has been developed (Caljon et al., 2015). This was realized by converting the indirect ELISA for Tsal antibody detection in host sera (Caljon et al., 2014b) into a competitive assay format using single epitope recognizing anti-Tsal nanobodies (Nbs). Nbs are single domain antigen-binding fragments of about 15 kDa derived from a nonconventional class of antibodies that can be found in camelids and sharks (Flajnik et al., 2011) and that can be selected through phage-display technology and panning methodologies. They are excellent affinity reagents with a small size, high stability, particular epitope range compared to conventional antibodies, and highly soluble in aqueous solutions. They are increasingly exploited in the development of medical diagnostic and therapeutic applications (Hassanzadeh-Ghassabeh et al., 2013). This anti-Tsal Nb-based competitive ELISA was able to detect exposure to a broad range of tsetse species (*G. morsitans morsitans, G. pallidipes, G. palpalis gambiensis,* and *G. fuscipes*) and did not cross-react with other hematophagous insects like *Stomoxys* or tabanids. Evaluation of the competitive ELISA test using a collection of plasmas of tsetse-exposed pigs revealed an increased sensitivity (>0.95) and accuracy (>0.95) compared to previously described indirect ELISA using the recombinant Tsal protein (Caljon et al., 2015). Moreover, the advantage of this assay is that it does not require adaptation to the sampled host species and can thus be applied to a wide range of hosts. A drawback of this anti-Tsal Nb-based competitive immunoassay is that currently the test only properly works with *G. morsitans morsitans* whole saliva extract as coating antigen. However, given that large *G. morsitans morsitans* colonies exist at several locations and that harvesting of whole saliva extracts is straightforward and with high yield, this is not imposing an important problem. So far, this test was demonstrated to be highly performant with plasmas of experimentally exposed pigs. Further testing is required to validate the assay for monitoring tsetse exposure in humans and cattle. The clear advantage of using the monoclonal Nbs for antisaliva antibody detection is that several Nbs directed to different immunogenic saliva proteins can be combined in one test in order to achieve a robust and sensitive multitarget immunoprofiling assay of the tsetse-exposed host.

In conclusion, it can be stated that immunogenic properties of tsetse saliva result in a strong IgG response in a variety of mammalian hosts that can be used as biomarker for the intensity of tsetse exposure. A highly abundant and immunogenic saliva protein of the Tsal family–proteins with an endonuclease signature and nucleic acid–binding activity (Caljon et al., 2012)—has shown to be the best salivary antigen candidate for the development of a highly specific biomarker to estimate host exposure to the important tsetse fly species of medical and veterinary importance. Different anti-Tsal antibody test formats are now developed and being validated for their use to estimate the tsetse bite exposure intensity especially in the human and bovine host among other relevant hosts. This tsetse saliva–based biomarker will be a valuable tool to measure in an easy way tsetse prevalence and tsetse–host contact in a defined study area allowing to monitor in time the impact of a tsetse control/eradication intervention and to detect reinvasion of previously cleared areas.

References

Alsan, M., 2015. The effect of the TseTse fly on African development. Am. Econ. Rev. 105, 382–410.

Alves-Silva, J., Ribeiro, J.M.C., Van Den Abbeele, J., Attardo, G., Hao, Z., Haines, L.R., Soares, M.B., Berriman, M., Aksoy, S., Lehane, M.J., 2010. An insight into the sialome of *Glossina morsitans morsitans*. BMC Genomics 11, 213.

Auty, H., Morrison, L.J., Torr, S.J., Lord, J., 2016. Transmission dynamics of rhodesian sleeping sickness at the interface of wildlife and livestock areas. Trends Parasitol. 32, 609–621.

Billingsley, P.F., Baird, J., Mitchell, J.A., Drakeley, C., 2006. Immune interactions between mosquitoes and their hosts. Parasite Immunol. 28, 143–153.

Bouyer, J., Bouyer, F., Donadeu, M., Rowan, T., Napier, G., 2013. Community- and farmer-based management of animal African trypanosomosis in cattle. Trends Parasitol. 29, 519–522.

Bouyer, F., Seck, M.T., Dicko, A.H., Sall, B., Lo, M., Vreysen, M.J.B., Chia, E., Bouyer, J., Wane, A., 2014. Ex-ante benefit-cost analysis of the elimination of a *Glossina palpalis gambiensis* population in the Niayes of Senegal. PLoS Negl. Trop. Dis. 8, e3112.

Caljon, G., Van Den Abbeele, J., Sternberg, J.M., Coosemans, M., De Baetselier, P., Magez, S., 2006. Tsetse fly saliva biases the immune response to Th2 and induces anti-vector antibodies that are a useful tool for exposure assessment. Int. J. Parasitol. 36, 1025–1035.

Caljon, G., Broos, K., De Goeyse, I., De Ridder, K., Sternberg, J.M., Coosemans, M., De Baetselier, P., Guisez, Y., Van Den Abbeele, J., 2009. Identification of a functional Antigen5-related allergen in the saliva of a blood feeding insect, the tsetse fly. Insect Biochem. Mol. Biol. 39, 332–341.

Caljon, G., De Ridder, K., De Baetselier, P., Coosemans, M., Van Den Abbeele, J., 2010. Identification of a tsetse fly salivary protein with dual inhibitory action on human platelet aggregation. PLoS One 5, e9671.

Caljon, G., De Ridder, K., Stijlemans, B., Coosemans, M., Magez, S., De Baetselier, P., Van Den Abbeele, J., 2012. Tsetse salivary gland proteins 1 and 2 are high affinity nucleic acid binding proteins with residual nuclease activity. PLoS One 7, e47233.

Caljon, G., De Vooght, L., Van Den Abbeele, J., 2014a. The biology of tsetse-trypanosome interactions. In: Magez, S., Radwanska, M. (Eds.), Trypanosomes and Trypanosomiasis. Springer-Verlag, Wien.

Caljon, G., Duguma, R., De Deken, R., Schauvliege, S., Gasthuys, F., Duchateau, L., Van Den Abbeele, J., 2014b. Serological responses and biomarker evaluation in mice and pigs exposed to tsetse fly bites. PLoS Negl. Trop. Dis. 8, e2911.

Caljon, G., Hussain, S., Vermeiren, L., Van Den Abbeele, J., 2015. Description of a nanobody-based competitive immunoassay to detect tsetse fly exposure. PLoS Negl. Trop. Dis. 9, e0003456.

Cappello, M., Bergum, P.W., Vlasuk, G.P., Furmidge, B.A., Pritchard, D.I., Aksoy, S., 1996. Isolation and characterization of the tsetse thrombin inhibitor: a potent antithrombotic peptide from the saliva of *Glossina morsitans morsitans*. Am. J. Trop. Med. Hyg. 54, 475–480.

Cappello, M., Li, S., Chen, X.O., Li, C.B., Harrison, L., Narashimhan, S., Beard, C.B., Aksoy, S., 1998. Tsetse thrombin inhibitor: bloodmeal-induced expression of an anticoagulant in salivary glands and gut tissue of *Glossina morsitans morsitans*. Proc. Natl. Acad. Sci. U.S.A. 95, 14290–14295.

Courtin, F., Camara, M., Rayaisse, J.-B., Kagbadouno, M., Dama, E., Camara, O., Traoré, I.S., Rouamba, J., Peylhard, M., Somda, M.B., Leno, M., Lehane, M.J., Torr, S.J., Solano, P., Jamonneau, V., Bucheton, B., 2015. Reducing human-tsetse contact significantly enhances the efficacy of sleeping sickness active screening campaigns: a promising result in the context of elimination. PLoS Negl. Trop. Dis. 9, e0003727.

Dama, E., Cornelie, S., Somda, B.M., Camara, M., Kambire, R., Courtin, F., Jamonneau, V., Demettre, E., Seveno, M., Bengaly, Z., Solano, P., Poinsignon, A., Remoue, F., Belem, A.M.G., Bucheton, B., 2013a. Identification of *Glossina palpalis gambiensis* specific salivary antigens: towards the development of a serologic biomarker of human exposure to tsetse flies in West Africa. Microbes Infect. 15, 416–427.

Dama, E., Cornelie, S., Camara, M., Somda, M.B., Poinsignon, A., Ilboudo, H., Ndille, E.E., Jamonneau, V., Solano, P., Remoue, F., Bengaly, Z., Belem, A.M.G., Bucheton, B., 2013b. In Silico identification of a candidate synthetic peptide (Tsgf1$_{18-43}$) to monitor human exposure to tsetse flies in West Africa. PLoS Negl. Trop. Dis. 7, e2455.

Delespaux, V., de Koning, H.P., 2007. Drugs and drug resistance in African trypanosomiasis. Drug Resist. Updat. 10, 30–50.

Desquesnes, M., Dia, M.L., 2003. *Trypanosoma vivax*: mechanical transmission in cattle by one of the most common African tabanids, *Atylotus agrestis*. Exp. Parasitol. 103, 35–43.

Feldmann, U., Dyck, V.A., Mattioli, R.C., Jannin, J., 2005. Potential impact of tsetse fly control involving the sterile insect technique. In: Dyck, V.A., Hendrichs, J., Robinson, A.S. (Eds.), Sterile Insect Technique. Principles and Practice in Area-wide Integrated Pest Management. Springer, Dordrecht.

Flajnik, M.F., Deschacht, N., Muyldermans, S., 2011. A case of convergence: why did a simple alternative to canonical antibodies arise in sharks and camels? PloS Biol. 9, e1001120.

Fontaine, A., Dioul, I., Bakkali, N., Missé, D., Pagès, F., Fusai, T., Rogier, C., Almeras, L., 2011. Implication of haematophagous arthropod salivary proteins in host-vector interactions. Parasit. Vectors 4, 187.

Franco, J.R., Simarro, P.P., Diarra, A., Ruiz-Postigo, J.A., Jannin, J.G., 2014. The journey towards elimination of gambiense human African trypanosomiasis: not far, nor easy. Parasitology 141, 748–760.

Gooding, R.H., Krafsur, E.S., 2005. Tsetse genetics: contributions to biology, systematics, and control of Tsetse Flies. Annu. Rev. Entomol. 50, 101–123.

Gouteux, J.P., Artzrouni, M., Jarry, M., 2001. A density-dependent model with reinvasion for estimating tsetse fly populations (Diptera: Glossinidae) through trapping. Bull. Entomol. Res. 91, 177–184.

Hassanzadeh-Ghassabeh, G., Devoogdt, N., De Pauw, P., Vincke, C., Muyldermans, S., 2013. Nanobodies and their potential applications. Nanomedicine (Lond) 8, 1013–1026.

Holmes, P., 2014. First WHO meeting of stakeholders on elimination of gambiense human African trypanosomiasis. PLoS Negl. Trop. Dis. 8, e3244.

International Glossina Genomics Initiative (IGGI), 2014. Genome sequence of the tsetse fly (*Glossina morsitans*): vector of African trypanosomiasis. Science 344, 380–386.

Jordan, A.M., 1986. Trypanosomiasis Control and African Rural Development. Longman, London.

Kabayo, J.P., 2002. Aiming to eliminate tsetse from Africa. Trends Parasitol. 18, 473–475.

Kgori, P.M., Modoa, S., Torr, S.J., 2006. The use of aerial spraying to eliminate tsetse from the Okavango Delta of Botswana. Acta Trop. 99, 184–199.

Krafsur, E.S., 2009. Tsetse flies: genetics, evolution, and role as vectors. Infection. Genet. Evol. 9, 124–141.

Leak, S.G.A., 1999. Tsetse biology and ecology: their role in the epidemiology and control of trypanosomosis, first ed. CABI, UK.

Lehane, M.J., 2005. The Biology of Blood-Sucking in Insects, second ed. Cambridge University Press, Cambridge.

Lehane, M., Alfaroukh, I., Bucheton, B., Camara, M., Harris, A., Kaba, D., Lumbala, C., Peka, M., Rayaisse, J.-B., Waiswa, C., Solano, P., Torr, S., 2016. Tsetse control and the elimination of Gambian sleeping sickness. PLoS Negl. Trop. Dis. 10, e0004437.

Li, S., Aksoy, S., 2000. A family of genes with growth factor and adenosine deaminase similarity are preferentially expressed in the salivary glands of *Glossina m. morsitans*. Gene 252, 83–93.

Li, S., Kwon, J., Aksoy, S., 2001. Characterization of genes expressed in the salivary glands of the tsetse fly, *Glossina morsitans morsitans*. Insect Mol. Biol. 10, 69–76.

Marzouki, S., Abdeladhim, M., Abdessalem, C.B., Oliveira, F., Ferjani, B., Gilmore, D., Louzir, H., Valenzuela, J.G., Ahmed, M.B., 2012. Salivary antigen SP32 is the immunodominant target of the antibody response to *Phlebotomus papatasi* bites in humans. PloS Negl. Trop. Dis. 11, e1911.

Morrison, L.J., Vezza, L., Rowan, T., Hope, J.C., 2016. Animal African trypanosomiasis: time to increase focus on clinically relevant parasite and host species. Trends Parasitol. 32, 599–607.

Orlandi-Pradines, E., Almeras, L., Denis de Senneville, L., Barbe, S., Remoue, F., Villard, C., Cornelie, S., Penhoat, K., Pascual, A., Bourgouin, C., Fontenille, D., Bonnet, J., Corre-Catelin, N., Reiter, P., Pages, F., Laffite, D., Boulanger, D., Simondon, F., Pradines, B., Fusai, T., Rogier, C., 2007. Antibody response against saliva antigens of *Anopheles gambiae* and *Aedes aegypti* in travellers in tropical Africa. Microbes Infect. 9, 1454–1462.

PAAT Technical and Scientific Series, 2010. Linking Sustainable Human and Animal African Trypanosomosis Control with Rural Development Strategies. PAAT Information Service Publications, WHO, Geneve and FAO, Rome.

Poinsignon, A., Cornelie, S., Remoue, F., Grebaut, P., Courtin, D., Garcia, A., Simondon, F., 2007. Human/

vector relationships during human African trypanosomiasis: initial screening of immunogenic salivary proteins of *Glossina* species. Am. J. Trop. Med. Hyg. 76, 327–333.

Poinsignon, A., Cornelie, S., Mestres-Simon, M., Lanfrancotti, A., Rossignol, M., Boulanger, D., Cisse, B., Sokhna, C., Arcà, B., Simondon, F., Remoue, F., 2008a. Novel peptide marker corresponding to salivary protein gSG6 potentially identifies exposure to *Anopheles* bites. PLoS One 3, e2472.

Poinsignon, A., Remoue, F., Rossignol, M., Cornelie, S., Courtin, D., Grébaut, P., Garcia, A., Simondon, F., 2008b. Human IgG antibody response to *Glossina* saliva: an epidemiologic marker of exposure to *Glossina* bites. Am. J. Trop. Med. Hyg. 78, 750–753.

Rizzo, C., Ronca, R., Fiorentino, G., Verra, F., Mangano, V., Poinsignon, A., Sirima, S.B., Nèbiè, I., Lombardo, F., Remoue, F., Coluzzi, M., Petrarca, V., Modiano, D., Arcà, B., 2011. Humoral response to the *Anopheles gambiae* salivary protein gSG6: a serological indicator of exposure to afrotropical malaria vectors. PLoS One 6, e17980.

Schofield, C.J., Kabayo, J.P., 2008. Trypanosomiasis vector control in Africa and Latin America. Parasit. Vectors 1, 24.

Schwarz, B.S., Ribeiro, J.M., Goldstein, M.D., 1990. Anti-tick antibodies: an epidemiological tool in Lyme disease research. Am. J. Epidemiol. 132, 58–66.

Schwarz, A., Sternberg, J.M., Johnston, V., Medrano-Mercado, N., Anderson, J.M., Humee, J.C.C., Valenzuela, J.G., Schaub, G.A., Billingsley, P.F., 2009. Antibody responses of domestic animals to salivary antigens of *Triatoma infestans* as biomarkers for low-level infestation of triatomines. Int. J. Parasitol. 39, 1021–1029.

Schwarz, A., Medrano-Mercado, N., Billingsley, P.F., Schaub, G.A., Sternberg, J.M., 2010. IgM-antibody responses of chickens to salivary antigens of *Triatoma infestans* as early biomarkers for low-level infestation of triatomines. Int. J. Parasitol. 40, 1295–1302.

Simarro, P.P., Cecchi, G., Franco, J.R., Paone, M., Diarra, A., Ruiz-Postigo, J.A., Fèvre, E.M., Mattioli, R.C., Jannin, J.G., 2012. Estimating and mapping the population at risk of sleeping sickness. PLoS Negl. Trop. Dis. 6, e1859.

Sinshaw, A., Abebe, G., Desquesnes, M., Yoni, W., 2006. Biting flies and *Trypanosoma vivax* infection in three highland districts bordering lake Tana, Ethiopia. Vet. Parasitol. 142, 35–46.

Solano, P., Ravel, S., de Meeus, T., 2010. How can tsetse population genetics contribute to African trypanosomiasis control? Trends Parasitol. 26, 255–263.

Somda, M.B., Bengaly, Z., Dama, E., Poinsignon, A., Dayo, G.K., Sidibe, I., Remoue, F., Sanon, A., Bucheton, B., 2013. First insights into the cattle serological response to tsetse salivary antigens: a promising direct biomarker of exposure to tsetse bites. Vet. Parasitol. 197, 332–340.

Somda, M.B., Cornelie, S., Bengaly, Z., Mathieu-Daudé, F., Poinsignon, A., Dama, E., Bouyer, J., Sidibé, I., Demettre, E., Seveno, M., Remoué, F., Sanon, A., Bucheton, B., 2016. Identification of a Tsal$_{152-75}$ salivary synthetic peptide to monitor cattle exposure to tsetse flies. Parasit. Vectors 9, 149.

Souza, A.P., Andrade, B.B., Aquino, D., Entringer, P., Miranda, J.C., Alcantara, R., Ruiz, D., Soto, M., Teixeira, C.R., Valenzuela, J.G., de Oliveira, C.I., Brodskyn, C.I., Barral-Netto, M., Barral, A., 2010. Using recombinant proteins from *Lutzomyia longipalpis* saliva to estimate human vector exposure in visceral leishmaniasis endemic areas. Plos Negl. Trop. Dis. 4, e649.

Sow, A., Sidibe, I., Bengaly, Z., Bance, A.Z., Sawadogo, G.J., Solano, P., Vreysen, M.J.B., Lancelot, R., Bouyer, J., 2012. Irradiated male tsetse from a 40-year-old colony are still competitive in a Riparian forest in Burkina Faso. PLoS One 7, e37124.

Torr, S.J., Hargrove, J.W., Vale, G.A., 2005. Towards a rational policy for dealing with tsetse. Trends Parasitol. 21, 537–541.

Van Den Abbeele, J., Caljon, G., Dierick, J.F., Moens, L., De Ridder, K., Coosemans, M., 2007. The *Glossina morsitans* tsetse fly saliva: general characteristics and identification of novel salivary proteins. Insect Biochem. Mol. Biol. 37, 1075–1085.

Van Den Abbeele, J., Caljon, G., De Ridder, K., De Baetselier, P., Coosemans, M., 2010. *Trypanosoma brucei* modifies the tsetse salivary composition, altering the fly feeding behavior that favors parasite transmission. PLoS Pathog. 6, e1000926.

Van den Bossche, P., de La Rocque, S., Hendrickx, G., Bouyer, J., 2010. A changing environment and the epidemiology of tsetse-transmitted livestock trypanosomiasis. Trends Parasitol. 26, 236–243.

Vinhas, V., Andrade, B.B., Paes, F., Bomura, A., Clarencio, J., Miranda, J.C., Bafica, A., Barral, A., Barral-Netto, M., 2007. Human anti-saliva immune response following experimental exposure to the visceral leishmaniasis vector, *Lutzomyia longipalpis*. Eur. J. Immunol. 37, 3111–3121.

Vreysen, M.J.B., Saleh, K.M., Ali, M.Y., Abdulla, A.M., Zhu, Z.-R., Juma, K.G., Dyck, V.A., Msangi, A.R., Mkonyi, P.A., Feldmann, H.U., 2000. *Glossina austeni* (Diptera: Glossinidae) eradicated on the island of Unguja, Zanzibar, using the sterile insect technique. J. Econ. Entomol. 93, 123–135.

Vreysen, M.J.B., Seck, M.T., Sall, B., Bouyer, J., 2013. Tsetse flies: their biology and control using area-wide integrated pest management approaches. J. Invertebr. Pathol. 112, S15–S25.

Yaro, M., Munyarda, K.A., Stearb, M.J., Grotha, D.M., 2016. Combatting African Animal Trypanosomiasis (AAT) in livestock: the potential role of trypanotolerance. Vet. Parasitol. 225, 43–52.

Zhao, X., Silva, T.L.A., Cronin, L., Savage, A.F., O'Neill, M., Nerima, B., Okedi, L.M., Aksoy, S., 2015. Immunogenicity and serological cross-reactivity of saliva proteins among different tsetse species. PLoS Negl. Trop. Dis. 9, e0004038.

Epidemiological Applications of Assessing Mosquito Exposure in a Malaria-Endemic Area

Andre Sagna[1,2], Anne Poinsignon[1,2], Franck Remoue[1,2]

[1]Institute of Research for Development (IRD), MIVEGEC Unit, Montpellier, France; [2]Institute Pierre Richet (IPR), Bouake, Ivory Coast

List of Abbreviations

Ab Antibody
gSG6 gambiae Salivary Gland protein six
gSG6-P1 gambiae Salivary Gland protein 6 - Peptide one
HI Holes Index
IRS Indoor Residual Spraying
ITN Insecticide Treated Net
LLIN Long Lasting Insecticide Treated Net
NMPC National Malaria Control Program
OD Optical Density
PI Physical Integrity
POC Point-of-Care
VCS Vector Control Strategies
WHO World Health Organization
WS Whole Saliva

INTRODUCTION

Mosquitoes are the most dangerous "animal" worldwide. They transmit a broad range of viral, protozoan, and metazoan pathogens responsible for devastating human and animal diseases (Gubler, 1998). Transmission occurs when an infected mosquito takes its blood meal the human host. Among mosquito-borne diseases, malaria represents the most widespread and serious infection in terms of heavy burden on health and economic development throughout the world. It causes nearly 500,000 deaths and 214 million clinical cases each year, children under 5 years of age and pregnant women being the most vulnerable (WHO, 2016). Malaria is caused by protozoa of the genus *Plasmodium* and transmitted to humans by infected *Anopheles* mosquitoes.

Entomological, parasitological, and clinical assessments are routinely used to evaluate the exposure of human populations to *Anopheles* vector bites and the risk of malaria transmission/infection. However, these methods are labor intensive and difficult to sustain on large scales, especially when transmission and *Anopheles* exposure levels are low (dry season, high altitude, urban settings, or after vector control) (Beier et al., 1999; Drakeley et al., 2005).

In particular, the entomological inoculation rate, the gold standard measure for mosquito–human transmission intensity of *Plasmodium*, is highly dependent on the density of human-biting *Anopheles* (Hay et al., 2000). This parameter is estimated using trapping methods such as human-landing catches of adult mosquitoes commonly used for sampling host-seeking mosquitoes and then determining human exposure levels. HLC may be limited because of ethical and logistical constraints especially in children because it estimates vector exposure at the population level and not at the individual level (Smith et al., 2005). Transmission estimates based on the prevalence or *Plasmodium* parasite densities of infection in human population (mainly by rapid diagnosis test and/or microscopy) are susceptible to microheterogeneity caused by climatic factors and the socioeconomic determinants impacting host-seeking behavior (Drakeley et al., 2005). Incidence of disease may be the closest indicator of the burden of disease in one focus (Smith et al., 2005). However, it can be subject to variability between sites and may not be appropriate for the evaluation of early phase studies of vector control or reliable for epidemic prediction (Smith et al., 2005). More recently, serological tool [antibody (Ab) responses to several *Plasmodium* antigens] correlates of transmission intensity have been described, yet they represent long-term rather than short-term exposure data (Drakeley et al., 2005). They are not then suitable in evaluating the short-term impact of vector control strategies (VCS).

To increase the impact of VCS, the current entomological methods should be improved by new complementary tools allowing to estimate the real human–*Anopheles* contact and consequently to evaluate accurately the risk of malaria transmission and the efficacy of VCS. The development of such tools needs to explore the close interactions between the human host and *Anopheles* mosquitoes. The physical contact between these two protagonists involves physiological interactions through salivary proteins of *Anopheles* mosquitoes, which are injected in human skin during the bite. Some of these salivary compounds are essential to the *Plasmodium* life cycle (Choumet et al., 2007). They have substantial antihemostatic, antiinflammatory, and immunomodulatory activities that assist the mosquito in the blood-feeding process by inhibiting several defense mechanisms of the human host (Ribeiro and Francischetti, 2003). Furthermore, many of them are immunogenic and elicit strong immune responses, evidenced by the swelling and itching that accompany a mosquito bite (Peng and Simons, 2004). Specific acquired cellular (Donovan et al., 2007; Schneider et al., 2011) or/and humoral (Ab) responses are developed by human individuals when exposed to bites of *Anopheles* mosquitoes (Andrade et al., 2009; Orlandi-Pradines et al., 2007; Remoue et al., 2006; Waitayakul et al., 2006). These immune responses may play several roles in pathogen transmission and disease evolution (Schneider and Higgs, 2008). In addition, recent studies have demonstrated that the intensity of the Ab response specific to whole salivary proteins could be a biomarker of the exposure level of human to *Anopheles* bites (Londono-Renteria et al., 2010; Remoue et al., 2006). Therefore, studying *Anopheles*–human immunological relationships can provide new promising tools for monitoring the real human–*Anopheles* contact and identifying individuals at risk of malaria transmission. It can also allow the development of novel methods for monitoring vector control and mosquito-release programs' effectiveness.

However, whole saliva could be inadequate as a biomarker tool, because it is a cocktail of various molecular components with different nature and biological functions (Poinsignon et al., 2008). Some of these molecules are ubiquitous and may potentially induce Ab cross-reactivities with common salivary epitopes of other hematophagous arthropods (Poinsignon et al., 2008). In addition, a lack of reproducibility between collected whole *Anopheles* saliva batches was observed and difficulties to obtain sufficient quantities needed for large-scale studies were highlighted (Poinsignon et al., 2008). Therefore, specific and antigenic proteins have been identified in the secretome of *Anopheles* mosquitoes (175, 115, 72, and 30 kDa bands; Cornelie et al., 2007) and one specific biomarker of *Anopheles* bites was developed by coupling bioinformatic and immunoepidemiological approaches. This promising candidate, namely, the gSG6-P1 salivary peptide (*Anopheles gambiae* Salivary Gland Protein-6 peptide 1), has been identified from gSG6-specific and antigenic protein and this specific peptide shown to be highly antigenic (Poinsignon et al., 2008). It has been then validated as a pertinent biomarker assessing specifically and reliably the exposure level to *Anopheles* bites (Drame et al., 2012; Poinsignon et al., 2009; Poinsignon et al., 2010) and/or the effectiveness of malaria vector control (Drame et al., 2010a) in all age-classes of human populations (newborns, infants, children, and adults) from several malaria endemic settings (rural, semiurban, and urban areas) throughout sub-Saharan Africa countries (Senegal, Angola, Kenya, Ivory Coast, and Benin) and in the Americas.

This chapter focuses on studies highlighting the approaches, techniques, and methods used to develop and validate specific candidate biomarkers of exposure to *Anopheles* bites. Potential applications of such salivary biomarkers of exposure to *Anopheles* vector bites in the field of operational research by National Malaria Control Programs (NMCP) are highlighted. Finally, the development of the salivary biomarker into rapid diagnostic tests, such as autoreactive dipstick "Point-of-Care" (POC) tests, that could be used at the operational level by NMCP is discussed.

DEVELOPMENT OF BIOMARKER OF HUMAN EXPOSURE TO ANOPHELES VECTOR BITES

The Concept of Immunological Marker of Exposure

Before obtaining a blood meal, the female *Anopheles*, like all hematophagous arthropods, must locate a blood vessel by introducing its mouthparts into the vertebrate host skin: this is the probing. Once a blood vessel is located, the insect can feed. The penetration of *Anopheles* mouthparts into the host skin results in hemostatic, inflammatory, and immune responses by the host (Ribeiro, 1987). The female *Anopheles* mosquito should be able to overcome these sophisticated barriers represented by human defense systems to successfully obtain a blood meal. To do so, they inject several times, into the host skin, some saliva while they are feeding. This saliva contains potent pharmacological and immunogenic components that inhibit inflammatory and hemostatic reactions and modulate the host immune system (Ribeiro, 1987, 2000). However, some salivary proteins are recognized by the host immune system and induce a specific Ab response. The first studies of salivary proteins of mosquitoes on the immune system concerned allergic reactions due to *Aedes* and *Culex* mosquito bites (Peng et al., 2004a, 2007; Peng and Simons, 2004). These mosquitoes express in their saliva a panel of allergens responsible for a production of immunoglobulin (Ig) E antibodies (Brummer-Korvenkontio et al., 1994; Reunala et al., 1994). These proteins can be used as tools for diagnosis and treatment of allergic reactions due to mosquito bites (Peng et al., 2004b, 2007; Remoue et al., 2007). In the *Anopheles* mosquitoes' saliva, no study has yet highlighted the presence and effect of allergens.

The first correlation between antisaliva IgG Ab and hematophagous arthropod exposure in human individuals was obtained using sera from outdoor workers exposed to *Ixodes damini* (*Ixodes scapularis*) tick bites in New Jersey, the United States (Schwartz et al., 1990). The authors reported a significant decrease of anti-tick IgG response levels after a period of no exposure to ticks. From this point, several serological studies demonstrated a relation between various hematophagous arthropod density and Ab response levels against their saliva (Drame et al., 2013a). For example, regarding *Aedes* mosquitoes, several studies in Finnish Lapland (Palosuo et al., 1997) or Canada (Peng et al., 2002) have reported an increasing anti-saliva IgG Ab after the summer peak exposure. Other studies showed that IgM antibodies against the saliva of triatomine bugs can be detected after one day of exposure and waned rapidly (18 days) (Schwarz et al., 2009, 2011).

Taken together, these serological studies suggested that the acquisition of an Ab response against mosquitoes' (and hematophagous arthropods') saliva is exposure dependent (Abdel-Naser et al., 2006; Peng et al., 1996; Reunala et al., 1994; Wilson et al., 2001). Then, a logical positive correlation between the human exposure level to *Anopheles* bites and human antimosquito saliva Ab level can be expected. The hypothesis is that the level of human IgG response against *Anopheles* saliva can represent the level of human exposure to *Anopheles* bites. In other words, an individual lowly exposed to *Anopheles* bites produces a low-level Ab response (<0.5 ΔOD) to *Anopheles* saliva while a highly exposed one produces a high-level Ab response (>1.5 ΔOD to *Anopheles* saliva). In this way, antimosquito saliva Ab response can be a pertinent epidemiological biomarker of human exposure to vector bites. The advantage is that this immunological tool is quantitative [values are expressed in optical density (OD)] and individual (one OD value represents the level of exposure of one individual).

Validation of the Concept

The hypothesis that the level of Ab response to *Anopheles* saliva could represent a biomarker showing the level of exposure to *Anopheles* bites are experienced in the field. *A. gambiae s.s.* whole saliva (WS) was collected and run on ELISA in sera samples from young children living in a seasonal malaria transmission region of Senegal (Remoue et al., 2006). Children were categorized into three groups according to their level of exposure to *Anopheles* bites assessed by CDC light trap method: low, medium, and high. Serological analyses indicated that the median levels of IgG responses to *A. gambiae s.s.* WS increased significantly according to the level of exposure of children to *A. gambiae s.s.* bites evaluated by a classical entomological method (Fig. 12.1). These findings confirmed that antisaliva IgG response could be a biomarker of exposure to *Anopheles* bites.

From this point, several studies were undertaken to evaluate the relevance of using salivary biomarker of exposure to *Anopheles* in malaria endemic areas. Orlandi-Pradine and colleagues reported that several French travelers developed specific Ab responses against *A. gambiae* and *Aedes aegypti* saliva after a 5-month stay in tropical Africa that strongly decreased several weeks after the end of their trip (Orlandi-Pradines et al., 2007). A rapid decrease of anti-*A. gambiae* saliva IgG levels was also reported 6 weeks (from initial 0.45 median ΔOD value to around 0) after insecticide-treated nets (ITNs) well-use in a semiurban population in Angola, before a new significant increase 2 months later following the stop of ITN use (Drame et al., 2010b). Another study reported considerable decreases in entomological (82.4%), parasitological (54.8%), and immunological criteria analyzed. In particular, the immunological data based on the level of antisaliva IgG Ab in children of all villages significantly dropped from 2008 to 2009, especially in villages using LLIN + ZF (long-lasting insecticide-treated nets + ZeroFly) and with IRS (insecticide residual spray) (Brosseau et al., 2012). The presence of anti-*Anopheles albimanus* WS Ab with exposure to mosquito bites has been recently

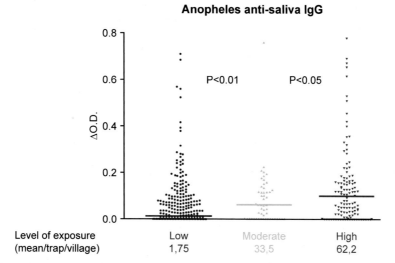

FIGURE 12.1   Antisaliva IgG responses according to the level of exposure to *Anopheles* **bites** (Remoue et al., 2006). The antisaliva IgG was measured in three villages with different level of exposure, low, medium, and high, to *A. gambiae* defined by entomological techniques. The level of antisaliva Ab increase according to the level of exposure to *Anopheles* bites. *Bars* indicate median value for each group. Statistical significance between two groups is indicated by a nonparametric Mann–Whitney *U*-test.

described in Haiti (Londono-Renteria et al., 2010).

Data on human exposure to anopheline saliva and its interaction with malaria were also provided. Remoue and colleagues showed, in Senegal, West Africa, that children who developed a malaria attack in December (end of the transmission season) had higher levels of anti-*A. gambiae* WS IgG in September (transmission season) of the same year, i.e., 3 months before they develop the disease (Remoue et al., 2006). Elsewhere in South-eastern Asia, it has been described that anti-*Anopheles dirus* salivary protein Ab occurred predominantly in patients with acute *Plasmodium falciparum* or *Plasmodium vivax* malaria; people from nonendemic areas do not carry such Abs (Waitayakul et al., 2006). In the Americas, the presence of anti-*Anopheles* saliva Ab has been also described. In adult volunteers from Brazil, an increase of anti-*Anopheles darlingi* WS Ab levels was associated with the presence of *P. vivax* infections (Andrade et al., 2009).

Taken together, these studies indicated that the estimation of human IgG Ab responses specific to *Anopheles* WS could provide a reliable biomarker for evaluating the *Anopheles* exposure level, the risk of malaria transmission, the disease outcomes, and the effectiveness of vector control strategies. However, the pertinence and the practical large-scale application of serological tests for epidemiological purposes have been hampered by several limitations. First, WS is a cocktail of various molecular components with different nature and biological functions. Some components are *Anopheles*-specific and others widely distributed within genus, families, orders, or classes of blood-sucking *Diptera* or arthropods (Ribeiro and Francischetti, 2003). Therefore, the evaluation of *Anopheles* exposure or vector control effectiveness based on the immunogenicity of WS could be skewed and over or underestimated by possible cross-reactivities between common epitopes between mosquito species or other organisms (Poinsignon et al., 2008). Second, the collection of saliva or salivary gland extracts is tedious and time consuming; therefore, it will be difficult or impossible to have an adequate production of mosquito saliva needed for large-scale epidemiological studies (Poinsignon et al., 2008). Third, saliva composition can be affected by several ecological parameters such as age, feeding status, or infectivity of *Anopheles* (Ribeiro and Francischetti, 2003), which in turn may influence the antisaliva immune response measured and may cause a lack of reproducibility between saliva batches. In addition, the standardization of immunological assay using whole saliva appeared difficult and time consuming. An alternative for optimizing the specificity of this immunological test would thus be to identify *Anopheles* genus-specific proteins/peptides (Lombardo et al., 2006).

From the Saliva of *Anopheles* Vectors to Synthetic Salivary Peptide

The increasing power of large-scale genomic, transcriptomic, and proteomic analyses allowed the accumulation of a considerable amount of information on the salivary secretions of blood-sucking arthropods (Ribeiro, 2003). As far as mosquitoes are concerned, the analysis of salivary transcriptomes of a number of *Anopheles* have allowed the discovery of a variety of genes that matched the sequence of various protein families, providing some clues on the evolution of blood feeding (Arca et al., 1999a; Arca et al., 1999b; 2005; Calvo et al., 2004; Calvo et al., 2007; Choumet et al., 2007; Das et al., 2010; Jariyapan et al., 2010, 2007, 2012; Lombardo et al., 2009; Valenzuela et al., 2003). Many of the salivary protein sequences are coded by genes related to intrinsic functions of the cell (housekeeping genes). However, large numbers of salivary proteins are secreted during plant or blood feeding (Ribeiro, 2003). Finally, some protein sequences present no similarities to known sequences deposited in databases (Arca et al., 1999a; 2005;

Calvo et al., 2009). This emphasizes how much still needs to be learned concerning the biological functions of salivary proteins in blood feeding, pathogen transmission, and manipulation of host responses.

The analysis of the adult *Anopheles* sialome has shown that secreted proteins and/or peptides (secretome) can be ubiquitous or specific to arthropod classes, orders, families, genus, or species (Calvo et al., 2006, 2009; Francischetti et al., 2002). In *A. gambiae* salivary gland females over 70 putative secreted salivary proteins have been identified (Arca et al., 2005). Among these specific proteins, only one protein appeared to be a pertinent candidate as biomarker, as demonstrated by several studies described below in this chapter: the gSG6 protein.

The gSG6 protein is a small protein first described in *A. gambiae* (Lanfrancotti et al., 2002) with a unique sequence that codes for a mature peptide/protein of ~10 kDa (116 amino acids) with 10 cysteine residues indicating the presence of potentially five disulfide bonds. A homologue was later found in the sialotranscriptome of *Anopheles stephensi* (Valenzuela et al., 2003) and *Anopheles funestus* (Calvo et al., 2007). *A. funestus SG6/fSG6 (f for funestus)* has 81% and 76% identities with *A. stephensi* and *A. gambiae* polypeptides, respectively. It is not found in the transcriptomes of the Culicinae subfamily members analyzed so far, i.e., *Culex pipiens quinquefasciatus*, *A. Aegypti*, and *Aedes albopictus* (Arca et al., 2007; Ribeiro et al., 2004, 2007). In *A. gambiae*, the transcript coding for gSG6 (g for *gambiae*) was found to be expressed 16 times greater in SGs of adult females than in males (Arca et al., 2005). The gSG6 protein appeared to play a role in blood feeding and was recruited in the anopheline subfamily most probably after the separation of the lineage, which gave origin to *Cellia* and *Anopheles* subgenera (Lombardo et al., 2009). The gSG6 protein, because immunogenic, can be therefore a reliable indicator of human exposure specific to *Anopheles* mosquito bites (Lombardo et al., 2009), vectors of malaria.

Identification of a Specific Salivary Biomarker of Exposure to Anopheles bites

The gSG6 salivary protein is reported to be immunogenic in travelers exposed for 5 months short periods to *Anopheles* bites (Orlandi-Pradines et al., 2007). In African children living in a malaria-endemic area, an immunoproteomic approach, suggested also the gSG6 antigenicity (Cornelie et al., 2007), which has been confirmed by ELISA using recombinant gSG6 protein (Poinsignon et al., 2008). Recently, its antigenicity has been confirmed in individuals from a malaria hyperendemic area of Burkina Faso (Rizzo et al., 2011a, 2011b). Indeed this study showed that individuals exposed to *Anopheles* bites developed specific Ab responses against a recombinant form of gSG6 protein expressed as purified N-terminal His-tagged recombinant protein in the *Escherichia coli* vector pET28b(+) (Lombardo et al., 2009; Rizzo et al., 2011a). In particular, increased anti-gSG6 IgG levels were observed in exposed individuals during the malaria transmission/rainy season (Rizzo et al., 2011a). In addition, anti-gSG6 IgG response appeared to be a reliable serological indicator of exposure to bites of the main African malaria vectors (*A. gambiae, Anopheles arabiensis*, and *A. funestus*) in the same area (Rizzo et al., 2011a). However, gSG6 recombinant protein has been described to result in a high background in control sera from individuals not exposed to *Anopheles* bites and considerable variations in specific Ab response among children supposed to be similarly exposed to *Anopheles* bites (Poinsignon et al., 2008). Therefore, with the objective of optimizing *Anopheles* specificity and reproducibility of the immunological assay, a peptide design approach was undertaken using bioinformatic tools (Poinsignon et al., 2008).

Peptide Design

Several algorithms were employed for prediction of potential immunogenic sites of the gSG6 protein by using bioinformatics (Poinsignon et al., 2008). The prediction of immunogenicity was based on the determination of physicochemical properties of the amino acid (AA) sequences with BcePred and FIMM databases and on the identification of MHC class 2 binding regions using the ProPred-2 online service. This led to define five gSG6 peptides (gSG6-P1 to gSG6-P5) of 20–27 AA residues in length, overlapping by at least three residues and spanning the entire sequence of the mature gSG6 protein. Both predictive methods for putative linear B-cell epitopes (FIMM and BcePred) assigned the highest immunogenicity to gSG6-P1, gSG6-P2, gSG6-P3, and then gSG6-P4 (Poinsignon et al., 2008).

Similarities were also searched for using the Blast family programs, including both the genome/EST libraries of other vector arthropods available in Vectorbase and of pathogens/organisms in nonredundant GenBank CDS databases (see Poinsignon et al., 2008). No relevant identity was found with proteins of other blood-sucking arthropods. Indeed, the longest perfect match was six AAs between a putative protein from *Pediculus humanus humanus* (?) and gSG6-P2 and gSG6-P3 peptides. In the case of gSG6-P1, the best match was four AAs in length with *C. pipiens quinquefasciatus* salivary adenosine deaminase. Moreover, no relevant similarity was found with sequences from pathogens or other organisms. The highest hits of gSG6-P1 were with the cyanobacterium *Microcystis aeruginosa* (3 AAs) and with *Ostreococcus* OsV5 virus (4 AAs). Altogether, this analysis confirmed the bona fide high specificity of the five selected gSG6 peptides for the *Anopheles* species. Peptides were then synthesized and purified (>90%) by specialized company.

Antigenicity of gSG6 Peptides

IgG Ab responses to the five gSG6 peptides were evaluated by enzyme-linked immunosorbent assay (ELISA) in a randomly selected subsample of children (n < 30) living in a rural area of Senegal (Poinsignon et al., 2008). All peptides were immunogenic, but the intensity of the

IgG level was clearly peptide dependent; weak immunogenicity was observed for gSG6-P3, gSG6-P4, and gSG6-P5, whereas gSG6-P1 and gSG6-P2 appeared highly immunogenic.

Validation as a Biomarker of Human Exposure

The specific IgG level to the two most antigenic gSG6 peptides (gSG6-P1 et gSG6-P2) was then evaluated according to the level of exposure (estimated by entomological data) in a larger sample (n = 241) of children living in a malaria seasonal area (Poinsignon et al., 2008). A positive trend was found for both peptides, but only significant for gSG6-P1. Altogether, these results indicated that only the IgG response to gSG6-P1 is suitable to be a pertinent biomarker of exposure to *Anopheles* bites and thus to risk of malaria.

Therefore, the gSG6-P1 was selected as the most pertinent candidate as marker of exposure. Indeed, this peptide appeared to satisfy several requirements that an exposure biomarker should fulfill. First, it thus far appears to be specific to *Anopheles* genus and therefore, no relevant cross-reactivity phenomena with epitopes from other proteins of arthropods or pathogens would be expected. Second, because it is of a synthetic nature, it guarantees high reproducibility of the immunological assay. Third, it elicits a specific Ab response, which correlates well with the level of exposure to *A. gambiae* bites. Now the question is what could be the applications of such a tool in the fight against malaria in endemic areas?

APPLICATIONS OF BIOMARKER OF HUMAN EXPOSURE IN EPIDEMIOLOGICAL CONTEXTS

Estimating mosquito-human contact rate in endemic areas for vector-transmitted diseases like malaria is an important tool to measure the risk of transmission of such diseases (Drame et al., 2013a). Since saliva from arthropods is injected at the time of transmission, efforts have been directed into discovering the salivary proteins/peptides able to elicit an immune response that correspond to the level of bites a person has received in a determined period of time. Recently, the gSG6-P1 has been validated as a biomarker of exposure to *A. gambiae s.l.* and to *A. funestus* (main malaria vectors in Africa) as previously described (Poinsignon et al., 2010). It could help identify individuals exposed to *Anopheles* bites and by consequences to identify individuals at risk of malaria transmission. In addition, such a marker of exposure may be useful to evaluate the effectiveness of vector control strategies deployed by NMCP.

Monitoring the Human Exposure to *Anopheles* Bites and Malaria Risk

Low-Level Exposure/Transmission Areas

There is an obvious need of new indicators and methods to evaluate, at individual and population levels, the exposure level to *Anopheles* bites especially in low exposure/transmission areas where current entomological and parasitological measurements present several limits of sensitivity. The sensitivity of the gSG6-P1 peptide biomarker was tested in such a setting in Senegal. Specific IgG responses to the gSG6-P1 peptide were detected in 36% of children living in villages where very few *A. gambiae* or none was collected by classical entomological methods (Poinsignon et al., 2009). This highlighted the high sensitivity and specificity of the gSG6-P1 epitope(s) after a low immunological boost induced by weak bites exposure (<3 *A. gambiae* bites/human/night). Recently, another study in an eligible area for malaria preelimination (North Senegal) reported the sensitivity of the biomarker to discriminate human exposure to *Anopheles* bites (Sagna et al., 2013b). In this area, a microgeographical heterogeneity of children exposure to *Anopheles* bites was described

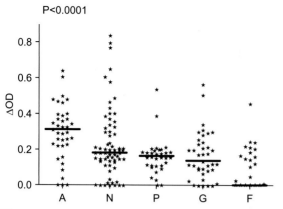

FIGURE 12.2 IgG Ab levels to gSG6-P1 according to the village (Sagna et al., 2013b) Individual IgG response levels are presented at the peak of malaria transmission (October 2008) in studied villages: Agniam (A), Niandane (N), Pendao (P), Guede (G), and Fanaye (F). *Bold lines* represent ΔOD median values. *P* value of the Kruskal–Wallis *U* test is indicated.

despite the general low level of exposure to *A. gambiae* bites (Fig. 12.2). These results point to the potential use of such an immunological tool as an epidemiological biomarker of *A. gambiae* bites for monitoring the risk of malaria transmission in very low exposure areas or the risk of reemergence after malaria interruption.

Other application of such salivary biomarker could be the evaluation of human exposure in urban areas, which represent sites with potentially low exposure to *Anopheles*. A study in Dakar city (Senegal) revealed considerable individual variation in anti-gSG6-P1 IgG levels between and within districts, in spite of a global low *Anopheles* exposure level and malaria transmission (Drame et al., 2012). Despite this individual heterogeneity, the median level of specific IgG and the percentage of immune responders differed significantly between districts. In addition, a positive association was observed between the exposure levels to *A. gambiae* bites, estimated by classical entomological methods, and the median IgG levels or the percentage of immune responders reflecting the real contact between human populations and *Anopheles*

mosquitoes (Drame et al., 2012). Differences in human exposure levels to *A. gambiae* bites could then partly explain district and/or group-variations in anti-gSG6-P1 IgG Ab response as previously described in a low-exposure rural area of Senegal (Poinsignon et al., 2009). Interestingly, in urban Dakar area, immunological parameters seemed to better discriminate the *Anopheles* exposure level between different groups compared to referent entomological data. Moreover, in this study, some discrepancies were observed in the correlation between immunological parameters and the exposure level to *A. gambiae* bites assessed by entomological data in districts. Even if the individual immune response capability could not be excluded between children and adults, this suggests the main role of the human behavior influencing the contact with vectors.

So far, all mentioned studies were conducted on subjects older than 1 year. However, to be more relevant in epidemiological surveys and studies on malaria, such a biomarker tool must pertinently be applicable to all human age classes, including new-borns and infants (<1 year old) who can also be bitten by *Anopheles* and at high risk of malaria transmission (Larru et al., 2009). In this way, a recent study has indicated that human Ab responses to gSG6-P1 biomarker help to assess *Anopheles* exposure level and the risk of malaria in younger than 1-year-old infants living in moderate to high transmission area of Benin (Drame et al., 2015). Indeed, the presence of anti-gSG6-P1 IgG and IgM levels in the blood of respectively 93.28% and 41.79% of 3-month-old infants (the majority of infants). Then, the specific Ab level gradually increased until 12 months of age (Fig. 12.3), whatever the *Anopheles* exposure level or the season.

These observations are consistent with the development and maturation patterns of the new-born immune system during the first months of life (Drame et al., 2015; Rogier, 2003). Indeed, the immature human immune system completes its maturation during infancy following exposition to antigens. Therefore,

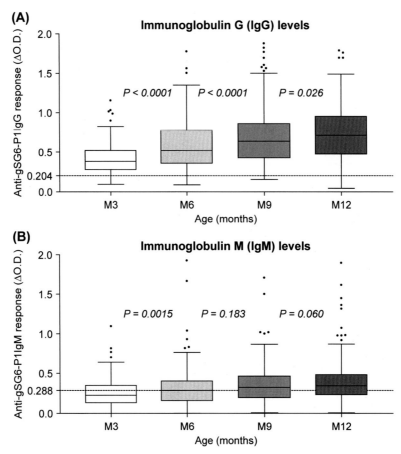

FIGURE 12.3 IgG and IgM responses to *Anopheles* **gSG6-P1 salivary peptide in the first year-life** (Drame et al., 2015). Individual IgG (A) and IgM (B) responses to the *Anopheles* gSG6-P1 are represented for infants in months 3 (white), 6 (light-gray), 9 (dark-gray), and 12 (black box) after their birth. *Horizontal lines* in the boxes indicate medians of the individual data. *Horizontal black dotted lines* represent the cut-off of IgG (0.204) and IgM (0.288) responder. Statistical significant differences between all age groups (multivariate linear mixed model analysis) are indicated.

new-borns are naive and increasingly susceptible to infectious agents; their immune system is not or insufficiently stimulated by antigens. In endemic malaria transmission areas, they are progressively exposed to salivary antigens of *Anopheles* (Rogier, 2003), probably explaining the progressive increase of anti-gSG6-P1 IgG and IgM from 3 to 12 months old. In conclusion of this study, it has been indicated that the exposure level to *Anopheles* bites, and then the risk of malaria infection, can be evaluated in infants by

assessing anti-gSG6-P1 IgM and IgG responses before and after 6 months of age, respectively. Individual or population factors and behaviors enhancing the level of the human–*Anopheles* contact with age can play a crucial role on accelerating this gradual acquisition (Carnevale et al., 1978; Geissbuhler et al., 2007).

In travelers, one study evaluated exposure of soldiers to American *Anopheles* vector bites using the gSG6-P1 peptide, previously validated as biomarker of exposure to African vectors bites (mainly *A. gambiae s.l.* and *A. funestus*). Results showed a significant increase of individual levels of anti-gSG6-P1 IgG after traveling to malaria areas (Haiti and the Dominican Republic) in comparison to those before entering those sites (Londono-Renteria et al., 2015). These results suggested that IgG responses to gSG6-P1 salivary peptide were closely associated with the degree of exposure to the bites of *A. albimanus*, *A. darling*, or *Anopheles punctimacula*, the malaria vectors described in these endemic areas in Latin America. This peptide biomarker can then be used to evaluate the level of exposure to bites of New World (Americas) species of *Anopheles* malaria vectors.

Risk of Malaria Infection

Previous studies provided data on human exposure to anopheline saliva and its relationship with malaria transmission in Africa (Remoue et al., 2006), in South-eastern Asia (Waitayakul et al., 2006), and in the Americas (Andrade et al., 2009; Londono-Renteria et al., 2010). *Anopheles* antisaliva IgG appeared to be a predictive tool or an indicator of clinical or asymptomatic malaria cases. Using the gSG6-P1 peptide of *Anopheles*, one study aimed to evaluate the risk of malaria infection in children living in a rural area of Senegal during the dry season, the season of the lowest exposure to malaria (Sagna et al., 2013a).

The study showed that children with *P. falciparum* infection in the dry season presented higher anti-gSG6-P1 IgG levels than uninfected children ($P < .01$). In addition, asymptomatic children showed higher specific IgG levels compared to uninfected ones (Fig. 12.4). IgG response to gSG6-P1 seems be a pertinent and reliable indicator for suspected *P. falciparum* infection in individuals living in a low transmission context. Another study in American soldiers demonstrated that the median level of specific IgG antibodies against gSG6-P1 was significantly higher in the malaria-infected group living in a low malaria-endemic area compared to the

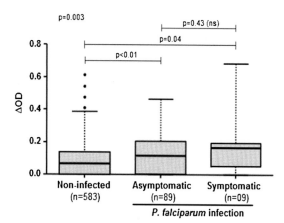

FIGURE 12.4 **IgG response levels to gSG6-P1 peptide according to malaria status** (Sagna et al., 2013a). Box plots show anti-gSG6-P1 IgG level (ΔOD) between uninfected, infected asymptomatic, and infected symptomatic individuals (cumulative data from January and June 2009). *Boxes* display the median ΔOD value, 25th and 75th percentiles. The whiskers show the 5th/95th percentiles and the dots indicate the outliers. Differences between two or three groups were tested using Mann–Whitney test and Kruskal–Wallis test, respectively.

uninfected group (Londono-Renteria et al., 2015). Interestingly, participants with "high" levels of IgG anti-gSG6-P1 antibodies had a 20 times higher risk to be infected by *P. falciparum* than individuals with lower Ab levels.

These data suggest that more mosquito bites elicit higher anti-gSG6-P1 antibodies and that higher Ab levels are also associated with a higher probability of being exposed to a bite from a *Plasmodium*-infected mosquito. Taken together, these different studies indicate that the level of gSG6-P1-specific IgG could be used as a biomarker for evaluating the risk of malaria infection or to suspect an infection in people with a weak exposure to *Anopheles*.

Regarding these data, one direct application of the salivary biomarker of exposure could be the elaboration of maps representing the risk of exposure to *Anopheles* bites. Such immunoepidemiological marker might represent a quantitative tool applied to field conditions and a complementary tool to those currently available, such as entomological, ecological, and environmental data (Billingsley et al., 2006; Kalluri et al., 2007). It could represent a geographic indicator of the risks of malaria transmission and thus a useful tool for predicting malaria morbidity risk as previously described (Remoue et al., 2006). Furthermore, it may represent a powerful tool for evaluation of vector control strategies (impregnated bed-net, intradomiciliary aspersion, etc.) and could here constitute a direct criterion for effectiveness and appropriate use (malaria control program) (Drame et al., 2010b).

Evaluation of Vector Control Strategies Effectiveness

Long and Short-Term Evaluation of Insecticide-Treated Net's Efficacy

A longitudinal study associating parasitological, entomological, and immunological assessments of the efficacy of ITN-based strategies using the gSG6-P1 biomarker has been conducted in a malaria-endemic area in Angola. Human IgG responses to gSG6-P1 peptide were evaluated in 105 individuals (adults and children) before and after the introduction of ITNs and compared to entomo-parasitological data (Drame et al., 2010a). A significant decrease of anti-gSG6-P1 IgG response was observed just after the effective use of ITNs (Fig. 12.5). The drop in specific IgG levels was associated with a considerable decrease of *P. falciparum* parasitaemia, the current WHO criterion for measuring the vector control efficacy (Smith et al., 2004). It was particularly marked in April–August 2006, corresponding to the season peak of *A. gambiae* exposure. Interestingly, the entomological data indicated that this season-dependent peak was of similar intensity before (2005) and after (2006) ITN use, suggesting ITN implementation had no impact on *A. gambiae* density, probably because of the low percentage of the overall human population covered in the studied area (Corran et al., 2007). This study indicated also

that the drop of anti-gSG6-P1 IgG response was associated with correct ITN use and not due to low *Anopheles* density. In addition, this was observed in all age groups studied (<7 years, 7–14 years, and >14 years), suggesting that this biomarker is relevant for ITN evaluation in all age groups. This rapid decrease after correct ITN usage appears to be a special property of anti-gSG6-P1 IgG, which is short-lived in the absence of ongoing antigenic stimulation, at/ for all age classes.

This specific Ab response does not seem to build up but wanes rapidly (around 6 weeks), when exposure failed (Drame et al., 2010a). This property represents a major strength when using such salivary biomarker of exposure for evaluating the efficacy of vector control. In addition, using a response threshold ($\Delta OD = 0.204$) combined with ΔOD_{ITNs}—the difference between April (after ITNs) and January 2006 (before)—makes possible the use of this operational biomarker at individual level. The threshold response (TR) represents the nonspecific background IgG response (the cut-off of immune response) and was calculated in non–*Anopheles*-exposed individuals (n = 14- neg; North of France) by using this formula: $TR = mean \quad (\Delta DO_{neg}) + 3SD = 0.204$. An exposed individual was then classified as an immune responder if its $\Delta OD > 0.204$. If the ΔOD_{ITNs} value is comprised between −0.204 and +0.204, no clear difference in exposure level to *Anopheles* bites can be defined (Drame et al., 2010a).

In contrast, if the individual ΔOD_{ITNs} value < −0.204, it could be concluded with a high level of confidence that this individual is benefiting from ITN installation. The ΔOD_{ITNs} parameter could therefore provide a measure of ITN efficacy at the individual level. An individual biomarker would also be relevant at the large-scale operational studies or surveillance in the field, i.e., in NMCP. In addition, the high sensitivity and specificity of the gSG6-P1 Ab response make it ideal for the evaluation of low-level exposure to *Anopheles*

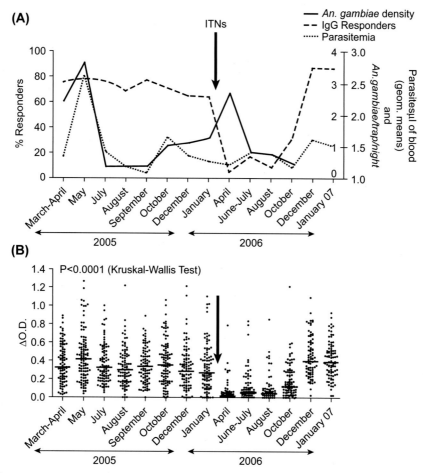

FIGURE 12.5 The salivary biomarker for evaluating the efficacy of *Anopheles* **control operation** (Drame et al., 2010b). The percentage (%) of anti-gSG6-P1 IgG immune responders (*thick-dotted line*) in the "immunological" subpopulation (n = 105), before (2005) and after (2006 and January 2007) the installation of ITNs (A). These results are presented together with the intensity of *P. falciparum* infection (mean parasitemia, *fine-dotted line*) measured in the same population and the mean of number of *Anopheles gambiae* (*solid line*) in the studied area (A). Entomological data were not available in December 2006 and January 2007 (the last 2 months of the study). *Arrows* indicate the installation of insecticide treated nets (ITNs) in February 2006. Individual anti-gSG6-P1 IgG levels (ΔOD) are presented before (2005) and after (2006) the installation of ITNs (B). *Bars* indicate the median value for each studied month. Statistically significant differences between months are indicated.

bites (Drame et al., 2012; Poinsignon et al., 2009), even when exposure or transmission is curtailed by NMCP efforts. Taken together, the estimation of human IgG responses to *Anopheles* gSG6-P1 could provide a reliable indicator for evaluating the efficacy of ITN-based strategies against malaria vectors, at individual and population levels, even after vector control, which generally induces particular low exposure/transmission contexts. This salivary biomarker is a relevant tool for the evaluation of short-term efficacy as well as longer-term monitoring of malaria VCSs.

Evaluation of Effectiveness of Diverse Vector Control Measures

A cross-sectional study conducted from October to December 2008 on 2774 residents (children and adults) of 45 districts of urban Dakar (Senegal) has validated IgG responses to gSG6-P1 as an epidemiological indicator evaluating the effectiveness of a range of VCSs (Drame et al., 2013b). Indeed, in this area, IgG levels to gSG6-P1 as well as the use of diverse malaria VCSs (ITNs, mosquito coils, spray bombs, ventilation, and/or incense) highly varied between districts (Drame et al., 2013b). This difference of use suggests some socioeconomic and cultural discrepancies between householders as described in large cities of the Ivory Coast (Doannio et al., 2006) and Tanzania (Geissbuhler et al., 2007). At the district level, specific IgG levels significantly decreased with VCS use in children as well as in adults. Among used VCS, ITNs, the first chosen preventive method (43.35% rate of use), by reducing drastically the human-Anopheles contact level in children as well as in adults, were by far the most efficient whatever age, period of sampling, or exposure level to mosquito bites. Spray bombs were secondarily associated to a decrease of specific IgG level, due certainly to their power and fast knock-down action on mosquitoes. But, their effects can be limited by the nonpersistence of used products and some socioeconomic considerations (Pages et al., 2007). In addition, they only have been recently adopted and are more expensive in the majority of sub-Saharan Africa cities (Pages et al., 2007), explaining their less frequent use (9.57% rate of use) in the Dakar area (Drame et al., 2013b). The noneffect of mosquito coil use is surprising, regardless to their well-adoption by residents (36.68% of rate of use), but it can be explained by their power deterrent effect, which tends to push Anopheles vectors outside where they can remain active (Pages et al., 2007). However, the protection ensured by ITN use seemed to be insufficient because anti-gSG6-P1 IgG levels in ITN users were specifically high in some periods of fairly high exposure to Anopheles bites. Changes in An. arabiensis behavior, the major malaria vector in the area, can also explain this lack of protection. This Anopheles species can bite outside the rooms/habitations with a maximal activity during the night (high mean around 10.00 p.m.), when people are not in bed and ITNs not hanged (Geissbuhler et al., 2007). Therefore, ITNs must be associated to a complementary VCS for an effective protection against Anopheles bites.

Taken together, these results suggest that the assessment of human IgG responses to Anopheles gSG6-P1 salivary peptide can provide a reliable evaluation of the effectiveness of malaria vector control in urban settings of Dakar whatever the age, sex, level of exposure to bites, or period of malaria transmission. Therefore, this salivary biomarker can be used to compare the effectiveness of different antimalaria vector strategies in order to identify the most suitable for application in a given area.

Comparing Effectiveness of Combined or Not Vector Control Strategies

In parallel to an entomological and parasitological evaluation, IgG responses to gSG6-P1 were also used to assess, in a randomized controlled trial in 28 villages in southern Benin, four malaria vector control interventions: LLIN targeted coverage to pregnant women and children younger than 6 years (TLLIN, reference group), LLIN universal coverage of all sleeping units (ULLIN), TLLIN plus full coverage of carbamate-IRS applied every 8 months (TLLIN + IRS), and ULLIN plus full coverage of carbamate-treated plastic sheeting (CTPS) lined up to the upper part of the household walls (ULLIN + CTPS) (Corbel et al., 2012). Results from this study have shown that specific IgG levels were similar in the four groups before intervention and only significantly lower in the ULLIN group compared to the others after intervention. In contrast to immunological data, clinical incidence density of malaria, the prevalence and parasite density of asymptomatic infections, and the density and aggressiveness of *Anopheles* mosquitoes, were not significantly different between the four groups before as well as after interventions (Corbel et al., 2012). These findings mean that LLIN used along by all the population of a given area may be more suitable in reducing the contact between human populations and the *Anopheles* vectors, even if any effect on malaria morbidity, infection, and transmission was not observed. Therefore, combining antivector tools do not undeniably reduce individual exposure to malaria vectors, even if significant effects on more rapidly reducing malaria transmission and burden have been reported (Okumu and Moore, 2011). These findings confirm that antivector saliva Ab response as a biomarker of exposure is also important for NMCPs and should help the design of more cost-effective strategies for malaria control and elimination.

Evaluation of Insecticide-Treated Nets' Physical Integrity

The biomarker of human exposure to *Anopheles* bites has also been used in operational conditions for evaluating LLINs physical integrity (PI) in the field (Noukpo et al., 2016). To do this, sera from 262 randomly selected children under 5 years living in Health Districts of Benin were analyzed and their LLINs removed for evaluating, in the laboratory, the PI, by using the new Holes Index (HI) recommended by WHO (WHO, 2011). The PI of the collected LLINs was then assessed by counting the number of holes (including tears and split seams), their location on the bed net, and their size (by measuring diameter in cm). Anti-gSG6-P1 IgG responses were assessed and compared to the PI of LLINs that these same children slept under and evaluated by the proportionate HI. Results showed a positive correlation between specific IgG levels and the HI. In other words, an increase of HI corresponding to a potential deterioration of LLINs efficacy (presence of holes) results in an increase of IgG level (exposure to anopheline bites). These data suggest that children sleeping under deteriorated LLINs were more exposed to *Anopheles* vector bites, thus are at high risk of malaria. These results demonstrate that the biomarker of exposure to *Anopheles* bites could be a pertinent indicator for monitoring the risk of malaria after LLINs mass-distribution campaigns in children under 5 years.

Importantly, the same authors used anti-gSG6-P1 IgG responses to help define a standardized threshold above which LLINs become ineffective to protect children sleeping under from *Anopheles* bites (Noukpo et al., 2016). Indeed, as true as there is a standardized HI developed so that the physical integrity of nets could be categorized (WHO, 2011), there is no standardized threshold at which an LLIN is considered as effective (offer substantial protection against mosquito bites) or ineffective (offer little or no protection against mosquito bites). Analyses of immunological data showed that children sleeping under LLINs with an HI higher than 100 were those who presented higher anti-gSG6-P1 IgG responses (Fig. 12.6). HI equal 100 seems to be the threshold above which an LLIN become ineffective for protecting the child sleeping under from mosquito bites.

However, the authors noticed that some children sleeping under LLINs considered as "effective" presented positive anti-gSG6-P1 IgG responses, which were quite high as children sleeping under LLINs considered as "ineffective." This suggests that they could also be exposed to *Anopheles* vector bites, when they were out of LLINs before sleeping or because the effective LLIN were not tucked properly underneath the mattress, so allowed mosquitoes' entrance. Anyway, findings reported here highlight the ability of using this new immunoepidemiological indicator based on human–vector contact to define categories of PI to include effective or ineffective (LLIN needs replacement) under field conditions.

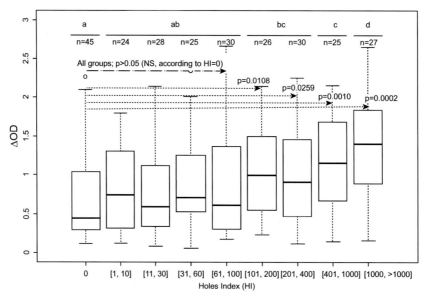

FIGURE 12.6 **IgG levels to gSG6-P1 salivary peptide according to Holes Index (HI) groups** (Noukpo et al., 2016). Box plots show anti-gSG6-P1 IgG level (ΔOD) of the different HI groups. The box plots display the median ΔOD value, 25th and 75th percentiles. The *dots* indicate the outliers and *bars* indicate the median value for each group. Differences between two groups were tested by using the nonparametric Mann–Whitney use test.

TOWARD THE DEVELOPMENT OF AN ANOPHELES DIPSTICK

All data presented above highlight the relevance of using an individual marker of exposure to arthropod vector bites in epidemiological studies on vector-borne diseases such as malaria. However, the detection of anti-gSG6-P1 IgG response levels was currently made by ELISA that is achieved within 2 days at the laboratory. By consequences, this technique needs a technical platform, experimented technicians, and gives not an immediate "response." Such method do not allow an operational and direct assessment in the field of human exposure to *Anopheles* bites, and thus the risk of malaria transmission and the efficacy of VCS in the field by NMCP. A simple, fast, and reliable device would be helpful to assess human-vector contact directly on the field bypassing the laboratory analysis. The IHVEC group, a section of the IRD-MIVEGEC unit and international

leader in salivary biomarker of exposure for the last decade, collaborates with the ACOBIOM Company (Montpellier, France) to develop a POC device for the rapid diagnostic of human exposure to *Anopheles* mosquito bites. The proposed POC device will be a lateral flow dipstick like malaria rapid diagnostic test (RDT) that will enable a rapid and easy detection of human IgG Ab response to *Anopheles* gSG6-P1 salivary peptide, in finger prick blood.

The first approach of this POC is to detect a color band before vector control, such as LLINs strategy, and this band disappears after effective VC (effective VC = reduction of the human-vector contact = decrease of the IgG specific to gSG6-P1). The first step of the development of such POC is the determination of threshold parameter (positive vs. negative) and of sensibility/specificity characteristics as all RDT. The second step will be then the global validation of such innovative diagnostic tools to assess the effectiveness of vector control against malaria.

Further studies will be necessary to validate the POC prototype in several studies in different malaria settings integrating phase 3 studies in the field. A second approach will be then the development of a lateral flow portable reader, which could be associated to optimize the quantification of the intensity of the color band indicating level of specific Ab response.

CONCLUSION

This chapter underlines the potential to use the *Anopheles* anti-saliva Ab response to get an insight of human population exposure to vector bites, to evaluate risk of transmission and to evaluate VCS efficacy in order to control the spread of malaria. To improve the use of salivary proteins as biomarkers, it appears necessary to clearly identify species-/genera-specific salivary antigen for *Anopheles*. One peptide, the gSG6-P1, was identified as specific to *Anopheles* genus. IgG Ab to this peptide was validated as biomarker of exposure to both of New World (Americas) and Old World (Africa) *Anopheles* species. The availability of such a tool allows to assess the risk of malaria transmission in low-level exposure/transmission areas and to accurately evaluate the effectiveness of current or experimental VCSs. For a direct impact in the fight of malaria, the biomarker is currently developed into a POC device for direct assessment (in the field) of the level of exposure and the effectiveness of vector control means by sanitary authorities for a better control of malaria.

References

Abdel-Naser, M.B., Lotfy, R.A., Al-Sherbiny, M.M., Sayed Ali, N.M., 2006. Patients with papular urticaria have IgG antibodies to bedbug (*Cimex lectularius*) antigens. Parasitol. Res. 98, 550–556.

Andrade, B.B., Rocha, B.C., Reis-Filho, A., Camargo, L.M., Tadei, W.P., Moreira, L.A., Barral, A., Barral-Netto, M., 2009. Anti-Anopheles darlingi saliva antibodies as marker of *Plasmodium vivax* infection and clinical immunity in the Brazilian Amazon. Malar. J. 8, 121.

Arca, B., Lombardo, F., Capurro, M., della Torre, A., Spanos, L., Dimopoulos, G., Louis, C., James, A.A., Coluzzi, M., 1999a. Salivary gland-specific gene expression in the malaria vector *Anopheles gambiae*. Parassitologia 41, 483–487.

Arca, B., Lombardo, F., de Lara Capurro, M., della Torre, A., Dimopoulos, G., James, A.A., Coluzzi, M., 1999b. Trapping cDNAs encoding secreted proteins from the salivary glands of the malaria vector *Anopheles gambiae*. Proc. Natl. Acad. Sci. U.S.A. 96, 1516–1521.

Arca, B., Lombardo, F., Francischetti, I.M., Pham, V.M., Mestres-Simon, M., Andersen, J.F., Ribeiro, J.M., 2007. An insight into the sialome of the adult female mosquito *Aedes albopictus*. Insect Biochem. Mol. Biol. 37, 107–127.

Arca, B., Lombardo, F., Valenzuela, J.G., Francischetti, I.M., Marinotti, O., Coluzzi, M., Ribeiro, J.M., 2005. An updated catalogue of salivary gland transcripts in the adult female mosquito, *Anopheles gambiae*. J. Exp. Biol. 208, 3971–3986.

Beier, J.C., Killeen, G.F., Githure, J.I., 1999. Short report: entomologic inoculation rates and *Plasmodium falciparum* malaria prevalence in Africa. Am. J. Trop. Med. Hyg. 61, 109–113.

Billingsley, P.F., Baird, J., Mitchell, J.A., Drakeley, C., 2006. Immune interactions between mosquitoes and their hosts. Parasite Immunol. 28, 143–153.

Brosseau, L., Drame, P.M., Besnard, P., Toto, J.C., Foumane, V., Le Mire, J., Mouchet, F., Remoue, F., Allan, R., Fortes, F., Carnevale, P., Manguin, S., 2012. Human antibody response to *Anopheles* saliva for comparing the efficacy of three malaria vector control methods in Balombo, Angola. PLoS One 7, e44189.

Brummer-Korvenkontio, H., Lappalainen, P., Reunala, T., Palosuo, T., 1994. Detection of mosquito saliva-specific IgE and IgG4 antibodies by immunoblotting. J. Allergy Clin. Immunol. 93, 551–555.

Calvo, E., Andersen, J., Francischetti, I.M., de, L.C.M., deBianchi, A.G., James, A.A., Ribeiro, J.M., Marinotti, O., 2004. The transcriptome of adult female *Anopheles darlingi* salivary glands. Insect Mol. Biol. 13, 73–88.

Calvo, E., Dao, A., Pham, V.M., Ribeiro, J.M., 2007. An insight into the sialome of *Anopheles funestus* reveals an emerging pattern in anopheline salivary protein families. Insect Biochem. Mol. Biol. 37, 164–175.

Calvo, E., Mans, B.J., Andersen, J.F., Ribeiro, J.M., 2006. Function and evolution of a mosquito salivary protein family. J. Biol. Chem. 281, 1935–1942.

Calvo, E., Pham, V.M., Marinotti, O., Andersen, J.F., Ribeiro, J.M., 2009. The salivary gland transcriptome of the neotropical malaria vector *Anopheles darlingi* reveals accelerated evolution of genes relevant to hematophagy. BMC Genomics 10, 57.

Carnevale, P., Frezil, J.L., Bosseno, M.F., Le Pont, F., Lancien, J., 1978. The aggressiveness of *Anopheles gambiae* A in relation to the age and sex of the human subjects. Bull. World Health Organ. 56, 147–154.

Choumet, V., Carmi-Leroy, A., Laurent, C., Lenormand, P., Rousselle, J.C., Namane, A., Roth, C., Brey, P.T., 2007. The salivary glands and saliva of Anopheles gambiae as an essential step in the Plasmodium life cycle: a global proteomic study. Proteomics 7, 3384–3394.

Corbel, V., Akogbeto, M., Damien, G.B., Djenontin, A., Chandre, F., Rogier, C., Moiroux, N., Chabi, J., Banganna, B., Padonou, G.G., Henry, M.C., 2012. Combination of malaria vector control interventions in pyrethroid resistance area in Benin: a cluster randomised controlled trial. Lancet Infect. Dis. 12, 617–626.

Cornelie, S., Remoue, F., Doucoure, S., Ndiaye, T., Sauvage, F.X., Boulanger, D., Simondon, F., 2007. An insight into immunogenic salivary proteins of Anopheles gambiae in African children. Malar. J. 6, 75.

Corran, P., Coleman, P., Riley, E., Drakeley, C., 2007. Serology: a robust indicator of malaria transmission intensity? Trends Parasitol. 23, 575–582.

Das, S., Radtke, A., Choi, Y.J., Mendes, A.M., Valenzuela, J.G., Dimopoulos, G., 2010. Transcriptomic and functional analysis of the Anopheles gambiae salivary gland in relation to blood feeding. BMC Genomics 11, 566.

Doannio, J.M., Doudou, D.T., Konan, L.Y., Djouaka, R., Pare Toe, L., Baldet, T., Akogbeto, M., Monjour, L., 2006. [Influence of social perceptions and practices on the use of bednets in the malaria control programme in Ivory Coast (West Africa)]. Med. Trop. (Mars) 66, 45–52.

Donovan, M.J., Messmore, A.S., Scrafford, D.A., Sacks, D.L., Kamhawi, S., McDowell, M.A., 2007. Uninfected mosquito bites confer protection against infection with malaria parasites. Infect. Immun. 75, 2523–2530.

Drakeley, C.J., Corran, P.H., Coleman, P.G., Tongren, J.E., McDonald, S.L., Carneiro, I., Malima, R., Lusingu, J., Manjurano, A., Nkya, W.M., Lemnge, M.M., Cox, J., Reyburn, H., Riley, E.M., 2005. Estimating medium- and long-term trends in malaria transmission by using serological markers of malaria exposure. Proc. Natl. Acad. Sci. U.S.A. 102, 5108–5113.

Drame, P.M., Poinsignon, A., Marie, A., Noukpo, H., Doucoure, S., Cornelie, S., Remoue, F., 2013a. New salivary biomarkers of human exposure to malaria vector bites. Book Chapter 23. In: Manguin, S. (Ed.), Anopheles mosquitoes- New Insights into Malaria Vectors. InTech open access Publisher, Rijeka, Croatia, pp. 755–795. ISBN: 978-953-51-1188-7.

Drame, P.M., Diallo, A., Poinsignon, A., Boussari, O., Dos Santos, S., Machault, V., Lalou, R., Cornelie, S., Le Hesran, J.Y., Remoue, F., 2013b. Evaluation of the effectiveness of malaria vector control measures in urban settings of Dakar by a specific Anopheles salivary biomarker. PLoS One. 8 (6), e66354.

Drame, P.M., Machault, V., Diallo, A., Cornelie, S., Poinsignon, A., Lalou, R., Sembene, M., Dos Santos, S., Rogier, C., Pages, F., Le Hesran, J.Y., Remoue, F., 2012. IgG responses to the gSG6-P1 salivary peptide for evaluating human exposure to Anopheles bites in urban areas of Dakar region. Senegal. Malar. J. 11, 72.

Drame, P.M., Poinsignon, A., Besnard, P., Cornelie, S., Le Mire, J., Toto, J.C., Foumane, V., Dos-Santos, M.A., Sembene, M., Fortes, F., Simondon, F., Carnevale, P., Remoue, F., 2010a. Human antibody responses to the Anopheles salivary gSG6-P1 peptide: a novel tool for evaluating the efficacy of ITNs in malaria vector control. PLoS One 5, e15596.

Drame, P.M., Poinsignon, A., Besnard, P., Le Mire, J., Dos-Santos, M.A., Sow, C.S., Cornelie, S., Foumane, V., Toto, J.C., Sembene, M., Boulanger, D., Simondon, F., Fortes, F., Carnevale, P., Remoue, F., 2010b. Human antibody response to Anopheles gambiae saliva: an immuno-epidemiological biomarker to evaluate the efficacy of insecticide-treated nets in malaria vector control. Am. J. Trop. Med. Hyg. 83, 115–121.

Drame, P.M., Poinsignon, A., Dechavanne, C., Cottrell, G., Farce, M., Ladekpo, R., Massougbodji, A., Cornelie, S., Courtin, D., Migot-Nabias, F., Garcia, A., Remoue, F., 2015. Specific antibodies to Anopheles gSG6-P1 salivary peptide to assess early childhood exposure to malaria vector bites. Malar. J. 14, 285.

Francischetti, I.M., Valenzuela, J.G., Pham, V.M., Garfield, M.K., Ribeiro, J.M., 2002. Toward a catalog for the transcripts and proteins (sialome) from the salivary gland of the malaria vector Anopheles gambiae. J. Exp. Biol. 205, 2429–2451.

Geissbuhler, Y., Chaki, P., Emidi, B., Govella, N.J., Shirima, R., Mayagaya, V., Mtasiwa, D., Mshinda, H., Fillinger, U., Lindsay, S.W., Kannady, K., de Castro, M.C., Tanner, M., Killeen, G.F., 2007. Interdependence of domestic malaria prevention measures and mosquito-human interactions in urban Dar es Salaam, Tanzania. Malar. J. 6, 126.

Gubler, D.J., 1998. Resurgent vector-borne diseases as a global health problem. Emerg. Infect. Dis. 4, 442–450.

Hay, S.I., Rogers, D.J., Toomer, J.F., Snow, R.W., 2000. Annual Plasmodium falciparum entomological inoculation rates (EIR) across Africa: literature survey, Internet access and review. Trans. R Soc. Trop. Med. Hyg. 94, 113–127.

Jariyapan, N., Baimai, V., Poovorawan, Y., Roytrakul, S., Saeung, A., Thongsahuan, S., Suwannamit, S., Otsuka, Y., Choochote, W., 2010. Analysis of female salivary gland proteins of the Anopheles barbirostris complex (Diptera: Culicidae) in Thailand. Parasitol. Res. 107, 509–516.

Jariyapan, N., Choochote, W., Jitpakdi, A., Harnnoi, T., Siriyasatein, P., Wilkinson, M.C., Junkum, A., Bates, P.A., 2007. Salivary gland proteins of the human malaria vector, *Anopheles dirus* B (Diptera: Culicidae). Rev. Inst. Med. Trop. Sao Paulo 49, 5–10.

Jariyapan, N., Roytrakul, S., Paemanee, A., Junkum, A., Saeung, A., Thongsahuan, S., Sor-Suwan, S., Phattanawiboon, B., Poovorawan, Y., Choochote, W., 2012. Proteomic analysis of salivary glands of female *Anopheles barbirostris* species A2 (Diptera: Culicidae) by two-dimensional gel electrophoresis and mass spectrometry. Parasitol Res. 111 (3), 1239–1249.

Kalluri, S., Gilruth, P., Rogers, D., Szczur, M., 2007. Surveillance of arthropod vector-borne infectious diseases using remote sensing techniques: a review. PLoS Pathog. 3, 1361–1371.

Lanfrancotti, A., Lombardo, F., Santolamazza, F., Veneri, M., Castrignano, T., Coluzzi, M., Arca, B., 2002. Novel cDNAs encoding salivary proteins from the malaria vector *Anopheles gambiae*. FEBS Lett. 517, 67–71.

Larru, B., Molyneux, E., Ter Kuile, F.O., Taylor, T., Molyneux, M., Terlouw, D.J., 2009. Malaria in infants below six months of age: retrospective surveillance of hospital admission records in Blantyre, Malawi. Malar. J. 8, 310.

Lombardo, F., Lanfrancotti, A., Mestres-Simon, M., Rizzo, C., Coluzzi, M., Arca, B., 2006. At the interface between parasite and host: the salivary glands of the African malaria vector *Anopheles gambiae*. Parassitologia 48, 573–580.

Lombardo, F., Ronca, R., Rizzo, C., Mestres-Simon, M., Lanfrancotti, A., Curra, C., Fiorentino, G., Bourgouin, C., Ribeiro, J.M., Petrarca, V., Ponzi, M., Coluzzi, M., Arca, B., 2009. The *Anopheles gambiae* salivary protein gSG6: an anopheline-specific protein with a blood-feeding role. Insect Biochem. Mol. Biol. 39, 457–466.

Londono-Renteria, B., Drame, P.M., Weitzel, T., Rosas, R., Gripping, C., Cardenas, J.C., Alvares, M., Wesson, D.M., Poinsignon, A., Remoue, F., Colpitts, T.M., 2015. An. gambiae gSG6-P1 evaluation as a proxy for human-vector contact in the Americas: a pilot study. Parasit Vectors 8, 533.

Londono-Renteria, B.L., Eisele, T.P., Keating, J., James, M.A., Wesson, D.M., 2010. Antibody response against *Anopheles albimanus* (Diptera: Culicidae) salivary protein as a measure of mosquito bite exposure in Haiti. J. Med. Entomol. 47, 1156–1163.

Noukpo, M.H., Damien, G.B., Elanga N'Dille, E., Sagna, A.B., Drame, P.M., Chaffa, E., Boussari, O., Corbel, V., Akogbeto, M., Remoue, F., 2016. Operational assessment of Long-Lasting Insecticidal Nets by using an *Anopheles* salivary biomarker of human-vector contact. Am. J. Trop. Med. Hyg. 95 (6), 1376–1382.

Okumu, F.O., Moore, S.J., 2011. Combining indoor residual spraying and insecticide-treated nets for malaria control in Africa: a review of possible outcomes and an outline of suggestions for the future. Malar. J. 10, 208.

Orlandi-Pradines, E., Almeras, L., Denis de Senneville, L., Barbe, S., Remoue, F., Villard, C., Cornelie, S., Penhoat, K., Pascual, A., Bourgouin, C., Fontenille, D., Bonnet, J., Corre-Catelin, N., Reiter, P., Pages, F., Laffite, D., Boulanger, D., Simondon, F., Pradines, B., Fusai, T., Rogier, C., 2007. Antibody response against saliva antigens of *Anopheles gambiae* and *Aedes aegypti* in travellers in tropical Africa. Microbes. Infect. 9, 1454–1462.

Pages, F., Orlandi-Pradines, E., Corbel, V., 2007. Vectors of malaria: biology, diversity, prevention, and individual protection. Med. Mal. Infect. 37, 153–161.

Palosuo, K., Brummer-Korvenkontio, H., Mikkola, J., Sahi, T., Reunala, T., 1997. Seasonal increase in human IgE and IgG4 antisaliva antibodies to Aedes mosquito bites. Int. Arch. Allergy Immunol. 114, 367–372.

Peng, Z., Beckett, A.N., Engler, R.J., Hoffman, D.R., Ott, N.L., Simons, F.E., 2004a. Immune responses to mosquito saliva in 14 individuals with acute systemic allergic reactions to mosquito bites. J. Allergy Clin. Immunol. 114, 1189–1194.

Peng, Z., Estelle, F., Simons, R., 2007. Mosquito allergy and mosquito salivary allergens. Protein Pept. Lett. 14, 975–981.

Peng, Z., Ho, M.K., Li, C., Simons, F.E., 2004b. Evidence for natural desensitization to mosquito salivary allergens: mosquito saliva specific IgE and IgG levels in children. Ann. Allergy Asthma Immunol. 93, 553–556.

Peng, Z., Li, H., Simons, F.E., 1996. Immunoblot analysis of IgE and IgG binding antigens in extracts of mosquitos *Aedes vexans*, *Culex tarsalis* and *Culiseta inornata*. Int. Arch. Allergy Immunol. 110, 46–51.

Peng, Z., Rasic, N., Liu, Y., Simons, F.E., 2002. Mosquito saliva-specific IgE and IgG antibodies in 1059 blood donors. J. Allergy Clin. Immunol. 110, 816–817.

Peng, Z., Simons, F.E., 2004. Mosquito allergy: immune mechanisms and recombinant salivary allergens. Int. Arch. Allergy Immunol. 133, 198–209.

Poinsignon, A., Cornelie, S., Ba, F., Boulanger, D., Sow, C., Rossignol, M., Sokhna, C., Cisse, B., Simondon, F., Remoue, F., 2009. Human IgG response to a salivary peptide, gSG6-P1, as a new immuno-epidemiological tool for evaluating low-level exposure to *Anopheles* bites. Malar. J. 8, 198.

Poinsignon, A., Cornelie, S., Mestres-Simon, M., Lanfrancotti, A., Rossignol, M., Boulanger, D., Cisse, B., Sokhna, C., Arca, B., Simondon, F., Remoue, F., 2008. Novel peptide marker corresponding to salivary protein gSG6 potentially identifies exposure to *Anopheles* bites. PLoS One 3, e2472.

Poinsignon, A., Samb, B., Doucoure, S., Drame, P.M., Sarr, J.B., Sow, C., Cornelie, S., Maiga, S., Thiam, C., Rogerie, F., Guindo, S., Hermann, E., Simondon, F., Dia, I., Riveau, G., Konate, L., Remoue, F., 2010. First attempt to validate the gSG6-P1 salivary peptide as an immuno-epidemiological tool for evaluating human exposure to *Anopheles funestus* bites. Trop. Med. Int. Health 15, 1198–1203.

Remoue, F., Alix, E., Cornelie, S., Sokhna, C., Cisse, B., Doucoure, S., Mouchet, F., Boulanger, D., Simondon, F., 2007. IgE and IgG4 antibody responses to *Aedes* saliva in African children. Acta Trop. 104, 108–115.

Remoue, F., Cisse, B., Ba, F., Sokhna, C., Herve, J.P., Boulanger, D., Simondon, F., 2006. Evaluation of the antibody response to *Anopheles* salivary antigens as a potential marker of risk of malaria. Trans. R Soc. Trop. Med. Hyg. 100, 363–370.

Reunala, T., Brummer-Korvenkontio, H., Palosuo, K., Miyanij, M., Ruiz-Maldonado, R., Love, A., Francois, G., Palosuo, T., 1994. Frequent occurrence of IgE and IgG4 antibodies against saliva of *Aedes communis* and *Aedes aegypti* mosquitoes in children. Int. Arch. Allergy Immunol. 104, 366–371.

Ribeiro, J.M., 1987. Role of saliva in blood-feeding by arthropods. Annu. Rev. Entomol. 32, 463–478.

Ribeiro, J.M., 2000. Blood-feeding in mosquitoes: probing time and salivary gland anti-haemostatic activities in representatives of three genera (*Aedes, Anopheles, Culex*). Med. Vet. Entomol. 14, 142–148.

Ribeiro, J.M., 2003. A catalogue of *Anopheles gambiae* transcripts significantly more or less expressed following a blood meal. Insect Biochem. Mol. Biol. 33, 865–882.

Ribeiro, J.M., Arca, B., Lombardo, F., Calvo, E., Phan, V.M., Chandra, P.K., Wikel, S.K., 2007. An annotated catalogue of salivary gland transcripts in the adult female mosquito, *Aedes aegypti*. BMC Genomics 8, 6.

Ribeiro, J.M., Charlab, R., Pham, V.M., Garfield, M., Valenzuela, J.G., 2004. An insight into the salivary transcriptome and proteome of the adult female mosquito *Culex pipiens quinquefasciatus*. Insect Biochem. Mol. Biol. 34, 543–563.

Ribeiro, J.M., Francischetti, I.M., 2003. Role of arthropod saliva in blood feeding: sialome and post-sialome perspectives. Annu. Rev. Entomol. 48, 73–88.

Rizzo, C., Ronca, R., Fiorentino, G., Mangano, V.D., Sirima, S.B., Nebie, I., Petrarca, V., Modiano, D., Arca, B., 2011a. Wide cross-reactivity between *Anopheles gambiae* and *Anopheles funestus* SG6 salivary proteins supports exploitation of gSG6 as a marker of human exposure to major malaria vectors in tropical Africa. Malar. J. 10, 206.

Rizzo, C., Ronca, R., Fiorentino, G., Verra, F., Mangano, V., Poinsignon, A., Sirima, S.B., Nebie, I., Lombardo, F., Remoue, F., Coluzzi, M., Petrarca, V., Modiano, D., Arca, B., 2011b. Humoral response to the *Anopheles gambiae* salivary protein gSG6: a serological indicator of exposure to Afrotropical malaria vectors. PLoS One 6, e17980.

Rogier, C., 2003. Childhood malaria in endemic areas: epidemiology, acquired immunity and control strategies. Med. Trop. (Mars) 63, 449–464.

Sagna, A.B., Gaayeb, L., Sarr, J.B., Senghor, S., Poinsignon, A., Boutouaba-Combe, S., Schacht, A.M., Hermann, E., Faye, N., Remoue, F., Riveau, G., 2013a. *Plasmodium falciparum* infection during dry season: IgG responses to *Anopheles gambiae* salivary gSG6-P1 peptide as sensitive biomarker for malaria risk in Northern Senegal. Malar. J. 12, 301.

Sagna, A.B., Sarr, J.B., Gaayeb, L., Drame, P.M., Ndiath, M.O., Senghor, S., Sow, C.S., Poinsignon, A., Seck, M., Hermann, E., Schacht, A.M., Faye, N., Sokhna, C., Remoue, F., Riveau, G., 2013b. gSG6-P1 salivary biomarker discriminates micro-geographical heterogeneity of human exposure to *Anopheles* bites in low and seasonal malaria areas. Parasit. Vectors 6, 68.

Schneider, B.S., Higgs, S., 2008. The enhancement of arbovirus transmission and disease by mosquito saliva is associated with modulation of the host immune response. Trans. R Soc. Trop. Med. Hyg. 102, 400–408.

Schneider, B.S., Mathieu, C., Peronet, R., Mecheri, S., 2011. *Anopheles stephensi* saliva enhances progression of cerebral malaria in a murine model. Vector Borne Zoonotic Dis. 11, 423–432.

Schwartz, B.S., Ribeiro, J.M., Goldstein, M.D., 1990. Anti-tick antibodies: an epidemiologic tool in Lyme disease research. Am. J. Epidemiol. 132, 58–66.

Schwarz, A., Juarez, J.A., Richards, J., Rath, B., Machaca, V.Q., Castro, Y.E., Malaga, E.S., Levy, K., Gilman, R.H., Bern, C., Verastegui, M., Levy, M.Z., 2011. Anti-triatomine saliva immunoassays for the evaluation of impregnated netting trials against Chagas disease transmission. Int. J. Parasitol. 41, 591–594.

Schwarz, A., Sternberg, J.M., Johnston, V., Medrano-Mercado, N., Anderson, J.M., Hume, J.C., Valenzuela, J.G., Schaub, G.A., Billingsley, P.F., 2009. Antibody responses of domestic animals to salivary antigens of *Triatoma infestans* as biomarkers for low-level infestation of triatomines. Int. J. Parasitol. 39, 1021–1029.

Smith, D.L., Dushoff, J., Snow, R.W., Hay, S.I., 2005. The entomological inoculation rate and *Plasmodium falciparum* infection in African children. Nature 438, 492–495.

Smith, T., Killeen, G., Lengeler, C., Tanner, M., 2004. Relationships between the outcome of *Plasmodium falciparum* infection and the intensity of transmission in Africa. Am. J. Trop. Med. Hyg. 71, 80–86.

Valenzuela, J.G., Francischetti, I.M., Pham, V.M., Garfield, M.K., Ribeiro, J.M., 2003. Exploring the salivary gland transcriptome and proteome of the *Anopheles stephensi* mosquito. Insect Biochem. Mol. Biol. 33, 717–732.

Waitayakul, A., Somsri, S., Sattabongkot, J., Looareesuwan, S., Cui, L., Udomsangpetch, R., 2006. Natural human humoral response to salivary gland proteins of *Anopheles* mosquitoes in Thailand. Acta Trop. 98, 66–73.

WHO, 2011. Guidelines for Monitoring the Durability of Long-Lasting Insecticidal Mosquito Nets under Operational Conditions. World Health Organization, p. 44.

WHO, 2016. Eliminating Malaria. World Health Organization, p. 24.

Wilson, A.D., Harwood, L.J., Bjornsdottir, S., Marti, E., Day, M.J., 2001. Detection of IgG and IgE serum antibodies to *Culicoides* salivary gland antigens in horses with insect dermal hypersensitivity (sweet itch). Equine Vet. J. 33, 707–713.

Ixodes Tick Saliva: A Potent Controller at the Skin Interface of Early Borrelia burgdorferi Sensu Lato Transmission

Sarah Bonnet[1], Nathalie Boulanger[2,3]

[1]UMR BIPAR 956 INRA-ANSES-ENVA, Maisons-Alfort, France; [2]Université de Strasbourg, Strasbourg, France; [3]Centre National de Référence Borrelia, Strasbourg, France

INTRODUCTION

Ticks are the most important vectors of pathogens affecting animals worldwide, and second after mosquitoes where humans are concerned (Dantas-Torres et al., 2012). They surpass all other arthropods with regard to the variety of pathogenic organisms they are able to transmit, including viruses, bacteria, and parasites (protozoa and helminths). Moreover, the risk of emergence or reemergence of tick-borne diseases (TBDs) has markedly increased with the intensification of human and animal mobility and due to socioeconomic and environmental changes (Kilpatrick and Randolph, 2012; Léger et al., 2013; Lindgren et al., 2012; Madder et al., 2011; Rizzoli et al., 2014).

Lyme borreliosis is the most important vector-borne disease of the Northern hemisphere (Stanek et al., 2012). It is transmitted by hard ticks that belong to the Ixodes genus. The infectious agent is a bacterium named Borrelia burgdorferi sensu lato, whereof 21 genospecies are currently identified. Human Lyme borreliosis is the most common tick-borne infection in Europe, as well as in the United States. The estimated incidence is 300,000 annual cases in the United States (Mead, 2015) and 65,500 cases in Europe (Rizzoli et al., 2011). Three Borrelia species have been identified as major causative agents in humans: B. burgdorferi sensu stricto (ss), Borrelia afzelii, and Borrelia garinii. B. garinii most often induces neurologic manifestations, while B. afzelii is mainly responsible for skin disorders and B. burgdorferi ss is the main cause of lyme arthritis (Radolf et al., 2012; Stanek et al., 2012). The bacteria typically circulate between the vector ticks and their wild animal reservoirs. The range of animal species that can become infected is large, including more than 300 different species (Humair and Gern, 2000). B. burgdorferi ss is the least host specialized, whereas B. afzelii and Borrelia bavariensis mainly infect rodents and B. garinii is found in

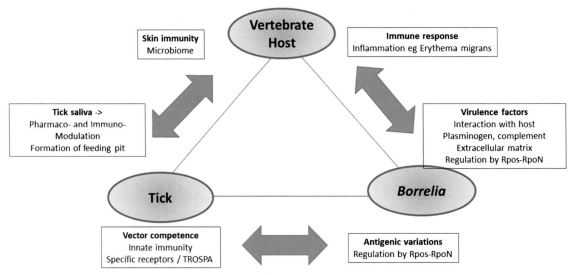

FIGURE 13.1 The complex interplay between the pathogen, *Borrelia burgdorferi* sensu lato, the vertebrate host, and the vector, *Ixodes* hard tick, and major factors influencing the tight interactions between the three actors.

birds. This specialization is linked to the host complement (Kurtenbach et al., 2006). Rodents and birds are the best reservoirs for *Borrelia* and the main blood source for larvae and nymphs, while adult ticks feed mainly on deer, which themselves are incompetent hosts for *Borrelia*. To analyze the physiopathology of the disease in the laboratory, the mouse model C3H/HeN is the choice to study Lyme borreliosis (Barthold et al., 1993; Wooten and Weis, 2001).

The skin constitutes a key interface in arthropod-borne diseases (ABD) (Bernard et al., 2014; Frischknecht, 2007); this organ has been considered for too long as a simple physical barrier. Since discovery of innate immunity and Toll-like receptors (TLRs) in mammals, the role of the skin as a major organ to protect organisms against infections has been recognized. The skin is an essential interface in the physiopathology of Lyme borreliosis. The tick inoculates the bacterium, *Borrelia*, which multiplies locally (Wikel, 2013). Erythema migrans (EM), the characteristic rash that can appear days to weeks after infection is the most common skin manifestation (about 60%–80% of cases) in humans

(Stanek et al., 2012). *Borrelia* then spreads to target organs: joints, heart, nervous system, and distant skin (Stanek et al., 2012). A number of bacteria, however, persists in skin for several months in absence of antibiotic treatment. This skin persistence is not confirmed yet in humans. As in all ABDs, the triad of host–arthropod–vector exhibit a very complex interplay (Fig. 13.1).

IXODES HARD TICK

Hard Tick Biology

Ticks possess many unusual features that contribute to their remarkable success as potential vectors. One of the most outstanding is their longevity and their reproductive potential, which makes them substantial pathogen reservoirs in the field. Another is the fact that for several species, they are able to feed on a very large host spectrum. Lastly, they consume a very large quantity of blood over a relatively long period of time, both increasing the chance of ingesting and transmitting a pathogen.

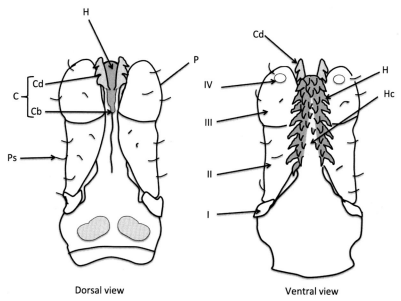

Dorsal view   Ventral view

FIGURE 13.2 Diagrammatic representation of the mouthparts of a representative hard tick. H: hypostome with a ventral side armed with rows of recurved teeth and a flat dorsal side with a narrow V-shaped channel (Hc) in the middle used for both saliva secretion and blood sucking. C: paired chelicerae comprising a body (Cb: a bulbous muscular basis and an elongated hollow shaft) and some cutting digits (Cd), each can be retracted in a protective sheath. The digits contain heavily sclerotized spines. When ticks attach to host skin, the chelicerae are protruded to expose the digits that cut the host skin. P: paired palps with several sensory sensilla (Ps) used to probe the host skin and identify chemical compounds. They comprise four segments (I, II, II, IV), the terminal one (article IV) and the first one (article I) being short.

The *Ixodidae* or hard ticks possess the most complex feeding biology of all hematophagous arthropods (Liu and Bonnet, 2014). Hard ticks have larval, nymphal, and adult forms, all requiring a blood meal to complete development and reproduction. A three-host life cycle, which includes host-seeking, feeding, and off-host molting (or egg laying) in each life stage, is the most common developmental pattern for the majority of hard ticks of medical and veterinary importance. After feeding of a rather substantial quantity of host blood, females drop off from their host to start oviposition in a sheltered microenvironment, laying up to several thousand eggs. Compared to other hematophagous arthropods, feeding in ixodid ticks is a slow and complex process, taking several days for repletion and detachment alone and necessitating extended control over the host's immune response. During this feeding process,

ticks inject saliva and absorb blood in an alternating pattern. Many proteins present in tick saliva dampen host defenses to assure adequate feeding (Kotal et al., 2015; Ribeiro, 1995), thereby creating a favorable environment for survival and propagation of tick-borne pathogens (TBP) transmitted via saliva (Kazimírová and Štibrániová, 2013; Wikel, 1999).

The main vectors of Lyme borreliosis belong to the *I. ricinus-persulcatus* complex: *Ixodes scapularis* in Eastern and Midwestern North America, *I. pacificus* in Western North America, *I. ricinus* in Western Europe, and *I. persulcatus* in Eastern Europe and Asia (Tsao, 2009).

Tick Feeding Process

Hard ticks are strictly hematophagous arthropods that stay attached to the host skin during several days of acquiring blood. Fig. 13.2 shows a diagram illustrating the mouthparts of a representative hard tick consisting of an unpaired hypostome, paired chelicerae, and paired palps, the latter being sensorial organs that do not enter into host skin. Ticks begin their attachment by cutting and tearing the host skin with their chelicerae, allowing the penetration of the hypostome, which will act as a harpoon to strengthen the anchoring of the tick in its host. Hard ticks also secrete cement into and above the gradually expanding lesion that will act as a kind of glue, which further strengthens skin anchoring by sealing the hypostome and the chelicerae in the skin. This first phase may take between 1 and 2 days before the tick begins to feed for several additional days, depending upon the life stage and the tick species. During the feeding process, ticks alternate blood sucking and secretion of saliva into the wound through the same channel (see Fig. 13.2). Injected saliva is essential to the success of the feeding process, facilitating the flow of blood (see later). Both saliva and cement are secreted by the salivary glands of the tick. Ticks are pool feeders, sucking fluids that are exuded into the wound generated by the bite. Thanks to fresh expandable cuticle synthesized, hard ticks are then able to imbibe enough blood to increase in size, as much as 100 times (Sonenshine and Anderson, 2014). Blood cells are lysed in the lumen of the midgut. In contrast to other hematophagous arthropods, the following digestion of the proteins and other molecules in the blood is an intracellular process which takes place within the epithelial cells of the midgut. Food storage in the midgut explains the tick's ability to wait for a vertebrate host sometimes for several years without feeding (Lara et al., 2005).

Pharmacological and Immunomodulatory Effect of Tick Saliva

The prolonged period of tick attachment to their host has sparked great interest in studying tick salivary gland secretions during feeding.

The evolutionary importance of salivary protein family repertoires has been previously recognized (Mans et al., 2008), and predicts that the molecular composition and function of tick saliva must mirror the interaction established with the host. Tick saliva is an essential biofluid for ticks, allowing ingestion of their obligate blood meal. In addition to the control of wound healing, all tick species are able to suppress the host's immune and inflammatory responses by secreting antiinflammatory and immune-modulatory molecules in their saliva (Francischetti et al., 2010; Kazimírová and Stibrániová, 2013; Ribeiro et al., 1985). Molecular characterization of tick saliva is thus a prerequisite for a full understanding of the mechanisms implicated in, not only tick feeding but also, pathogen transmission (Liu and Bonnet, 2014). Global advances in transcript and protein profiling have had a profound impact on the number of tick salivary components that may be studied at the same time. Indeed, transcript and protein profiling in tick saliva have been applied to several species of hard and soft ticks and led to the discovery of several factors which contribute to successful feeding and evasion of the host immune and hemostatic defenses. The first proteomic studies focusing on tick saliva were conducted in the 2000s (Madden et al., 2004). Since then, the salivary transcriptomes and proteomes of diverse species of hard ticks have been characterized including *I. scapularis* (McNally et al., 2012; Ribeiro et al., 2006; Valenzuela, 2002), *I. pacificus* (Francischetti et al., 2005), *I. ricinus* (Chmelar et al., 2008; Liu et al., 2014; Schwarz et al., 2013; Vennestrøm and Jensen, 2007). However, less than 5% of the identified proteins were expressed and their function verified (Francischetti et al., 2010).

Antihemostatic activities: Hemostasis is essential in mammals to control any wound and the possible loss of blood by vascular damage. During formation of a feeding lesion, platelets interact with various macromolecules at the level of the vascular endothelium and adhere to it to form the clot. Thrombin transforms the soluble fibrinogen into insoluble fibrin, which strengthens the formation of the clot. To take its blood meal and ensure a continuous blood flow, the tick thus needs to avoid this process of coagulation. In addition, as ticks are pool feeders, they require strong vasodilatory substances to increase blood flow perfusion in superficial regions of the skin. Several antihemostatic factors acting on platelet adhesion/aggregation, activation of the intrinsic and extrinsic pathways of coagulation, and thrombin formation have been identified in tick saliva from several genera including both soft and hard ticks (Maritz-Olivier et al., 2007). Among them are the vasodilatory prostaglandins (Bowman et al., 1996) and prostacyclin (Ribeiro et al., 1988). In *I. scapularis*, two tissue factor pathway inhibitors, ixolaris and penthalaris (Francischetti et al., 2004, 2002), and one inhibitor of the factor Xa, Salp14 (Narasimhan et al., 2004), have been identified. Ticks also secrete factors that counteract platelet aggregation like apyrases, described among others for *I. scapularis* (Ribeiro et al., 2006). Lastly, some serine protease inhibitors (serpins) with anticoagulant activity have been identified both in *I. scapularis* (Dai et al., 2012; Pichu et al., 2014) and *I. ricinus* (Chmelar et al., 2011; Liu et al., 2014).

Antiinflammatory activities: To allow for their long blood meal, hard ticks must also avoid their rejection by the vertebrate host. For that purpose, certain molecules present in the saliva target the attenuation of inflammation and pain perception. Histamine-binding proteins able to bind and neutralize histamine, an essential mediator of inflammation, have been described in several tick species (Wikel, 1982a), including lipocalins in *I. ricinus* (Beaufays et al., 2008). Lipocalins, small extracellular proteins that bind to histamine, serotonin, and prostaglandin, are implicated in the regulation of cell homeostasis and vertebrate immune responses (Flower, 1996). In *I. dammini* (previous name for *I. scapularis*), a kinase able to neutralize the effect of bradykinin has also been described (Ribeiro et al., 1985).

Modulators of host immunity: Like other hematophagous arthropods, ticks face host immunity during the blood-feeding process, and have consequently evolved a complex and sophisticated immunological armament against this potentially harmful process. Accordingly, tick saliva contains immunomodulatory components that allow ticks to successfully feed and favor pathogen transmission by interfering with host immune responses.

Concerning innate immunity, tick saliva targets neutrophils and macrophages and inhibits their migration (Wikel, 1999). It includes factors able to decrease killing activity of natural killer cells (Kubes et al., 2002). It also inhibits the activation of TLRs, especially TLR2 and TLR1, which are involved in the recognition of the major surface lipoprotein of *B. burgdorferi* sl, OspC. Salp15, the most studied immunomodulatory tick molecule, interacts with OspC and inhibits Toll activation. Then, alarmins (chemokines, antimicrobial peptides such as defensin and cathelicidin) are neutralized and many danger signals potentially secreted by skin cells are blocked (Marchal et al., 2011). When transmitted to host, *Borrelia* has also to evade another important part of the innate immune system, the complement pathway, which constitutes one of the first defense lines involved during an infection in vertebrate hosts (Nielsen and Leslie, 2002). Findings from a number of studies suggest that ticks have evolved mechanisms to impair complement activity. Salp15 has also been implicated in the protection of *Borrelia* species from complement and antibody-mediated killing by the host. *Ixodes scapularis* anticomplement (Valenzuela et al., 2000) and some homologous proteins identified in *I. ricinus* (Daix et al., 2007) target the alternate pathway of the complement cascade by inhibiting the deposition of C3b and C5b and the subsequent generation of anaphylatoxin C3a preventing the degranulation of mast cells and neutrophils. *Ixodes* anticomplement has been also identified as an anticomplement protein in *I. ricinus* that binds to properdin, leading to the inhibition of C3 convertase and the alternative complement pathway (Couvreur et al., 2008).

Adaptive immunity is similarly targeted by tick saliva. It suppresses the production of pro-inflammatory cytokines and induces a polarized type 2 immune response (Oliveira et al., 2010; Skallova et al., 2008). Early studies showed that the saliva inhibits the CD4[+] T cells in vitro (Wikel, 1982b). Salp15 was then identified in *I. scapularis* as a salivary protein with multiple functions on adaptive immunity. This protein was shown to be responsible for the inhibition of CD4[+] T-cell activation, specifically binding to the T-cell co-receptor CD4 (Anguita et al., 2002; Garg et al., 2006; Juncadella et al., 2007), but also for the inhibition of cytokine expression by dendritic cells (Hovius et al., 2008). Iris isolated from *I. ricinus* is able to suppress in vitro proliferation of ConA-stimulated naïve spleen cells and to downregulate the production of INF-γ by T-lymphocytes and antigen-presenting cells (Leboulle et al., 2002). The impact of *I. scapularis* saliva on CD8[+] lymphocytes has also been reported with the discovery of sialostatin that inhibits cathepsin (Schwarz et al., 2012). In *I. scapularis* as well, salivary prostaglandin E_2 was found to be responsible for the inhibition of IL-12, IL-2, and TNF-α production via dendritric cells (Sa-Nunes et al., 2007). Similarly, B-lymphocyte proliferation was affected by tick saliva in vitro (Wikel, 1982b). Later, BIP (B-cell inhibitory protein) has been identified in *I. ricinus*, to be involved in inhibition of B-lymphocytes proliferation (Hannier et al., 2004).

BORRELIA BURGDORFERI SENSU LATO

The Bacterium

Spirochetes belong to the order Spirochaetales, which includes the causative agents of borrelioses, syphilis, and leptospirosis. Tick-borne borrelioses are divided into two groups. Those responsible for relapsing fevers occur in both temperate and tropical zones and are mainly transmitted by soft ticks, whereas those responsible for Lyme borreliosis are transmitted by the hard tick, *Ixodes* sp., being the most common tick-borne zoonosis in the Northern hemisphere (Ogden et al., 2014). Spirochetes are extracellular bacteria, helically shaped with multiple flagella and measuring 10–30 μm in length. Up to now, around 21 genospecies causing Lyme borreliosis were described in the group of *B. burgdorferi* (sl), but new species have been described now (Ivanova et al., 2014; Pritt et al., 2016). Only few species are responsible for Lyme borreliosis in humans: *B. burgdorferi* (ss) in North America; *B. afzelii*, *B. garinii*, *B. burgdorferi* ss, *B. bavariensis*, and *Borrelia spielmanii* in Europe. In Asia, eight species are described and three are human pathogens: *B. afzelii*, *B. garinii*, and *B. bavariensis* (Stanek et al., 2012). As an extracellular pathogen, *B. burgdorferi* sl interacts with host cells, tissues, and extracellular matrix.

Borrelia Development Within the Tick

During transmission from the vertebrate host to the tick, the bacteria undergo drastic changes in their enzootic cycle, namely temperature, pH, immune system, and osmotic pressure. To control these modifications, *Borrelia* proteins are regulated by a complex system of gene regulation, the RpoS-RpoN system (reviewed by Norris et al., 2010). When *Borrelia* is transmitted by the tick to a vertebrate host during the blood meal, several bacterial molecules including OspB and OspD are downregulated (reviewed in Ogden et al., 2014). One of them, the surface lipoprotein OspA, allows the binding of the bacteria to a specific receptor, tick receptor for outer surface protein A (TROSPA), on gut cells. Bacteria persist upon the tick molt, until a blood meal on a new vertebrate host triggers the migration of *Borrelia* from the midgut to the salivary glands via the hemolymph

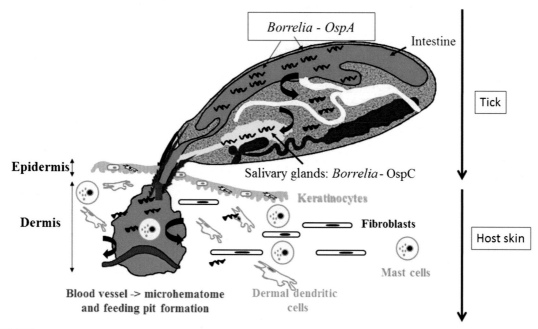

FIGURE 13.3 Mechanisms of *Borrelia* transmission. Tick mouthparts dilacerate the epidermis and the dermis, producing the formation of a feeding pool where saliva is inoculated, as well as *Borrelia burgdorferi* sensu lato. *From McCoy, K., Boulanger, N., 2016. Tiques et Maladies à tiques: Biologie, écologie évolutive, épidémiologie. IRD Editions.*

(Fig. 13.3). Two tick proteins, ISDLP (*Ixodes scapularis* dystroglycan-like protein) and TRE31, have been described also on gut cells. These proteins, upregulated during tick blood meal and expressed when the vertebrate host is infected by *Borrelia*, may help the bacteria to go through the midgut epithelium (Coumou et al., 2016; Zhang et al., 2011). The migration of the spirochetes within the tick might also be facilitated by the mammal plasminogen, present in the blood meal (Fuchs et al., 1994).

Within the salivary glands, the bacteria mainly express the lipoprotein OspC that binds to the tick salivary protein Salp15 (Ramamoorthi et al., 2005). Salp15, as well as Salp20 and TSLPI, will further help the bacteria to evade the immune mechanisms of the vertebrate host (Hourcade et al., 2016; Ramamoorthi et al., 2005; Schuijt et al., 2011a; Wagemakers et al., 2016). The delineation in bacterial population is however not as strict in reality, and the lipoproteins can be very

heterogeneous on bacteria during its development within the tick. Some will express only OspA, some only OspC, some OspA and OspC, and for others neither of the two (Ohnishi et al., 2001). In addition, Osp expression and dissemination dynamics vary among strains as observed for *B. afzelii* and *B. garinii* (Fingerle et al., 2002) and *B. burgdorferi* ss (Ohnishi et al., 2001) due to different *Ixodes* species and *Borrelia* genospecies in Europe and Asia. Therefore, the delay of *Borrelia* migration from the gut to the salivary glands and its subsequent transmission from the tick to the vertebrate host may significantly vary between the geographic areas where different genospecies are found (Fingerle et al., 2002).

In the United States, where *I. scapularis* transmits only *B. burgdorferi* ss to humans, the highest risk of transmission occurs at 36 h after initiation of the blood meal. In Europe, where additional *Borrelia* species are transmitted like *B. afzelii* and *B. garinii*, the risk of transmission appears already at 16h (Kahl et al., 1998). No precise study has been done with *B. burgdorferi* ss in *I. ricinus*, as this *Borrelia* species is much less prevalent in humans in Europe (Stanek et al., 2012). *I. ricinus* found in Europe seems to be a better transmitter than *I. scapularis* at least in murine models (Crippa et al., 2002). A quantification of *Borrelia* in the midgut compared to the salivary glands in *Ixodes* by polymerase chain reaction (PCR) showed that the bacteria multiply intensively during the uptake of the blood meal in the midgut where thousands of spirochetes were detected. However, spirochetes detected in the salivary glands were only in the range of hundreds (Piesman et al., 2001). During *Borrelia* development within the vector, tick innate immunity has also been studied and a group of six putative defensins has been identified (Tonk et al., 2015), one at least being clearly expressed in the gut upon *Borrelia* infection (Rudenko et al., 2005). The precise role of these defensins is not known, especially whether they regulate *Borrelia* infection within the tick.

Early Development of *Borrelia* in the Skin

Ticks are telmophages, i.e., they lacerate host tissues inducing the formation of a microhematoma, followed by a feeding lesion into which the motile bacteria will be inoculated together with tick saliva (Bockenstedt et al., 2014). The number of *Borrelia* inoculated during the overall blood meal is estimated to be in the hundreds (Kern et al., 2011). At this time point of transmission, OspC is essential since its absence leads to an abortion of the infection in the vertebrate host (Tilly et al., 2006). Its expression is enhanced while infected ticks feed and transmit spirochetes, and it is downregulated after establishment of the infection in the vertebrate host and the presence of *Borrelia* in the skin. Indeed, the use of *Borrelia* mutants for OspC revealed that OspC is essential during the first 24h of the transmission, since *Borrelia* mutants are cleared

from the skin (Tilly et al., 2007). *Borrelia* is an extracellular bacterium and it is going to multiply in the skin. The peak is at around day 7 upon intradermal inoculation (Kern et al., 2015) and at day 10 upon a tick bite (Kern et al., 2011). In a rabbit model, OspC has also been shown to be essential for the early transmission, before VlsE, a variable surface protein, that starts to appear at day 5 after intradermal inoculation (Crother et al., 2004). *Borrelia* also expresses different proteins that interact with the extracellular matrix which can facilitate the migration through the dermis or the persistence in the skin as DbpA and DbpB with decorin, a dermatan sulfate proteoglycan (Guo et al., 1995), or BBK32 with fibronectin (Brissette and Gaultney, 2014; Hovius et al., 2007; Probert and Johnson, 1998).

As observed in the tick, *Borrelia* also binds to plasminogen and complement regulator via different adhesins that help *Borrelia* to better disseminate through host tissue (Brissette and Gaultney, 2014; Coburn et al., 2013). It has also been reported that the bacterial load in wild-type mice infected by ticks fed on plasminogen-deficient mice is significantly reduced (Coleman et al., 1997). Once in the skin, spirochetes can spread contiguously through the dermis and/or disseminate through the blood to the target organs, in particular the heart, the nervous system, joints and/or distant skin. The spreading of *Borreliae* within the skin will enhance uptake of *Borrelia* by the unfed tick. This other aspect of transmission, from the infected host to the tick, which is essential to maintain *Borrelia* in the environment, is still poorly investigated. Due to the persistence of *Borreliae* for months in the skin as shown in mouse models and in dogs, the transmission is relatively quick. It takes around 8h to be detected in ticks in the process of the blood meal (Shih et al., 1992; N. Boulanger, personal communication). This quick transmission process ensures an efficient maintenance of infectious agents in the enzoonotic cycle. The role of tick saliva in this specific process is not clear. However, a tick saliva protein, Salp25D plays a critical role in the acquisition of *Borrelia* from the infected vertebrate host. Its silencing in the salivary gland impairs specifically the tick infection with *B. burgdorferi* (Narasimhan et al., 2007). In addition, a specific phenomenon of transmission from infected to uninfected tick has been observed when they feed close to each other, described as the co-feeding phenomenon (Gern and Rais, 1996). It occurs in absence of bacteremia in the host and is suspected to be facilitated by tick saliva (Kazimírová and Štibrániová, 2013; Nuttall and Labuda, 2004). The occurrence of a local immunosuppression by tick saliva that might locally reactivate the bacterial multiplication is possible, but it has not been confirmed by quantitative PCR (N. Boulanger-unpublished data).

THE VERTEBRATE HOST: THE SKIN, A KEY INTERFACE

Skin Structure

The skin is the primary interface in ABDs, where pathogens and arthropod saliva are inoculated. It will provide the first line of defense against the transmitted pathogens, if cutaneous immune reactions are efficient (Bernard et al., 2014). The skin is a complex immune organ made of an upper layer, the epidermis, made with 90% keratinocytes with few Langerhans cells (resident dendritic cells of the epidermal layer). CD8+ T cells can also be found in this layer. Then below, the dermis has a more complex structure and a greater cell diversity with more specialized immune cells (dendritic cells, mast cells, CD4+ T cells, natural killer cells, and macrophages) within a network made by resident cells, the fibroblasts, and the extracellular matrix. Therefore, the skin is going to provide an efficient immune surveillance by trafficking of cells between the local cutaneous immune system, draining lymph nodes, and circulation (Di Meglio et al., 2011; Nestle et al., 2009). This organ

might also potentially provide immune-privileged sites for pathogens in the context of ABDs. The hair follicles seem to be an organ of choice for malaria parasites (Ménard et al., 2013) and neutrophils, cells essential for the further development of leishmaniasis (Peters et al., 2008) and trypanosomiasis (Stijlemans et al., 2014). Up to now, for *Borrelia* no such immune-privileged site has been identified, although its persistence in the skin has been clearly demonstrated.

Effect of Tick Bite and Saliva on the Vertebrate Host Skin

The tick mouthparts disrupt the skin and the hematophagous tick engorges for 3–10 days. Mouse model of being bitten by infected nymphs is the best model to study the process of *Borrelia* transmission (Barthold et al., 1990). If infected, the tick inoculates *Borreliae* that interact with TLR2 on immune cells (Wooten et al., 2002). Using in vitro and in vivo models of skin injury, it has been shown that the lacerations of the skin likely release double-strand RNA from damaged cells that activate TLR3. This TLR3 activation is increased by *Borrelia*–TLR2 interaction. However, the tick saliva inhibits this inflammation, pointing out the strong control of the saliva on all skin processes (Bernard et al., 2016).

Skin acts as a physical barrier but also contains different resident cells, keratinocytes and fibroblasts, and specific immune cells (eosinophils, mast cells, dendritic cells). Tick saliva targets all the skin resident cells and produces a favorable site for pathogen inoculation like *Borrelia*. It exerts an antialarmin effect on keratinocytes blocking all the danger signals, notably antimicrobial peptides secretion (Marchal et al., 2011). It also induces fibroblast lysis that might be responsible for formation of the feeding pool (Schramm et al., 2012).

Histological studies in BALB/c mice bitten by uninfected *I. scapularis* revealed that a first tick bite induces a vascular dilatation and few inflammatory cells recruitment at the site of tick attachment that likely facilitates the transmission and dissemination of pathogens (Krause et al., 2009). However, mice repeatedly exposed to tick bite present less vasodilatation and more inflammatory cells. Similar observations were made in people exposed regularly to tick bites. The authors conclude that repeated exposures to ticks likely protect animals and humans from TBDs (Burke et al., 2005; Krause et al., 2009). More generally, animals repeatedly exposed to tick bites develop an immunity with specific production of antibodies against the different tick saliva proteins (Vu Hai et al., 2013), impairing tick attachment and the length of the blood meal (Brossard and Wikel, 1997; Wikel and Alarcon-Chaidez, 2001). A microarray analysis performed early after the tick bite on skin lesions showed that neutrophils are major infiltrating cells by 12 h of tick feeding (Heinze et al., 2012). However, very few studies have investigated persistence of tick saliva and to what extent the immunosuppressive effect acts in situ in the host skin. Only one study demonstrated the presence of Salp15 in the skin at least for 2 days after the nymph attachment on mouse skin (Anguita et al., 1998).

Early Development of *Borrelia* in the Skin: Physiopathology

Histology of tick bite infected with *Borrelia* reveals a cellular infiltrate made of lymphocytes and macrophages with few plasma cells (Mullegger, 2004). Activated dendritic cells are also observed. Very few *Borreliae* are present in the skin at the site of inoculation. However, the skin is a site of persistence for *Borrelia* since they can be detected for months in rodent skin (Shih et al., 1995), as well as in dog skin (Krupka and Straubinger, 2010). In humans, specific clinical manifestation observed is EM, 3–30 days after the infectious tick bite in approximately 80% of patients (Stanek et al., 2012). No real convincing animal model has

been able to reproduce up to now this clinical manifestation, although such manifestation has been described in rabbit (Foley et al., 1995) and in nonhuman primate (Philipp et al., 1993). These two models are rarely used and mice do not present specific skin manifestations. This absence of good models impairs our understanding of the specific skin physiopathology of Lyme borreliosis in humans.

APPLICATIONS FOR DISEASE UNDERSTANDING AND CONTROL

Diagnostic Tools

In veterinary medicine, clinical manifestations are the most reliable tool for Lyme diagnosis. In human, the current diagnosis is based mainly on a patient's clinical examination coupled to a two-step serology (ELISA and Western Blot). Direct diagnosis relies on bacterial culture, often long and tedious, and PCR. Generally, early diagnosis will pose any problem because of the pathognomonic sign of the disease: the EM. However, this event is not present in all patients: only 80% of the population has this clinical manifestation (Stanek et al., 2012). Late diagnosis is more complex given the multitude of organs targeted by the bacteria. In some patients are described chronic Lyme disease or disease refractory to antibiotics; but these two concepts are controversial and seem to be very subjective since no live organisms are detected (Stanek et al., 2012).

To try to solve the problem of diagnosis, a targeted proteomic approach using Selected Reaction Monitoring mass spectrometry (SRM-MS/MS) has been developed to detect bacterial proteins in the skin during early infection of mouse model (Schnell et al., 2015). The authors first used the technique of nontargeted proteomics in mouse model to identify bacterial proteins in the skin. Different *Borrelia* proteins have been detected, among those OspC and flagellin. Then, the technique of targeted proteomics SRM-MS/MS has been performed and further tested in patients with EM in which *Borrelia* proteins have also been identified. The persistence of the bacteria in animal skin for months should allow to detect markers of late infection in mice that could be used in patients with disseminated infections as a new diagnostic tool.

Vaccine Development

In humans, vaccine prevention of TBD is at present far from satisfactory. While vaccination against tick-borne encephalitis (TBE), a potentially fatal neurological disease, provides a high level of protection (95%–99% efficacy), and even cross-protection between the three TBE subtypes (Loew-Baselli et al., 2011); no vaccine exists for the most prevalent TBD, Lyme borreliosis. In animals, babesiosis is currently controlled by antiprotozoal drugs and acaricides or, for *Babesia bovis* and *B. bigemina*, via unsatisfactory vaccination with attenuated parasites that can result in the risk of contamination and adverse reactions (Vos de and Bock, 2000). Then, in light of limited understanding of immunity to TBP, TBP strain diversity and, more generally, the transmission of multiple TBP by the same tick species, vaccine strategies that target tick antigens holding the promise of affording broad protection against TBD are increasingly being sought (Nuttall et al., 2006; Willadsen, 2004). These antigens might be implicated in conserved processes in vector infestation and vector capacity. Antitick vaccines could potentially reduce transmission of TBD indirectly, through reduction in tick burden, or directly, through interference with tick components that enhance TBP transmission.

Currently, vaccines derived from Bm86 (a midgut protein of *Rhipicephalus microplus*), are the only vaccines directed against ectoparasites that are commercially available (in Australia and Cuba) (de la Fuente et al., 1999; Willadsen et al., 1995). Vaccine eliciting antibodies are believed to lyze the tick's gut wall, thus interfering with feeding and subsequent egg production. When applied within an integrated management strategy, such vaccines are effective in reducing tick burden, and may thus indirectly diminish transmission of certain TBP. Such vaccination strategy works for the so-called cattle tick *R. microplus* that fed only on cattle for which the vaccine is intended. However, for several species of ticks responsible for important TBD as Lyme disease, such as *Ixodes* sp. which feed on multiple hosts and notably on wildlife, vaccines that exert a direct effect on tick blood meal acquisition or vector competence must be sought.

Several studies have reported that ticks produce differentially expressed transcripts in response to pathogen infection; some of them corresponding to factors implicated in vector competence (reviewed in Liu and Bonnet, 2014). Some of these factors can elicit protective immune responses when used to vaccinate against TBD, as demonstrated in studies involving TROSPA, SILK, Q38, and subolesin proteins (Merino et al., 2013). Proteins that play a key role in tick feeding, also represent good vaccine candidates, as immunity against such proteins may block the feeding process prior to pathogen transmission, which typically occurs many hours or even days after tick attachment. Indeed, vaccines that provide immunity against salivary proteins are likely to reduce pathogen transmission, as has been reported for Salp15 in the transmission of *Borrelia* sp. to mice by *I. scapularis* (Dai et al., 2009) or 64TRP (a cement protein) in the transmission of TBE to mice by *I. ricinus* (Labuda et al., 2006). Lastly, it was reported that vaccination against metalloproteases or the serpin iris derived from *I. ricinus* salivary glands interfered with completion of the blood meal and subsequently affected their reproductive fitness (Decrem et al., 2008; Prevot et al., 2007).

Regarding anti-*Borrelia* vaccines, no vaccine against Lyme borreliosis is currently available. One recombinant vaccine, Lymerix, was commercialized by SmithKline Beecham in the United States in 1999 but was withdrawn in 2002 for potential side effects, need of several injections, and little efficiency in Europe (Moyer, 2015). Therefore, new developments are urgently needed in this field (Schuijt et al., 2011b). Vaccine candidates are mainly the surface lipoproteins of *Borrelia*, OspA, and OspC. When OspA is used, it is tested as a transmission blocking vaccine (Poljak et al., 2012; Wressnigg et al., 2013). OspC should constitute a good candidate since it is essential in the early transmission of the bacteria when inoculated into the skin; however, this lipoprotein presents too variable sequences to be used as vaccine candidate. *Borrelia*-infected skin should also be explored by proteomics to detect and identify new vaccine candidates, essential in the early transmission and persistence of *Borrelia*.

Finally, the elaboration of a vaccine combining tick and pathogen antigens could increase the vaccine efficiency like it has been demonstrated in separate studies with Salp15 (Dai et al., 2009) and OspA on the transmission of *Borrelia* sp. by *I. scapularis* (Wressnigg et al., 2013).

CURRENT ADVANCES AND FUTURE DIRECTIONS

In addition to being an important public health problem, ticks, both as vectors and pests, generate substantial economic loss worldwide. Current control methods, consisting largely of the application of chemical acaricides, are poorly effective and/or unsustainable (development of resistance, environmental impact, cost). New approaches that are environmentally sustainable and that provide broad protection against current and future TBP are thus urgently needed. While innovative strategies like development of good diagnostic tools and vaccine development for management of TBD are urgently needed, it is unlikely that these

will emerge unless significant progress can be made in understanding the tripartite relationship between ticks, pathogens, and hosts. In this context, the focus on skin interface for the development of diagnostic tools should be developed. As in Lyme borreliosis, *Borrelia* has been shown to persist for months in the skin (N. Boulanger and M. Wooten, personal communication); mass spectrometry has been tested as potential tool to detect bacterial proteins in this organ (Schnell et al., 2015). This technique constitutes a promising tool for the detection of markers of infection in ABDs, using the skin as biological material.

Concerning vaccines, the primary rate-limiting step in development of antitick vaccines is identification of protective antigenic targets (Mulenga et al., 2000). The utilization of so-called "exposed" antigens present in saliva, rather than "concealed" tick antigens to which the host is never naturally exposed, may allow natural boosting of the host response upon exposure to ticks (Nuttall et al., 2006). Anti-tick vaccines are expected to provide broad protection against diverse TBP. They may be combined with vaccines against individual TBP, when these are characterized and cognate vaccines are available, to afford greater protection, but are also—and most significantly—expected to protect against uncharacterized TBP, whether these are currently in circulation or as yet to emerge.

Further investigation of the skin interface to detect additional tick saliva candidates for Lyme borreliosis control deserves further investigation. To improve our understanding of the immune response in the skin in response to pathogens is also essential. The discovery of the skin innate immunity in late 90s (Bernard and Gallo, 2011), and the role of the skin microbiome in inflammation, should help to understand the role of this organ in the control of infections (Grice and Segre, 2011; Sanford and Gallo, 2013). The skin seems to harbor different immune-privileged sites for pathogens as shown for malaria parasites. These pathogens persisting in the skin might be responsible of the development of a specific immunity against pathogens (Gueirard et al., 2010). As long as we do not understand the mechanisms underlying the development of pathogens transmitted by arthropods, vaccine and diagnosis success will be compromised. Studies, using intravital microscopy (Ménard et al., 2013; Peters et al., 2008) and mass spectrometry (Schnell et al., 2015), already allowed great improvements. The skin definitely represents a key organ in ABDs that deserves further investigation (Bernard et al., 2015, 2014).

References

Anguita, J., Ramamoorthi, N., Hovius, J.W.R., Das, S., Thomas, V., Persinski, R., Conze, D., Askenase, P.W., Rincón, M., Kantor, F.S., Fikrig, E., 2002. Salp15, an *Ixodes scapularis* salivary protein, inhibits CD4+ T cell activation. Immunity 16, 849–859.

Anguita, J., Rincón, M., Samanta, S., Barthold, S.W., Flavell, R.A., Fikrig, E., 1998. *Borrelia burgdorferi*-infected, interleukin-6-deficient mice have decreased Th2 responses and increased lyme arthritis. J. Infect. Dis. 178, 1512–1515.

Barthold, S., Beck, D., Hansen, G., Terwilliger, G., Moody, K., 1990. Lyme borreliosis in selected strains and ages of laboratory mice. J. Infect. Dis. 162, 133–138.

Barthold, S., de Souza, M., Janotka, J., Smith, A., Persing, D., 1993. Chronic Lyme borreliosis in the laboratory mouse. Am. J. Pathol. 143, 959–971.

Beaufays, J., Adam, B., Menten-Dedoyart, C., Fievez, L., Grosjean, A., Decrem, Y., Prévôt, P.-P., Santini, S., Brasseur, R., Brossard, M., Vanhaeverbeek, M., Bureau, F., Heinen, E., Lins, L., Vanhamme, L., Godfroid, E., 2008. Ir-LBP, an *Ixodes ricinus* tick salivary LTB4-binding lipocalin, interferes with host neutrophil function. PLoS One 3, e3987.

Bernard, J.J., Gallo, R.L., 2011. Protecting the boundary: the sentinel role of host defense peptides in the skin. CMLS 68, 2189–2199.

Bernard, Q., Gallo, R., Jaulhac, B., Nakatsuji, T., Luft, B., Yang, X., Boulanger, N., 2016. *Ixodes* tick saliva suppresses the keratinocyte cytokine response to TLR2/TLR3 ligands during early exposure to Lyme borreliosis. Exp. Dermatol. 25, 26–31.

Bernard, Q., Jaulhac, B., Boulanger, N., 2015. Skin and arthropods: An effective interaction used by pathogens in vector-borne diseases. Eur. J. Dermatol. 25 (Suppl. 1), 18–22. http://dx.doi.org/10.1684/ejd.2015.2550.

Bernard, Q., Jaulhac, B., Boulanger, N., 2014. Smuggling across the border: how arthropod-borne pathogens evade and exploit the host defense system of the skin. J. Invest. Dermatol. 134, 1211–1219.

Bockenstedt, L., Gonzalez, D., Mao, J., Li, M., Belperron, A., Haberman, A., 2014. What ticks do under your skin: two-photon intravital imaging of *Ixodes scapularis* feeding in the presence of the lyme disease spirochete. Yale J. Biol. Med. 87, 3–13.

Bowman, A., Dillwith, J., Sauer, J., 1996. Tick salivary prostaglandins: presence, origin and significance. Parasitol. Today 12, 388–396.

Brissette, C., Gaultney, R., 2014. That's my story, and I'm sticking to it–an update on *B. burgdorferi* adhesins. Front Cell Infect. Microbiol. 4, 41.

Brossard, M., Wikel, S.K., 1997. Immunology of interactions between ticks and hosts. Med. Vet. Entomol. 11, 270–276.

Burke, G., Wikel, S.K., Spielman, A., Telford, S.R., McKay, K., Krause, P.J., 2005. Hypersensitivity to ticks and Lyme disease risk. Emerg. Infect. Dis. 11, 36–41.

Chmelar, J., Anderson, J., Mu, J., Jochim, R., Valenzuela, J., Kopecky, J., 2008. Insight into the sialome of the castor bean tick, *Ixodes ricinus*. BMC Genomics 9, 233.

Chmelar, J., Oliveira, C.J., Rezacova, P., Francischetti, I.M.B., Kovarova, Z., Pejler, G., Kopacek, P., Ribeiro, J.M.C., Mares, M., Kopecky, J., Kotsyfakis, M., 2011. A tick salivary protein targets cathepsin G and chymase and inhibits host inflammation and platelet aggregation. Blood 117, 736–744.

Coburn, J., Leong, J., Chaconas, G., 2013. Illuminating the roles of the *Borrelia burgdorferi* adhesins. Trends Microbiol. 21, 372–379.

Coleman, J., Gebbia, J., Piesman, J., Degen, J., Bugge, T., Benach, J., 1997. Plasminogen is required for efficient dissemination of *B. burgdorferi* in ticks and for enhancement of spirochetemia in mice. Cell 89, 1111–1119.

Coumou, J., Narasimhan, S., Trentelman, J., Wagemakers, A., Koetsveld, J., Ersoz, J., Oei, A., Fikrig, E., Hovius, J., 2016. *Ixodes scapularis* dystroglycan-like protein promotes *Borrelia burgdorferi* migration from the gut. J. Mol. Med. 94, 361–370.

Couvreur, B., Beaufays, J., Charon, C., Lahaye, K., Gensale, F., Denis, V., Charloteaux, B., Decrem, Y., Prévôt, P.P., Brossard, M., Vanhamme, L., Godfroid, E., 2008. Variability and action mechanism of a family of anticomplement proteins in *Ixodes ricinus*. PLoS One 3, e1400.

Crippa, M., Rais, O., Gern, L., 2002. Investigations on the mode and dynamics of transmission and infectivity of *Borrelia burgdorferi* sensu stricto and *Borrelia afzelii* in *Ixodes ricinus* ticks. Vector Borne Zoonotic Dis. 2, 3–9.

Crother, T., Champion, C., Whitelegge, J., Aguilera, R., Wu, X., Blanco, D., Miller, J., Lovett, M., 2004. Temporal analysis of the antigenic composition of *Borrelia burgdorferi* during infection in rabbit skin. Infect. Immun. 72, 5063–5072.

Dai, J., Wang, P., Adusumilli, S., Booth, C.J., Narasimhan, S., Anguita, J., Fikrig, E., 2009. Antibodies against a tick protein, Salp15, protect mice from the Lyme disease agent. Cell. Host Microbe 6, 482–492.

Dai, S.X., Zhang, A.D., Huang, J.F., 2012. Evolution, expansion and expression of the Kunitz/BPTI gene family associated with long-term blood feeding in *Ixodes scapularis*. BMC Evol. Biol. 12, 4.

Daix, V., Schroeder, H., Praet, N., Georgin, J.-P., Chiappino, I., Gillet, L., De Fays, K., Decrem, Y., Leboulle, G., Godfroid, E., Bollen, A., Pastoret, P.-P., Gern, L., Sharp, P.M., Vanderplasschen, A., 2007. *Ixodes* ticks belonging to the *Ixodes ricinus* complex encode a family of anticomplement proteins. Insect Mol. Biol. 16, 155–166.

Dantas-Torres, F., Chomel, B.B., Otranto, D., 2012. Ticks and tick-borne diseases: a one health perspective. Trends Parasitol. 28, 437–446.

de la Fuente, J., Rodriguez, M., Montero, C., Redondo, M., Garcia-Garcia, J.C., Mendez, L., Serrano, E., Valdes, M., Enriquez, A., Canales, M., Ramos, E., Boue, O., Machado, H., Lleonart, R., 1999. Vaccination against ticks (*Boophilus* spp.): the experience with the Bm86-based vaccine Gavac. Genet. Anal. 15, 143–148.

Decrem, Y., Beaufays, J., Blasioli, V., Lahaye, K., Brossard, M., Vanhamme, L., Godfroid, E., 2008. A family of putative metalloproteases in the salivary glands of the tick *Ixodes ricinus*. FEBS J. 275, 1485–1499.

Di Meglio, P., Perera, G.K., Nestle, F.O., 2011. The multitasking organ: recent insights into skin immune function. Immunity 35, 857–869.

Fingerle, V., Rauser, S., Hammer, B., Kahl, O., Heimerl, C., Schulte-Spechtel, U., Gern, L., Wilske, B., April 2002. Dynamics of dissemination and outer surface protein expression of different European *Borrelia burgdorferi* sensu lato strains in artificially infected *Ixodes ricinus* nymphs. J. Clin. Microbiol. 40 (4), 1456–1463.

Flower, D.R., 1996. The lipocalin protein family: structure and function. Biochem. J. 318 (Pt. 1), 1–14.

Foley, D., Gayek, R., Skare, J., Wagar, E., Champion, C., Blanco, D., Lovett, M., Miller, J., 1995. Rabbit model of Lyme borreliosis: erythema migrans, infection-derived immunity, and identification of *Borrelia burgdorferi* proteins associated with virulence and protective immunity. J. Clin. Invest. 96, 965–975.

Francischetti, I., Sa-Nunes, A., Mans, B., Santos, I., Ribeiro, J., 2010. The role of saliva in tick feeding. Front. Biosci. 14, 2051–2088.

Francischetti, I.M.B., Mather, T.N., Ribeiro, J.M.C., 2004. Penthalaris, a novel recombinant five-Kunitz tissue factor pathway inhibitor (TFPI) from the salivary gland of the tick vector of Lyme disease, *Ixodes scapularis*. Thromb. Haemost. 91, 886–898.

Francischetti, I.M.B., My Pham, V., Mans, B.J., Andersen, J.F., Mather, T.N., Lane, R.S., Ribeiro, J.M.C., 2005. The transcriptome of the salivary glands of the female western black-legged tick *Ixodes pacificus* (Acari: Ixodidae). Insect Biochem. Mol. Biol. 35, 1142–1161.

Francischetti, I.M.B., Valenzuela, J.G., Andersen, J.F., Mather, T.N., Ribeiro, J.M.C., 2002. Ixolaris, a novel recombinant tissue factor pathway inhibitor (TFPI) from the salivary gland of the tick, *Ixodes scapularis*: identification of factor X and factor Xa as scaffolds for the inhibition of factor VIIa/tissue factor complex. Blood 99, 3602–3612.

Frischknecht, F., 2007. The skin as interface in the transmission of arthropod-borne pathogens. Cell. Microbiol. 9, 1630–1640.

Fuchs, H., Wallich, R., Simon, M., Kramer, M., 1994. The outer surface protein A of the spirochete *Borrelia burgdorferi* is a plasmin(ogen) receptor. Proc. Natl. Acad. Sci. U. S. A. 91, 12594–12598.

Garg, R., Juncadella, I.J., Ramamoorthi, N., Ashish, Ananthanarayanan, S.K., Thomas, V., Rincón, M., Krueger, J.K., Fikrig, E., Yengo, C.M., Anguita, J., 2006. Cutting edge: CD4 is the receptor for the tick saliva immunosuppressor, Salp15. J. Immunol. 177, 6579–6583.

Gern, L., Rais, O., 1996. Efficient transmission of *Borrelia burgdorferi* between cofeeding *Ixodes ricinus* ticks (Acari: Ixodidae). J. Med. Entomol. 33, 189–192.

Grice, E.A., Segre, J.A., 2011. The skin microbiome. Nat. Rev. Microbiol. 9, 244–253.

Gueirard, P., Tavares, J., Thiberge, S., Bernex, F., Ishino, T., Milon, G., Franke-Fayard, B., Janse, C.J., Ménard, R., Amino, R., 2010. Development of the malaria parasite in the skin of the mammalian host. Proc. Natl. Acad. Sci. U. S. A. 107, 18640–18645.

Guo, B., Norris, S., Rosenberg, L., Höök, M., September 1995. Adherence of *Borrelia burgdorferi* to the proteoglycan decorin. Infect. Immun. 63 (9), 3467–3472.

Hannier, S., Liversidge, J., Sternberg, J.M., Bowman, A.S., 2004. Characterization of the B-cell inhibitory protein factor in *Ixodes ricinus* tick saliva: a potential role in enhanced *Borrelia burgdoferi* transmission. Immunology 113, 401–408.

Heinze, D.M., Carmical, J.R., Aronson, J.F., Thangamani, S., 2012. Early immunologic events at the tick-host interface. PLoS One 7, e47301.

Hourcade, D., Akk, A., Mitchell, L., Zhou, H., Hauhart, R., Pham, C., 2016. Anti-complement activity of the *Ixodes scapularis* salivary protein Salp20. Mol. Immunol. 69, 62–69.

Hovius, J.W.R., de Jong, M.A.W.P., den Dunnen, J., Litjens, M., Fikrig, E., van der Poll, T., Gringhuis, S.I., Geijtenbeek, T.B.H., 2008. Salp15 binding to DC-sign inhibits cytokine expression by impairing both nucleosome remodeling and mRNA stabilization. PLoS Pathog. 4, e31.

Hovius, J.W.R., van Dam, A.P., Fikrig, E., 2007. Tick-host-pathogen interactions in Lyme borreliosis. Trends Parasitol. 23, 434–438.

Humair, P., Gern, L., 2000. The wild hidden face of Lyme borreliosis in Europe. Microbe. Infect. 2, 915–922.

Ivanova, L., Tomova, A., González-Acuña, D., Murúa, R., Moreno, C., Hernández, C., Cabello, J., Cabello, C., Daniels, T., Godfrey, H., Cabello, F., 2014. *Borrelia chilensis*, a new member of the *Borrelia burgdorferi* sensu lato complex that extends the range of this genospecies in the Southern Hemisphere. Env. Microbiol. 16, 1069–1080.

Juncadella, I.J., Garg, R., Ananthnarayanan, S.K., Yengo, C.M., Anguita, J., 2007. T-cell signaling pathways inhibited by the tick saliva immunosuppressor, Salp15. FEMS Immunol. Med. Microbiol. 49, 433–438.

Kahl, O., Janetzki-Mittmann, C., Gray, J., Jonas, R., Stein, J., de Boer, R., 1998. Risk of infection with *Borrelia burgdorferi* sensu lato for a host in relation to the duration of nymphal *Ixodes ricinus* feeding and the method of tick removal. X. Zentralbl Bakteriol 287, 41–52.

Kazimírová, M., Štibrániová, I., 2013. Tick salivary compounds: their role in modulation of host defences and pathogen transmission. Front. Cell. Infect. Microbiol. 3, 43.

Kern, A., Collin, E., Barthel, C., Michel, C., Jaulhac, B., Boulanger, N., 2011. Tick saliva represses innate immunity and cutaneous inflammation in a murine model of lyme disease. Vector Borne Zoonotic Dis. 11, 1343–1350.

Kern, A., Schnell, G., Bernard, Q., Bœuf, A., Jaulhac, B., Collin, E., Barthel, C., Ehret-Sabatier, L., Boulanger, N., 2015. Heterogeneity of *Borrelia burgdorferi* sensu stricto population and its involvement in *Borrelia* pathogenicity: study on murine model with specific emphasis on the skin interface. PLoS One 10, e0133195.

Kilpatrick, A., Randolph, S., 2012. Drivers, dynamics, and control of emerging vector-borne zoonotic diseases. Lancet 380, 1946–1955.

Kotal, J., Langhansova, H., Lieskovska, J., Andersen, J.F., Francischetti, I.M., Chavakis, T., Kopecky, J., Pedra, J.H., Kotsyfakis, M., Chmelar, J., 2015. Modulation of host immunity by tick saliva. J. Proteomic. 128, 58–68.

Krause, P.J., Grant-Kels, J.M., Tahan, S.R., Dardick, K.R., Alarcon-Chaidez, F., Bouchard, K., Visini, C., Deriso, C., Foppa, I.M., Wikel, S., 2009. Dermatologic changes induced by repeated *Ixodes scapularis* bites and implications for prevention of tick-borne infection. Vector Borne Zoonotic Dis. 9, 603–610.

Krupka, I., Straubinger, R., 2010. Lyme borreliosis in dogs and cats: background, diagnosis, treatment and prevention of infections with *Borrelia burgdorferi* sensu stricto. Vet. Clin. North Am. Small Anim. Pr. 40, 1103–1119.

Kubes, M., Kocakova, P., Slovak, M., Slavikova, M., Fuchsberger, N., Nuttall, P.A., 2002. Heterogeneity in the effect of different Ixodid tick species on human natural killer cell activity. Parasite Immunol. 24, 23–28.

Kurtenbach, K., Hanincová, K., Tsao, J., Margos, G., Fish, D., Ogden, N., 2006. Fundamental processes in the evolutionary ecology of Lyme borreliosis. Nat. Rev. Microbiol. 4, 660–669.

Labuda, M., Trimnell, A.R., Licková, M., Kazimírová, M., Davies, G.M., Lissina, O., Hails, R.S., Nuttall, P.A., 2006. An antivector vaccine protects against a lethal vector-borne pathogen. PLoS Pathog. 2, e27.

Lara, F.A., Lins, U., Bechara, G.H., Oliveira, P.L., 2005. Tracing heme in a living cell: hemoglobin degradation and heme traffic in digest cells of the cattle tick *Boophilus microplus*. J. Exp. Biol. 208, 3093–3101.

Leboulle, G., Crippa, M., Decrem, Y., Mejri, N., Brossard, M., Bollen, A., Godfroid, E., 2002. Characterization of a novel salivary immunosuppressive protein from *Ixodes ricinus* ticks. J. Biol. Chem. 277, 10083–10089.

Léger, E., Vourc'h, G., Vial, L., Chevillon, C., McCoy, K., 2013. Changing distributions of ticks: causes and consequences. Exp. Appl. Acarol. 59, 219–244.

Lindgren, E., Andersson, Y., Suk, J.E., Sudre, B., Semenza, J.C., 2012. Public health. Monitoring EU emerging infectious disease risk due to climate change. Science 336, 418–419.

Liu, X.Y., Bonnet, S.I., 2014. Hard tick factors implicated in pathogen transmission. PLoS Negl. Trop. Dis. 8, e2566.

Liu, X.Y., de la Fuente, J., Cote, M., Galindo, R.C., Moutailler, S., Vayssier-Taussat, M., Bonnet, S.I., 2014. IrSPI, a tick serine protease inhibitor involved in tick feeding and *Bartonella henselae* infection. PLoS Negl. Trop. Dis. 8, e2993.

Loew-Baselli, A., Poellabauer, E.M., Pavlova, B.G., Fritsch, S., Firth, C., Petermann, R., Barrett, P.N., Ehrlich, H.J., 2011. Prevention of tick-borne encephalitis by FSME-IMMUN vaccines: review of a clinical development programme. Vaccine 29, 7307–7319.

Madden, R.D., Sauer, J.R., Dillwith, J.W., 2004. A proteomics approach to characterizing tick salivary secretions. Exp. Appl. Acarol. 32, 77–87.

Madder, M., Thys, E., Achi, L., Toure, A., De Deken, R., 2011. *Rhipicephalus (Boophilus) microplus*: a most successful invasive tick species in West-Africa. Exp. Appl. Acarol. 53, 139–145.

Mans, B.J., Andersen, J.F., Francischetti, I.M.B., Valenzuela, J.G., Schwan, T.G., Pham, V.M., Garfield, M.K., Hammer, C.H., Ribeiro, J.M.C., 2008. Comparative sialomics between hard and soft ticks: implications for the evolution of blood-feeding behavior. Insect Biochem. Mol. Biol. 38, 42–58.

Marchal, C., Schramm, F., Kern, A., Luft, B., Yang, X., Schuijt, T., Hovius, J., Jaulhac, B., Boulanger, N., 2011. Antialarmin effect of tick saliva during the transmission of Lyme.

Maritz-Olivier, C., Stutzer, C., Jongejan, F., Neitz, A.W., Gaspar, A.R., 2007. Tick anti-hemostatics: targets for future vaccines and therapeutics. Trends Parasitol. 23, 397–407.

McNally, K.L., Mitzel, D.N., Anderson, J.M., Ribeiro, J.M., Valenzuela, J.G., Myers, T.G., Godinez, A., Wolfinbarger, J.B., Best, S.M., Bloom, M.E., 2012. Differential salivary gland transcript expression profile in *Ixodes scapularis* nymphs upon feeding or flavivirus infection. Ticks Tick Borne Dis. 3, 18–26.

Mead, P.S., 2015. Epidemiology of Lyme disease. Infect. Dis. Clin. North Am. 29, 187–210.

Ménard, R., Tavares, J., Cockburn, I., Markus, M., Zavala, F., Amino, R., 2013. Looking under the skin: the first steps in malarial infection and immunity. Nat. Rev. Microbiol. 11, 701–712.

Merino, O., Antunes, S., Mosqueda, J., Moreno-Cid, J.A., Perez de la Lastra, J.M., Rosario-Cruz, R., Rodriguez, S., Domingos, A., de la Fuente, J., 2013. Vaccination with proteins involved in tick-pathogen interactions reduces vector infestations and pathogen infection. Vaccine 31, 5889–5896.

Moyer, M., 2015. The growing global battle against blood-sucking ticks. Nature 524, 406–408.

Mulenga, A., Sugimoto, C., Onuma, M., 2000. Issues in tick vaccine development: identification and characterization of potential candidate vaccine antigens. Microbe. Infect. 2, 1353–1361.

Mullegger, R., 2004. Dermatological manifestations of Lyme borreliosis. Eur. J. Dermatol. 14, 296–309.

Narasimhan, S., Montgomery, R.R., DePonte, K., Tschudi, C., Marcantonio, N., Anderson, J.F., Sauer, J.R., Cappello, M., Kantor, F.S., Fikrig, E., 2004. Disruption of *Ixodes scapularis* anticoagulation by using RNA interference. Proc. Natl. Acad. Sci. U. S. A. 101, 1141–1146.

Narasimhan, S., Sukumaran, B., Bozdogan, U., Thomas, V., Liang, X., DePonte, K., Marcantonio, N., Koski, R., Anderson, J., Kantor, F., Fikrig, E., 2007. A tick antioxidant facilitates the Lyme disease agent's successful migration from the mammalian host to the arthropod vector. Cell Host Microbe. 2, 7–18.

Nestle, F.O., Di Meglio, P., Qin, J.-Z., Nickoloff, B.J., 2009. Skin immune sentinels in health and disease. Nat. Rev. Immunol. 9, 679–691.

Nielsen, C.H., Leslie, R.G., 2002. Complement's participation in acquired immunity. J. Leukoc. Biol. 72, 249–261.

Norris, S., Coburn, J., Leong, J., Hu, L.T., Hook, M., 2010. Pathobiology of lyme disease Borrelia. Borrelia Mol. Biol. Host Interact. Pathog. 299–331.

Nuttall, P., Labuda, M., 2004. Tick-host interactions: saliva-activated transmission. Parasitology 129 (Suppl.), S177–S189.

Nuttall, P.A., Trimnell, A.R., Kazimirova, M., Labuda, M., 2006. Exposed and concealed antigens as vaccine targets for controlling ticks and tick-borne diseases. Parasite Immunol. 28, 155–163.

Ogden, N., Artsob, H., Margos, G., Tsao, J., 2014. Non-ricketsial tick-borne bacteria and the diseases they cause. In: Sonenshine, D., Roe, M.R. (Eds.), Biology of eTicks. Oxford University Press, pp. 278–312.

Ohnishi, J., Piesman, J., de Silva, A.M., 2001. Antigenic and genetic heterogeneity of *Borrelia burgdorferi* populations transmitted by ticks. PNAS 98, 670–675.

Oliveira, C.J., Carvalho, W.A., Garcia, G.R., Gutierrez, F.R., de Miranda Santos, I.K., Silva, J.S., Ferreira, B.R., 2010. Tick saliva induces regulatory dendritic cells: MAP-kinases and Toll-like receptor-2 expression as potential targets. Vet. Parasitol. 167, 288–297.

Peters, N., Egen, J., Secundino, N., Debrabant, A., Kimblin, N., Kamhawi, S., Lawyer, P., Fay, M., Germain, R., Sacks, D., 2008. In vivo imaging reveals an essential role for neutrophils in leishmaniasis transmitted by sand flies. Science 322, 1634.

Philipp, M., Aydintug, M., Bohm, R.J., Cogswell, F., Dennis, V., Lanners, H., Lowrie, R.J., Roberts, E., Conway, M., Karaçorlu, M., Peyman, G., Gubler, D., Johnson, B., Piesman, J., Gu, Y., 1993. Early and early disseminated phases of Lyme disease in the rhesus monkey: a model for infection in humans. Infect. Immun. 61, 3047–3059.

Pichu, S., Ribeiro, J.M., Mather, T.N., Francischetti, I.M., 2014. Purification of a serine protease and evidence for a protein C activator from the saliva of the tick, *Ixodes scapularis*. Toxicon 77, 32–39.

Piesman, J., Schneider, B., Zeidner, N., 2001. Use of quantitative PCR to measure density of *Borrelia burgdorferi* in the midgut and salivary glands of feeding tick vectors. J. Clin. Microbiol. 39, 4145–4148.

Poljak, A., Comstedt, P., Hanner, M., Schüler, W., Meinke, A., Wizel, B., Lundberg, U., 2012. Identification and characterization of *Borrelia* antigens as potential vaccine candidates against Lyme borreliosis. Vaccine 30, 4398–4406.

Prevot, P., Couvreur, B., Denis, V., Brossard, M., Vanhamme, L., Godfroid, E., 2007. Protective immunity against *Ixodes ricinus* induced by a salivary serpin. Vaccine 25, 3284–3292.

Pritt, B., Mead, P., Johnson, D., Neitzel, D., Respicio-Kingry, L., Davis, J., Schiffman, E., Sloan, L., Schriefer, M., Replogle, A., Paskewitz, S., Ray, J., Bjork, J., Steward, C., Deedon, A., Lee, X., Kingry, L., Miller, T., Feist, M., Theel, E., Patel, R., Irish, C., Petersen, J., 2016. Identification of a novel pathogenic *Borrelia* species causing Lyme borreliosis with unusually high spirochaetaemia: a descriptive study. Lancet Infect. Dis. S1473–3099, 464–468.

Probert, W., Johnson, B., 1998. Identification of a 47 kDa fibronectin-binding protein expressed by *Borrelia burgdorferi* isolate B31. Mol. Microbiol. 30, 1003–1015.

Radolf, J.D., Caimano, M.J., Stevenson, B., Hu, L.T., 2012. Of ticks, mice and men: understanding the dual-host lifestyle of Lyme disease spirochaetes. Nat. Rev. Microbiol. 10, 87–99.

Ramamoorthi, N., Narasimhan, S., Pal, U., Bao, F.K., Yang, X.F.F., Fish, D., Anguita, J., Norgard, M.V., Kantor, F.S., Anderson, J.F., Koski, R.A., Fikrig, E., 2005. The Lyme disease agent exploits a tick protein to infect the mammalian host. Nature 436, 573–577.

Ribeiro, J.M., Makoul, G., Levine, J., Robinson, D., Spielman, A., 1985. Antihemostatic, antiinflammatory, and immunosuppressive properties of the saliva of a tick, *Ixodes dammini*. J. Exp. Med. 161, 332–344.

Ribeiro, J., Alarcon-Chaidez, F., Francischetti, I., Mans, B., Mather, T., Valenzuela, J., Wikel, S., 2006. An annotated catalog of salivary gland transcripts from *Ixodes scapularis* ticks. Insect Biochem. Mol. Biol. 36, 111–129.

Ribeiro, J., Makoul, G., Robinson, D., 1988. *Ixodes dammini*: evidence for salivary prostacyclin secretion. J. Parasitol. 74, 1068–1069.

Ribeiro, J.M., 1995. Blood-feeding arthropods: live syringes or invertebrate pharmacologists? Infect. Agents Dis. 4, 143–152.

Rizzoli, A., Hauffe, H., Carpi, G., Vourc, H.G., Neteler, M., Rosa, R., 2011. Lyme borreliosis in Europe. Euro Surveill. 16.

Rizzoli, A., Silaghi, C., Obiegala, A., Rudolf, I., Hubálek, Z., Földvári, G., Plantard, O., Vayssier-Taussat, M., Bonnet, S., Spitalská, E., Kazimírová, M., 2014. *Ixodes ricinus* and its transmitted pathogens in urban and peri-urban areas in Europe: new hazards and relevance for public health. Front Public Heal. 2, 251.

Rudenko, N., Golovchenko, M., Edwards, M., Grubhoffer, L., 2005. Differential expression of *Ixodes ricinus* tick genes induced by blood feeding or *Borrelia burgdorferi* infection. J. Med. Entomol. 42, 36–41.

Sanford, J.A., Gallo, R.L., 2013. Functions of the skin microbiota in health and disease. Semin. Immunol. 25, 370–377.

Sa-Nunes, A., Bafica, A., Lucas, D.A., Conrads, T.P., Veenstra, T.D., Andersen, J.F., Mather, T.N., Ribeiro, J.M., Francischetti, I.M., 2007. Prostaglandin E2 is a major inhibitor of dendritic cell maturation and function in *Ixodes scapularis* saliva. J. Immunol. 179, 1497–1505.

Schnell, G., Boeuf, A., Westermann, B., Jaulhac, B., Carapito, C., Boulanger, N., Ehret-Sabatier, L., 2015. Discovery and targeted proteomics on cutaneous biopsies: a promising work toward an early diagnosis of Lyme disease. Mol. Cell. Proteomic. 14, 1254–1264.

Schramm, F., Kern, A., Barthel, C., Nadaud, S., Meyer, N., Jaulhac, B., Boulanger, N., 2012. Microarray analyses of inflammation response of human dermal fibroblasts to different strains of *Borrelia burgdorferi* sensu stricto. PLoS One 7, e40046.

Schuijt, T.J., Coumou, J., Narasimhan, S., Dai, J., Deponte, K., Wouters, D., Brouwer, M., Oei, A., Roelofs, J.J., van Dam, A.P., van der Poll, T., Van't Veer, C., Hovius, J.W., Fikrig, E., 2011a. A tick mannose-binding lectin inhibitor interferes with the vertebrate complement cascade to enhance transmission of the Lyme disease agent. Cell Host Microbe 10, 136–146.

Schuijt, T.J., Hovius, J.W., van der Poll, T., Van Dam, A.P., Fikrig, E., 2011b. Lyme borreliosis vaccination: the facts, the challenge, the future. Trends Parasitol. 27, 40–47.

Schwarz, A., Valdes, J.J., Kotsyfakis, M., 2012. The role of cystatins in tick physiology and blood feeding. Ticks Tick Borne Dis. 3, 117–127.

Schwarz, A., von Reumont, B., Erhart, J., Chagas, A., Ribeiro, J., Kotsyfakis, M., 2013. De novo *Ixodes ricinus* salivary gland transcriptome analysis using two next-generation sequencing methodologies. FASEB J. 27, 4745–4756.

Shih, C., Liu, L., Spielman, A., December 1995. Differential spirochetal infectivities to vector ticks of mice chronically infected by the agent of Lyme disease. J. Clin. Microbiol. 33 (12), 3164–3168.

Shih, C.M., Pollack, R.J., Telford, S.R., Spielman, A., 1992. Delayed dissemination of Lyme disease spirochetes from the site of deposition in the skin of mice. J. Infect. Dis. 166, 827–831.

Skallova, A., Iezzi, G., Ampenberger, F., Kopf, M., Kopecky, J., 2008. Tick saliva inhibits dendritic cell migration, maturation, and function while promoting development of Th2 responses. J. Immunol. 180, 6186–6192.

Sonenshine, D.E., Anderson, J.M., 2014. Mouthparts and digestive system. In: Sonenshine, D.E., Michael Rao, R. (Eds.), Biology of Ticks. Oxford University Press, New York, pp. 122–162.

Stanek, G., Wormser, G., Gray, J., Strle, F., 2012. Lyme borreliosis. Lancet 379, 461–473.

Stijlemans, B., Leng, L., Brys, L., Sparkes, A., Vansintjan, L., Caljon, G., Raes, G., Van Den Abbeele, J., Van Ginderachter, J.A., Beschin, A., Bucala, R., De Baetselier, P., 2014. MIF contributes to *Trypanosoma brucei* associated immunopathogenicity development. PLoS Pathog. 10, e1004414.

Tilly, K., Bestor, A., Jewett, M.W., Rosa, P., 2007. Rapid clearance of Lyme disease spirochetes lacking OspC from skin. Infect. Immun. 75, 1517–1519.

Tilly, K., Krum, J.G., Bestor, A., Jewett, M.W., Grimm, D., Bueschel, D., Byram, R., Dorward, D., Vanraden, M.J., Stewart, P., Rosa, P., 2006. *Borrelia burgdorferi* OspC protein required exclusively in a crucial early stage of mammalian infection. Infect. Immun. 74, 3554–3564.

Tonk, M., Cabezas-Cruz, A., Valdés, J.J., Rego, R., Grubhoffer, L., Estrada-Peña, A., Vilcinskas, A., Kotsyfakis, M., Rahnamaeian, M., 2015. *Ixodes ricinus* defensins attack distantly-related pathogens. Dev. Comp. Immunol. 53, 358–365.

Tsao, J., 2009. Reviewing molecular adaptations of Lyme borreliosis spirochetes in the context of reproductive fitness in natural transmission cycles. Vet. Res. 40, 36.

Valenzuela, J.G., 2002. High-throughput approaches to study salivary proteins and genes from vectors of disease. Insect Biochem. Mol. Biol. 32, 1199–1209.

Valenzuela, J.G., Charlab, R., Mather, T.N., Ribeiro, J.M.C., 2000. Purification, cloning, and expression of a novel salivary anticomplement protein from the tick, *Ixodes scapularis*. J. Biol. Chem. 275, 18717–18723.

Vennestrøm, J., Jensen, P.M., 2007. Ixodes ricinus: the potential of two-dimensional gel electrophoresis as a tool for studying host-vector-pathogen interactions. Exp. Parasitol. 115, 53–58.

Vos de, A.J., Bock, R.E., 2000. Vaccination against bovine babesiosis. Ann. N. Y Acad. Sci. 916, 540–545.

Vu Hai, V., Pages, F., Boulanger, N., Audebert, S., Parola, P., Almeras, L., 2013. Immunoproteomic identification of antigenic salivary biomarkers detected by *Ixodes ricinus*-exposed rabbit sera. Ticks Tick. Borne. Dis. 4, 459–468.

Wagemakers, A., Coumou, J., Schuijt, T., Oei, A., Nijhof, A., van't, V.C., van der Poll, T., Bins, A., Hovius, J., 2016. An Ixodes ricinus tick salivary lectin pathway inhibitor protects *Borrelia burgdorferi* sensu lato from human complement. Vector Borne Zoonotic Dis. 16, 223–228.

Wikel, S.K., 1999. Tick modulation of host immunity: an important factor in pathogen transmission. Int. J. Parasitol. 29, 851–859.

Wikel, S., 1982a. Histamine content of tick attachment sites and the effects of H1 and H2 histamine antagonists on the expression of resistance. Ann. Trop. Med. Parasitol. 76, 179–185.

Wikel, S., 1982b. Influence of *Dermacentor andersoni* infestation on lymphocyte responsiveness to mitogens. Ann. Trop. Med. Parasitol. 76, 627–632.

Wikel, S.K., 2013. Ticks and tick-borne pathogens at the cutaneous interface: host defenses, tick countermeasures, and a suitable environment for pathogen establishment. Front. Microbiol. 4, 337.

Wikel, S.K., Alarcon-Chaidez, F.J., 2001. Progress toward molecular characterization of ectoparasite modulation of host immunity. Vet. Parasitol. 101, 275–287.

Willadsen, P., 2004. Anti-tick vaccines. Parasitology 129 (Suppl.), S367–S387.

Willadsen, P., Bird, P., Cobon, G.S., Hungerford, J., 1995. Commercialisation of a recombinant vaccine against *Boophilus microplus*. Parasitology 110 (Suppl.), S43–S50.

Wooten, R., Weis, J., 2001. Host-pathogen interactions promoting inflammatory Lyme arthritis: use of mouse models for dissection of disease processes. Curr. Opin. Microbiol. 4, 274–279.

Wooten, R.M., Ma, Y., Yoder, R.A., Brown, J.P., Weis, J.H., Zachary, J.F., Kirschning, C.J., Weis, J.J., 2002. Toll-like receptor 2 is required for innate, but not acquired, host defense to *Borrelia burgdorferi*. J. Immunol. 168, 348–355.

Wressnigg, N., Pöllabauer, E., Aichinger, G., Portsmouth, D., Löw-Baselli, A., Fritsch, S., Livey, I., Crowe, B., Schwendinger, M., Brühl, P., Pilz, A., Dvorak, T., Singer, J., Firth, C., Luft, B., Schmitt, B., Zeitlinger, M., Müller, M., Kollaritsch, H., Paulke-Korinek, M., Esen, M., Kremsner, P., Ehrlich, H., Barrett, P., 2013. Safety and immunogenicity of a novel multivalent OspA vaccine against Lyme borreliosis in healthy adults: a double-blind, randomised, dose-escalation phase 1/2 trial. Lancet Infect. Dis. 13, 680–689.

Zhang, L., Zhang, Y., Adusumilli, S., Liu, L., Narasimhan, S., Dai, J., Zhao, Y., Fikrig, E., 2011. Molecular interactions that enable movement of the Lyme disease agent from the tick gut into the hemolymph. PLoS Pathog. 7, e1002079.

Translation of Saliva Proteins Into Tools to Prevent Vector-Borne Disease Transmission

Sukanya Narasimhan[1], Tyler R. Schleicher[1], Erol Fikrig[1,2]

[1]Yale University School of Medicine, New Haven, CT, United States; [2]Howard Hughes Medical Institute, Chevy Chase, MD, United States

INTRODUCTION

Blood-feeding arthropods are vectors of human diseases worldwide (Russel et al., 2013). There are at least 12,000 species of blood-feeding arthropods (Lehane, 2005) with approximately 1000 arthropod species capable of transmitting human pathogens (Krenn and Aspock, 2012). Listed in Table 14.1 are some of the major human disease agents transmitted by hematophagous arthropod vectors. Vector-borne diseases, which account for about 20% of the global burden of infectious diseases, annually result in greater than 1 million deaths and nearly 1 billion disability adjusted life years (Murray et al., 2012). Industrialization and urbanization of the developed world has changed the dynamics of many vector-borne diseases by increasing the spread of vectors and chances of pathogen transmission (Knudsen and Slooff, 1992; Neiderud, 2015). As climate changes, market globalization, and concomitant increase in travel pose the potential of shifting the geographic ranges of arthropods (Aderson, 2010; Sutherst, 2004; Vora, 2008), new threats have emerged in unexpected places by bringing arthropod vectors and vector-borne disease agents such as West Nile virus (Roehrig et al., 2002), chikungunya (Fredericks and Fernandez-Sesma, 2014), and Zika virus (Weaver et al., 2016) to the developed world. Emergence of vectors and pathogens in non endemic areas is significant, as an immunologically naïve population is likely to be more severely affected by these newly encountered pathogens (Kilpatrick and Randolph, 2012; Sutherst, 2004). There exists an urgency to control the prevalence and spread of vector-borne diseases worldwide. Management of vector-borne diseases has traditionally relied on insecticides and acaricides (Roberts and Andre, 1994; Wilkinson, 1976). Development of resistance to these chemical treatments and the collateral damage to nontarget species and the environment (Brogdon and McAllister, 2004; Roberts and Andre, 1994;

TABLE 14.1 Transmission Patterns of Major Vector-Borne Diseases That Threaten Humans

Vector	Pathogen	Transmission	Disease	Cycle
MOSQUITOES				
Anopheles spp.	*Plasmodium* spp.	Saliva	Malaria	Anthroponotic
Aedes aegypti	Dengue virus	Saliva	Dengue fever	Anthroponotic
Aedes	Yellow fever virus	Saliva	Yellow fever	Anthroponotic
Aedes	Chikungunya virus	Saliva	Chikungunya	Anthroponotic
Aedes	Zika virus	Saliva		Anthroponotic
Aedes	Rift Valley fever virus	Saliva	Rift Valley fever	Zoonotic
Ae. triseriatus	La Crosse virus	Saliva	Encephalitis	Zoonotic
Aedes, Culex	Encephalitis viruses (Eastern, Western, Venezuelan Equine, Japanese, St. Louis)	Saliva	Encephalitis	Zoonotic
Culex	Sindbis virus	Saliva	Sindbis fever	Zoonotic
Anopheles	O'nyong–nyong virus	Saliva	O'nyong–nyong fever	Anthroponotic
Culex spp.	West Nile virus	Saliva	Encephalitis	Zoonotic
Anopheles, Aedes, Culex	*Wuchereria bancrofti*	Saliva	Lymphatic filariasis	Anthroponotic
TICKS				
Ixodes scapularis	*Borrelia burgdorferi*	Saliva	Lyme disease	Zoonotic
Dermacentor, Rhipicephalus	*Rickettsia rickettsii*	Saliva	Rocky mountain spotted fever	Zoonotic
I. scapularis	*Babesia microti*	Saliva	Babesiosis	Zoonotic
I. scapularis	Powassan virus	Saliva	Powassan disease	Zoonotic
I. scapularis	*Anaplasma phagocytophilum*	Saliva	Anaplasmosis	Zoonotic
Dermacentor	Colorado tick fever virus	Saliva	Colorado tick fever	Zoonotic
Ambylomma americanum	*Ehrlichia* spp.	Saliva	Ehrlichiosis	Zoonotic
Ornithodoros spp.	*Borrelia* spp.	Saliva	Relapsing fever	Zoonotic
A. americanum	Unknown	Saliva	Southern tick-associated Rash Illness (STARI)	Unknown
Dermacentor, Ambylomma	*Francisella tularensis*	Saliva	Tularemia	Zoonotic
Hyalomma spp.	Crimean-Congo Hemorrhagic fever virus	Saliva	Hemorrhagic fever	Zoonotic
TSETSE FLIES				
Glossina morsitans morsitans	*Trypanosoma brucei*	Saliva	African sleeping sickness	Both

TABLE 14.1 Transmission Patterns of Major Vector-Borne Diseases That Threaten Humans—cont'd

Vector	Pathogen	Transmission	Disease	Cycle
REDUVIID BUGS				
Rhodnius prolixus	*Trypanosoma cruzi*	Feces	Chagas disease	Both
SANDFLIES				
Lutzomyia and *Phlebotomus* spp.	*Leishmania* spp.	Saliva	Leishmaniasis	Both
FLEAS				
Xenopsylla cheopis	*Yersinia pestis*	Saliva	Plague	Zoonotic
BLACK FLIES				
Simulium spp.	*Onchocerca volvulus*	Saliva	River blindness	Anthroponotic
LICE				
Pediculus humanus	*Rickettsia prowazekii*	Feces	Typhus fever	Anthroponotic

Roberts et al., 1997) emphasizes the need to develop effective, economical, safe, and environment friendly alternatives to control disease vectors. Scientific insights gained in the last two decades have resulted in the emergence of new paradigms to better understand and provide novel approaches for control of vector-borne diseases (Table 14.1).

EVOLUTION OF HEMATOPHAGY

Evolution of arthropod hematophagy is closely linked to the evolution of the vertebrate hemostatic system that is thought to have initiated almost 400 million years ago (Davidson et al., 2003) and established in mammals and birds just over 200 million years ago (Doolittle and Feng, 1987). Since blood served as a means of transporting nutrients such as proteins, sugars, lipids, and minerals within the vertebrate host, the stage was set for arthropods to devise ways to obtain this nutrient-rich medium for their own survival. A comparison of the salivary proteomes of current day blood-feeding arthropods suggests that arthropods evolved this unique ectoparasitic lifestyle on several

independent occasions (Balashov Yu, 1984; Law et al., 1992; Ribeiro, 1995) (Grimaldi, 2010; Mans et al., 2008a; Schofield and Galvao, 2009). Closely related lineages evolved hematophagy independently and on different occasions as seen by related protein families with differing functional mechanisms geared toward hematophagy (Mans, 2011; Ribeiro and Francischetti, 2003).

In insects, the adaptation to hematophagy seems to have occurred at least 5 times at the order level (Mans, 2011) and even possibly up to 20 times within dipterans (true flies) (Grimaldi, 2010). Blood feeding in insects is thought to have evolved somewhere in the late Jurassic–early Cretaceous period (~200–150 MYA), and in mites and specifically in ticks around 100 MYA (Grimaldi, 2010; Mans et al., 2008a). In arachnids, hematophagy is restricted to the subclass Acari that includes mites and ticks (Walter and Proctor, 2013). Although only about 10% of Acari are obligate blood feeders, blood feeding seems to have evolved in different lineages (including Astigmata and Prostigmata, suborders of Acari) multiple times (Mans et al., 2008a; Radovsky, 1969). This might also underlie the observation that while all 900 plus species of ticks are obligate hematophagous arthropods (Walter and Proctor, 2013) distinct differences in the biology of blood-feeding is observed between Argasid and Ixodid ticks, the two major familes in ticks (Mans et al., 2002a; Mans and Neitz, 2004). Argasid or soft ticks feed for less than an hour and feed multiple times in each developmental stage, while Ixodid or hard ticks feed once in each developmental stage for several days. Even within the Ixodid ticks, significant differences, particularly in salivary gland transcriptomes, have been observed among prostriate (that includes the *Ixodes* ticks) and metastriate (that includes *Amblyomma*, *Dermacentor*, and *Rhipicephalus*) ticks (Francischetti et al., 2009).

Prior to blood feeding, many of these arthropods were predators (such as the reduviid bugs), saprophagous/scavengers (such as ticks), or sap suckers of plants and flowers (such as mosquitoes and flies), and these specific lifestyles had already equipped the arthropod with a scaffold of proteins that could be easily modified and fine-tuned to hematophagy over time. Close association with vertebrates during these ancestral activities potentially allowed the preexisting functions of digestive enzymes and toxins of a predator arthropod, proteases of a scavenger arthropod, and lipid and protein digesting enzymes of a plant feeding arthropod to be utilized for probing and obtaining blood (Lehane, 2005; Mans, 2011). For example, ancestors of hematophagous arthropods that likely fed on dead or dying arthropod remains would have utilized extracellular matrix digesting enzymes such as hyaluronidase to access the nutrients. Such an enzymatic activity has indeed been described in tick saliva, potentially facilitating the formation of hematoma during feeding (Ribeiro et al., 2000). It is conceivable that odorant proteins such as the D7 family of proteins abundant in the saliva of many insects including mosquitoes (Anderson et al., 2006; Arca et al., 2007; Mans et al., 2008a; Valenzuela et al., 2003) might have served to locate flowers and plants by mosquito ancestors and later transformed to

locate vertebrate hosts and to facilitate hematophagy by binding to biogenic amines such as histamine and serotonin to prevent vasoconstriction and platelet aggregation (Calvo et al., 2006a). The lipocalins in ticks have similarly evolved to bind histamines and serotonins to facilitate blood feeding (Paesen et al., 2000). Thus protein families specific to lineages likely expanded to converge on functions critical for successful hematophagy (Mans, 2011; Ribeiro and Francischetti, 2003).

STRATEGIES OF VERTEBRATE HOST HEMOSTASIS

It is only expected that as the blood circulatory system became established in the vertebrate system, strategies to prevent blood loss or hemostasis imperative for life, would have evolved around the same window of time (Doolittle and Feng, 1987; Munoz-Chapuli et al., 2005). Evolution of the blood coagulation network is thought to have occurred just before the divergence of teleosts (about 450 MYA) and possibly evolved through two rounds of global gene duplication (Davidson et al., 2003). A critical first step in mammalian blood coagulation is the binding of platelets, small anucleated cells derived from megakaryocytes, to collagen exposed upon tissue damage (Clemetson, 2012; Jennings, 2009) via their surface glycoprotein receptors resulting in platelet aggregation and activation. Subsequently, in conjunction with five serine proteases (factor VII [FVII]; factor IX [FIX]; factor X [FX]; protein C [PC]; and prothrombin [PT]) and five cofactors (tissue factor [TF]; factor V [FV]; factor VIII [FVIII]; thrombomodulin; and protein S) a stable fibrin clot is generated to plug the injured blood vessel. The coagulation system as a whole is a complex network of interactions between proteases and protease inhibitors regulated elegantly in a positive and negative feedback loop so that the fibrin clot that is formed to prevent blood loss is also eventually dissolved and tissue damage repaired. These processes are reviewed in detail elsewhere (Palta et al., 2014; Triplett, 2000; Versteeg et al., 2013) and are briefly summarized in Fig. 14.3 in the interest of clarity. Briefly, two pathways lead to the formation of a fibrin clot—the extrinsic and intrinsic pathways. Initiated by two different triggers, the two pathways converge on a common pathway that leads to the rapid amplification of the response to injury and formation of the clot. The extrinsic pathway is activated when vascular injury exposes TF to coagulation factors in the blood (Versteeg et al., 2013). TF binds factor VII and converts it to activated factor VIIa. The TF/VIIa complex then proteolytically activates factor IX and X to IXa and Xa. The binding of factor Xa to Va on TF-expressing cells forms the prothrombinase complex that generates increasing amounts of activated thrombin. This thrombin further activates coagulation factors and platelets that have adhered on collagen that is exposed upon tissue injury. When there is no injury, the endothelial cells lining the vascular wall prevent coagulation by the secretion of platelet inhibitors and coagulation inhibitors. Tissue injury exposes the subendothelial layer that contains collagen and Von Willebrand factor (vWF) that facilitate platelet adhesion and aggregation. Both events are critical for hemostasis. Platelet adhesion leads to degranulation of α and δ granules of platelets, which release various factors including calcium, adenosine triphosphate (ATP), adenosine diphosphate (ADP), serotonin, histamine, and epinephrine that are key mediators of pain, inflammation, vasoconstriction, and vasodilation (Assoian and Sporn, 1986). Platelet aggregation provides the surface that enables the assembly of activated coagulation factors required for the formation of a stable fibrin clot. The formation of the TF/VIIa complex and the platelet aggregation thus initiates the secondary steps of hemostasis that lead to activation of prothrombin to thrombin required for the formation of the fibrin clot. Thrombin generated by the extrinsic pathway alone is not sufficient to generate a robust response to prevent blood loss. The intrinsic pathway helps amplify the amounts of activated thrombin by the activation of factor VIII and V by thrombin to form the prothrombinase complex that swiftly accelerates the activity of factor Xa to form increasing amounts of thrombin required to form a stable clot (Palta et al., 2014). The emerging integrated model of coagulation reveals that the intrinsic and extrinsic phases are really synchronized events encompassing an initiation phase or extrinsic phase that drives the amplification and propagation phase of coagulation (Versteeg et al., 2013).

STRATEGIES OF THE ARTHROPOD TO IMPAIR HOST HEMOSTASIS

For blood feeding to commence, the arthropod must locate it's chosen vertebrate host. This in itself is a difficult task driven by odor/scent, CO_2 gradients, and touch (Bowen, 1991; Day, 2005; Gillespie et al., 2004). Once a suitable host is located, the vector penetrates the skin with coarse, fine piercing or tearing mouthparts to access the blood meal within. Regardless of whether the arthropod feeds from a cannulated vessel (like mosquitoes) or from a pool of hemorrhage (like ticks and biting flies), the first phase of exploring or probing for the blood meal is accompanied by lacerations of the blood vessels that lie in the dermis (Ribeiro and Francischetti, 2003). From this stage on, the vector has the task of circumventing host hemostasis. It is now confirmed that almost all hematophagous arthropods encode for at least one inhibitor of vasodilation, one inhibitor of platelet aggregation/adhesion, and one inhibitor of serine proteases that accelerate coagulation. But we must bear in mind that while insects such as mosquitoes, sand flies, and tsetse flies take minutes (Lehane, 2005), Argasid ticks take hours and Ixodid ticks take many days to feed to repletion (Mans and

Neitz, 2004). In addition, some vectors such as ticks feed on multiple hosts to complete their life cycles (Sonenshine and Roe, 2014). Therefore, the vast molecular diversities and repertoires of salivary proteins encoded by different arthropods are shaped by innate differences in their biology of feeding and host preferences.

When the vector damages host skin tissue and associated capillaries, vertebrate hemostasis commences immediately and is marked by: (1) vasoconstriction to decrease blood flow to the site; (2) attachment of platelets to exposed collagen, followed by aggregation and activation; (3)

engagement of tissue factor to factor VIIa to generate thrombin; and (4) further activation of platelets, thrombin, and coagulation factors, culminating in clot formation (Ribeiro and Francischetti, 2003). Arthropod saliva elaborates diverse components to impair host hemostasis starting with vasodilators to counter vasoconstriction (Fig. 14.1), the vertebrate host's first response to blood loss (Ribeiro and Francischetti, 2003). Vasodilators help enlarge the blood vessel thus making it easier to locate blood and also help increase blood flow allowing rapid engorgement. The vector of Chagas disease, *Rhodnius prolixus*, encodes a novel strategy to prevent

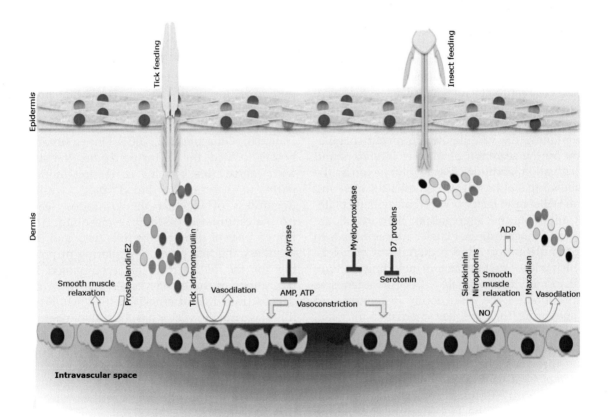

FIGURE 14.1 Schematic representation of arthropod strategies to circumvent vasoconstriction. Tissue damage concomitant with blood loss triggers vasoconstriction. Arthropod saliva circumvents this by targeting critical components that lead to vasoconstriction. The different salivary molecules depicted here represent a summary of work cited in Champagne (2005), Francischetti et al. (2009), Iwanaga et al. (2014), Kazimirova and Stibraniova (2013), Ribeiro and Francischetti (2003).

vasoconstriction by secreting at least four nitric oxide (NO)-binding heme proteins or nitrophorins (Ribeiro et al., 1993). X-ray crystallographic analysis of these proteins revealed a traditional lipocalin-fold (Andersen et al., 1997, 1998) and showed that these were indeed derivatives of lipocalins that had evolved to bind heme and serve as NO storage depots which release NO once saliva is deposited in the more neutral pH of the host dermis (Andersen et al., 2000; Weichsel et al., 1998). NO has been shown to have several physiological functions (Kubes et al., 1991) and, in this context of hematophagy, functions to increase smooth muscle relaxation leading to vasodilation (Palmer et al., 1987). The strategy of NO production to counter vasoconstriction is also used by other hematophagous arthropods. *Aedes* saliva contains sialokinins that bind to endothelial tachykinin receptors that stimulate NO production at the feeding site (Champagne and Ribeiro, 1994; Ribeiro, 1992). The saliva of the bed bug, *Cimex lectularius*, also contains a protein that has the ability to store and release NO at the feeding site (Valenzuela and Ribeiro, 1998). While the function of this protein is analogous to that of *R. prolixus* nitrophorins, the structure reveals an inositol polyphosphate 5-phosphatase fold with a novel mechanism for the reversible binding of NO (Weichsel et al., 2005), and represents a classic example of convergent evolution. Maxadilan (MAX), a peptide in the saliva of phlebotomine sandflies of the genus *Lutzomyia* (Lerner et al., 1991), has been shown to promote vasodilation by binding to the pituitary adenylate cyclase activating peptide receptor that is involved in modulating vascular tone (Moro and Lerner, 1997). MAX is about 500 times more potent than the vertebrate vasodilator, calcitonin gene-related peptide (Moro and Lerner, 1997). Old World sandflies lack MAX and use vasodilatory amines such as adenosine and AMP that are accumulated in the saliva to impair vasoconstriction (Ribeiro et al., 1999). *Anopheles* mosquitoes use yet another strategy to thwart vasoconstriction. They utilize the activity of salivary myeloperoxidases to destroy noradrenalin and serotonin, potent vasoconstrictors that are released from platelets (Ribeiro, 1996; Ribeiro and Valenzuela, 1999).

The D7 group of proteins (related to odorant-binding proteins), abundant in both mosquito and sandfly saliva, has been shown to bind to small molecules (biogenic amines such as histamine and serotonin) and impair hemostasis by inhibiting vasoconstriction (Calvo et al., 2006a; Valenzuela et al., 2002a). Ixodid ticks have been shown to release prostaglandins to promote vasodilation (Bowman et al., 1995, 1996). Soft ticks of the genus *Ornithodoros* have been shown to encode a novel vasodilator, the tick adrenomedullin (TAM), that is homolgous to vertebrate adrenomedullins (a vasodilator peptide hormone) and is thought to have been acquired from their vertebrate host by horizontal gene transfer several million years ago (Iwanaga et al., 2014).

Platelet aggregation, the critical first step in forming the plug to prevent blood loss (Ware and Suva, 2011), is dependent on ADP, collagen, and thrombin (Fig. 14.2). Several hematophagous arthropods of the genus *Aedes* (Champagne et al., 1995), *Anopheles* (Cupp et al., 1994), and *Triatomes* (Faudry et al., 2006) have been shown to rely on apyrase, a homolog of vertebrate 5'-nucleotidase, to destroy ADP and ATP at the feeding lesion (Fig. 14.2) (Faudry et al., 2004). Apyrase-like functions on novel proteins have also been identified in other arthropods (Mans et al., 1998, 2000; Valenzuela et al., 2001b). Aegyptin a collagen binding salivary protein from *Aedes aegypti* was shown to inhibit platelet aggregation by impairing the ability of collagen to interact with platelet glycoproteins (Calvo et al., 2007) (Fig. 14.2). Nitrophorin-like proteins that bind biogenic amines such as serotonin and epinephrine have also been identified in reduviid bugs (Andersen et al., 2004) and impair platelet aggregation. More recently, dipetalodipin, a protein in the saliva of *Triatoma pallidipennis* was shown to inhibit platelet aggregation

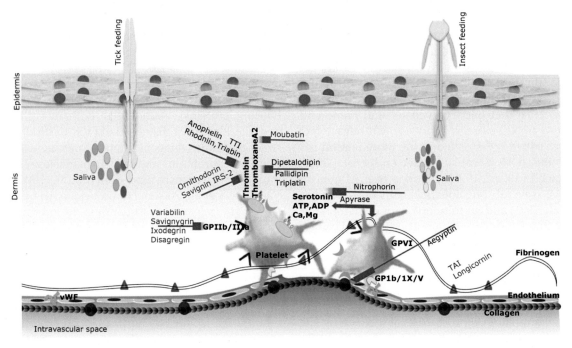

FIGURE 14.2 Overview of arthropod salivary proteins that impair platelet aggregation. Upon tissue damage platelets bind to exposed collagen and get activated releasing ATP, serotonin, ADP, epinephrine, and various prothrombotic factors including thrombin and thromboxane A2 that promote platelet aggregation and fibrin polymerization to form the clot. Tick salivary proteins (shown in *blue font*), and insect salivary proteins (shown in *black fonts*) impair various steps in platelet aggregation. The schematic summarizes findings detailed in the following review articles (Fontaine et al., 2011; Francischetti, 2010; Kazimirova and Stibraniova, 2013; Mans et al., 2008b).

by binding to thromboxane A2 (TXA2), a critical mediator of platelet aggregation (Assumpcao et al., 2010). A similar strategy is also used by triplatin, a salivary protein from the triatomine vector of Chagas disease, which binds to and prevents thromboxane A2-mediated platelet aggregation (Ma et al., 2012). Ticks encode a variety of platelet aggregation inhibitors (Fig. 14.2) and impair platelet aggregation by different strategies (Mans et al., 2008a). For instance, moubatin from *Ornithodoros moubata* prevents collagen-mediated platelet activation by binding to TXA2 (Mans and Ribeiro, 2008) while the Tick adhesion inhibitor from *O. moubata* prevents platelet binding to collagen (Karczewski et al., 1995; Mans et al., 2002b, 2003). Platelet disaggregation activity has also been observed in the saliva of diverse ticks wherein platelet

disaggregation is enabled by proteolysis of fibrinogen by apyrase itself or by blocking the interaction of fibrinogen with its platelet receptor (Francischetti, 2010; Kazimirova and Stibraniova, 2013).

Arthropod saliva also elaborates anticoagulants that target key components of the coagulation pathway including thrombin and factor X/Xa (Fig. 14.3). Both intrinsic and extrinsic pathways converge at the activation of factor X and the subsequent activation of thrombin (Fig. 14.3) and have therefore been a preferred target of hematophagous arthropods. While *Anopheles* mosquito (Calvo et al., 2004; Figueiredo et al., 2012; Ronca et al., 2012; Watanabe et al., 2011) and tsetse (Cappello et al., 1998) saliva encode inhibitors of thrombin, *Culex* mosquito saliva predominantly

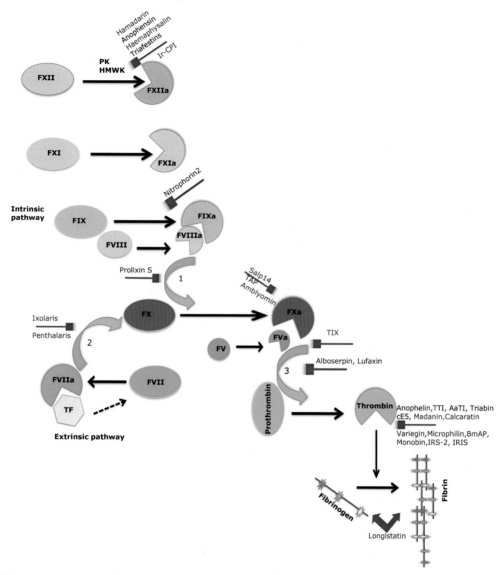

FIGURE 14.3 Schematic representation of specific steps of the vertebrate coagulation cascade targeted by arthropod salivary proteins. Vertebrate hemostasis is achieved by a series of proteases that finally culminates in the activation of thrombin and polymerization of fibrin to form a stable clot. Tick salivary proteins (shown in *blue font*) and insect salivary proteins (shown in *black font*) impair various steps of this pathway to impair coagulation. *Curved arrows* 1, 2, and 3 represent the intrinsic Xase, extrinsic Xase, and prothrombinase respectively. The schematic is a summary of literature on arthropod salivary anticoagulants (Calvo et al., 2011; Chmelar et al., 2012; Collin et al., 2012; Kazimirova and Stibraniova, 2013; Mizurini et al., 2010; Watanabe et al., 2011).

encodes inhibitors of factor X (Stark and James, 1998). In *R. prolixus*, the reduviid bug, a novel anticoagulant, nitrophorin 2 binds with high affinity to factors X and to Xa and impairs factor Xa's ability to form the Xase complex and Prolixin S blocks the intrinsic Xase complex (Hellmann and Hawkins, 1965; Mizurini et al., 2010) (Fig. 14.3).

Both soft and hard ticks stay attached to their host for hours and days, respectively (Sonenshine and Roe, 2014). Therefore, their arsenal has to be more diverse and potent. A diverse array of serine protease inhibitors or serpins are elaborated by a wide variety of ticks and thought to play a role in hemostasis (Chmelar et al., 2012; Dai et al., 2012; Francischetti et al., 2008a,b; Mans and Neitz, 2004). Serpins with the ability to block coagulation have been identified from *Ixodes*, *Amblyomma*, *Rhipicephalus*, and *Hemophysalis* species (Ibelli et al., 2014; Kim et al., 2015; Prevot et al., 2006; Rodriguez-Valle et al., 2015; Tirloni et al., 2016). Inhibitors of factor Xa have been identified from a variety of tick species and all seem to share a common Kunitz-type protein fold (Corral-Rodriguez et al., 2009; Valenzuela et al., 2002b; Ware and Suva, 2011). Ixolaris and penthalaris, two and five Kunitz domain containing proteins respectively, identified in *Ixodes. scapularis*, are potent inhibitors of the Xase complex of the extrinsic pathway (Francischetti et al., 2004; Monteiro et al., 2008) (Fig. 14.3). Salivary proteins with novel structures elaborated by *I. scapularis* saliva have also been shown to inhibit coagulation as shown by the inhibition of factor Xa by Salp14 (Narasimhan et al., 2002), and the specific inhibition of factor Xa-mediated activation of factor V by TIX (Schuijt et al., 2013) (Fig. 14.3). Further, iterating the diversity of the hemostatic strategies of ticks is the identification of several inhibitors of thrombin activity from the saliva of hard and soft ticks (Figs. 14.2 and 14.3) (Ciprandi et al., 2006; Fontaine et al., 2011; Iwanaga et al., 2003; Jablonka et al., 2015; Koh et al., 2007; Mans et al., 2008b; Xu et al., 2016; Zhu et al., 1997).

Since hemostasis occurs within minutes of vascular injury, impairment of hemostasis is imperative for both rapid and slow feeders. Therefore, hematophagous arthropods have devised multiple and even redundant strategies to impair platelet aggregation and vasoconstriction, and the propagation and amplification phases of hemostasis (Champagne, 2005; Francischetti, 2010).

Redundancy is critical to ensure that hemostasis is effectively achieved. Over the last two decades, the vector biology field has made vast progress in deciphering a molecular understanding of arthropod anticoagulation strategies (Fontaine et al., 2011; Francischetti et al., 2009; Mans, 2011). However, as genomes and salivary transcriptomes of various hematophagous arthropods unravel, it is evident that we understand only a small proportion of the functional genome (Gulia-Nuss et al, 2016; International Glossina Genome Initiative, 2014). Structural similarities do not always predict functional similarities (Valenzuela et al., 2002b); for instance, the triabin/lipocalin family of salivary proteins identified from the reduviids encompass a family of structurally related proteins with diverse functions ranging from carrier proteins to protease inhibitors to receptor binding proteins (Hernandez-Vargas et al., 2016). It is quite likely that the saliva of arthropods contains several more anticoagulant strategies that remain to be identified.

HOST IMMUNE RESPONSES TO ARTHROPOD ATTACHMENT AND FEEDING

Immune modulation at the vector–host skin interface is critical to the success of arthropod feeding. Regardless of whether it is a pool (telmophage) or vessel (solenophage) feeder, skin is breached during feeding and cutaneous innate immune responses at the epidermis and dermis represent the first line of host defense against arthropod bite and salivary components (Wikel, 2013). As mouthparts of the arthropod are inserted into the skin penetrating epidermis and even dermis, vector saliva contacts nerve endings, blood vessels, fibroblasts, and innate immune cells such as natural killer (NK) cells, neutrophils, macrophages, mast cells, basophils, eosinophils, and lymphocytes that are recruited to the bite site by an array of soluble immune mediators such as complement components, prostaglandins, cytokines, and chemokines generated in response to tissue and vascular injury (Clark, 2010; Kupper and Fuhlbrigge, 2004; Nestle et al., 2009). Cytokines and chemokines, released by various innate immune cells in response to tissue injury, complement activation and/or pathogen recognition by pathogen recognition receptors such as Toll-like receptors that mediate communication between the innate and adaptive immune cells (Lacy and Stow, 2011) orchestrate the immune response.

NK cells in addition to killing microbial pathogens regulate the functions of various innate and adaptive immune cells by secretion of cytokines including IFN-λ and TNF and thus influence the development of the immune response (O'Sullivan et al., 2015). Neutrophils function in multiple ways including by phagocytosis of microbes (Ribeiro et al., 1990) and by releasing reactive oxygen species that is likely detrimental to the vector and pathogen that is entering or exiting the host (Narasimhan et al., 2007b). Neutrophils also secrete chemoattractants/cytokines such as IL-8, IL-6, TNF, and IL-12 for further recruitment of neutrophils and initiation of wound healing (Francischetti et al., 2005). Macrophages function in presentation of vector salivary antigens and secretion of cytokines such as TNF-α and IFN-λ that direct the T_H1/T_H2 adaptive immune responses of the host (Hall and Titus, 1995; Mbow et al., 1998; Norsworthy et al., 2004). Macrophages also function in the phagocytosis of pathogens that are deposited along with vector saliva (Ribeiro and Francischetti, 2003). Basophils, and mast cells, secrete biogenic amines such as histamine and serotonins and cytokines such as IL-4 that contribute to the allergic response (Askenase et al., 1982; Kazimirova and Stibraniova, 2013; Wikel, 1999) and drive T_H2-centric immune responses (Kim et al., 2014; Otsuka et al., 2013; Siracusa et al., 2013). Eosinophils, traditionally a hallmark of helminth infections, are recruited to the vector bite site by chemokines such as CCL11/eotaxin secreted by various immune cells including T and B cells (Conroy and Williams, 2001). Eosinophils are known to secrete both T_H1 and T_H2 driving cytokines such as IFN-λ, and IL-12, IL-4, and IL-13 (Klion and Nutman, 2004; Spencer et al., 2009).

Salivary antigens are also seen by innate immune cells including keratinocytes, Langerhans cells, and dendritic cells that serve as immune sentinels of the skin and take up salivary antigens and present them to T helper cells in the lymph nodes (Merad et al., 2008; Wikel, 2013) to initiate the adaptive immune response. The innate immune responses and adaptive immune responses together direct the extent and nature of the inflammatory responses at the bite site upon primary and secondary encounters of the vector with the host. Since pathogens are deposited along with saliva at the bite site, modulation of host immune responses by the vector are paramount to pathogen survival and infection of the host (Brossard and Wikel, 2004; Nuttall and Labuda, 2004). Complement activation for instance, especially alternative and lectin pathways are critical innate immune responses and contribute to vasoactive and chemotactic functions that ultimately promote, in the case of tick feeding, acquired tick resistance resulting in tick rejection (Gordon and Allen, 1991; Wikel, 1979). Deposition of the membrane attack complexes as part of the final stages of complement activation is also detrimental to pathogens that arrive at the bite site (Schuijt et al., 2011a; Tyson et al., 2008).

Tissue injury triggers hemostatic mechanisms and in turn signals inflammation and pain sensations (Verhamme and Hoylaerts, 2009) and the two pathways intersect at many points. For example, ATP released from tissue injury, in addition to histamine and serotonin released from activated platelets, are pain inducers (Cook and McCleskey, 2002) and also increase vascular permeability (Jenne et al., 2013). Activation of factor XII in response to tissue damage results in the generation of bradykinin—another inducer of pain (Kaplan and

Ghebrehiwet, 2010). Pain would alert the host of the arthropod intruder and promote immediate removal of the arthropod by the host. In addition, ATP released upon tissue injury, as well as activated thrombin mediate neutrophil recruitment (Kaplan et al., 2015; Kaur et al., 2001). Activation and release of neutrophil contents modify the tissue matrix, exacerbate pain and vascular permeability, and escalate inflammation (Butterfield et al., 2006). Tissue injury and activation of the coagulation cascade also triggers activation of the complement cascade (Markiewski and Lambris, 2007; Neher et al., 2011). If the arthropod is infected, the feeding process also presents a septic injury resulting in activation of the complement pathway (Markiewski and Lambris, 2007), activation of leukocytes, and increased inflammatory responses (Mollnes et al., 2002; Muller, 2013). Further, within a few hours of tissue injury and the accompanying host reaction tissue repair responses such as angiogenesis, fibroblast proliferation and the regeneration of extracellular matrix is initiated (Browder et al., 2000; Li et al., 2003). Tissue repair processes are probably irrelevant to fast blood feeders such as mosquitoes and tsetse flies. However, for hard ticks that feed for 5–6 days, wound repair would prevent engorgement. Therefore, tick saliva has to circumvent host repair mechanisms as well (Francischetti et al., 2009).

Host immune responses to arthropod saliva are varied and complex and depend both on the host and vector species. Mellanby (1946) has suggested that upon encounter with mosquito salivary antigens the cutaneous immune response likely evolves from no response to delayed type hypersensitivity (DTH), to immediate type hypersensitivity (ITH), and finally to desensitization. Conceivably, in endemic areas, an individual might be at the delayed hypersensitivity stage for saliva of one mosquito species and at the immediate hypersensitivity stage for the bites of another mosquito species. At the first encounter, the dendritic cells

process the salivary antigens and this leads to the activation and expansion of antigen-specific T cells. At the second encounter, these antigen-specific T cells engage with the antigens, secrete proinflammatory cytokines resulting in the activation and recruitment of macrophages in the vicinity of the deposited antigens (Ferreira et al., 1988, 1993). It was shown, using human volunteers from sandfly endemic regions, that sandfly saliva facilitates the recruitment and infiltration of the inflammatory cells, predominantly macrophages, to the bite site within 6–12 h and inflammation peaks around 48–72 h. Biopsies of the skin bite sites showed increased presence of T_H1 cells producing the classic T_H1 cytokine IFN-γ suggesting a typical T_H1-driven DTH (Oliveira et al., 2013b). DTH and the concomitant vasodilation have been shown to be advantageous to sandflies and bed bugs (Belkaid et al., 2000) by increasing blood flow and decreasing feeding time. But, not all humans demonstrate a T_H1 response to sandfly bites—some demonstrate a T_H2 or a mixed T_H1/T_H2 response to sandfly bites iterating host-dependent modulation of immune responses (Oliveira et al., 2013b).

Mosquito bites, on the other hand, are suggested to suppress antigen-specific DTH responses (Depinay et al., 2006). In some instances, continuous exposure to salivary antigens of mosquitoes (Cornelie et al., 2007; Shen et al., 1989), ticks (Gauci et al., 1988; Parmar et al., 1996), or reduviid bugs (Chapman et al., 1986) results in the expansion of IgE-producing B cells. When tissue resident mast cells loaded with antigen-specific IgE engage with specific salivary antigens, mast cells are activated and release vasoactive amines such as histamine and diverse cytokines, especially IL-4 that allows the development of a T_H2/antibody-mediated immune response (Ribeiro and Francischetti, 2003; Wikel, 1996). Activated mast cells also produce nerve growth factor that can act on nociceptive neurons to increase their sensitivity to pain-inducing molecules such as bradykinin, histamine, and serotonin (Julius and Basbaum, 2001). Histamine is also an important mediator of the itch response (Shim and Oh, 2008) that alerts the host to the presence of the feeding arthropod. These events are immediate, hence known as ITH. The effect of histamine and bradykinin lasts for about 20–30 min and wanes once the molecules are metabolized (Ribeiro and Francischetti, 2003). Continued exposure to the salivary antigens could finally lead to the maturation of IgG subtypes with increased affinity and avidity to the specific antigens. Neutralization of critical salivary functions by host IgGs would likely deter feeding. It should be noted that not all arthropod salivary antigens induce IgE antibodies and that some arthropod saliva induce IgM followed by IgG (Schwarz et al., 2010).

An unusual and striking response to tick feeding is observed upon repeated tick infestations of guinea pigs and was originally described by Trager (1939). Seminal work by several groups (Allen and Kemp, 1982; Askenase et al., 1982; Wikel and Allen, 1976) has since shown that the immune response to repeated tick infestations of guinea pigs is characterized by the infiltration predominantly of basophils, mast cells, and eosinophils (Wikel, 1996). McLaren et al. (1983) demonstrated by elegant temporal electron microscopic examination of the tick feeding sites that secondary tick infestations of guinea pigs result in the rapid recruitment of basophils within 6–18 h of tick attachment and degranulation of basophils peaks around 18 h. Basophils were shown to comprise almost 90% of the infiltrate at the tick feeding site (Brown and Knapp, 1981) and work using basophil-depleted mice has emphasized the critical and non redundant role of basophils in mediating tick resistance (Wada et al., 2010).

The phenomenon of cutaneous basophil hypersensitivity (CBH), resembling the Jones–Mote reaction (Richerson et al., 1970), accompanied by degranulation of basophils, leads to the release of noxious components including histamine at the bite site. It has been speculated that these components are detrimental to tick feeding and additionally also leads to increased awareness of the host to tick attachment provoking host behavior such as scratching and removal of the tick (Brown et al., 1984; Wikel, 1996, 2013). Acquired resistance to tick feeding develops within a week after the primary tick infestation (Askenase et al., 1982) and is long-lasting. The basophilic recruitment and degranulation was shown to be dependent on T cells, B cells, and IgG_1 or IgE antibodies to specific tick salivary antigens (Graziano et al., 1983; Mitchell et al., 1982). Basophils have been shown to have the ability to serve as antigen presenting cells (Sokol and Medzhitov, 2010) and thus might promote Th2 memory response upon tick infestations of animals such as guinea pigs. CBH is specific only to certain tick–host relationships. For instance, *I. scapularis* infestations of guinea pigs elicit CBH (Narasimhan et al., 2007a) but not *I.scapularis* infestations of mice, its natural host (Wikel et al., 1997). *I. scapularis* infestations of mice promotes a predominantly neutrophilic recruitment at the tick bite site (Krause et al., 2009) and does not result in recruitment of basophils nor tick rejection. This dichotomy in immune response is not mechanistically understood. But, it suggests that the evolution of specific host–vector associations is likely determined by how well the vector can evade specific host immune responses (Ribeiro and Francischetti, 2003; Ribeiro et al., 1998) and that the array of functions present in arthropod saliva might be best adapted to the immune responses of certain hosts akin to a "lock-and-key" and this might drive vector–host specificity (Ribeiro et al., 1985).

Most vector saliva is believed to skew the host responses toward an anti inflammatory T_H2 immune response that is less detrimental to the vector and often beneficial to the transmitted pathogen/s (Brossard and Wikel, 2004; Caljon et al., 2006a,b; Mbow et al., 1998; Mejri and Brossard, 2007; Schoeler et al., 1999). Host immune responses to vector bites vary in the context of the host and vector species, and vectors have evolved to evade, subvert (Fontaine et al., 2011), and even co-opt the immune responses to their advantage (Belkaid et al., 2000), as outlined in the following paragraphs.

STRATEGIES OF THE ARTHROPOD TO IMPAIR HOST INFLAMMATION

Inflammation at the bite site is a continuum of interactions between the vector and innate and adaptive immune responses of the host triggered by the piercing or tearing of the skin and the salivary components spit into the bite site. By virtue of the fact that host hemostasis and inflammation are intertwined (Verhamme and Hoylaerts, 2009), the anti hemostatic mechanisms of saliva proteins often provide dual functions by impairing both hemostasis and consequently inflammation (Fig. 14.4). For instance, removal of histamine by histamine-binding lipocalins (Calvo et al., 2006a; Mans, 2005; Paesen et al., 2000) impairs vasoconstriction, as well as nociception, itching, and inflammation mediated by histamine (Shim and Oh, 2008). Impairment of platelet aggregation and

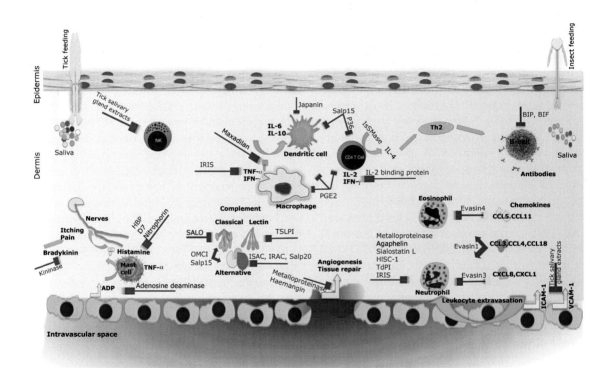

FIGURE 14.4   Schematic representation of modulation of host immune responses by arthropod salivary components. Breach of the skin barrier and rupture of capillaries in the dermis that accompany arthropod feeding triggers various immune responses in the host. Tick salivary proteins (shown in *blue font*), and insect salivary proteins (shown in *black font*) impair several components of the immune responses in order to dampen host immunity. Detailed descriptions of these salivary components and their mode of actions are available in the following references (Alarcon-Chaidez et al., 2009; Andrade et al., 2005; Charlab et al., 2000; Deruaz et al., 2008; Fontaine et al., 2011; Waisberg et al., 2014; Wikel, 2013).

thrombin activation also impair associated inflammatory signaling (Kaur et al., 2001; Verhamme and Hoylaerts, 2009). Tissue damage leads to the extracellular increase in ADP from ATP degradation and is an important mediator of platelet aggregation (Francischetti, 2010). Adenosine is also pronociceptive (Sawynok, 1998), and this can result in host awareness to the arthropod. Therefore, saliva from various arthropods express enzymes such as apyrase (Hughes, 2013) and adenosine deaminase (Kato et al., 2007; Ribeiro et al., 2001) that hydrolyze adenosine and allows the arthropod to feed on the vertebrate host.

As the innate immune response leads to an adaptive immune response (Litman et al., 2010), arthropod saliva has evolved functions that impair the ability of macrophages and dendritic cells to prevent antigen presentation to T helper cells (Anguita et al., 2002; Cross et al., 1994; Hovius et al., 2008a; Mbow et al., 1998; Ramachandra and Wikel, 1992; Wasserman et al., 2004). Ticks remain attached to the host for days, hence an elaborate, complex array of immunomodulators and tick salivary immunomodulators have been examined in greater detail compared to that from other hematophagous arthropod saliva (Fontaine et al., 2011). It is not surprising that a single species of tick (*I. scapularis*) elaborates approximately 500 or so salivary proteins (Francischetti et al., 2009; Ribeiro et al., 2006), while mosquito (*Aedes* and *Anopheles*) saliva contains about 55–100 proteins (Almeras et al., 2010; Arca et al., 2005; Calvo et al., 2006b; Ribeiro et al., 2007, 2016). Tick salivary proteins include an array of proteases, protease inhibitors with potential anti hemostatic functions (Diaz-Martin et al., 2013; Kim et al., 2016; Mans et al., 2008a; McNally et al., 2012; Prevot et al., 2006; Valenzuela et al., 2002b), anti complement proteins (Schroeder et al., 2007; Schuijt et al., 2011a; Tyson et al., 2008; Valenzuela et al., 2000), proteins that bind and inactivate cytokines (Gillespie et al., 2001; Kocakova et al., 2003), proteins that bind chemokines and impair chemotaxis of leukocytes (Deruaz et al., 2008), proteins

that prevent angiogenesis and tissue repair (Francischetti et al., 2005; Hajnicka et al., 2011), and histamine binding proteins that bind histamine (Paesen et al., 2000) so as to prevent itch responses that would alert the host of the vector.

Further, the influx of neutrophils to the tick bite site and neutrophil activation is impaired by tick salivary components that bind chemokines and cytokines involved in neutrophil migration, disintegrin metalloprotease-mediated reduction in the expression of β2 integrin adhesion molecules on neutrophils (Beaufays et al., 2008; Guo et al., 2009), and by reducing expression of adhesion molecules such as VCAM-1, ICAM-1, and P-Selectins on vascular endothelial cells required for leukocyte migration to the bite site (Kazimirova and Stibraniova, 2013; Maxwell et al., 2005). Neutrophil migration is also inhibited by the elaboration of proteins that block IL-8/CXCL8—chemokines that attract neutrophils to the bite site (Hajnicka et al., 2011; Vancova et al., 2010b). Wound healing associated with neutrophil migration is impaired by tick salivary proteins that prevent fibroblast migration and proliferation, reduce angiogenesis, and by proteins such as metalloproteinases that block remodeling of the extracellular matrix (Francischetti et al., 2009; Fukumoto et al., 2006; Islam et al., 2009). Dendritic cells (DC) that serve as a key link between innate and adaptive immunity are also impaired by saliva from diverse tick species using diverse strategies including inhibition of DC differentiation (Cavassani et al., 2005), impeding DC migration in response to chemoattractants (Oliveira et al., 2008a), impairing the elaboration of DC cytokines and reducing DC stimulation (Sa-Nunes et al., 2007), reducing the secretion of DC cytokines such as INF-α, TNF-α, INF-γ, IL-1, and IL-6, and polarizing the T-helper response toward a T_H2 response and away from a proinflammatory T_H1 and T_H17 adaptive response (Alarcon-Chaidez et al., 2009; Andrade et al., 2005; Brossard and Wikel, 2004; Fontaine et al., 2011; Heinze et al., 2012b; Kotal et al., 2015;

Preston et al., 2013; Wikel, 2013). Further, tick saliva has been shown to modulate chemokines and cytokines to effectively downregulate both innate and adaptive immune repsonses (Kazimirova and Stibraniova, 2013). Tick saliva from several *Ixodes* species has been shown to block several chemokines including the chemokine CCL3 that serves to attract neutrophils, basophils, NK cells, monocytes, and T cells; CCL8 that attracts neutrophils; CCL5 that attracts basophils and eosinophils; CCL11 that attracts eosinophils; and CCL18 that attracts T cells (Deruaz et al., 2008; Oliveira et al., 2008a; Peterkova et al., 2008; Vancova et al., 2010b). Tick saliva has also been shown to directly impact B and T cells by inhibiting cytokines such as IL-2 and impairing T-cell proliferation (Gillespie et al., 2001; Ramachandra and Wikel, 1992) and by directing a T_H2 response by suppressing expression of IL-2 and IFN-γ (T_H1 cytokines) and increasing expression of cytokines such as IL-4 (Alarcon-Chaidez et al., 2009). T-cell proliferation (Anguita et al., 2002; Leboulle et al., 2002) and CD4-mediated (Juncadella and Anguita, 2009) signaling are also impaired by *Ixodes* saliva.

It must also be noted that often the salivary strategies are counterintuitive. For instance, while it seems logical that arthropods would secrete salivary vasodilators to counter vasoconstriction and provide increased blood flow to the feeding site, this also brings to the feeding site increased numbers of leukocytes potentially detrimental to hematophagy. Perhaps, the tradeoffs are in favor of the arthropod. For example, PGE2, a prostaglandin secreted in *Ixodes* saliva impairs vasoconstriction, but also increases pain induced by bradykinin (Ribeiro et al., 1985). However, a salivary kininase activity elaborated in tick saliva helps destroy bradykinin and counters the negative influence of PGE2 in pain sensation (Ribeiro et al., 1985) (Fig. 14.4). PGE2 also modulates the inflammatory functions of macrophages, dendritic cells, and consequently of T lymphocytes (Poole et al., 2013; Ribeiro

et al., 1985; Wikel, 2013). *I. scapularis* remains attached to the host for several days and feeding proceeds in phases of slow and rapid engorgement (Anderson and Magnarelli, 2008). A tick histamine release factor is secreted by *I. scapularis* saliva during the rapid phase of engorgement and would potentially result in vasodilation (Dai et al., 2010). This increased histamine might also indirectly facilitate the onset of tissue repair (Yang et al., 2014) and likely allow retraction of tick mouthparts in preparation for tick detachment. As advances in genomics and proteomics promote a greater understanding of the fascinating functions of arthropod saliva (Almeras et al., 2010; Anderson et al., 2006; Chmelar et al., 2008, 2016a; Choumet et al., 2007; Cotte et al., 2014; Gulia-Nuss et al., 2016; Montandon et al., 2016; Valenzuela et al., 2002b), it is also becoming increasingly clear that we have only achieved a partial understanding of the functional sialome.

In general, saliva from various arthropods have been shown to suppress T_H1 cytokines such as IFN-γ and IL-2 (Fontaine et al., 2011; Wikel, 2013) and drive antibody-mediated T_H2 immune responses that are more conducive to vector feeding. These collective observations are summarized in Fig. 14.4 to provide a brief overview of the preemptive and proactive strategies of arthropod saliva to circumvent host immune responses. Not surprisingly, the pathogens transmitted by these vectors collude with the vector and exploit the salivary functions to ensure successful infection and dissemination in the host (Jones et al., 1992; Pingen et al., 2016; Ramamoorthi et al., 2005; Scholl et al., 2016; Styer et al., 2011; Thangamani et al., 2010; Titus and Ribeiro, 1988). This phenomenon of saliva-assisted/activated pathogen transmission (Nuttall and Labuda, 2004) is described later and provides the thrust for research efforts aimed at identifying and vaccine targeting transmission-promoting salivary antigens in the hope that this would impair the successful transmission of vector-borne pathogens to the vertebrate host.

SALIVA-ASSISTED PATHOGEN TRANSMISSION

The attachment of the arthropod vector to the host provides the conduit for pathogen transmission. Yet, the vector provides more than just the physical conduit—vector saliva helps shape the host immune responses in favor of pathogen survival especially in the initial phase of pathogen arrival in the host (Fig. 14.4). As noted earlier, laceration of blood vessels and tissue injury caused by the arthropod vector during feeding trigger host defense responses and the pathogen is deposited amidst this hostile environment (Chong et al., 2013; Heinze et al., 2014; Oliveira et al., 2013a; Ribeiro and Francischetti, 2003; Sakhon et al., 2013). The anti hemostatic and anti inflammatory molecules of vector saliva inadvertently facilitate pathogen survival (Figs. 14.1–14.4). Indeed, several studies have demonstrated that an inoculum of pathogen with the cognate vector saliva exacerbates infection (Table 14.2) and underscores the role of saliva in assisting pathogen transmission (Hovius et al., 2008b; Leitner et al., 2013; Nuttall and Labuda, 2004). Salivary immunomodulators that suppress vasoconstriction; complement activation, T-cell activation, inflammasome activation, dendritic cell functions, and neutrophil and macrophage functions (Abdeladhim et al., 2014; Boppana et al., 2009; Gillespie et al., 2000; Hovius et al., 2008b; Kotal et al., 2015; Moro and Lerner, 1997; Oliveira et al., 2011; Pingen et al., 2016; Schuijt et al., 2011a) passively assist the pathogen during transmission (Wang et al., 2016)

In contrast to this bystander effect of salivary proteins on pathogen transmission, Salp15, an *Ixodes* tick salivary protein shown to inhibit CD4+ T-cell activation (Anguita et al., 2002) and dendritic cell activation (Hovius et al., 2008a), additionally facilitates survival of *Borrelia burgdorferi,* the agent of Lyme disease, in the host in the early phase of transmission by directly binding to the spirochete (Ramamoorthi et al., 2005) via the spirochete outer surface protein C or OspC (Fig. 14.5). OspC is a spirochete protein critical for the spirochete early in infection (Grimm et al., 2004) and the spirochete would benefit from hiding OspC from the host's cellular and humoral responses. Binding of Salp15 to OspC thus prevented OspC from being "seen" by the host immune responses, especially when the host also has anti-OspC antibodies from prior *B. burgdorferi* infections (Ramamoorthi et al., 2005). Immunity against Salp15 resulted in rapid clearance of the spirochete in mice that had anti-*B. burgdorferi* antibodies (Dai et al., 2009). This revealed a new facet of saliva-assisted transmission and demonstrated that vector saliva–pathogen interactions were more intimate than envisaged. Further, it showed that pathogen-interacting salivary antigens could be vaccine targeted to enhance pathogen clearance (Fig. 14.5).

Whether mosquito salivary proteins can directly or indirectly influence the transmission and survival efficiency of *Plasmodium* is not understood. *Plasmodium* sporozoites deposited in the dermis by a *Plasmodium*-infected mosquito apparently remain in the skin for several hours and are vulnerable to host innate immune responses recruited in response to mosquito bite-induced tissue and vascular injury (Sinnis and Zavala, 2012). Further, parasites draining into the lymph nodes and parasites that traverse dendritic cells are thought to present parasite epitopes to T cells to initiate adaptive immune responses against the parasite and CD11c+ DCs are key players in the priming of *Plasmodium*-specific CD8+ T cells (Jung et al., 2002). Previous research on whether mosquito saliva can alter the transmission of *Plasmodium* has focused on immunization of mice with whole salivary gland extract (SGE) or repeated exposure to uninfected mosquito bites, followed by *Plasmodium* sporozoite inoculations via mosquito bite or intravenous/intradermal injections (Table 14.3). The results of these studies have been inconsistent and conflicting. Two studies have described protection from whole SGE immunization and prior exposure to

TABLE 14.2   Studies Assessing Saliva-Induced Potentiation of the Disease

Vector	Pathogen	Response	Reference(s)
MOSQUITOES			
Anopheles stephensi	*Plasmodium berghei*	Enhancement	Schneider et al. (2011)
	P. berghei	No effect	Kebaier et al. (2010)
	Plasmodium yoelii	No effect	Kebaier et al. (2010)
Aedes triseriatus	Cache Valley virus	Enhancement	Edwards et al. (1998)
	La cross virus	Enhancement	Schneider et al. (2004)
	Vesicular Stomatitis virus	Enhancement	Limesand et al. (2000)
Aedes aegypti	Cache Valley virus	Enhancement	Edwards et al. (1998)
	Sindbis virus	Enhancement	Schneider et al. (2004)
	Dengue virus	Enhancement	Conway et al. (2014)
	Rift Valley fever virus	Enhancement	Le Coupanec et al. (2013)
Culex pipiens	Cache Valley virus	Enhancement	Edwards et al. (1998)
Culex tarsalis	West Nile virus	Enhancement	Styer et al. (2011)
TSETSE FLIES			
Glossina morsitans morsitans	*Trypanosoma brucei brucei*	Enhancement	Caljon et al. (2006a,b)
REDUVIID BUGS			
Rhodnius prolixus	*Trypanosoma cruzi*	Enhancement	Mesquita et al. (2008)
SANDFLIES			
Lutzomyia longipalpis	*Leishmania infantum chagasi*	Enhancement	Warburg et al. (1994)
	Le. infantum chagasi	No effect	Gomes et al. (2008) **and** Paranhos-Silva et al. (2003)
	Leishmania amazonensis	Enhancement	Morris et al. (2001), Laurenti et al. (2009a,b), Francesquini et al. (2014), Thiakaki et al. (2005), Norsworthy et al. (2004), Theodos et al. (1991a,b)
	Leishmania braziliensis	Enhancement	Donnelly et al. (1998), Carregaro et al. (2013), Lima and Titus (1996), Samuelson et al. (1991)
	Leishmania major	Enhancement	Morris et al. (2001), Theodos et al. (1991a,b), Theodos and Titus (1993), Titus and Ribeiro (1988)
Lutzomyia flaviscutellata	*Le. amazonensis*	Inhibition	Francesquini et al. (2014)
	Le. braziliensis	Inhibition	Francesquini et al. (2014)
Lutzomyia complexus	*Le. amazonensis*	Inhibition	Francesquini et al. (2014)
	Le. braziliensis	Inhibition	Francesquini et al. (2014)

TABLE 14.2 Studies Assessing Saliva-Induced Potentiation of the Disease—cont'd

Vector	Pathogen	Response	Reference(s)
Phlebotomus papatasi	*Le. major*	Enhancement	Belkaid et al. (1998), Theodos et al. (1991a,b), Mbow et al. (1998)
FLEAS			
Xenopsylla cheopis	*Yersinia pestis*	No effect	Bosio et al. (2014)

For complete citations refer to McDowell (2015).
Table modified from McDowell, M.A., 2015. Vector-transmitted disease vaccines: targeting salivary proteins in transmission (SPIT). Trends Parasitol. 31, 363–372.

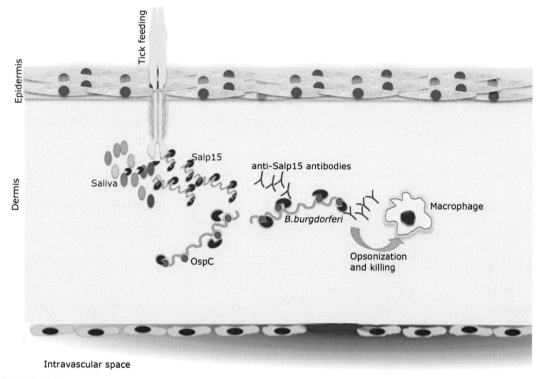

FIGURE 14.5 Vaccine targeting of arthropod salivary proteins that bind to the pathogen. Tick salivary protein, Salp15, binds to *Borrelia burgdorferi* outer surface protein OspC to prevent host immune responses against OspC. Antibodies to Salp15 deposited on the pathogen-bound Salp15 facilitate *B. burgdorferi* clearance by macrophages (Ramamoorthi et al., 2005).

mosquito bites (Alger et al., 1972; Donovan et al., 2007), while a separate study utilizing a similar approach was unable to support those findings, suggesting that antibodies to mosquito saliva do not have a significant impact on the ability of an animal to resist malaria (Kebaier et al., 2010). In addition, another study found that mosquito feeding could even enhance the severity of malaria (Schneider et al., 2011). Overall, it seems that natural mosquito bites do not provide a significant amount of protection from malaria, as individuals that live in endemic regions do not

TABLE 14.3   Studies Assessing Saliva-Induced Protection

Vector	Pathogen	Response	Reference(s)
MOSQUITOES			
Anopheles stephensi	*Plasmodium yoelii*	Protection	Donovan et al. (2007)
	Plasmodium berghei	No effect	Kebaier et al. (2010)
Unreported mosquito species	*P. berghei*	Protection	Alger et al. (1972), Alger and Harant (1976)
Anopheles fluviatilis	*Plasmodium gallinaceum*	Protection	Rocha et al. (2004)
Aedes aegypti	West Nile virus	Enhancement	Schneider et al. (2006)
Culex tarsalis	West Nile virus	Enhancement	Reagan et al. (2012)
TSETSE FLIES			
Glossina morsitans morsitans	*Trypanosoma brucei brucei*	Enhancement	Caljon et al. (2006a,b)
SAND FLIES			
Lutzomyia longipalpis	*Leishmania amazonensis*	Enhancement	Thiakaki et al. (2005)
	Leishmania major	Protection	Gomes et al. (2012), Rogers et al. (2002)
Lutzomyia intermedia	*Leishmania braziliensis*	Enhancement	de Moura et al. (2007)
Phlebotomus papatasi	*Le. major*	Protection	Kamhawi et al. (2000), Belkaid et al. (1998), Ahmed et al. (2010), Lestinova et al. (2015)
Ph. duboscqi	*Le. major*	Protection	Rohousova et al. (2011)
TICKS			
Dermacentor andersoni	*Francisella tularensis*	Protection	Bell et al. (1979)
Ixodes scapularis	*Borrelia burgdorferi*	Protection	Wikel et al. (1997)
B. burgdorferi	*B. burgdorferi*	Protection	Nazario et al. (1998)

For complete citations refer to McDowell (2015).
Table modified from McDowell, M.A., 2015. Vector-transmitted disease vaccines: targeting salivary proteins in transmission (SPIT). Trends Parasitol. 31, 363–372.

develop high levels of immunity. Immunization with SGE or the introduction of saliva by a mosquito bite could lead to the production of host antibodies to vector proteins with contrasting biological functions, thus eliminating or reducing the potential for a protective environment. It has been shown earlier that infection of mice with purified *Plasmodium yoelii* sporozoites elicits robust adaptive CD8+ T-cell responses that are detrimental to sporozoite survival (Schofield et al., 1987; Weiss et al., 1988) reducing liver infection and subsequent disease. A study has demonstrated that infection of mice with unpurified *Plasmodium* sporozoites containing *Anopheles* salivary proteins did not provide a protective immune response (Billman et al., 2016) and garners further evidence that mosquito saliva might modulate the immune response that is more favorable to parasite survival, especially in the initial phase of infection. Putative immunomodulatory functions have indeed been identified in *Anopheles* salivary transcriptomes (Andrade et al., 2005). Ascertaining the biochemical functions of these salivary proteins and targeting a subset of salivary proteins that modulate immune responses in favor of *Plasmodium* transmission might be a viable approach to impair *Plasmodium* transmission to the vertebrate host.

To date, most of the proteins in mosquito saliva have no known function. It was shown that infection of the mosquito salivary gland with *P. falciparum* induces a mosquito protein, agaphelin, which inhibits neutrophil elastase (Waisberg et al., 2014). Inhibition of a neutrophil function is an example of an immunomodulatory function of a mosquito saliva protein that might facilitate *Plasmodium* parasite survival during transmission to the host. However, the researchers did not examine whether inhibiting agaphelin actually altered transmission of *P. falciparum*. RNA silencing of some select salivary gland genes involved in inhibition of host hemostasis (*D7L2, 5′ nucleotidase, anophelin*) has been shown to disrupt feeding of *Anopheles* mosquitoes (Das et al., 2010); however, the impact on pathogen transmission has not been addressed.

Some groups of proteins are specific to particular species of mosquitoes such as the salivary gland 1 (SG1) family only found in anopheline mosquitoes (Ribeiro et al., 2010). There are seven proteins within the SG1 group of proteins, all of which currently have no predicted function or similarity to other known proteins in public databases (Arca and Lombardo, 1999; Arca et al., 2005; Ribeiro et al., 2010). A study performed a global analysis of salivary gland proteomes of several different *Anopheles* vectors and found some molecular diversity between these closely related mosquitoes (Fontaine et al., 2012). Interestingly, secreted salivary proteins were more divergent across different *Anopheles* mosquito species compared to house-keeping genes suggesting rapid evolution of salivary proteins

that likely interact with the vertebrate host (Fontaine et al., 2012). In particular, this study compared *Anopheles darlingi* and *An. albimanus* (the primary malaria vectors or North and South America) and *An. gambaie, An. arabiensis,* and *An. funestus* (the primary malaria vectors in Africa) (Fontaine et al., 2012). Nevertheless, several protein families were also commonly represented among all the anopheline species compared. The common protein families included the apyrase/5′nucleotidase, GE-rich 30 kDa antiplatelet family, long and short form of D7 proteins, SG3, SG7, SG10 or hypothetical 6.2 kDa proteins, antigen 5, and mucin/13.5 kDa proteins. Apyrase and anti platelet family proteins are known to have a role in facilitating hematophagy. SDS-PAGE analysis of protein extracts from *An. gambiae* and *An. arabiensis* belonging to the *An. gambiae* sensu lato complex showed significant similarities in protein band profiles compared to that of *An. stephensi* and *An. albimanus* that belong to *Cellia* and *Nyssorynchus* subgenera, respectively (Fontaine et al., 2012). Whether differences in salivary gland proteins may have an impact on how a vector can efficiently transmit *Plasmodium* or different species of this parasite remains to be deciphered.

Mosquito saliva has also been shown to enhance the ability of virus to cause disease (Table 14.2). A majority of these studies have focused on the co-inoculation of viruses with saliva or whole SGE. Saliva and SGE have been shown to enhance the infection of multiple viruses, including dengue, West Nile, and Rift Valley fever virus (Table 14.2). In addition, prior exposure to mosquito bites can enhance the progression of viral pathogenesis in the case of West Nile virus (Schneider et al., 2006, 2007). Very few studies have focused on the contribution of individual salivary proteins during vector transmission of a virus. One study identified a serine protease that can amplify dengue virus infection (Conway et al., 2014). This serine protease modified the environment through degradation of extracellular matrix proteins, enabling better proliferation of the virus (Conway et al., 2014). Proteins that enhance transmission might therefore be vaccine targeted to derail transmission. A study by Thangamani et al. (2010) showed that murine dermal immune responses to chikungunya virus (CHIKV) delivered by mosquito bite was significantly different from the immune responses elicited to CHIKV delivered by needle inoculation. The study showed that in presence of mosquito saliva the IL-4 levels were significantly increased and IL-2, IFN-γ, and TLR3 transcript levels decreased when compared to that seen upon needle inoculation (Thangamani et al., 2010). More interestingly, the study observed that CHIKV-infected mosquito bites elicited an almost 150-fold increase in IL-4 transcripts at the bite sites when compared to that in uninfected mosquito bite sites. These observations iterate that anti inflammatory and immunomodulatory effects of vector saliva might favor pathogen survival in the host (Wikel, 2013), and underscore the intimate interactions between the pathogen and the vector saliva during pathogen transmission.

Analyses of mosquito saliva have identified several sporozoite–mosquito interactions that occur within the salivary gland. Currently it is known that TRAP, an important sporozoite surface protein interacts with saglin, an *Anopheles* salivary gland protein (Ghosh et al., 2001, 2009; Wengelnik et al., 1999) to facilitate *Plasmodium* invasion of the salivary gland. Another interaction shown to facilitate sporozoite invasion of the mosquito salivary gland was between the circumsporozoite protein (CSP) and the mosquito CSP-binding protein (CSPBP) (Wang et al., 2013). Blocking these interactions resulted in mosquitoes with lower levels of *Plasmodium* parasites within the salivary glands. However, there are currently no proteins in saliva that have been identified to bind to *Plasmodium* sporozoites to facilitate survival in the host during the initial phase of transmission to the host. If such *Plasmodium*-binding salivary proteins are identified in future, it would extend the Salp15-OspC paradigm of a tick

salivary protein directly binding to the surface protein of the Lyme disease agent to facilitate pathogen transmission (Ramamoorthi et al., 2005) and such *Plasmodium* interacting salivary proteins may additionally serve as vaccine targets to block malaria transmission.

Saliva from the major mosquito vectors (*Aedes*, *Anopheles*, and *Culex*) is known to elicit an antibody response in the host (Ali et al., 2012; Cantillo et al., 2014; Doucoure et al., 2012; Malafronte Rdos et al., 2003; Remoue et al., 2006). For example, gSG6 an abundant salivary gland protein found in *Anopheles* mosquitoes (the major vectors of malaria) (Ali et al., 2012) was shown to induce IgG4 type antibodies (Rizzo et al., 2014). *An. gambiae* salivary protein, anophelin (also known as cE5), has been shown to induce a longer lasting IgG1-type immune response. The distinct IgG subtypes are important to note because they can highlight differences in host immune outcomes. In regards to *anophelin* and gSG6, both IgG1 and IgG4 are associated with Th2-type immune tolerant responses (Bretscher, 2014; Stevens et al., 1988) and the physiological significance of the different immune responses elicited by individual saliva proteins is not well understood. It will be important to identify the antibodies stimulated by distinct saliva antigens and determine whether the antigens lead to the development of sustained protective antibody responses, especially after exposure to different levels of natural mosquito bites. These complex interactions may have implications in the design of vaccines for saliva-transmitted pathogens.

Sandfly saliva and salivary gland lysates have been shown to facilitate the infection of mice with *Leishmania* species (Araujo-Santos et al., 2014; Norsworthy et al., 2004; Prates et al., 2011; Theodos et al., 1991a,b). Immunity to sandfly saliva was shown to protect against *Leishmania* infection by infected sandfly bites (Kamhawi et al., 2000) and offered hope that salivary proteins may be targeted to impair parasite transmission (Kamhawi et al., 2014). One of the molecules shown to be responsible for enhancing *Leishmania* infection of mice is maxadilan or MAX (Morris et al., 2001). In addition to its role as a vasodilator (Lerner and Shoemaker, 1992; Ribeiro et al., 1989), MAX was shown to dampen the immune responses of macrophages that the *Leishmania* parasites infect (Soares et al., 1998). MAX was shown to inhibit the production of NO and TNF-α by macrophages and these are required for macrophages to kill the parasite (Qureshi et al., 1996). Further, MAX and sandfly saliva modulate the cytokine responses of the macrophages (increased IL-10 and IL-6) (Fig. 14.4) in favor of *Leishmania* infection (Abdeladhim et al., 2014; Reddy et al., 2008). It is then logical to expect that vaccine targeting of critical salivary molecules could potentially impair infection. Mice vaccinated against MAX and challenged with sandflies showed reduced feeding (Milleron et al., 2004). Further when MAX-vaccinated mice were infected with a cocktail of *Leishmania major* and SGE the mice developed less severe lesions and less parasite burden (Morris et al., 2001). Thus a MAX-based vaccine would potentially decrease parasite transmission by impairing feeding, as well as impairing survival of the parasite in the host.

Repeated feeding of *Phlebotomus papatasi* (Old World sandflies) on mice resulted in a generation of antibodies against several salivary proteins including a protein called SP15 (Valenzuela et al., 2001a). SP15-based DNA vaccine protected mice from *L. major* infections and lesions. Now, using a non human primate (NHP) model, it has been further iterated that repeated exposure of animals to uninfected sandflies results in immune responses that target several salivary proteins including SP15 (Oliveira et al., 2015). Immunization of NHPs against SP15 or repeated exposure to sandfly bites provided significant protection against cutaneous leishmaniasis (Oliveira et al., 2015). When pathogens are delivered along with small doses of vector saliva (doses that mimic a natural vector bite), saliva promotes pathogen transmission (Morris et al.,

2001; Titus et al., 2006). However, when larger doses of saliva are delivered by repeated vector bites or by pre exposure to SGE, the host develops an immune response. Thus in endemic areas, where a person is likely to be repeatedly exposed to even 50–100 vector bites a day (Coleman et al., 2011), resistance to infection is likely to develop. Consistent with this posit, a study by Gomes et al. (2002) showed that children in endemic areas who develop anti sandfly salivary antibodies develop protective Th1 responses to *Leishmania* infection. In contrast, children who did not have anti sand fly salivary antibodies developed non protective Th2 responses to *Leishmania* infection (Barral et al., 2000) and this is consistent with the laboratory observations made by Oliveiro et al. (2015) using an NHP animal model. Taken together, these observations also emphasize that the most realistic approach to examine and understand vector–host–pathogen interactions and to ultimately exploit this understanding to develop strategies to block pathogen transmission is to utilize animal models of vector transmission of pathogens.

Seminal observations by William Trager (1939) showed that guinea pigs or rabbits repeatedly infested with *Dermacentor variabilis*, the dog tick, develop robust acquired immunity to *D. variabilis* salivary antigens resulting in rapid tick rejection. The link between acquired resistance to tick bite and blocking of tick-borne pathogen transmission was established when it was demonstrated that acquired resistance to uninfected *D. andersoni* induced resistance to transmission of *Francisella tularensis* by the same tick species (Bell et al., 1979). Acquired resistance to tick feeding also resulted in decreased transmission of *B. burgdorferi* to the host (Nazario et al., 1998) and bolstered the presumption that immunity to arthropod salivary antigens is a viable option to block tick feeding and pathogen transmission. Indeed, there is accumulating evidence that people exposed to frequent tick bites in natural situations develop a noticeable immune response to tick attachment (Burke et al., 2005). The human immune response to repeated tick bites was the development of cutaneous hypersensitivity and itch response that arguably resulted in avoidance behavior, removal of the tick, and a lower likelihood of acquiring tick-borne disease.

Although tick rejection is not observed on the natural/murine host upon repeated infestations of mice, repeated infestations of laboratory strains of mice with clean *I.scapularis* followed by challenge with *B. burgdorferi*-infected ticks resulted in decreased pathogen transmission (Wikel et al., 1997) and passive transfer of tick-immune rabbit sera into naïve laboratory strains of mice also impaired tick transmission of *B. burgdorferi* to the murine host without affecting tick feeding (Narasimhan et al., 2007a). Similarly, in sandfly endemic areas, it has been shown that some individuals do develop antibodies to sandfly salivary proteins with no obvious impact on sandfly feeding but with a negative impact on *Leishmania* parasite survival and severity of infection (Oliveira et al., 2013b). Taken together, these observations uncouple vector feeding from pathogen transmission and infection of the host. Immune responses of the host to the salivary components of the vector might not impair the ability of the vector to obtain its blood meal, but indirectly influence the survival ability of the pathogens deposited at the feeding site during probing and modify the disease outcome.

An interesting example of both the vector and the pathogen modulating a critical component of the immune response is the modulation of the complement pathway by tick salivary proteins such as ISAC (Valenzuela et al., 2000), Salp20 (Tyson et al., 2008), and TSLPI (Schuijt et al., 2011a). If ticks are infected with *B. burgdorferi*, the complement regulator acquiring surface proteins (CRASPs) of *B. burgdorferi* bind to host complement regulator proteins and escape complement-mediated killing (Kraiczy and Stevenson, 2013). It is then likely that a vaccine targeting a combination of tick anticomplements and *B. burgdroferi* CRASPS might be a more potent strategy to prevent *B. burgdorferi* transmission.

We must also bear in mind that not all vector-borne pathogens take the salivary route. *Trypanosoma cruzii*, the agent of Chagas disease, transmitted by the reduviid bug is excreted from the gut and deposited on the host skin when the reduviid bug takes a blood meal (Table 14.1) and enters the host via abrasions in the skin and mucosal surfaces (Rassi ct al., 2010). However, during the process of feeding, salivary components are secreted into the host and modulate the feeding site (Montandon et al., 2016; Montfort et al., 2000). Since the excreted *T. cruzii* are deposited in close proximity to the feeding site, the parasites are likely to enter the host at or near the immunomodulated site. Thus, salivary components of the reduviid bug might still serve as targets to impair *T. cruzii* transmission. Transmission of *Rickettsia prowazakii*, the agent of typhus, by the human body louse occurs in a similar manner (Table 14.1). The bacteria are shed/excreted from the gut and deposited on the skin when the louse takes a blood meal (Bechah et al., 2008). It is likely that salivary components of the louse dampen and modulate the feeding site, providing an advantage to the bacteria that enter the host following the blood-feeding event. In the rather bizarre and adventurous experiments conducted by Dr. Weigl in the 1940s, a vaccine against typhus was developed using ground up fed body lice *Pediculus humanus humanus* (Arthur, 2015). It is conceivable that salivary components included in the extract might have additionally provided immunity to prevent bacterial transmission and infection. Along the same lines, a vaccination approach currently being examined to prevent malaria transmission is the immunization of humans with whole irradiated sporozoites obtained from the dissection of mosquito salivary glands. Pilot studies have demonstrated partial protection in human volunteers (Hoffman et al., 2002). This irradiated sporozoite-based vaccine developed by Sanaria (Epstein et al., 2011) might potentially include mosquito salivary proteins bound to or co purified with the sporozoites resulting in enhanced protection.

However, the laborious nature of salivary gland dissections required to generate sporozoite vaccines for large-scale immunization of at risk populations makes this approach difficult (Menard et al., 2013).

STRATEGIES FOR IDENTIFICATION OF SALIVARY VACCINE TARGETS TO BLOCK PATHOGEN TRANSMISSION

The complexity of the saliva and the functional redundancy of the arthropod vector's salivary transcriptome suggests that a vaccine based on a single salivary protein might not be able to effectively block the ability of a vector to obtain its blood meal and concomitantly transmit harbored pathogens. In fact, vaccine targeting of individual proteins involved in preventing host hemostasis might increase probing time of fast-feeders like mosquitoes as suggested by RNA interference-mediated silencing of *An. gambiae* salivary apyrase (Boisson et al., 2006). Increasing the probing time might inadvertently increase parasite deposition in the host (Rossignol et al., 1984). Vaccine targeting of a tick gut anticoagulant protein, Ixophilin, resulted in delayed repletion without blocking tick feeding and did indeed result in increased pathogen burden in the murine host (Narasimhan et al., 2013), presumably by increasing the window of transmission time. Another study revealed that immunization of mice with recombinant D7 from *Culex tarsalis* (Reagan et al., 2012) enhanced West Nile infection, providing a good example of a protein that is abundant and antigenic, but does not provide protection. While no functional studies have been reported, the *Culex* D7 used in this study has sequence homology to other mosquito D7 proteins that bind biogenic amines and alter hemostasis as discussed earlier. Due to the anti hemostatic properties of D7, immunization and induction of antibodies may have impaired the mosquito from obtaining an efficient blood meal, leading to an increase in probing and the consequent inoculation of the host with a higher viral titer. Furthermore, in the context of pathogen transmission, it is important to bear in mind that immunity against different salivary proteins might polarize the immune response differently. For instance, immunization against the sandfly protein Sp44 provided a predominant T_H2 polarization and therefore exacerbated *Leishmania* infection. In contrast, immunization against the sandfly protein Sp15 polarized the response toward a T_H1 response and protected against *Leishmania* infection (Oliveira et al., 2008b). Therefore, salivary antigens have to be chosen wisely.

Various molecular strategies have been employed to identify potential salivary vaccine targets (Table 14.4). Screening for antigenic proteins using sera from humans or animals exposed to multiple vector bites has provided a viable first-line strategy to short list potential targets (Das et al., 2001; Lewis et al., 2015; Oliveira et al., 2008b; Rizzo et al., 2014; Schuijt et al., 2011b). Larval and nymphal hard ticks feed for several days and feeding proceeds in phases defined grossly as slow in the first 1–2 days and then rapid in the 3rd and 4th day (Anderson and Magnarelli, 2008). It is now understood that the salivary proteome changes to meet feeding phase-specific requirements (Kim et al., 2016; Narasimhan et al., 2007a; Somda et al., 2013) and is consistent with the histopathological and molecular examination of the dermis at the tick bite site that suggests distinct differences in the composition of the inflammatory milieu that accumulates in the early and late stages of feeding (Chmelar et al., 2016b; Heinze et al., 2012a; Heinze et al., 2012b; Krause et al., 2009). In *I. scapularis* nymphs, transcripts encoding for histamine binding proteins were expressed early in feeding to potentially suppress a proinflammatory milieu, pain, and itch (Dai et al., 2010), while transcripts for a histamine release factor were expressed late in feeding (Dai et al., 2010) to

TABLE 14.4   Approaches to Identify Vaccine Targets in Vector Saliva

Method	Vector	Interaction(s)	Screen	Reference(s)
IMMUNOSCREENS				
Yeast Display				
	Tick	TIX, TSLPI	Tick immune serum—SG library	Schuijt et al. (2011a,b, 2013)
	Tick	Ixofin3D	*Borrelia* OMPs—MG library	Narasimhan et al. (2014)
	Mosquito	CSP-CSPBP	*P. falciparum* CSP—SG library	Wang et al. (2013)
Phage Display				
	Mosquito	TRAP-Saglin	In vivo selection—SG and MG phage	Ghosh et al. (2001, 2009)
	Tick	Salp14, Salp15	Tick immune serum—SG library	Das et al. (2001)
Western Blot	Sandfly	SP15	SGE probed with serum from mice exposed to sandfly bites	Valenzuela et al. (2001a,b)
SILENCING TECHNIQUES				
RNA Interference				
	Tick	Subolesin	dsRNA injection and selection	de la Fuente et al. (2006)
CRISPR/Cas9				
	Mosquito (*Aedes*)	None	Method papers	Kistler et al. (2015) and Dong et al. (2015)
	Mosquito (*Anopheles*)	None	Reproductive genes	Hammond et al. (2016)

CSP, circumsporozoite protein; *MG*, midgut; *OMPs*, outer membrane proteins; *SG*, salivary gland; *SGE*, salivary gland extract.

increase vasodilation and facilitate engorgment. These findings were also consistent with the observation that tick sensitivity to histamine diminished in the late stages of feeding (Kemp and Bourne, 1980; Paine et al., 1983). Expressions of several Kunitz-domain containing protease inhibitors and anti chemokine activity in salivary glands of several Ixodid ticks have also shown temporal differences in protein expression (Schwarz et al., 2014; Vancova et al., 2010a). Ibelli et al. (2013) showed that tick saliva encodes several cystatins with some expressed preferentially early in feeding suggesting a putative function in attachment to the host or creation of a feeding lesion uptake and some expressed in the late phase suggesting a role in blood meal uptake. Acquired resistance to tick feeding results in rapid tick rejection within the first 12–24h of tick attachment. Presumably, host immunity is directed against salivary proteins expressed in the early phase and it might be critical to identify this subset of salivary proteins (Kim et al., 2016; Narasimhan et al., 2007a; Wikel, 2013).

Therefore, it has been proposed that tick salivary antigens secreted early in feeding might be more suitable vaccine targets to prevent tick feeding and the concomitant pathogen transmission at the very outset (Lewis et al., 2015; Narasimhan et al., 2007a) and studies are aimed at identifying immunogenic tick salivary antigens expressed early in feeding (Lewis et al., 2015; Radulovic et al., 2014).

Approaches based solely on antigenicity are likely to overlook critical salivary proteins that have evolved to escape host immune-detection or secreted in small yet potent amounts. For example, sialostatin L2, an *I. scapularis* salivary cystatin inhibits papain-like proteases, cathepsins L and C, key enzymes in vertebrate immunity, at micromolar concentrations (Kotsyfakis et al., 2006). Sialostatin L inhibition of cathepsin C prevents the activation of granule serine proteases in cytotoxic T lymphocytes and NK cells, mast cells, and neutrophils, and serves as an immunomodulator upon tick infestation. The potency of sialostatin L2 showed that the tick is likely to secrete small amounts of the protein to achieve sufficient inhibition of its enzymatic targets (Kotsyfakis et al., 2006). Expectedly, sera coming from guinea pigs exposed to four successive tick infestations did not recognize sialostatin L2 (Kotsyfakis et al., 2008). However, vaccination of guinea pigs against sialostatin L2 followed by tick challenge provided impaired tick feeding and decreased engorgement weights demonstrating the utility of "silent" antigens as vaccine targets to impair feeding and transmission (Kotsyfakis et al., 2008).

Reverse screens using an RNA interference approach (Table 14.4) to identify salivary proteins critical for pathogen transmission has also identified several tick salivary proteins including subolesin (de la Fuente et al., 2006, 2010; Karim et al., 2010). Subolesin functions as a transcription factor in several arthropods and in ticks regulating NF-kB-dependent and independent genes involved in development and innate immunity (de la Fuente et al., 2008; Galindo et al., 2009). Using different Ixodid tick models including *I. scapularis*, *Am. americanum*, *D. variabilis*, and *R. microplus* (Almazan et al., 2005, 2010; de la Fuente et al., 2010, 2011), it was shown that immunization against tick subolesin provided multiple phenotypes including decreased feeding, decreased survival, decreased oviposition, and decreased vectorial capacity. Subolesin orthologs known as akirins represent a family of evolutionarily conserved proteins with similar functions in vertebrates and insects (de la Fuente et al., 2011) and interestingly immunization of mice against *Ae. albopictus* akirin followed by challenge with mosquitoes, or sandflies or mites provided a significant decrease in the survival and fertility of these vectors (Canales et al., 2009; Harrington et al., 2009) suggesting that subolesin or subolesin-based vaccine could serve as a broad-spectrum vaccine against multiple arthropod vectors of pathogens (de la Fuente et al., 2011). Subolesin and akirins are intracellular proteins and challenge the traditional view that anti arthropod vaccines target extracellular or secreted antigens of the vector (Lal et al., 2001; Milleron et al., 2004; Tellam et al., 2001; Trimnell et al., 2005; Willadsen, 2004). The mechanism by which host antibodies against subolesin might enter arthropod salivary gland or midgut cells remains to be deciphered.

The CRISPR/Cas9 might offer a more stable approach for knocking down/out arthropod salivary genes and has been applied to *Ae. aegypti* and *An. gambiae* to generate knockout mosquitoes to better understand the physiological functions of non saliva proteins (Dong et al., 2015; Hammond et al., 2016; Kistler et al., 2015) and further enhance our ability to understand the functions of the salivary transcriptome and identify relevant vaccine targets to impair pathogen transmission.

Prompted by the finding that Salp15 binds directly to spirochetes and facilitates transmission (Ramamoorthi et al., 2005), the search for additional salivary proteins that might directly engage with pathogens is also being examined. In this regard, yeast surface display approach has been used to study *Plasmodium–Anopheles salivary gland* interactions (Table 14.4) (Wang et al., 2013). Briefly, a library of salivary gland peptides/proteins from the vector was expressed on the surface of yeast cells and screened with fluorescently labeled inactivated pathogens or recombinant surface antigens from the pathogen to identify potential interactions between pathogen and vector. Wang et al. (2013) identified a *Plasmodium* CSPBP that facilitated *Plasmodium* invasion of the salivary glands. While the impact of vaccine targeting this protein on *Plasmodium* transmission to the vertebrate host is not known, antibody-mediated neutralization of CSPBP reduced parasite burden in mosquito salivary glands post blood feeding (Wang et al., 2013). An interesting study by Ghosh et al. (2001) utilized a phage display peptide library to identify *Plasmodium–Anopheles* interactions and showed that a 12 amino acid peptide (SM1) specifically bound to a ligand was present in salivary gland and midguts and that this ligand was a critical *Plasmodium* interactant that promoted *Plasmodium* development in the midgut and salivary glands. Identification of the SMI-binding mosquito ligand would be essential to exploit the ligand as a potential vaccine target.

Using the rationale that the cement protein secreted into the tick feeding site ensures tick attachment, a cement protein from *R. appendiculatus* was identified as a potential salivary vaccine candidate to prevent tick-borne encephalitis virus (TBEV) transmission (Labuda et al., 2006). The possibility that regions of the cement protein conserved in several tick species might be exploited to develop a pan-vaccine against ticks has been addressed (Trimnell et al., 2005), but not come to fruition yet. The possibility of using the carbohydrate epitopes of salivary proteins has also been suggested as an approach to develop a pan-arthropod vaccine (Dinglasan et al., 2005). The N-linked glycans that posttranslationally modify many arthropod salivary proteins are immunogenic (Hoffmann-Sommergruber et al., 2011). Unlike salivary proteins, the limited diversity of these N-linked structures presents an easier route to define immunogenic glycan epitopes that might be vaccine targeted (Seppo and Tiemeyer, 2000). Nevertheless, cross-reactivity with host glycan decorations would also have to be factored into the selection criteria as with all salivary protein antigens.

DEVELOPMENT OF SALIVARY PROTEIN-BASED VACCINES—LIMITATIONS

The efforts over the last two decades driven by technological advancements in genomics (Aksoy et al., 2014; Gulia-Nuss et al., 2016; Holt et al., 2002; Valenzuela et al., 2002b), proteomics (Choumet et al., 2007; Grabowski et al., 2016; Kalume et al., 2005), and functional genetics (Dong et al., 2015; Hall et al., 2015) have expanded our understanding of the vector saliva in the context of feeding and pathogen transmission. While we are poised to translate this knowledge to strategies to control the spread of vector-borne diseases, several hurdles remain to be overcome. There are many individual and environmental factors that likely influence the development of vector-borne diseases including age and immune status of the host, in addition to the history of the host's prior vector exposure (Kilpatrick and Randolph, 2012). In endemic areas, individuals are continuously exposed to mosquito and sandfly bites (Coleman et al., 2011; Trung et al., 2004). One would expect that host immune responses to salivary proteins must present a hostile environment for the vector and the pathogen(s) it harbors. Yet, people in endemic areas do get infected and would argue against the utility of salivary proteins as vaccine targets. However, people in endemic areas tend to have less severe clinical manifestations of the disease, due potentially to a combination of immunity to vector salivary proteins and to the proteins of the pathogen (Kamhawi, 2000; Struik and Riley, 2004). Human immunity to arthropod saliva is not well characterized and the few published studies do not reveal a clear picture, with some suggesting that vector saliva elicits a T_H2 response and some suggesting that it is mixed and that it is this ratio that determines infection and disease severity (Abdeladhim et al., 2011; Andrade et al., 2009; Carvalho et al., 2015; McDowell, 2015; Oliveira et al., 2013b; Vinhas et al., 2007). The potent salivary proteins of the vector might have naturally evolved to prevent eliciting a robust immune response in the natural host. Repeated bites could induce tolerance (Feingold et al., 1968; Pasare and Medzhitov, 2003), as suggested by Mellanby (1946), to salivary antigens and compromise the efficacy of salivary antigen-based vaccine. On the other hand, continuous exposure to vector salivary antigen might also serve to boost the immune responses elicited by vaccination. But, studies on the immune response to sandfly bites in humans in sandfly endemic areas has shown that some bitten individuals maintain a T_H2 response profile that is not protective against leishmaniasis (Oliveira et al., 2013b) and might present an immunological hurdle toward developing an efficient vaccine. Therefore, an appropriate adjuvant will be an essential component of an efficient vaccine design (Petrovsky and Aguilar, 2004) to coax a protective immune response.

Experiments conducted in controlled laboratory environments using inbred mice strains and laboratory-reared arthropods do not factor in the many real-life variables and animal models might not fully recapitulate human host–vector interactions. For example, while saliva from lab-reared sandflies significantly exacerbated leishmaniasis in animal models (Ben Hadj Ahmed et al., 2010; Laurenti et al., 2009b), saliva from wild-caught flies provided only a modest effect (Laurenti et al., 2009a). In nature, the host is continuously exposed to the vector in contrast to experimental animals and could additionally influence immune responses at the bite site (Rohousova et al., 2011).

Further, the immune response of salivary proteins varies from host to host and immune responses can be protective in one host and not in the other as seen with LJM19, a sandfly salivary antigen that elicited a protective immune response against *L. infantum chagasi* and *L. braziliensis* in hamsters (Tavares et al., 2011), and in dogs (Collin et al., 2009), but not in mice (Tavares et al., 2011). Oliveira et al. (2015) have shown, using a non human primate model, that immunization of NHPs against the sandfly protein SP15 provided protection against cutaneous leishmaniasis when challenged with infected sand flies. *I. scapularis* infestations of the murine host does not elicit resistance to tick salivary antigens while guinea pigs mount a robust response to tick salivary antigens (Brown and Askenase, 1985; Wikel and Allen, 1976; Wikel et al., 1997). Such vastly differing host immune responses to vector bites remain a confounding factor in the development of vector-based vaccines. The development of humanized mouse models now offers a technique to better characterize and develop transmission-blocking vaccines for human use. Mouse strains engrafted with human hepatic cells and human red blood cells have been successfully used to support the development of *P. falciparum* and may help further advance therapeutic approaches to prevent malaria and other vector transmitted diseases (Foquet et al., 2013; Soulard et al., 2015; Vaughan et al., 2012).

CONCLUSION

At the core of hematophagy is the reward of a meal rich in various nutrients essential for the vector to mate, reproduce, and increase its population. It would be a gross oversight if we ignore the gut where the blood meal arrives bringing the vector in intimate contact with the various immune components of the host. The gut where the blood meal is processed also represents the first barrier that the pathogen surmounts when it infects the vector.

Therefore, the gut presents equally potent opportunities to derail feeding, interrupt the life cycle of the vector, and block pathogen transmission. Hence, transmission-blocking vaccines based on vector gut proteins are also being explored as a viable strategy to decrease prevalence and transmission of vector-borne pathogens (Dostalova and Volf, 2012; Kamhawi et al., 2004; Narasimhan et al., 2014; Neelakanta and Sultana, 2015; Nunes et al., 2014). Naturally, gut antigens are less likely to be deposited into the host and thus an anamnestic memory response to boost the vaccine is not available and a combination of concealed and exposed vector antigens might be more robust (Trimnell et al., 2002).

A successful tick vaccine against *R. (Boophilus) microplus* based on the gut protein, Bm86, was first developed in Australia (Willadsen et al., 1989) and further studied (de la Fuente et al., 1999; Rodriguez-Mallon, 2016) setting a precedent that validates a gut protein-based approach. Immunization of cattle against Bm86 using *Pichia pastoris*-generated recombinant Bm86 effectively inhibited *R. microplus* infestations and consequently thwarted pathogen transmission to the cattle (De La Fuente et al., 2000; de la Fuente et al., 1999; Garcia-Garcia et al., 1998; Rodriguez et al., 1995). Cattle can be infested with different *Rhipicephalus* species, and vaccination against Bm86 (commercial vaccine named Gavac) also provided efficient (>99%) protection against *B. annulatus* and *B.decoloratus* infestations suggesting that Bm86 could serve as a vaccine against multiple tick species (Fragoso et al., 1998). While Bm86 was not as effective against some strains of *B. microplus*, immunization against another gut antigen Bm95 was shown to circumvent this limitation and provide robust protection against Bm86 vaccine-resistant strains (Garcia-Garcia et al., 2000). It is not clear if Bm86 homologs of other ticks might serve as protective vaccine targets as immunization of rabbits against Bm86 homologs from *I. ricinus* did not provide protection against *I. ricinus* infestations (Coumou et al., 2014).

Traditionally, subunit vaccines against vector-borne infectious agents have targeted components of the pathogen such as OspA of *B. burgdorferi* against Lyme disease (Embers and Narasimhan, 2013; Moyer, 2015), the CSP of *P. falciparum* against malaria (Greenwood and Doumbo, 2016), or heat killed/attenuated virus particles against TBEV (Morozova et al., 2014). The OspA vaccine against Lyme disease (also known as Lymerix) is an unusual transmission-blocking vaccine that targets the spirochete in the tick gut prior to its transmission to the host (de Silva et al., 1996; Fikrig et al., 1992). *B. burgdorferi* outer surface protein OspA was shown to facilitate spirochete colonization of the tick gut and was not required for transmission to the host (Embers and Narasimhan, 2013; Radolf et al., 2012). However, *B. burgdorferi* spirochetes express OspA in the tick gut prior to the commencement of transmission (Radolf et al., 2012) and vaccination against OspA generates *Borrelicidal* anti-OspA antibodies in the host and these enter the tick gut along with the blood meal and effectively target OspA expressing spirochetes in the gut and subsequently prevent transmission (Thanassi and Schoen, 2000). The vaccine has however been removed from the market due to poor sales and due to a perceived notion of side effects related to the OspA vaccine (Plotkin, 2016).

The current malaria vaccine, RTS,S/AS01 (also known as Mosquirix), targets the CSP, an abundant protein on the surface of *P. falciparum* sporozoites. CSP has several important functions including binding to heparan sulfate proteoglycans on the surface of hepatic cells, allowing the sporozoites to efficiently invade the host liver (Hollingdale et al., 1984; Menard, 2000; Yoshida et al., 1980). Therefore, the vaccine has only a short window of time to control or kill the parasite—from the time of deposition of *Plasmodium* in the skin to the infection of the liver. Phase III clinical trials have demonstrated that the RTS,S/AS01 vaccine affords partial protection from clinical malaria in children, however, with waning immunity (Gosling and von Seidlein, 2016; Olotu et al., 2016; White et al., 2015). The World Health Organization has recommended large-scale pilot studies in several high-risk African countries to further assess the efficacy of this vaccine (Greenwood and Doumbo, 2016). It is speculated that the RTS,S/AS01 vaccine formulation induces high titres of antigen-specific CD4[+] T-cell response (Moreno et al., 1993) and increased pro-inflammatory cytokines such as IFN-γ (Di Rosa and Matzinger, 1996) and this could contribute to the killing of the sporozoites prior to entry into the liver or the schizont stages in the liver—both stages of the *Plasmodium* express CSP on their surface (Mota and Rodriguez, 2002; Yoshida et al., 1980). Further, the vaccine was shown to provide maximum protection against *Plasmodium* strains that carried CSP protein alleles that matched those represented in the RTS,S/AS01 vaccine (Neafsey et al., 2015).

Vaccine targets based on antigens of the arthropod vector have the advantage of impairing multiple pathogens transmitted by the same vector and certainly a desired approach from a public health perspective. A viable approach to develop successful vaccines against vector-borne diseases might therefore lie in a holistic design of multivalent vaccines targeting critical components of the vector saliva, gut, and the pathogen/s. We must also recognize that the expanding understanding of arthropod vectors and the microbial world within these vectors reveal nontraditional opportunities to control vector-borne diseases such as the utility of *Wolbachia* strains to block dengue virus transmission by *Ae. aegypti* mosquitoes (McGraw and O'Neill, 2013). Findings also draw attention to gut microbiota of the vertebrate host and show that colonization of the human gut by ∝-Gal-expressing bacteria elicit ∝-Gal-specific natural antibodies and these provide protection against malaria potentially by targeting the glycans on the *Plasmodium* cell surface (Yilmaz et al., 2014).

Technological advances in research areas not directly related to vector biology have really propelled vector biology research toward a better understanding of the physiological significance and functions of the vector salivary proteins and their interactions with the pathogen and the host. For instance, massively parallel sequencing tools for RNA and DNA (Tucker et al., 2009) have made it possible to sequence vector genomes (Gulia-Nuss et al., 2016; Holt et al., 2002; International Glossina Genome Initiative, 2014), genomes of the pathogens they transmit and endosymbionts they harbor (Akman et al., 2002; Barbour, 2014; Berriman et al., 2005; Fraser et al., 1997), and their host genomes (McPherson et al., 2001; Venter et al., 2001). These tools have also helped us to define the salivary transcriptomes of the vector (Abdeladhim et al., 2012; Alves-Silva et al., 2010; Garcia et al., 2014; Zivkovic et al., 2010). Similarly, mass spectroscopy and proteomics tools to define even femtogram levels of proteins (Aebersold and Mann, 2003; Yates et al., 2009) have led to an increased understanding of the composition of salivary proteomes (Fontaine et al., 2012; Francischetti et al., 2008a; Mastronunzio et al., 2012; Patramool et al., 2012; Ribeiro and Francischetti, 2003; Vijay et al., 2014), in the context of the pathogens they transmit. RNA interference (Barnard et al., 2012) and CRISPR (Hammond et al., 2016) techniques to generate transient and germ line knock-outs and knock-ins respectively now make it possible to infer the biological significance of novel genes and their encoded proteins. Microscopy/imaging techniques that enable in vivo visualization of host–pathogen and vector–pathogen interactions have revealed a real-time understanding of vector–pathogen and host–pathogen interactions during pathogen transmission (Bockenstedt et al., 2012; Dunham-Ems et al., 2009; Menard et al., 2013). Armed with these rapidly evolving and potent tools to address vector–pathogen–host interactions, the field is poised to exploit a combination of traditional and novel strategies to identify arthropod salivary vaccine targets that may be exploited to effectively control and manage vector-borne diseases.

Acknowledgments

We wish to thank Dr. Madan K Anant for assistance with the illustrations in this chapter. While we have strived to incorporate significant findings in the field that relate to this chapter, we recognize that the diversity of the arthropod vectors and the strategies they use to impair host defense responses cannot be exhaustively addressed within the confines of this chapter. We express our sincere apologies for any inadvertent omissions of work pertinent to this chapter. Work from our group cited in this chapter was funded by grants from the NIAID (NIH) and in part by a gift from the John Monsky and Jennifer Weis Monsky Lyme Disease Research Fund. E.F is an investigator of the Howard Hughes Medical Institute.

References

Abdeladhim, M., Ben Ahmed, M., Marzouki, S., Belhadj Hmida, N., Boussoffara, T., Belhaj Hamida, N., Ben Salah, A., Louzir, H., 2011. Human cellular immune response to the saliva of *Phlebotomus papatasi* is mediated by IL-10-producing CD8+ T cells and Th1-polarized CD4+ lymphocytes. PLoS Negl. Trop. Dis. 5, e1345.

Abdeladhim, M., Jochim, R.C., Ben Ahmed, M., Zhioua, E., Chelbi, I., Cherni, S., Louzir, H., Ribeiro, J.M., Valenzuela, J.G., 2012. Updating the salivary gland transcriptome of *Phlebotomus papatasi* (Tunisian strain): the search for sand fly-secreted immunogenic proteins for humans. PLoS One 7, e47347.

Abdeladhim, M., Kamhawi, S., Valenzuela, J.G., 2014. What's behind a sand fly bite? The profound effect of sand fly saliva on host hemostasis, inflammation and immunity. Infect. Genet. Evol. 28, 691–703.

Aderson, A., 2010. Contributions in the first 21st century decade to environmental health vector borne disease research. Infect. Dis. Res. Treat. 2, 17–24.

Aebersold, R., Mann, M., 2003. Mass spectrometry-based proteomics. Nature 422, 198–207.

Ahmed, S.B., Kaabi, B., Chelbi, I., Derbali, M., Cherni, S., Laouini, D., Zhioua, E., 2010. Lack of protection of pre-immunization with saliva of long-term colonized *Phlebotomus papatasi* against experimental challenge with *Leishmania major* and saliva of wild-caught *P. papatasi*. Am. J. Trop. Med. Hyg 83, 512–514.

Akman, L., Yamashita, A., Watanabe, H., Oshima, K., Shiba, T., Hattori, M., Aksoy, S., 2002. Genome sequence of the endocellular obligate symbiont of tsetse flies, *Wigglesworthia glossinidia*. Nat. Genet. 32, 402–407.

Aksoy, S., Attardo, G., Berriman, M., Christoffels, A., Lehane, M., Masiga, D., Toure, Y., 2014. Human African trypanosomiasis research gets a boost: unraveling the tsetse genome. PLoS Negl. Trop. Dis. 8, e2624.

Alarcon-Chaidez, F.J., Boppana, V.D., Hagymasi, A.T., Adler, A.J., Wikel, S.K., 2009. A novel sphingomyelinase-like enzyme in Ixodes scapularis tick saliva drives host CD4 T cells to express IL-4. Parasite Immunol. 31, 210–219.

Alger, N.E., Harant, J., 1976. *Plasmodium berghei*: sporozoite challenge, protection, and hypersensitivity in mice. Exp. Parasitol 40, 273–280.

Alger, N.E., Harant, J.A., Willis, L.C., Jorgensen, G.M., 1972. Sporozoite and normal salivary gland induced immunity in malaria. Nature 238, 341.

Ali, Z.M., Bakli, M., Fontaine, A., Bakkali, N., Vu Hai, V., Audebert, S., Boublik, Y., Pages, F., Remoue, F., Rogier, C., Fraisier, C., Almeras, L., 2012. Assessment of Anopheles salivary antigens as individual exposure biomarkers to species-specific malaria vector bites. Malar. J. 11, 439.

Allen, J.R., Kemp, D.H., 1982. Observations on the behaviour of *Dermacentor andersoni* larvae infesting normal and tick resistant Guinea-pigs. Parasitology 84, 195–204.

Almazan, C., Blas-Machado, U., Kocan, K.M., Yoshioka, J.H., Blouin, E.F., Mangold, A.J., de la Fuente, J., 2005. Characterization of three Ixodes scapularis cDNAs protective against tick infestations. Vaccine 23, 4403–4416.

Almazan, C., Lagunes, R., Villar, M., Canales, M., Rosario-Cruz, R., Jongejan, F., de la Fuente, J., 2010. Identification and characterization of *Rhipicephalus (Boophilus) microplus* candidate protective antigens for the control of cattle tick infestations. Parasitol. Res. 106, 471–479.

Almeras, L., Fontaine, A., Belghazi, M., Bourdon, S., Boucomont-Chapeaublanc, E., Orlandi-Pradines, E., Baragatti, M., Corre-Catelin, N., Reiter, P., Pradines, B., Fusai, T., Rogier, C., 2010. Salivary gland protein repertoire from *Aedes aegypti* mosquitoes. Vector Borne Zoonotic Dis. 10, 391–402.

Alves-Silva, J., Ribeiro, J.M., Van Den Abbeele, J., Attardo, G., Hao, Z., Haines, L.R., Soares, M.B., Berriman, M., Aksoy, S., Lehane, M.J., 2010. An insight into the sialome of *Glossina morsitans morsitans*. BMC Genomics 11, 213.

Andersen, J.F., Champagne, D.E., Weichsel, A., Ribeiro, J.M., Balfour, C.A., Dress, V., Montfort, W.R., 1997. Nitric oxide binding and crystallization of recombinant nitrophorin I, a nitric oxide transport protein from the blood-sucking bug *Rhodnius prolixus*. Biochemistry 36, 4423–4428.

Andersen, J.F., Ding, X.D., Balfour, C., Shokhireva, T.K., Champagne, D.E., Walker, F.A., Montfort, W.R., 2000. Kinetics and equilibria in ligand binding by nitrophorins 1-4: evidence for stabilization of a nitric oxide-ferriheme complex through a ligand-induced conformational trap. Biochemistry 39, 10118–10131.

Andersen, J.F., Gudderra, N.P., Francischetti, I.M., Valenzuela, J.G., Ribeiro, J.M., 2004. Recognition of anionic phospholipid membranes by an antihemostatic protein from a blood-feeding insect. Biochemistry 43, 6987–6994.

Andersen, J.F., Weichsel, A., Balfour, C.A., Champagne, D.E., Montfort, W.R., 1998. The crystal structure of nitrophorin 4 at 1.5 A resolution: transport of nitric oxide by a lipocalin-based heme protein. Structure 6, 1315–1327.

Anderson, J.F., Magnarelli, L.A., 2008. Biology of ticks. Infect. Dis. Clin. North Am. 22, 195–215 (v).

Anderson, J.M., Oliveira, F., Kamhawi, S., Mans, B.J., Reynoso, D., Seitz, A.E., Lawyer, P., Garfield, M., Pham, M., Valenzuela, J.G., 2006. Comparative salivary gland transcriptomics of sandfly vectors of visceral leishmaniasis. BMC Genomics 7, 52.

Andrade, B.B., Rocha, B.C., Reis-Filho, A., Camargo, L.M., Tadei, W.P., Moreira, L.A., Barral, A., Barral-Netto, M., 2009. Anti-*Anopheles darlingi* saliva antibodies as marker of *Plasmodium vivax* infection and clinical immunity in the Brazilian Amazon. Malar. J. 8, 121.

Andrade, B.B., Teixeira, C.R., Barral, A., Barral-Netto, M., 2005. Haematophagous arthropod saliva and host defense system: a tale of tear and blood. An. Acad. Bras. Cienc. 77, 665–693.

Anguita, J., Ramamoorthi, N., Hovius, J.W., Das, S., Thomas, V., Persinski, R., Conze, D., Askenase, P.W., Rincon, M., Kantor, F.S., Fikrig, E., 2002. Salp15, an ixodes scapularis salivary protein, inhibits CD4$^+$ T cell activation. Immunity 16, 849–859.

Araujo-Santos, T., Prates, D.B., Franca-Costa, J., Luz, N.F., Andrade, B.B., Miranda, J.C., Brodskyn, C.I., Barral, A., Bozza, P.T., Borges, V.M., 2014. Prostaglandin E$_2$/leukotriene B$_4$ balance induced by *Lutzomyia longipalpis* saliva favors *Leishmania infantum* infection. Parasit. Vectors 7, 601.

Arca, B., Lombardo, F., de Lara Capurro, M., della Torre, A., Dimopoulos, G., James, A.A., Coluzzi, M., 1999. Trapping cDNAs encoding secreted proteins from the salivary glands of the malaria vector *Anopheles gambiae*. Proc. Natl. Acad. Sci. U. S. A. 96, 1516–1521.

Arca, B., Lombardo, F., Francischetti, I.M., Pham, V.M., Mestres-Simon, M., Andersen, J.F., Ribeiro, J.M., 2007. An insight into the sialome of the adult female mosquito *Aedes albopictus*. Insect Biochem. Mol. Biol. 37, 107–127.

Arca, B., Lombardo, F., Valenzuela, J.G., Francischetti, I.M., Marinotti, O., Coluzzi, M., Ribeiro, J.M., 2005. An updated catalogue of salivary gland transcripts in the adult female mosquito, *Anopheles gambiae*. J. Exp. Biol. 208, 3971–3986.

Arthur, A., 2015. The Fantastic Laboratory of Dr. Weigl. W.W Norton and Company, New York.

Askenase, P.W., Bagnall, B.G., Worms, M.J., 1982. Cutaneous basophil-associated resistance to ectoparasites (ticks). I. Transfer with immune serum or immune cells. Immunology 45, 501–511.

Assoian, R.K., Sporn, M.B., 1986. Type beta transforming growth factor in human platelets: release during platelet degranulation and action on vascular smooth muscle cells. J. Cell Biol. 102, 1217–1223.

Assumpcao, T.C., Alvarenga, P.H., Ribeiro, J.M., Andersen, J.F., Francischetti, I.M., 2010. Dipetalodipin, a novel multifunctional salivary lipocalin that inhibits platelet aggregation, vasoconstriction, and angiogenesis through unique binding specificity for TXA$_2$, PGF$_2\alpha$, and 15(S)-HETE. J. Biol. Chem. 285, 39001–39012.

Balashov Yu, S., 1984. Interaction between blood-sucking arthropods and their hosts, and its influence on vector potential. Annu. Rev. Entomol. 29, 137–156.

Barbour, A.G., 2014. Phylogeny of a relapsing fever *Borrelia* species transmitted by the hard tick *Ixodes scapularis*. Infect. Genet. Evol. 27, 551–558.

Barnard, A.C., Nijhof, A.M., Fick, W., Stutzer, C., Maritz-Olivier, C., 2012. RNAi in arthropods: insight into the machinery and applications for understanding the pathogen-vector interface. Genes (Basel) 3, 702–741.

Barral, A., Honda, E., Caldas, A., Costa, J., Vinhas, V., Rowton, E.D., Valenzuela, J.G., Charlab, R., Barral-Netto, M., Ribeiro, J.M., 2000. Human immune response to sand fly salivary gland antigens: a useful epidemiological marker? Am. J. Trop. Med. Hyg. 62, 740–745.

Beaufays, J., Adam, B., Menten-Dedoyart, C., Fievez, L., Grosjean, A., Decrem, Y., Prevot, P.P., Santini, S., Brasseur, R., Brossard, M., Vanhaeverbeek, M., Bureau, F., Heinen, E., Lins, L., Vanhamme, L., Godfroid, E., 2008. Ir-LBP, an ixodes ricinus tick salivary LTB4-binding lipocalin, interferes with host neutrophil function. PLoS One 3, e3987.

Bechah, Y., Capo, C., Mege, J.L., Raoult, D., 2008. Epidemic typhus. Lancet Infect. Dis. 8, 417–426.

Belkaid, Y., Kamhawi, S., Modi, G., Valenzuela, J., Noben-Trauth, N., Rowton, E., Ribeiro, J., Sacks, D.L., 1998. Development of a natural model of cutaneous leishmaniasis: powerful effects of vector saliva and saliva preexposure on the long-term outcome of *Leishmania major* infection in the mouse ear dermis. J. Exp. Med 188, 1941–1953.

Belkaid, Y., Valenzuela, J.G., Kamhawi, S., Rowton, E., Sacks, D.L., Ribeiro, J.M., 2000. Delayed-type hypersensitivity to *Phlebotomus papatasi* sand fly bite: an adaptive response induced by the fly? Proc. Natl. Acad. Sci. U. S. A. 97, 6704–6709.

Bell, J.F., Stewart, S.J., Wikel, S.K., 1979. Resistance to tick-borne *Francisella tularensis* by tick-sensitized rabbits: allergic klendusity. Am. J. Trop. Med. Hyg. 28, 876–880.

Ben Hadj Ahmed, S., Chelbi, I., Kaabi, B., Cherni, S., Derbali, M., Zhioua, E., 2010. Differences in the salivary effects of wild-caught versus colonized *Phlebotomus papatasi* (Diptera: Psychodidae) on the development of zoonotic cutaneous leishmaniasis in BALB/c mice. J. Med. Entomol. 47, 74–79.

Berriman, M., Ghedin, E., Hertz-Fowler, C., Blandin, G., Renauld, H., Bartholomeu, D.C., Lennard, N.J., Caler, E., Hamlin, N.E., Haas, B., Bohme, U., Hannick, L., Aslett, M.A., Shallom, J., Marcello, L., Hou, L., Wickstead, B., Alsmark, U.C., Arrowsmith, C., Atkin, R.J., Barron, A.J., Bringaud, F., Brooks, K., Carrington, M., Cherevach, I., Chillingworth, T.J., Churcher, C., Clark, L.N., Corton, C.H., Cronin, A., Davies, R.M., Doggett, J., Djikeng, A., Feldblyum, T., Field, M.C., Fraser, A., Goodhead, I., Hance, Z., Harper, D., Harris, B.R., Hauser, H., Hostetler, J., Ivens, A., Jagels, K., Johnson, D., Johnson, J., Jones, K., Kerhornou, A.X., Koo, H., Larke, N., Landfear, S., Larkin, C., Leech, V., Line, A., Lord, A., Macleod, A., Mooney, P.J., Moule, S., Martin, D.M., Morgan, G.W., Mungall, K., Norbertczak, H., Ormond, D., Pai, G., Peacock, C.S., Peterson, J., Quail, M.A., Rabbinowitsch, E., Rajandream, M.A., Reitter, C., Salzberg, S.L., Sanders, M., Schobel, S., Sharp, S., Simmonds, M., Simpson, A.J., Tallon, L., Turner, C.M., Tait, A., Tivey, A.R., Van Aken, S., Walker, D., Wanless, D., Wang, S., White, B., White, O., Whitehead, S., Woodward, J., Wortman, J., Adams, M.D., Embley, T.M., Gull, K., Ullu, E., Barry, J.D., Fairlamb, A.H., Opperdoes, F., Barrell, B.G., Donelson, J.E., Hall, N., Fraser, C.M., Melville, S.E., El-Sayed, N.M., 2005. The genome of the African trypanosome *Trypanosoma brucei*. Science 309, 416–422.

Billman, Z.P., Seilie, A.M., Murphy, S.C., 2016. Purification of *Plasmodium* sporozoites enhances parasite-specific CD8+ T cell responses. Infect. Immun. 84, 2233–2242.

Bockenstedt, L.K., Gonzalez, D.G., Haberman, A.M., Belperron, A.A., 2012. Spirochete antigens persist near cartilage after murine Lyme borreliosis therapy. J. Clin. Invest. 122, 2652–2660.

Boisson, B., Jacques, J.C., Choumet, V., Martin, E., Xu, J., Vernick, K., Bourgouin, C., 2006. Gene silencing in mosquito salivary glands by RNAi. FEBS Lett. 580, 1988–1992.

Boppana, V.D., Thangamani, S., Adler, A.J., Wikel, S.K., 2009. SAAG-4 is a novel mosquito salivary protein that programmes host CD4 T cells to express IL-4. Parasite Immunol. 31, 287–295.

Bosio, C.F., Viall, A.K., Jarrett, C.O., Gardner, D., Rood, M.P., Hinnebusch, B.J., 2014. Evaluation of the murine immune response to *Xenopsylla cheopis* flea saliva and its effect on transmission of *Yersinia pestis*. PLoS Negl. Trop. Dis 8, e3196.

Bowen, M.F., 1991. The sensory physiology of host-seeking behavior in mosquitoes. Annu. Rev. Entomol. 36, 139–158.

Bowman, A.S., Dillwith, J.W., Sauer, J.R., 1996. Tick salivary prostaglandins: presence, origin and significance. Parasitol. Today 12, 388–396.

Bowman, A.S., Sauer, J.R., Zhu, K., Dillwith, J.W., 1995. Biosynthesis of salivary prostaglandins in the lone star tick, *Amblyomma americanum*. Insect Biochem. Mol. Biol. 25, 735–741.

Bretscher, P.A., 2014. On the mechanism determining the TH1/TH2 phenotype of an immune response, and its pertinence to strategies for the prevention, and treatment, of certain infectious diseases. Scand. J. Immunol. 79, 361–376.

Brogdon, W.G., McAllister, J.C., 2004. Insecticide resistance and vector control. J. Agromedicine 9, 329–345.

Brossard, M., Wikel, S.K., 2004. Tick immunobiology. Parasitology (129 Suppl.), S161–S176.

Browder, T., Folkman, J., Pirie-Shepherd, S., 2000. The hemostatic system as a regulator of angiogenesis. J. Biol. Chem. 275, 1521–1524.

Brown, S.J., Askenase, P.W., 1985. Rejection of ticks from guinea pigs by anti-hapten-antibody-mediated degranulation of basophils at cutaneous basophil hypersensitivity sites: role of mediators other than histamine. J. Immunol. 134, 1160–1165.

Brown, S.J., Barker, R.W., Askenase, P.W., 1984. Bovine resistance to *Amblyomma americanum* ticks: an acquired immune response characterized by cutaneous basophil infiltrates. Vet. Parasitol. 16, 147–165.

Brown, S.J., Knapp, F.W., 1981. Response of hypersensitized guinea pigs to the feeding of *Amblyomma americanum* ticks. Parasitology 83, 213–223.

Burke, G., Wikel, S.K., Spielman, A., Telford, S.R., McKay, K., Krause, P.J., Tick-borne Infection Study, G., 2005. Hypersensitivity to ticks and Lyme disease risk. Emerg. Infect. Dis. 11, 36–41.

Butterfield, T.A., Best, T.M., Merrick, M.A., 2006. The dual roles of neutrophils and macrophages in inflammation: a critical balance between tissue damage and repair. J. Athl. Train. 41, 457–465.

Caljon, G., Van Den Abbeele, J., Sternberg, J.M., Coosemans, M., De Baetselier, P., Magez, S., 2006a. Tsetse fly saliva biases the immune response to Th2 and induces anti-vector antibodies that are a useful tool for exposure assessment. Int. J. Parasitol. 36, 1025–1035.

Caljon, G., Van Den Abbeele, J., Stijlemans, B., Coosemans, M., De Baetselier, P., Magez, S., 2006b. Tsetse fly saliva accelerates the onset of *Trypanosoma brucei* infection in a mouse model associated with a reduced host inflammatory response. Infect. Immun 74, 6324–6330.

Calvo, E., Andersen, J., Francischetti, I.M., deL Capurro, M., deBianchi, A.G., James, A.A., Ribeiro, J.M., Marinotti, O., 2004. The transcriptome of adult female *Anopheles darlingi* salivary glands. Insect Mol. Biol. 13, 73–88.

Calvo, E., Mans, B.J., Andersen, J.F., Ribeiro, J.M., 2006a. Function and evolution of a mosquito salivary protein family. J. Biol. Chem. 281, 1935–1942.

Calvo, E., Mizurini, D.M., Sa-Nunes, A., Ribeiro, J.M., Andersen, J.F., Mans, B.J., Monteiro, R.Q., Kotsyfakis, M., Francischetti, I.M., 2011. Alboserpin, a factor Xa inhibitor from the mosquito vector of yellow fever, binds heparin and membrane phospholipids and exhibits antithrombotic activity. J. Biol. Chem. 286, 27998–28010.

Calvo, E., Pham, V.M., Lombardo, F., Arca, B., Ribeiro, J.M., 2006b. The sialotranscriptome of adult male *Anopheles gambiae* mosquitoes. Insect Biochem. Mol. Biol. 36, 570–575.

Calvo, E., Tokumasu, F., Marinotti, O., Villeval, J.L., Ribeiro, J.M., Francischetti, I.M., 2007. Aegyptin, a novel mosquito salivary gland protein, specifically binds to collagen and prevents its interaction with platelet glycoprotein VI, integrin $\alpha 2\beta 1$, and von Willebrand factor. J. Biol. Chem. 282, 26928–26938.

Canales, M., Naranjo, V., Almazan, C., Molina, R., Tsuruta, S.A., Szabo, M.P., Manzano-Roman, R., Perez de la Lastra, J.M., Kocan, K.M., Jimenez, M.I., Lucientes, J., Villar, M., de la Fuente, J., 2009. Conservation and immunogenicity of the mosquito ortholog of the tick-protective antigen, subolesin. Parasitol. Res. 105, 97–111.

Cantillo, J.F., Fernandez-Caldas, E., Puerta, L., 2014. Immunological aspects of the immune response induced by mosquito allergens. Int. Arch. Allergy Immunol. 165, 271–282.

Cappello, M., Li, S., Chen, X., Li, C.B., Harrison, L., Narashimhan, S., Beard, C.B., Aksoy, S., 1998. Tsetse thrombin inhibitor: bloodmeal-induced expression of an anticoagulant in salivary glands and gut tissue of *Glossina morsitans morsitans*. Proc. Natl. Acad. Sci. U. S. A. 95, 14290–14295.

Carregaro, V., Costa, D.L., Brodskyn, C., Barral, A.M., Barral-Netto, M., Cunha, F.Q., Silva, J.S., 2013. Dual effect of *Lutzomyia longipalpis* saliva on *Leishmania braziliensis* infection is mediated by distinct saliva-induced cellular recruitment into BALB/c mice ear. BMC Microbiol. 13, 102.

Carvalho, A.M., Cristal, J.R., Muniz, A.C., Carvalho, L.P., Gomes, R., Miranda, J.C., Barral, A., Carvalho, E.M., de Oliveira, C.I., 2015. Interleukin 10-dominant immune response and increased risk of cutaneous leishmaniasis after natural exposure to *Lutzomyia intermedia* sand flies. J. Infect. Dis. 212, 157–165.

Cavassani, K.A., Aliberti, J.C., Dias, A.R., Silva, J.S., Ferreira, B.R., 2005. Tick saliva inhibits differentiation, maturation and function of murine bone-marrow-derived dendritic cells. Immunology 114, 235–245.

Champagne, D.E., 2005. Antihemostatic molecules from saliva of blood-feeding arthropods. Pathophysiol. Haemost. Thromb. 34, 221–227.

Champagne, D.E., Ribeiro, J.M., 1994. Sialokinin I and II: vasodilatory tachykinins from the yellow fever mosquito *Aedes aegypti*. Proc. Natl. Acad. Sci. U. S. A. 91, 138–142.

Champagne, D.E., Smartt, C.T., Ribeiro, J.M., James, A.A., 1995. The salivary gland-specific apyrase of the mosquito *Aedes aegypti* is a member of the 5′-nucleotidase family. Proc. Natl. Acad. Sci. U. S. A. 92, 694–698.

Chapman, M.D., Marshall, N.A., Saxon, A., 1986. Identification and partial purification of species-specific allergens from *Triatoma protracta* (Heteroptera:Reduviidae). J. Allergy Clin. Immunol. 78, 436–442.

Charlab, R., Rowton, E.D., Ribeiro, J.M., 2000. The salivary adenosine deaminase from the sand fly Lutzomyia longipalpis. Exp. Parasitol. 95, 45–53.

Chmelar, J., Anderson, J.M., Mu, J., Jochim, R.C., Valenzuela, J.G., Kopecky, J., 2008. Insight into the sialome of the castor bean tick, *Ixodes ricinus*. BMC Genomics 9, 233.

Chmelar, J., Calvo, E., Pedra, J.H., Francischetti, I.M., Kotsyfakis, M., 2012. Tick salivary secretion as a source of antihemostatics. J. Proteomics 75, 3842–3854.

Chmelar, J., Kotal, J., Karim, S., Kopacek, P., Francischetti, I.M., Pedra, J.H., Kotsyfakis, M., 2016a. Sialomes and Mialomes: a systems-biology view of tick tissues and tick-host interactions. Trends Parasitol. 32, 242–254.

Chmelar, J., Kotal, J., Kopecky, J., Pedra, J.H., Kotsyfakis, M., 2016b. All for one and one for all on the tick-host battlefield. Trends Parasitol. 32, 368–377.

Chong, S.Z., Evrard, M., Ng, L.G., 2013. Lights, camera, and action: vertebrate skin sets the stage for immune cell interaction with arthropod-vectored pathogens. Front. Immunol. 4, 286.

Choumet, V., Carmi-Leroy, A., Laurent, C., Lenormand, P., Rousselle, J.C., Namane, A., Roth, C., Brey, P.T., 2007. The salivary glands and saliva of *Anopheles gambiae* as an essential step in the *Plasmodium* life cycle: a global proteomic study. Proteomics 7, 3384–3394.

Ciprandi, A., de Oliveira, S.K., Masuda, A., Horn, F., Termignoni, C., 2006. *Boophilus microplus*: its saliva contains microphilin, a small thrombin inhibitor. Exp. Parasitol. 114, 40–46.

Clark, R.A., 2010. Skin-resident T cells: the ups and downs of on site immunity. J. Invest. Dermatol. 130, 362–370.

Clemetson, K.J., 2012. Platelets and primary haemostasis. Thromb. Res. 129, 220–224.

Coleman, R.E., Burkett, D.A., Sherwood, V., Caci, J., Dennett, J.A., Jennings, B.T., Cushing, R., Ploch, J., Hopkins, G., Putnam, J.L., 2011. Impact of phlebotomine sand flies on United State military operations at Tallil Air Base, Iraq: 6. Evaluation of insecticides for the control of sand flies. J. Med. Entomol. 48, 584–599.

Collin, N., Assumpcao, T.C., Mizurini, D.M., Gilmore, D.C., Dutra-Oliveira, A., Kotsyfakis, M., Sa-Nunes, A., Teixeira, C., Ribeiro, J.M., Monteiro, R.Q., Valenzuela, J.G., Francischetti, I.M., 2012. Lufaxin, a novel factor Xa inhibitor from the salivary gland of the sand fly *Lutzomyia longipalpis* blocks protease-activated receptor 2 activation and inhibits inflammation and thrombosis in vivo. Arterioscler. Thromb. Vasc. Biol. 32, 2185–2198.

Collin, N., Gomes, R., Teixeira, C., Cheng, L., Laughinghouse, A., Ward, J.M., Elnaiem, D.E., Fischer, L., Valenzuela, J.G., Kamhawi, S., 2009. Sand fly salivary proteins induce strong cellular immunity in a natural reservoir of visceral leishmaniasis with adverse consequences for Leishmania. PLoS Pathog. 5, e1000441.

Conroy, D.M., Williams, T.J., 2001. Eotaxin and the attraction of eosinophils to the asthmatic lung. Respir. Res. 2, 150–156.

Conway, M.J., Watson, A.M., Colpitts, T.M., Dragovic, S.M., Li, Z., Wang, P., Feitosa, F., Shepherd, D.T., Ryman, K.D., Klimstra, W.B., Anderson, J.F., Fikrig, E., 2014. Mosquito saliva serine protease enhances dissemination of dengue virus into the mammalian host. J. Virol. 88, 164–175.

Cook, S.P., McCleskey, E.W., 2002. Cell damage excites nociceptors through release of cytosolic ATP. Pain 95, 41–47.

Cornelie, S., Remoue, F., Doucoure, S., Ndiaye, T., Sauvage, F.X., Boulanger, D., Simondon, F., 2007. An insight into immunogenic salivary proteins of Anopheles gambiae in African children. Malar. J. 6, 75.

Corral-Rodriguez, M.A., Macedo-Ribeiro, S., Barbosa Pereira, P.J., Fuentes-Prior, P., 2009. Tick-derived Kunitz-type inhibitors as antihemostatic factors. Insect Biochem. Mol. Biol. 39, 579–595.

Cotte, V., Sabatier, L., Schnell, G., Carmi-Leroy, A., Rousselle, J.C., Arsene-Ploetze, F., Malandrin, L., Sertour, N., Namane, A., Ferquel, E., Choumet, V., 2014. Differential expression of Ixodes ricinus salivary gland proteins in the presence of the Borrelia burgdorferi sensu lato complex. J. Proteomics 96, 29–43.

Coumou, J., Wagemakers, A., Trentelman, J.J., Nijhof, A.M., Hovius, J.W., 2014. Vaccination against Bm86 homologues in rabbits does not impair Ixodes ricinus feeding or oviposition. PLoS One 10, e0123495.

Cross, M.L., Cupp, E.W., Enriquez, F.J., 1994. Differential modulation of murine cellular immune responses by salivary gland extract of Aedes aegypti. Am. J. Trop. Med. Hyg. 51, 690–696.

Cupp, E.W., Cupp, M.S., Ramberg, F.B., 1994. Salivary apyrase in African and New World vectors of Plasmodium species and its relationship to malaria transmission. Am. J. Trop. Med. Hyg. 50, 235–240.

Dai, J., Narasimhan, S., Zhang, L., Liu, L., Wang, P., Fikrig, E., 2010. Tick histamine release factor is critical for Ixodes scapularis engorgement and transmission of the lyme disease agent. PLoS Pathog. 6, e1001205.

Dai, J., Wang, P., Adusumilli, S., Booth, C.J., Narasimhan, S., Anguita, J., Fikrig, E., 2009. Antibodies against a tick protein, Salp15, protect mice from the Lyme disease agent. Cell Host Microbe 6, 482–492.

Dai, S.X., Zhang, A.D., Huang, J.F., 2012. Evolution, expansion and expression of the Kunitz/BPTI gene family associated with long-term blood feeding in Ixodes Scapularis. BMC Evol. Biol. 12, 4.

Das, S., Banerjee, G., DePonte, K., Marcantonio, N., Kantor, F.S., Fikrig, E., 2001. Salp25D, an Ixodes scapularis antioxidant, is 1 of 14 immunodominant antigens in engorged tick salivary glands. J. Infect. Dis. 184, 1056–1064.

Das, S., Radtke, A., Choi, Y.J., Mendes, A.M., Valenzuela, J.G., Dimopoulos, G., 2010. Transcriptomic and functional analysis of the Anopheles gambiae salivary gland in relation to blood feeding. BMC Genomics 11, 566.

Davidson, C.J., Tuddenham, E.G., McVey, J.H., 2003. 450 million years of hemostasis. J. Thromb. Haemost. 1, 1487–1494.

Day, J.F., 2005. Host-seeking strategies of mosquito disease vectors. J. Am. Mosq. Control Assoc. 21, 17–22.

de la Fuente, J., Almazan, C., Blas-Machado, U., Naranjo, V., Mangold, A.J., Blouin, E.F., Gortazar, C., Kocan, K.M., 2006. The tick protective antigen, 4D8, is a conserved protein involved in modulation of tick blood ingestion and reproduction. Vaccine 24, 4082–4095.

de la Fuente, J., Manzano-Roman, R., Naranjo, V., Kocan, K.M., Zivkovic, Z., Blouin, E.F., Canales, M., Almazan, C., Galindo, R.C., Step, D.L., Villar, M., 2010. Identification of protective antigens by RNA interference for control of the lone star tick, Amblyomma americanum. Vaccine 28, 1786–1795.

de la Fuente, J., Maritz-Olivier, C., Naranjo, V., Ayoubi, P., Nijhof, A.M., Almazan, C., Canales, M., Perez de la Lastra, J.M., Galindo, R.C., Blouin, E.F., Gortazar, C., Jongejan, F., Kocan, K.M., 2008. Evidence of the role of tick subolesin in gene expression. BMC Genomics 9, 372.

de la Fuente, J., Moreno-Cid, J.A., Canales, M., Villar, M., de la Lastra, J.M., Kocan, K.M., Galindo, R.C., Almazan, C., Blouin, E.F., 2011. Targeting arthropod subolesin/akirin for the development of a universal vaccine for control of vector infestations and pathogen transmission. Vet. Parasitol. 181, 17–22.

De La Fuente, J., Rodriguez, M., Garcia-Garcia, J.C., 2000. Immunological control of ticks through vaccination with Boophilus microplus gut antigens. Ann. N. Y. Acad. Sci. 916, 617–621.

de la Fuente, J., Rodriguez, M., Montero, C., Redondo, M., Garcia-Garcia, J.C., Mendez, L., Serrano, E., Valdes, M., Enriquez, A., Canales, M., Ramos, E., Boue, O., Machado, H., Lleonart, R., 1999. Vaccination against ticks (Boophilus spp.): the experience with the Bm86-based vaccine Gavac. Genet. Anal. 15, 143–148.

de Moura, T.R., Oliveira, F., Novais, F.O., Miranda, J.C., Clarencio, J., Follador, I., Carvalho, E.M., Valenzuela, J.G., Barral-Netto, M., Barral, A., Brodskyn, C., de Oliveira, C.I., 2007. Enhanced Leishmania braziliensis infection following pre-exposure to sandfly saliva. PLoS Negl. Trop. Dis 1, e84.

de Silva, A.M., Telford 3rd, S.R., Brunet, L.R., Barthold, S.W., Fikrig, E., 1996. Borrelia burgdorferi OspA is an arthropod-specific transmission-blocking Lyme disease vaccine. J. Exp. Med. 183, 271–275.

Depinay, N., Hacini, F., Beghdadi, W., Peronet, R., Mecheri, S., 2006. Mast cell-dependent down-regulation of antigen-specific immune responses by mosquito bites. J. Immunol. 176, 4141–4146.

Deruaz, M., Frauenschuh, A., Alessandri, A.L., Dias, J.M., Coelho, F.M., Russo, R.C., Ferreira, B.R., Graham, G.J., Shaw, J.P., Wells, T.N., Teixeira, M.M., Power, C.A., Proudfoot, A.E., 2008. Ticks produce highly selective chemokine binding proteins with antiinflammatory activity. J. Exp. Med. 205, 2019–2031.

Di Rosa, F., Matzinger, P., 1996. Long-lasting CD8 T cell memory in the absence of CD4 T cells or B cells. J. Exp. Med. 183, 2153–2163.

Diaz-Martin, V., Manzano-Roman, R., Valero, L., Oleaga, A., Encinas-Grandes, A., Perez-Sanchez, R., 2013. An insight into the proteome of the saliva of the argasid tick Ornithodoros moubata reveals important differences in saliva protein composition between the sexes. J. Proteomics 80, 216–235.

Dinglasan, R.R., Valenzuela, J.G., Azad, A.F., 2005. Sugar epitopes as potential universal disease transmission blocking targets. Insect Biochem. Mol. Biol. 35, 1–10.

Dong, S., Lin, J., Held, N.L., Clem, R.J., Passarelli, A.L., Franz, A.W., 2015. Heritable CRISPR/Cas9-mediated genome editing in the yellow fever mosquito, Aedes aegypti. PLoS One 10, e0122353.

Donnelly, K.B., Lima, H.C., Titus, R.G., 1998. Histologic characterization of experimental cutaneous leishmaniasis in mice infected with Leishmania braziliensis in the presence or absence of sand fly vector salivary gland lysate. J. Parasitol 84, 97–103.

Donovan, M.J., Messmore, A.S., Scrafford, D.A., Sacks, D.L., Kamhawi, S., McDowell, M.A., 2007. Uninfected mosquito bites confer protection against infection with malaria parasites. Infect. Immun. 75, 2523–2530.

Doolittle, R.F., Feng, D.F., 1987. Reconstructing the evolution of vertebrate blood coagulation from a consideration of the amino acid sequences of clotting proteins. Cold Spring Harb. Symp. Quant. Biol. 52, 869–874.

Dostalova, A., Volf, P., 2012. Leishmania development in sand flies: parasite-vector interactions overview. Parasit. Vectors 5, 276.

Doucoure, S., Mouchet, F., Cornelie, S., DeHecq, J.S., Rutee, A.H., Roca, Y., Walter, A., Herve, J.P., Misse, D., Favier, F., Gasque, P., Remoue, F., 2012. Evaluation of the human IgG antibody response to Aedes albopictus saliva as a new specific biomarker of exposure to vector bites. PLoS Negl. Trop. Dis. 6, e1487.

Dunham-Ems, S.M., Caimano, M.J., Pal, U., Wolgemuth, C.W., Eggers, C.H., Balic, A., Radolf, J.D., 2009. Live imaging reveals a biphasic mode of dissemination of Borrelia burgdorferi within ticks. J. Clin. Invest. 119, 3652–3665.

Edwards, J.F., Higgs, S., Beaty, B.J., 1998. Mosquito feeding-induced enhancement of Cache valley virus (Bunyaviridae) infection in mice. J. Med. Entomol 35, 261–265.

Embers, M.E., Narasimhan, S., 2013. Vaccination against Lyme disease: past, present, and future. Front. Cell. Infect. Microbiol. 3, 6.

Epstein, J.E., Tewari, K., Lyke, K.E., Sim, B.K., Billingsley, P.F., Laurens, M.B., Gunasekera, A., Chakravarty, S., James, E.R., Sedegah, M., Richman, A., Velmurugan, S., Reyes, S., Li, M., Tucker, K., Ahumada, A., Ruben, A.J., Li, T., Stafford, R., Eappen, A.G., Tamminga, C., Bennett, J.W., Ockenhouse, C.F., Murphy, J.R., Komisar, J., Thomas, N., Loyevsky, M., Birkett, A., Plowe, C.V., Loucq, C., Edelman, R., Richie, T.L., Seder, R.A., Hoffman, S.L., 2011. Live attenuated malaria vaccine designed to protect through hepatic CD8+ T cell immunity. Science 334, 475–480.

Faudry, E., Lozzi, S.P., Santana, J.M., D'Souza-Ault, M., Kieffer, S., Felix, C.R., Ricart, C.A., Sousa, M.V., Vernet, T., Teixeira, A.R., 2004. Triatoma infestans apyrases belong to the 5′-nucleotidase family. J. Biol. Chem. 279, 19607–19613.

Faudry, E., Santana, J.M., Ebel, C., Vernet, T., Teixeira, A.R., 2006. Salivary apyrases of Triatoma infestans are assembled into homo-oligomers. Biochem. J. 396, 509–515.

Feingold, B., Benjamini, E., Michaeli, D., 1968. The allergic responses to insect bites. Annu. Rev. Entomol. 13, 137–158.

Ferreira, S.H., Lorenzetti, B.B., Bristow, A.F., Poole, S., 1988. Interleukin-1 beta as a potent hyperalgesic agent antagonized by a tripeptide analogue. Nature 334, 698–700.

Ferreira, S.H., Lorenzetti, B.B., Poole, S., 1993. Bradykinin initiates cytokine-mediated inflammatory hyperalgesia. Br. J. Pharmacol. 110, 1227–1231.

Figueiredo, A.C., de Sanctis, D., Gutierrez-Gallego, R., Cereija, T.B., Macedo-Ribeiro, S., Fuentes-Prior, P., Pereira, P.J., 2012. Unique thrombin inhibition mechanism by anophelin, an anticoagulant from the malaria vector. Proc. Natl. Acad. Sci. U. S. A. 109, E3649–E3658.

Fikrig, E., Telford 3rd, S.R., Barthold, S.W., Kantor, F.S., Spielman, A., Flavell, R.A., 1992. Elimination of Borrelia burgdorferi from vector ticks feeding on OspA-immunized mice. Proc. Natl. Acad. Sci. U. S. A. 89, 5418–5421.

Fontaine, A., Diouf, I., Bakkali, N., Misse, D., Pages, F., Fusai, T., Rogier, C., Almeras, L., 2011. Implication of haematophagous arthropod salivary proteins in host-vector interactions. Parasit. Vectors 4, 187.

Fontaine, A., Fusai, T., Briolant, S., Buffet, S., Villard, C., Baudelet, E., Pophillat, M., Granjeaud, S., Rogier, C., Almeras, L., 2012. Anopheles salivary gland proteomes from major malaria vectors. BMC Genomics 13, 614.

Foquet, L., Hermsen, C.C., van Gemert, G.J., Libbrecht, L., Sauerwein, R., Meuleman, P., Leroux-Roels, G., 2013. Molecular detection and quantification of *Plasmodium falciparum*-infected human hepatocytes in chimeric immune-deficient mice. Malar. J. 12, 430.

Fragoso, H., Rad, P.H., Ortiz, M., Rodriguez, M., Redondo, M., Herrera, L., de la Fuente, J., 1998. Protection against *Boophilus annulatus* infestations in cattle vaccinated with the *B. microplus* Bm86-containing vaccine Gavac. off. Vaccine 16, 1990–1992.

Francesquini, F.C., Silveira, F.T., Passero, L.F., Tomokane, T.Y., Carvalho, A.K., Corbett, C.E., Laurenti, M.D., 2014. Salivary gland homogenates from wild-caught sand flies *Lutzomyia flaviscutellata* and *Lutzomyia* (*Psychodopygus*) *complexus* showed inhibitory effects on *Leishmania* (*Leishmania*) *amazonensis* and *Leishmania* (*Viannia*) *braziliensis* infection in BALB/c mice. Int. J. Exp. Pathol 95, 418–426.

Francischetti, I.M., 2010. Platelet aggregation inhibitors from hematophagous animals. Toxicon 56, 1130–1144.

Francischetti, I.M., Mans, B.J., Meng, Z., Gudderra, N., Veenstra, T.D., Pham, V.M., Ribeiro, J.M., 2008a. An insight into the sialome of the soft tick, *Ornithodorus parkeri*. Insect Biochem. Mol. Biol. 38, 1–21.

Francischetti, I.M., Mather, T.N., Ribeiro, J.M., 2004. Penthalaris, a novel recombinant five-Kunitz tissue factor pathway inhibitor (TFPI) from the salivary gland of the tick vector of Lyme disease, *Ixodes scapularis*. Thromb. Haemost. 91, 886–898.

Francischetti, I.M., Mather, T.N., Ribeiro, J.M., 2005. Tick saliva is a potent inhibitor of endothelial cell proliferation and angiogenesis. Thromb. Haemost. 94, 167–174.

Francischetti, I.M., Meng, Z., Mans, B.J., Gudderra, N., Hall, M., Veenstra, T.D., Pham, V.M., Kotsyfakis, M., Ribeiro, J.M., 2008b. An insight into the salivary transcriptome and proteome of the soft tick and vector of epizootic bovine abortion, *Ornithodoros coriaceus*. J. Proteomics 71, 493–512.

Francischetti, I.M., Sa-Nunes, A., Mans, B.J., Santos, I.M., Ribeiro, J.M., 2009. The role of saliva in tick feeding. Front. Biosci. (Landmark Ed.) 14, 2051–2088.

Fraser, C.M., Casjens, S., Huang, W.M., Sutton, G.G., Clayton, R., Lathigra, R., White, O., Ketchum, K.A., Dodson, R., Hickey, E.K., Gwinn, M., Dougherty, B., Tomb, J.F., Fleischmann, R.D., Richardson, D., Peterson, J., Kerlavage, A.R., Quackenbush, J., Salzberg, S., Hanson, M., van Vugt, R., Palmer, N., Adams, M.D., Gocayne, J., Weidman, J., Utterback, T., Watthey, L., McDonald, L., Artiach, P., Bowman, C., Garland, S., Fuji, C., Cotton, M.D., Horst, K., Roberts, K., Hatch, B., Smith, H.O., Venter, J.C., 1997. Genomic sequence of a Lyme disease spirochaete, *Borrelia burgdorferi*. Nature 390, 580–586.

Fredericks, A.C., Fernandez-Sesma, A., 2014. The burden of dengue and chikungunya worldwide: implications for the southern United States and California. Ann. Glob. Health 80, 466–475.

Fukumoto, S., Sakaguchi, T., You, M., Xuan, X., Fujisaki, K., 2006. Tick troponin I-like molecule is a potent inhibitor for angiogenesis. Microvasc. Res. 71, 218–221.

Galindo, R.C., Doncel-Perez, E., Zivkovic, Z., Naranjo, V., Gortazar, C., Mangold, A.J., Martin-Hernando, M.P., Kocan, K.M., de la Fuente, J., 2009. Tick subolesin is an ortholog of the akirins described in insects and vertebrates. Dev. Comp. Immunol. 33, 612–617.

Garcia, G.R., Gardinassi, L.G., Ribeiro, J.M., Anatriello, E., Ferreira, B.R., Moreira, H.N., Mafra, C., Martins, M.M., Szabo, M.P., de Miranda-Santos, I.K., Maruyama, S.R., 2014. The sialotranscriptome of *Amblyomma triste*, *Amblyomma parvum* and *Amblyomma cajennense* ticks, uncovered by 454-based RNA-seq. Parasit. Vectors 7, 430.

Garcia-Garcia, J.C., Montero, C., Redondo, M., Vargas, M., Canales, M., Boue, O., Rodriguez, M., Joglar, M., Machado, H., Gonzalez, I.L., Valdes, M., Mendez, L., de la Fuente, J., 2000. Control of ticks resistant to immunization with Bm86 in cattle vaccinated with the recombinant antigen Bm95 isolated from the cattle tick, *Boophilus microplus*. Vaccine 18, 2275–2287.

Garcia-Garcia, J.C., Soto, A., Nigro, F., Mazza, M., Joglar, M., Hechevarria, M., Lamberti, J., de la Fuente, J., 1998. Adjuvant and immunostimulating properties of the recombinant Bm86 protein expressed in *Pichia pastoris*. Vaccine 16, 1053–1055.

Gauci, M., Stone, B.F., Thong, Y.H., 1988. Isolation and immunological characterisation of allergens from salivary glands of the Australian paralysis tick *Ixodes holocyclus*. Int. Arch. Allergy Appl. Immunol. 87, 208–212.

Ghosh, A.K., Devenport, M., Jethwaney, D., Kalume, D.E., Pandey, A., Anderson, V.E., Sultan, A.A., Kumar, N., Jacobs-Lorena, M., 2009. Malaria parasite invasion of the mosquito salivary gland requires interaction between the *Plasmodium* TRAP and the *Anopheles saglin* proteins. PLoS Pathog. 5, e1000265.

Ghosh, A.K., Ribolla, P.E., Jacobs-Lorena, M., 2001. Targeting *Plasmodium* ligands on mosquito salivary glands and midgut with a phage display peptide library. Proc. Natl. Acad. Sci. U. S. A. 98, 13278–13281.

Gillespie, R.D., Dolan, M.C., Piesman, J., Titus, R.G., 2001. Identification of an IL-2 binding protein in the saliva of the Lyme disease vector tick, *Ixodes scapularis*. J. Immunol. 166, 4319–4326.

Gillespie, R.D., Mbow, M.L., Titus, R.G., 2000. The immunomodulatory factors of bloodfeeding arthropod saliva. Parasite Immunol. 22, 319–331.

Gillespie, S.H., Smith, G.L., Osbourn, A.E., 2004. Microbe-Vector Interactions in Vector-Borne Diseases. Cambridge University Press, Cambridge Eng., New York.

Gomes, R., Oliveira, F., Teixeira, C., Meneses, C., Gilmore, D.C., Elnaiem, D.E., Kamhawi, S., Valenzuela, J.G., 2012. Immunity to sand fly salivary protein LJM11 modulates host response to vector-transmitted *Leishmania* conferring ulcer-free protection. J. Invest. Dermatol 132, 2735–2743.

Gomes, R., Teixeira, C., Teixeira, M.J., Oliveira, F., Menezes, M.J., Silva, C., de Oliveira, C.I., Miranda, J.C., Elnaiem, D.E., Kamhawi, S., Valenzuela, J.G., Brodskyn, C.I., 2008. Immunity to a salivary protein of a sand fly vector protects against the fatal outcome of visceral leishmaniasis in a hamster model. Proc. Natl. Acad. Sci. U.S.A 105, 7845–7850.

Gomes, R.B., Brodskyn, C., de Oliveira, C.I., Costa, J., Miranda, J.C., Caldas, A., Valenzuela, J.G., Barral-Netto, M., Barral, A., 2002. Seroconversion against *Lutzomyia longipalpis* saliva concurrent with the development of anti-*Leishmania chagasi* delayed-type hypersensitivity. J. Infect. Dis. 186, 1530–1534.

Gordon, J.R., Allen, J.R., 1991. Nonspecific activation of complement factor 5 by isolated *Dermacentor andersoni* salivary antigens. J. Parasitol. 77, 296–301.

Gosling, R., von Seidlein, L., 2016. The future of the RTS,S/AS01 malaria vaccine: an alternative development plan. PLoS Med. 13, e1001994.

Grabowski, J.M., Perera, R., Roumani, A.M., Hedrick, V.E., Inerowicz, H.D., Hill, C.A., Kuhn, R.J., 2016. Changes in the proteome of Langat-infected *Ixodes scapularis* ISE6 Cells: metabolic pathways associated with flavivirus infection. PLoS Negl. Trop. Dis. 10, e0004180.

Graziano, F.M., Gunderson, L., Larson, L., Askenase, P.W., 1983. IgE antibody-mediated cutaneous basophil hypersensitivity reactions in guinea pigs. J. Immunol. 131, 2675–2681.

Greenwood, B., Doumbo, O.K., 2016. Implementation of the malaria candidate vaccine RTS,S/AS01. Lancet 387, 318–319.

Grimaldi, D., 2010. Evolution of the Insects, first ed. Cambridge University Press, New York.

Grimm, D., Tilly, K., Byram, R., Stewart, P.E., Krum, J.G., Bueschel, D.M., Schwan, T.G., Policastro, P.F., Elias, A.F., Rosa, P.A., 2004. Outer-surface protein C of the Lyme disease spirochete: a protein induced in ticks for infection of mammals. Proc. Natl. Acad. Sci. U. S. A. 101, 3142–3147.

Gulia-Nuss, M., Nuss, A.B., Meyer, J.M., Sonenshine, D.E., Roe, R.M., Waterhouse, R.M., Sattelle, D.B., de la Fuente, J., Ribeiro, J.M., Megy, K., Thimmapuram, J., Miller, J.R., Walenz, B.P., Koren, S., Hostetler, J.B., Thiagarajan, M., Joardar, V.S., Hannick, L.I., Bidwell, S., Hammond, M.P., Young, S., Zeng, Q., Abrudan, J.L., Almeida, F.C., Ayllon, N., Bhide, K., Bissinger, B.W., Bonzon-Kulichenko, E., Buckingham, S.D., Caffrey, D.R., Caimano, M.J., Croset, V., Driscoll, T., Gilbert, D., Gillespie, J.J., Giraldo-Calderon, G.I., Grabowski, J.M.,

Jiang, D., Khalil, S.M., Kim, D., Kocan, K.M., Koci, J., Kuhn, R.J., Kurtti, T.J., Lees, K., Lang, E.G., Kennedy, R.C., Kwon, H., Perera, R., Qi, Y., Radolf, J.D., Sakamoto, J.M., Sanchez-Gracia, A., Severo, M.S., Silverman, N., Simo, L., Tojo, M., Tornador, C., Van Zee, J.P., Vazquez, J., Vieira, F.G., Villar, M., Wespiser, A.R., Yang, Y., Zhu, J., Arensburger, P., Pietrantonio, P.V., Barker, S.C., Shao, R., Zdobnov, E.M., Hauser, F., Grimmelikhuijzen, C.J., Park, Y., Rozas, J., Benton, R., Pedra, J.H., Nelson, D.R., Unger, M.F., Tubio, J.M., Tu, Z., Robertson, H.M., Shumway, M., Sutton, G., Wortman, J.R., Lawson, D., Wikel, S.K., Nene, V.M., Fraser, C.M., Collins, F.H., Birren, B., Nelson, K.E., Caler, E., Hill, C.A., 2016. Genomic insights into the *Ixodes scapularis* tick vector of Lyme disease. Nat. Commun. 7, 10507.

Guo, X., Booth, C.J., Paley, M.A., Wang, X., DePonte, K., Fikrig, E., Narasimhan, S., Montgomery, R.R., 2009. Inhibition of neutrophil function by two tick salivary proteins. Infect. Immun. 77, 2320–2329.

Hajnicka, V., Vancova-Stibraniova, I., Slovak, M., Kocakova, P., Nuttall, P.A., 2011. Ixodid tick salivary gland products target host wound healing growth factors. Int. J. Parasitol. 41, 213–223.

Hall, A.B., Basu, S., Jiang, X., Qi, Y., Timoshevskiy, V.A., Biedler, J.K., Sharakhova, M.V., Elahi, R., Anderson, M.A., Chen, X.G., Sharakhov, I.V., Adelman, Z.N., Tu, Z., 2015. SEX DETERMINATION. A male-determining factor in the mosquito *Aedes aegypti*. Science 348, 1268–1270.

Hall, L.R., Titus, R.G., 1995. Sand fly vector saliva selectively modulates macrophage functions that inhibit killing of Leishmania major and nitric oxide production. J. Immunol. 155, 3501–3506.

Hammond, A., Galizi, R., Kyrou, K., Simoni, A., Siniscalchi, C., Katsanos, D., Gribble, M., Baker, D., Marois, E., Russell, S., Burt, A., Windbichler, N., Crisanti, A., Nolan, T., 2016. A CRISPR-Cas9 gene drive system targeting female reproduction in the malaria mosquito vector *Anopheles gambiae*. Nat. Biotechnol. 34, 78–83.

Harrington, D., Canales, M., de la Fuente, J., de Luna, C., Robinson, K., Guy, J., Sparagano, O., 2009. Immunisation with recombinant proteins subolesin and Bm86 for the control of *Dermanyssus gallinae* in poultry. Vaccine 27, 4056–4063.

Heinze, D.M., Carmical, J.R., Aronson, J.F., Alarcon-Chaidez, F., Wikel, S., Thangamani, S., 2014. Murine cutaneous responses to the rocky mountain spotted fever vector, *Dermacentor andersoni*, feeding. Front. Microbiol. 5, 198.

Heinze, D.M., Carmical, J.R., Aronson, J.F., Thangamani, S., 2012a. Early immunologic events at the tick-host interface. PLoS One 7, e47301.

Heinze, D.M., Wikel, S.K., Thangamani, S., Alarcon-Chaidez, F.J., 2012b. Transcriptional profiling of the murine cutaneous response during initial and subsequent infestations with *Ixodes scapularis* nymphs. Parasit. Vectors 5, 26.

Hellmann, K., Hawkins, R.I., 1965. Prolixins-S and prolixin-G; two anticoagulants from *Rhodnius prolixus* Stal. Nature 207, 265–267.

Hernandez-Vargas, M.J., Santibanez-Lopez, C.E., Corzo, G., 2016. An insight into the triabin protein family of american hematophagous reduviids: functional, structural and phylogenetic analysis. Toxins (Basel) 8, 44.

Hoffman, S.L., Goh, L.M., Luke, T.C., Schneider, I., Le, T.P., Doolan, D.L., Sacci, J., de la Vega, P., Dowler, M., Paul, C., Gordon, D.M., Stoute, J.A., Church, L.W., Sedegah, M., Heppner, D.G., Ballou, W.R., Richie, T.L., 2002. Protection of humans against malaria by immunization with radiation-attenuated *Plasmodium falciparum* sporozoites. J. Infect. Dis. 185, 1155–1164.

Hoffmann-Sommergruber, K., Paschinger, K., Wilson, I.B., 2011. Glycomarkers in parasitic infections and allergy. Biochem. Soc. Trans. 39, 360–364.

Hollingdale, M.R., Nardin, E.H., Tharavanij, S., Schwartz, A.L., Nussenzweig, R.S., 1984. Inhibition of entry of *Plasmodium falciparum* and *P. vivax* sporozoites into cultured cells; an in vitro assay of protective antibodies. J. Immunol. 132, 909–913.

Holt, R.A., Subramanian, G.M., Halpern, A., Sutton, G.G., Charlab, R., Nusskern, D.R., Wincker, P., Clark, A.G., Ribeiro, J.M., Wides, R., Salzberg, S.L., Loftus, B., Yandell, M., Majoros, W.H., Rusch, D.B., Lai, Z., Kraft, C.L., Abril, J.F., Anthouard, V., Arensburger, P., Atkinson, P.W., Baden, H., de Berardinis, V., Baldwin, D., Benes, V., Biedler, J., Blass, C., Bolanos, R., Boscus, D., Barnstead, M., Cai, S., Center, A., Chaturverdi, K., Christophides, G.K., Chrystal, M.A., Clamp, M., Cravchik, A., Curwen, V., Dana, A., Delcher, A., Dew, I., Evans, C.A., Flanigan, M., Grundschober-Freimoser, A., Friedli, L., Gu, Z., Guan, P., Guigo, R., Hillenmeyer, M.E., Hladun, S.L., Hogan, J.R., Hong, Y.S., Hoover, J., Jaillon, O., Ke, Z., Kodira, C., Kokoza, E., Koutsos, A., Letunic, I., Levitsky, A., Liang, Y., Lin, J.J., Lobo, N.F., Lopez, J.R., Malek, J.A., McIntosh, T.C., Meister, S., Miller, J., Mobarry, C., Mongin, E., Murphy, S.D., O'Brochta, D.A., Pfannkoch, C., Qi, R., Regier, M.A., Remington, K., Shao, H., Sharakhova, M.V., Sitter, C.D., Shetty, J., Smith, T.J., Strong, R., Sun, J., Thomasova, D., Ton, L.Q., Topalis, P., Tu, Z., Unger, M.F., Walenz, B., Wang, A., Wang, J., Wang, M., Wang, X., Woodford, K.J., Wortman, J.R., Wu, M., Yao, A., Zdobnov, E.M., Zhang, H., Zhao, Q., Zhao, S., Zhu, S.C., Zhimulev, I., Coluzzi, M., della Torre, A., Roth, C.W., Louis, C., Kalush, F., Mural, R.J., Myers, E.W., Adams, M.D., Smith, H.O., Broder, S., Gardner, M.J., Fraser, C.M., Birney, E., Bork, P., Brey, P.T., Venter, J.C., Weissenbach, J., Kafatos, F.C., Collins, F.H., Hoffman, S.L., 2002. The genome sequence of the malaria mosquito *Anopheles gambiae*. Science 298, 129–149.

Hovius, J.W., de Jong, M.A., den Dunnen, J., Litjens, M., Fikrig, E., van der Poll, T., Gringhuis, S.I., Geijtenbeek, T.B., 2008a. Salp15 binding to DC-SIGN inhibits cytokine expression by impairing both nucleosome remodeling and mRNA stabilization. PLoS Pathog. 4, e31.

Hovius, J.W., Levi, M., Fikrig, E., 2008b. Salivating for knowledge: potential pharmacological agents in tick saliva. PLoS Med. 5, e43.

Hughes, A.L., 2013. Evolution of the salivary apyrases of blood-feeding arthropods. Gene 527, 123–130.

Ibelli, A.M., Hermance, M.M., Kim, T.K., Gonzalez, C.L., Mulenga, A., 2013. Bioinformatics and expression analyses of the *Ixodes scapularis* tick cystatin family. Exp. Appl. Acarol. 60, 41–53.

Ibelli, A.M., Kim, T.K., Hill, C.C., Lewis, L.A., Bakshi, M., Miller, S., Porter, L., Mulenga, A., 2014. A blood meal-induced *Ixodes scapularis* tick saliva serpin inhibits trypsin and thrombin, and interferes with platelet aggregation and blood clotting. Int. J. Parasitol. 44, 369–379.

International Glossina Genome Initiative, I., 2014. Genome sequence of the tsetse fly (*Glossina morsitans*): vector of African trypanosomiasis. Science 344, 380–386.

Islam, M.K., Tsuji, N., Miyoshi, T., Alim, M.A., Huang, X., Hatta, T., Fujisaki, K., 2009. The Kunitz-like modulatory protein haemangin is vital for hard tick blood-feeding success. PLoS Pathog. 5, e1000497.

Iwanaga, S., Isawa, H., Yuda, M., 2014. Horizontal gene transfer of a vertebrate vasodilatory hormone into ticks. Nat. Commun. 5, 3373.

Iwanaga, S., Okada, M., Isawa, H., Morita, A., Yuda, M., Chinzei, Y., 2003. Identification and characterization of novel salivary thrombin inhibitors from the ixodidae tick, *Haemaphysalis longicornis*. Eur. J. Biochem. 270, 1926–1934.

Jablonka, W., Kotsyfakis, M., Mizurini, D.M., Monteiro, R.Q., Lukszo, J., Drake, S.K., Ribeiro, J.M., Andersen, J.F., 2015. Identification and mechanistic analysis of a novel tick-derived inhibitor of thrombin. PLoS One 10, e0133991.

Jenne, C.N., Urrutia, R., Kubes, P., 2013. Platelets: bridging hemostasis, inflammation, and immunity. Int. J. Lab. Hematol. 35, 254–261.

Jennings, L.K., 2009. Mechanisms of platelet activation: need for new strategies to protect against platelet-mediated atherothrombosis. Thromb. Haemost. 102, 248–257.

Jones, L.D., Hodgson, E., Williams, T., Higgs, S., Nuttall, P.A., 1992. Saliva activated transmission (SAT) of Thogoto virus: relationship with vector potential of different haematophagous arthropods. Med. Vet. Entomol. 6, 261–265.

Julius, D., Basbaum, A.I., 2001. Molecular mechanisms of nociception. Nature 413, 203–210.

Juncadella, I.J., Anguita, J., 2009. The immunosuppresive tick salivary protein, Salp15. Adv. Exp. Med. Biol. 666, 121–131.

Jung, S., Unutmaz, D., Wong, P., Sano, G., De los Santos, K., Sparwasser, T., Wu, S., Vuthoori, S., Ko, K., Zavala, F., Pamer, E.G., Littman, D.R., Lang, R.A., 2002. In vivo depletion of CD11c+ dendritic cells abrogates priming of CD8+ T cells by exogenous cell-associated antigens. Immunity 17, 211–220.

Kalume, D.E., Okulate, M., Zhong, J., Reddy, R., Suresh, S., Deshpande, N., Kumar, N., Pandey, A., 2005. A proteomic analysis of salivary glands of female *Anopheles gambiae* mosquito. Proteomics 5, 3765–3777.

Kamhawi, S., 2000. The biological and immunomodulatory properties of sand fly saliva and its role in the establishment of Leishmania infections. Microbes Infect. 2, 1765–1773.

Kamhawi, S., Aslan, H., Valenzuela, J.G., 2014. Vector saliva in vaccines for visceral leishmaniasis: a brief encounter of high consequence? Front. Public Health 2, 99.

Kamhawi, S., Belkaid, Y., Modi, G., Rowton, E., Sacks, D., 2000. Protection against cutaneous leishmaniasis resulting from bites of uninfected sand flies. Science 290, 1351–1354.

Kamhawi, S., Ramalho-Ortigao, M., Pham, V.M., Kumar, S., Lawyer, P.G., Turco, S.J., Barillas-Mury, C., Sacks, D.L., Valenzuela, J.G., 2004. A role for insect galectins in parasite survival. Cell 119, 329–341.

Kaplan, A.P., Ghebrehiwet, B., 2010. The plasma bradykinin-forming pathways and its interrelationships with complement. Mol. Immunol. 47, 2161–2169.

Kaplan, Z.S., Zarpellon, A., Alwis, I., Yuan, Y., McFadyen, J., Ghasemzadeh, M., Schoenwaelder, S.M., Ruggeri, Z.M., Jackson, S.P., 2015. Thrombin-dependent intravascular leukocyte trafficking regulated by fibrin and the platelet receptors GPIb and PAR4. Nat. Commun. 6, 7835.

Karczewski, J., Waxman, L., Endris, R.G., Connolly, T.M., 1995. An inhibitor from the argasid tick *Ornithodoros moubata* of cell adhesion to collagen. Biochem. Biophys. Res. Commun. 208, 532–541.

Karim, S., Troiano, E., Mather, T.N., 2010. Functional genomics tool: gene silencing in *Ixodes scapularis* eggs and nymphs by electroporated dsRNA. BMC Biotechnol. 10, 1.

Kato, H., Jochim, R.C., Lawyer, P.G., Valenzuela, J.G., 2007. Identification and characterization of a salivary adenosine deaminase from the sand fly *Phlebotomus duboscqi*, the vector of Leishmania major in sub-Saharan Africa. J. Exp. Biol. 210, 733–740.

Kaur, J., Woodman, R.C., Ostrovsky, L., Kubes, P., 2001. Selective recruitment of neutrophils and lymphocytes by thrombin: a role for NF-kappaB. Am. J. Physiol. Heart Circ. Physiol. 281, H784–H795.

Kazimirova, M., Stibraniova, I., 2013. Tick salivary compounds: their role in modulation of host defences and pathogen transmission. Front. Cell. Infect. Microbiol. 3, 43.

Kebaier, C., Voza, T., Vanderberg, J., 2010. Neither mosquito saliva nor immunity to saliva has a detectable effect on the infectivity of *Plasmodium* sporozoites injected into mice. Infect. Immun. 78, 545–551.

Kemp, D.H., Bourne, A., 1980. *Boophilus microplus*: the effect of histamine on the attachment of cattle-tick larvae– studies in vivo and in vitro. Parasitology 80, 487–496.

Kilpatrick, A.M., Randolph, S.E., 2012. Drivers, dynamics, and control of emerging vector-borne zoonotic diseases. Lancet 380, 1946–1955.

Kim, B.S., Wang, K., Siracusa, M.C., Saenz, S.A., Brestoff, J.R., Monticelli, L.A., Noti, M., Tait Wojno, E.D., Fung, T.C., Kubo, M., Artis, D., 2014. Basophils promote innate lymphoid cell responses in inflamed skin. J. Immunol. 193, 3717–3725.

Kim, T.K., Tirloni, L., Pinto, A.F., Moresco, J., Yates 3rd, J.R., da Silva Vaz Jr., I., Mulenga, A., 2016. *Ixodes scapularis* tick saliva proteins sequentially secreted every 24 h during blood feeding. PLoS Negl. Trop. Dis. 10, e0004323.

Kim, T.K., Tirloni, L., Radulovic, Z., Lewis, L., Bakshi, M., Hill, C., da Silva Vaz Jr., I., Logullo, C., Termignoni, C., Mulenga, A., 2015. Conserved *Amblyomma americanum* tick Serpin19, an inhibitor of blood clotting factors Xa and XIa, trypsin and plasmin, has anti-haemostatic functions. Int. J. Parasitol. 45, 613–627.

Kistler, K.E., Vosshall, L.B., Matthews, B.J., 2015. Genome engineering with CRISPR-Cas9 in the mosquito *Aedes aegypti*. Cell Rep. 11, 51–60.

Klion, A.D., Nutman, T.B., 2004. The role of eosinophils in host defense against helminth parasites. J. Allergy Clin. Immunol. 113, 30–37.

Knudsen, A.B., Slooff, R., 1992. Vector-borne disease problems in rapid urbanization: new approaches to vector control. Bull. World Health Organ. 70, 1–6.

Kocakova, P., Slavikova, M., Hajnicka, V., Slovak, M., Gasperik, J., Vancova, I., Fuchsberger, N., Nuttall, P.A., 2003. Effect of fast protein liquid chromatography fractionated salivary gland extracts from different ixodid tick species on interleukin-8 binding to its cell receptors. Folia Parasitol. (Praha) 50, 79–84.

Koh, C.Y., Kazimirova, M., Trimnell, A., Takac, P., Labuda, M., Nuttall, P.A., Kini, R.M., 2007. Variegin, a novel fast and tight binding thrombin inhibitor from the tropical bont tick. J. Biol. Chem. 282, 29101–29113.

Kotal, J., Langhansova, H., Lieskovska, J., Andersen, J.F., Francischetti, I.M., Chavakis, T., Kopecky, J., Pedra, J.H., Kotsyfakis, M., Chmelar, J., 2015. Modulation of host immunity by tick saliva. J. Proteomics 128, 58–68.

Kotsyfakis, M., Anderson, J.M., Andersen, J.F., Calvo, E., Francischetti, I.M., Mather, T.N., Valenzuela, J.G., Ribeiro, J.M., 2008. Cutting edge: immunity against a "silent" salivary antigen of the Lyme vector *Ixodes scapularis* impairs its ability to feed. J. Immunol. 181, 5209–5212.

Kotsyfakis, M., Sa-Nunes, A., Francischetti, I.M., Mather, T.N., Andersen, J.F., Ribeiro, J.M., 2006. Antiinflammatory and immunosuppressive activity of sialostatin L, a salivary cystatin from the tick *Ixodes scapularis*. J. Biol. Chem. 281, 26298–26307.

Kraiczy, P., Stevenson, B., 2013. Complement regulator-acquiring surface proteins of *Borrelia burgdorferi*: structure, function and regulation of gene expression. Ticks Tick Borne Dis. 4, 26–34.

Krause, P.J., Grant-Kels, J.M., Tahan, S.R., Dardick, K.R., Alarcon-Chaidez, F., Bouchard, K., Visini, C., Deriso, C., Foppa, I.M., Wikel, S., 2009. Dermatologic changes induced by repeated *Ixodes scapularis* bites and implications for prevention of tick-borne infection. Vector Borne Zoonotic Dis. 9, 603–610.

Krenn, H.W., Aspock, H., 2012. Form, function and evolution of the mouthparts of blood-feeding Arthropoda. Arthropod Struct. Dev. 41, 101–118.

Kubes, P., Suzuki, M., Granger, D.N., 1991. Nitric oxide: an endogenous modulator of leukocyte adhesion. Proc. Natl. Acad. Sci. U. S. A. 88, 4651–4655.

Kupper, T.S., Fuhlbrigge, R.C., 2004. Immune surveillance in the skin: mechanisms and clinical consequences. Nat. Rev. Immunol. 4, 211–222.

Labuda, M., Trimnell, A.R., Lickova, M., Kazimirova, M., Davies, G.M., Lissina, O., Hails, R.S., Nuttall, P.A., 2006. An antivector vaccine protects against a lethal vector-borne pathogen. PLoS Pathog. 2, e27.

Lacy, P., Stow, J.L., 2011. Cytokine release from innate immune cells: association with diverse membrane trafficking pathways. Blood 118, 9–18.

Lal, A.A., Patterson, P.S., Sacci, J.B., Vaughan, J.A., Paul, C., Collins, W.E., Wirtz, R.A., Azad, A.F., 2001. Antimosquito midgut antibodies block development of *Plasmodium falciparum* and *Plasmodium vivax* in multiple species of *Anopheles* mosquitoes and reduce vector fecundity and survivorship. Proc. Natl. Acad. Sci. U. S. A. 98, 5228–5233.

Laurenti, M.D., da Matta, V.L., Pernichelli, T., Secundino, N.F., Pinto, L.C., Corbett, C.E., Pimenta, P.P., 2009a. Effects of salivary gland homogenate from wild-caught and laboratory-reared *Lutzomyia longipalpis* on the evolution and immunomodulation of *Leishmania (Leishmania) amazonensis* infection. Scand. J. Immunol. 70, 389–395.

Laurenti, M.D., Silveira, V.M., Secundino, N.F., Corbett, C.E., Pimenta, P.P., 2009b. Saliva of laboratory-reared *Lutzomyia longipalpis* exacerbates *Leishmania (Leishmania) amazonensis* infection more potently than saliva of wild-caught *Lutzomyia longipalpis*. Parasitol. Int. 58, 220–226.

Law, J.H., Ribeiro, J.M., Wells, M.A., 1992. Biochemical insights derived from insect diversity. Annu. Rev. Biochem. 61, 87–111.

Le Coupanec, A., Babin, D., Fiette, L., Jouvion, G., Ave, P., Misse, D., Bouloy, M., Choumet, V., 2013. *Aedes* mosquito saliva modulates Rift Valley fever virus pathogenicity. PLoS Negl. Trop. Dis 7, e2237.

Leboulle, G., Crippa, M., Decrem, Y., Mejri, N., Brossard, M., Bollen, A., Godfroid, E., 2002. Characterization of a novel salivary immunosuppressive protein from *Ixodes ricinus* ticks. J. Biol. Chem. 277, 10083–10089.

Lehane, 2005. The Biology of Blood-sucking Insects. Cambridge University Press.

Leitner, W.W., Wali, T., Costero-Saint Denis, A., 2013. Is arthropod saliva the achilles' heel of vector-borne diseases? Front. Immunol. 4, 255.

Lerner, E.A., Ribeiro, J.M., Nelson, R.J., Lerner, M.R., 1991. Isolation of maxadilan, a potent vasodilatory peptide from the salivary glands of the sand fly *Lutzomyia longipalpis*. J. Biol. Chem. 266, 11234–11236.

Lerner, E.A., Shoemaker, C.B., 1992. Maxadilan. Cloning and functional expression of the gene encoding this potent vasodilator peptide. J. Biol. Chem. 267, 1062–1066.

Lestinova, T., Vlkova, M., Votypka, J., Volf, P., Rohousova, I., 2015. *Phlebotomus papatasi* exposure cross-protects mice against *Leishmania major* co-inoculated with *Phlebotomus duboscqi* salivary gland homogenate. Acta Trop 144, 9–18.

Lewis, L.A., Radulovic, Z.M., Kim, T.K., Porter, L.M., Mulenga, A., 2015. Identification of 24 h *Ixodes scapularis* immunogenic tick saliva proteins. Ticks Tick Borne Dis. 6, 424–434.

Li, J., Zhang, Y.P., Kirsner, R.S., 2003. Angiogenesis in wound repair: angiogenic growth factors and the extracellular matrix. Microsc. Res. Tech. 60, 107–114.

Lima, H.C., Titus, R.G., 1996. Effects of sand fly vector saliva on development of cutaneous lesions and the immune response to *Leishmania braziliensis* in BALB/c mice. Infect. Immun 64, 5442–5445.

Limesand, K.H., Higgs, S., Pearson, L.D., Beaty, B.J., 2000. Potentiation of vesicular stomatitis New Jersey virus infection in mice by mosquito saliva. Parasite Immunol 22, 461–467.

Litman, G.W., Rast, J.P., Fugmann, S.D., 2010. The origins of vertebrate adaptive immunity. Nat. Rev. Immunol. 10, 543–553.

Ma, D., Assumpcao, T.C., Li, Y., Andersen, J.F., Ribeiro, J., Francischetti, I.M., 2012. Triplatin, a platelet aggregation inhibitor from the salivary gland of the triatomine vector of Chagas disease, binds to TXA$_2$ but does not interact with glycoprotein PVI. Thromb. Haemost. 107, 111–123.

Malafronte Rdos, S., Calvo, E., James, A.A., Marinotti, O., 2003. The major salivary gland antigens of *Culex quinquefasciatus* are D7-related proteins. Insect Biochem. Mol. Biol. 33, 63–71.

Mans, B.J., 2005. Tick histamine-binding proteins and related lipocalins: potential as therapeutic agents. Curr. Opin. Investig. Drugs 6, 1131–1135.

Mans, B.J., 2011. Evolution of vertebrate hemostatic and inflammatory control mechanisms in blood-feeding arthropods. J. Innate Immun. 3, 41–51.

Mans, B.J., Andersen, J.F., Francischetti, I.M., Valenzuela, J.G., Schwan, T.G., Pham, V.M., Garfield, M.K., Hammer, C.H., Ribeiro, J.M., 2008a. Comparative sialomics between hard and soft ticks: implications for the evolution of blood-feeding behavior. Insect Biochem. Mol. Biol. 38, 42–58.

Mans, B.J., Andersen, J.F., Schwan, T.G., Ribeiro, J.M., 2008b. Characterization of anti-hemostatic factors in the argasid, *Argas monolakensis*: implications for the evolution of blood-feeding in the soft tick family. Insect Biochem. Mol. Biol. 38, 22–41.

Mans, B.J., Coetzee, J., Louw, A.I., Gaspar, A.R., Neitz, A.W., 2000. Disaggregation of aggregated platelets by apyrase from the tick, *Ornithodoros savignyi* (Acari: Argasidae). Exp. Appl. Acarol. 24, 271–282.

Mans, B.J., Gaspar, A.R., Louw, A.I., Neitz, A.W., 1998. Apyrase activity and platelet aggregation inhibitors in the tick *Ornithodoros savignyi* (Acari: Argasidae). Exp. Appl. Acarol. 22, 353–366.

Mans, B.J., Louw, A.I., Neitz, A.W., 2002a. Evolution of hematophagy in ticks: common origins for blood coagulation and platelet aggregation inhibitors from soft ticks of the genus *Ornithodoros*. Mol. Biol. Evol. 19, 1695–1705.

Mans, B.J., Louw, A.I., Neitz, A.W., 2002b. Savignygrin, a platelet aggregation inhibitor from the soft tick *Ornithodoros savignyi*, presents the RGD integrin recognition motif on the Kunitz-BPTI fold. J. Biol. Chem. 277, 21371–21378.

Mans, B.J., Louw, A.I., Neitz, A.W., 2003. The influence of tick behavior, biotope and host specificity on concerted evolution of the platelet aggregation inhibitor savignygrin, from the soft tick *Ornithodoros savignyi*. Insect Biochem. Mol. Biol. 33, 623–629.

Mans, B.J., Neitz, A.W., 2004. Adaptation of ticks to a blood-feeding environment: evolution from a functional perspective. Insect Biochem. Mol. Biol. 34, 1–17.

Mans, B.J., Ribeiro, J.M., 2008. Function, mechanism and evolution of the moubatin-clade of soft tick lipocalins. Insect Biochem. Mol. Biol. 38, 841–852.

Markiewski, M.M., Lambris, J.D., 2007. The role of complement in inflammatory diseases from behind the scenes into the spotlight. Am. J. Pathol. 171, 715–727.

Mastronunzio, J.E., Kurscheid, S., Fikrig, E., 2012. Postgenomic analyses reveal development of infectious *Anaplasma phagocytophilum* during transmission from ticks to mice. J. Bacteriol. 194, 2238–2247.

Maxwell, S.S., Stoklasek, T.A., Dash, Y., Macaluso, K.R., Wikel, S.K., 2005. Tick modulation of the in-vitro expression of adhesion molecules by skin-derived endothelial cells. Ann. Trop. Med. Parasitol. 99, 661–672.

Mbow, M.L., Bleyenberg, J.A., Hall, L.R., Titus, R.G., 1998. *Phlebotomus papatasi* sand fly salivary gland lysate down-regulates a Th1, but up-regulates a Th2, response in mice infected with Leishmania major. J. Immunol. 161, 5571–5577.

McDowell, M.A., 2015. Vector-transmitted disease vaccines: targeting salivary proteins in transmission (SPIT). Trends Parasitol. 31, 363–372.

McGraw, E.A., O'Neill, S.L., 2013. Beyond insecticides: new thinking on an ancient problem. Nat. Rev. Microbiol. 11, 181–193.

McLaren, D.J., Worms, M.J., Brown, S.J., Askenase, P.W., 1983. Ornithodorus tartakovskyi: quantitation and ultrastructure of cutaneous basophil responses in the Guinea pig. Exp. Parasitol. 56, 153–168.

McNally, K.L., Mitzel, D.N., Anderson, J.M., Ribeiro, J.M., Valenzuela, J.G., Myers, T.G., Godinez, A., Wolfinbarger, J.B., Best, S.M., Bloom, M.E., 2012. Differential salivary gland transcript expression profile in Ixodes scapularis nymphs upon feeding or flavivirus infection. Ticks Tick Borne Dis. 3, 18–26.

McPherson, J.D., Marra, M., Hillier, L., Waterston, R.H., Chinwalla, A., Wallis, J., Sekhon, M., Wylie, K., Mardis, E.R., Wilson, R.K., Fulton, R., Kucaba, T.A., Wagner-McPherson, C., Barbazuk, W.B., Gregory, S.G., Humphray, S.J., French, L., Evans, R.S., Bethel, G., Whittaker, A., Holden, J.L., McCann, O.T., Dunham, A., Soderlund, C., Scott, C.E., Bentley, D.R., Schuler, G., Chen, H.C., Jang, W., Green, E.D., Idol, J.R., Maduro, V.V., Montgomery, K.T., Lee, E., Miller, A., Emerling, S., Kucherlapati, Gibbs, R., Scherer, S., Gorrell, J.H., Sodergren, E., Clerc-Blankenburg, K., Tabor, P., Naylor, S., Garcia, D., de Jong, P.J., Catanese, J.J., Nowak, N., Osoegawa, K., Qin, S., Rowen, L., Madan, A., Dors, M., Hood, L., Trask, B., Friedman, C., Massa, H., Cheung, V.G., Kirsch, I.R., Reid, T., Yonescu, R., Weissenbach, J., Bruls, T., Heilig, R., Branscomb, E., Olsen, A., Doggett, N., Cheng, J.F., Hawkins, T., Myers, R.M., Shang, J., Ramirez, L., Schmutz, J., Velasquez, O., Dixon, K., Stone, N.E., Cox, D.R., Haussler, D., Kent, W.J., Furey, T., Rogic, S., Kennedy, S., Jones, S., Rosenthal, A., Wen, G., Schilhabel, M., Gloeckner, G., Nyakatura, G., Siebert, R., Schlegelberger, B., Korenberg, J., Chen, X.N., Fujiyama, A., Hattori, M., Toyoda, A., Yada, T., Park, H.S., Sakaki, Y., Shimizu, N., Asakawa, S., Kawasaki, K., Sasaki, T., Shintani, A., Shimizu, A., Shibuya, K., Kudoh, J., Minoshima, S., Ramser, J., Seranski, P., Hoff, C., Poustka, A., Reinhardt, R., Lehrach, H., International Human Genome Mapping, C., 2001. A physical map of the human genome. Nature 409, 934–941.

Mejri, N., Brossard, M., 2007. Splenic dendritic cells pulsed with *Ixodes ricinus* tick saliva prime naive CD4[+]T to induce Th2 cell differentiation in vitro and in vivo. Int. Immunol. 19, 535–543.

Mellanby, K., 1946. Man's reaction to mosquito bites. Nature 158, 554.

Menard, R., 2000. The journey of the malaria sporozoite through its hosts: two parasite proteins lead the way. Microbes Infect. 2, 633–642.

Menard, R., Tavares, J., Cockburn, I., Markus, M., Zavala, F., Amino, R., 2013. Looking under the skin: the first steps in malarial infection and immunity. Nat. Rev. Microbiol. 11, 701–712.

Merad, M., Ginhoux, F., Collin, M., 2008. Origin, homeostasis and function of Langerhans cells and other langerin-expressing dendritic cells. Nat. Rev. Immunol. 8, 935–947.

Mesquita, R.D., Carneiro, A.B., Bafica, A., Gazos-Lopes, F., Takiya, C.M., Souto-Padron, T., Vieira, D.P., Ferreira-Pereira, A., Almeida, I.C., Figueiredo, R.T., Porto, B.N., Bozza, M.T., Graca-Souza, A.V., Lopes, A.H., Atella, G.C., Silva-Neto, M.A., 2008. *Trypanosoma cruzi* infection is enhanced by vector saliva through immunosuppressant mechanisms mediated by lysophosphatidylcholine. Infect. Immun 76, 5543–5552.

Milleron, R.S., Ribeiro, J.M., Elnaime, D., Soong, L., Lanzaro, G.C., 2004. Negative effect of antibodies against maxadilan on the fitness of the sand fly vector of American visceral leishmaniasis. Am. J. Trop. Med. Hyg. 70, 278–285.

Mitchell, E.B., Brown, S.J., Askenase, P.W., 1982. IgG1 antibody-dependent mediator release after passive systemic sensitization of basophils arriving at cutaneous basophil hypersensitivity reactions. J. Immunol. 129, 1663–1669.

Mizurini, D.M., Francischetti, I.M., Andersen, J.F., Monteiro, R.Q., 2010. Nitrophorin 2, a factor IX(a)-directed anticoagulant, inhibits arterial thrombosis without impairing haemostasis. Thromb. Haemost. 104, 1116–1123.

Mollnes, T.E., Song, W.C., Lambris, J.D., 2002. Complement in inflammatory tissue damage and disease. Trends Immunol. 23, 61–64.

Montandon, C.E., Barros, E., Vidigal, P.M., Mendes, M.T., Anhe, A.C., de Oliveira Ramos, H.J., de Oliveira, C.J., Mafra, C., 2016. Comparative proteomic analysis of the saliva of the *Rhodnius prolixus*, *Triatoma lecticularia* and *Panstrongylus herreri* triatomines reveals a high interespecific functional biodiversity. Insect Biochem. Mol. Biol. 71, 83–90.

Monteiro, R.Q., Rezaie, A.R., Bae, J.S., Calvo, E., Andersen, J.F., Francischetti, I.M., 2008. Ixolaris binding to factor X reveals a precursor state of factor Xa heparin-binding exosite. Protein Sci. 17, 146–153.

Montfort, W.R., Weichsel, A., Andersen, J.F., 2000. Nitrophorins and related antihemostatic lipocalins from *Rhodnius prolixus* and other blood-sucking arthropods. Biochim. Biophys. Acta 1482, 110–118.

Moreno, A., Clavijo, P., Edelman, R., Davis, J., Sztein, M., Sinigaglia, F., Nardin, E., 1993. CD4+ T cell clones obtained from *Plasmodium falciparum* sporozoite-immunized volunteers recognize polymorphic sequences of the circumsporozoite protein. J. Immunol. 151, 489–499.

Moro, O., Lerner, E.A., 1997. Maxadilan, the vasodilator from sand flies, is a specific pituitary adenylate cyclase activating peptide type I receptor agonist. J. Biol. Chem. 272, 966–970.

Morozova, O.V., Bakhvalova, V.N., Potapova, O.F., Grishechkin, A.E., Isaeva, E.I., Aldarov, K.V., Klinov, D.V., Vorovich, M.F., 2014. Evaluation of immune response and protective effect of four vaccines against the tick-borne encephalitis virus. Vaccine 32, 3101–3106.

Morris, R.V., Shoemaker, C.B., David, J.R., Lanzaro, G.C., Titus, R.G., 2001. Sandfly maxadilan exacerbates infection with *Leishmania major* and vaccinating against it protects against *L. major* infection. J. Immunol. 167, 5226–5230.

Mota, M.M., Rodriguez, A., 2002. Invasion of mammalian host cells by *Plasmodium* sporozoites. Bioessays 24, 149–156.

Moyer, M.W., 2015. The growing global battle against blood-sucking ticks. Nature 524, 406–408.

Muller, W.A., 2013. Getting leukocytes to the site of inflammation. Vet. Pathol. 50, 7–22.

Munoz-Chapuli, R., Carmona, R., Guadix, J.A., Macias, D., Perez-Pomares, J.M., 2005. The origin of the endothelial cells: an evo-devo approach for the invertebrate/vertebrate transition of the circulatory system. Evol. Dev. 7, 351–358.

Murray, C.J., Vos, T., Lozano, R., Naghavi, M., Flaxman, A.D., Michaud, C., Ezzati, M., Shibuya, K., Salomon, J.A., Abdalla, S., Aboyans, V., Abraham, J., Ackerman, I., Aggarwal, R., Ahn, S.Y., Ali, M.K., Alvarado, M., Anderson, H.R., Anderson, L.M., Andrews, K.G., Atkinson, C., Baddour, L.M., Bahalim, A.N., Barker-Collo, S., Barrero, L.H., Bartels, D.H., Basanez, M.G., Baxter, A., Bell, M.L., Benjamin, E.J., Bennett, D., Bernabe, E., Bhalla, K., Bhandari, B., Bikbov, B., Bin Abdulhak, A., Birbeck, G., Black, J.A., Blencowe, H., Blore, J.D., Blyth, F., Bolliger, I., Bonaventure, A., Boufous, S., Bourne, R., Boussinesq, M., Braithwaite, T., Brayne, C., Bridgett, L., Brooker, S., Brooks, P., Brugha, T.S., Bryan-Hancock, C., Bucello, C., Buchbinder, R., Buckle, G., Budke, C.M., Burch, M., Burney, P., Burstein, R., Calabria, B., Campbell, B., Canter, C.E., Carabin, H., Carapetis, J., Carmona, L., Cella, C., Charlson, F., Chen, H., Cheng, A.T., Chou, D., Chugh, S.S., Coffeng, L.E., Colan, S.D., Colquhoun, S., Colson, K.E., Condon, J., Connor, M.D., Cooper, L.T., Corriere, M., Cortinovis, M., de Vaccaro, K.C., Couser, W., Cowie, B.C., Criqui, M.H., Cross, M., Dabhadkar, K.C., Dahiya, M., Dahodwala, N., Damsere-Derry, J., Danaei, G., Davis, A., De Leo, D., Degenhardt, L., Dellavalle, R., Delossantos, A., Denenberg, J., Derrett, S., Des Jarlais, D.C., Dharmaratne, S.D., Dherani, M., Diaz-Torne, C., Dolk, H., Dorsey, E.R., Driscoll, T., Duber, H., Ebel, B., Edmond, K., Elbaz, A., Ali, S.E., Erskine, H., Erwin, P.J., Espindola, P., Ewoigbokhan, S.E., Farzadfar, F., Feigin, V., Felson, D.T., Ferrari, A., Ferri, C.P., Fevre,

E.M., Finucane, M.M., Flaxman, S., Flood, L., Foreman, K., Forouzanfar, M.H., Fowkes, F.G., Fransen, M., Freeman, M.K., Gabbe, B.J., Gabriel, S.E., Gakidou, E., Ganatra, H.A., Garcia, B., Gaspari, F., Gillum, R.F., Gmel, G., Gonzalez-Medina, D., Gosselin, R., Grainger, R., Grant, B., Groeger, J., Guillemin, F., Gunnell, D., Gupta, R., Haagsma, J., Hagan, H., Halasa, Y.A., Hall, W., Haring, D., Haro, J.M., Harrison, J.E., Havmoeller, R., Hay, R.J., Higashi, H., Hill, C., Hoen, B., Hoffman, H., Hotez, P.J., Hoy, D., Huang, J.J., Ibeanusi, S.E., Jacobsen, K.H., James, S.L., Jarvis, D., Jasrasaria, R., Jayaraman, S., Johns, N., Jonas, J.B., Karthikeyan, G., Kassebaum, N., Kawakami, N., Keren, A., Khoo, J.P., King, C.H., Knowlton, L.M., Kobusingye, O., Koranteng, A., Krishnamurthi, R., Laden, F., Lalloo, R., Laslett, L.L., Lathlean, T., Leasher, J.L., Lee, Y.Y., Leigh, J., Levinson, D., Lim, S.S., Limb, E., Lin, J.K., Lipnick, M., Lipshultz, S.E., Liu, W., Loane, M., Ohno, S.L., Lyons, R., Mabweijano, J., MacIntyre, M.F., Malekzadeh, R., Mallinger, L., Manivannan, S., Marcenes, W., March, L., Margolis, D.J., Marks, G.B., Marks, R., Matsumori, A., Matzopoulos, R., Mayosi, B.M., McAnulty, J.H., McDermott, M.M., McGill, N., McGrath, J., Medina-Mora, M.E., Meltzer, M., Mensah, G.A., Merriman, T.R., Meyer, A.C., Miglioli, V., Miller, M., Miller, T.R., Mitchell, P.B., Mock, C., Mocumbi, A.O., Moffitt, T.E., Mokdad, A.A., Monasta, L., Montico, M., Moradi-Lakeh, M., MoraN, A., Morawska, L., Mori, R., Murdoch, M.E., Mwaniki, M.K., Naidoo, K., Nair, M.N., Naldi, L., Narayan, K.M., Nelson, P.K., Nelson, R.G., Nevitt, M.C., Newton, C.R., Nolte, S., Norman, P., Norman, R., O'Donnell, M., O'Hanlon, S., Olives, C., Omer, S.B., Ortblad, K., Osborne, R., Ozgediz, D., Page, A., Pahari, B., Pandian, J.D., Rivero, A.P., Patten, S.B., Pearce, N., Padilla, R.P., Perez-ruiz, F., Perico, N., Pesudovs, K., Phillips, D., Phillips, M.R., Pierce, K., Pion, S., Polanczyk, G.V., Polinder, S., Pope 3rd, C.A., Popova, S., Porrini, E., Pourmalek, F., Prince, M., Pullan, R.L., Ramaiah, K.D., Ranganathan, D., Razavi, H., Regan, M., Rehm, J.T., Rein, D.B., Remuzzi, G., Richardson, K., Rivara, F.P., Roberts, T., Robinson, C., De Leon, F.R., Ronfani, L., Room, R., Rosenfeld, L.C., Rushton, L., Sacco, R.L., Saha, S., Sampson, U., Sanchez-riera, L., Sanman, E., Schwebel, D.C., Scott, J.G., Segui-gomez, M., Shahraz, S., Shepard, D.S., Shin, H., Shivakoti, R., Singh, D., Singh, G.M., Singh, J.A., Singleton, J., Sleet, D.A., Sliwa, K., Smith, E., Smith, J.L., Stapelberg, N.J., Steer, A., Steiner, T., Stolk, W.A., Stovner, L.J., Sudfeld, C., Syed, S., Tamburlini, G., Tavakkoli, M., Taylor, H.R., Taylor, J.A., Taylor, W.J., Thomas, B., Thomson, W.M., Thurston, G.D., Tleyjeh, I.M., Tonelli, M., Towbin, J.A., Truelsen, T., Tsilimbaris, M.K., Ubeda, C., Undurraga, E.A., van der Werf, M.J., van Os, J., Vavilala, M.S., Venketasubramanian, N., Wang, M., Wang, W., Watt, K., Weatherall, D.J., Weinstock, M.A., Weintraub, R., Weisskopf, M.G., Weissman, M.M., White, R.A., Whiteford, H., Wiebe, N., Wiersma, S.T., Wilkinson, J.D., Williams, H.C., Williams, S.R., Witt, E., Wolfe, F., Woolf, A.D., Wulf, S., Yeh, P.H., Zaidi, A.K., Zheng, Z.J., Zonies, D., Lopez, A.D., AlMazroa, M.A., Memish, Z.A., 2012. Disability-adjusted life years (DALYs) for 291 diseases and injuries in 21 regions, 1990–2010: a systematic analysis for the Global Burden of Disease Study 2010. Lancet 380, 2197–2223.

Narasimhan, S., Coumou, J., Schuijt, T.J., Boder, E., Hovius, J.W., Fikrig, E., 2014. A tick gut protein with fibronectin III domains aids *Borrelia burgdorferi* congregation to the gut during transmission. PLoS Pathog. 10, e1004278.

Narasimhan, S., Deponte, K., Marcantonio, N., Liang, X., Royce, T.E., Nelson, K.F., Booth, C.J., Koski, B., Anderson, J.F., Kantor, F., Fikrig, E., 2007a. Immunity against *Ixodes scapularis* salivary proteins expressed within 24 hours of attachment thwarts tick feeding and impairs Borrelia transmission. PLoS One 2, e451.

Narasimhan, S., Koski, R.A., Beaulieu, B., Anderson, J.F., Ramamoorthi, N., Kantor, F., Cappello, M., Fikrig, E., 2002. A novel family of anticoagulants from the saliva of *Ixodes scapularis*. Insect Mol. Biol. 11, 641–650.

Narasimhan, S., Perez, O., Mootien, S., DePonte, K., Koski, R.A., Fikrig, E., Ledizet, M., 2013. Characterization of Ixophilin, a thrombin inhibitor from the gut of *Ixodes scapularis*. PLoS One 8, e68012.

Narasimhan, S., Sukumaran, B., Bozdogan, U., Thomas, V., Liang, X., DePonte, K., Marcantonio, N., Koski, R.A., Anderson, J.F., Kantor, F., Fikrig, E., 2007b. A tick anti-oxidant facilitates the Lyme disease agent's successful migration from the mammalian host to the arthropod vector. Cell Host Microbe 2, 7–18.

Nazario, S., Das, S., de Silva, A.M., Deponte, K., Marcantonio, N., Anderson, J.F., Fish, D., Fikrig, E., Kantor, F.S., 1998. Prevention of *Borrelia burgdorferi* transmission in guinea pigs by tick immunity. Am. J. Trop. Med. Hyg. 58, 780–785.

Neafsey, D.E., Juraska, M., Bedford, T., Benkeser, D., Valim, C., Griggs, A., Lievens, M., Abdulla, S., Adjei, S., Agbenyega, T., Agnandji, S.T., Aide, P., Anderson, S., Ansong, D., Aponte, J.J., Asante, K.P., Bejon, P., Birkett, A.J., Bruls, M., Connolly, K.M., D'Alessandro, U., Dobano, C., Gesase, S., Greenwood, B., Grimsby, J., Tinto, H., Hamel, M.J., Hoffman, I., Kamthunzi, P., Kariuki, S., Kremsner, P.G., Leach, A., Lell, B., Lennon, N.J., Lusingu, J., Marsh, K., Martinson, F., Molel, J.T., Moss, E.L., Njuguna, P., Ockenhouse, C.F., Ogutu, B.R., Otieno, W., Otieno, L., Otieno, K., Owusu-Agyei, S., Park, D.J., Pelle, K., Robbins, D., Russ, C., Ryan, E.M., Sacarlal, J., Sogoloff, B., Sorgho, H., Tanner, M., Theander, T., Valea, I., Volkman, S.K., Yu, Q., Lapierre, D., Birren, B.W., Gilbert, P.B., Wirth, D.F., 2015. Genetic diversity and protective efficacy of the RTS,S/AS01 malaria vaccine. N. Engl. J. Med. 373, 2025–2037.

Neelakanta, G., Sultana, H., 2015. Transmission-blocking vaccines: focus on anti-vector vaccines against tick-borne diseases. Arch. Immunol. Ther. Exp. (Warsz) 63, 169–179.

Neher, M.D., Weckbach, S., Flierl, M.A., Huber-Lang, M.S., Stahel, P.F., 2011. Molecular mechanisms of inflammation and tissue injury after major trauma–is complement the "bad guy". J. Biomed. Sci. 18, 90.

Neiderud, C.J., 2015. How urbanization affects the epidemiology of emerging infectious diseases. Infect. Ecol. Epidemiol. 5, 27060.

Nestle, F.O., Di Meglio, P., Qin, J.Z., Nickoloff, B.J., 2009. Skin immune sentinels in health and disease. Nat. Rev. Immunol. 9, 679–691.

Norsworthy, N.B., Sun, J., Elnaiem, D., Lanzaro, G., Soong, L., 2004. Sand fly saliva enhances Leishmania amazonensis infection by modulating interleukin-10 production. Infect. Immun. 72, 1240–1247.

Nunes, J.K., Woods, C., Carter, T., Raphael, T., Morin, M.J., Diallo, D., Leboulleux, D., Jain, S., Loucq, C., Kaslow, D.C., Birkett, A.J., 2014. Development of a transmission-blocking malaria vaccine: progress, challenges, and the path forward. Vaccine 32, 5531–5539.

Nuttall, P.A., Labuda, M., 2004. Tick-host interactions: saliva-activated transmission. Parasitology (129 Suppl.), S177–S189.

O'Sullivan, T.E., Sun, J.C., Lanier, L.L., 2015. Natural killer cell memory. Immunity 43, 634–645.

Oliveira, C.J., Cavassani, K.A., More, D.D., Garlet, G.P., Aliberti, J.C., Silva, J.S., Ferreira, B.R., 2008a. Tick saliva inhibits the chemotactic function of MIP-1α and selectively impairs chemotaxis of immature dendritic cells by down-regulating cell-surface CCR5. Int. J. Parasitol. 38, 705–716.

Oliveira, C.J., Sa-Nunes, A., Francischetti, I.M., Carregaro, V., Anatriello, E., Silva, J.S., Santos, I.K., Ribeiro, J.M., Ferreira, B.R., 2011. Deconstructing tick saliva: non-protein molecules with potent immunomodulatory properties. J. Biol. Chem. 286, 10960–10969.

Oliveira, F., de Carvalho, A.M., de Oliveira, C.I., 2013a. Sand-fly saliva-leishmania-man: the trigger trio. Front. Immunol. 4, 375.

Oliveira, F., Lawyer, P.G., Kamhawi, S., Valenzuela, J.G., 2008b. Immunity to distinct sand fly salivary proteins primes the anti-Leishmania immune response towards protection or exacerbation of disease. PLoS Negl. Trop. Dis. 2, e226.

Oliveira, F., Rowton, E., Aslan, H., Gomes, R., Castrovinci, P.A., Alvarenga, P.H., Abdeladhim, M., Teixeira, C., Meneses, C., Kleeman, L.T., Guimaraes-Costa, A.B., Rowland, T.E., Gilmore, D., Doumbia, S., Reed, S.G., Lawyer, P.G., Andersen, J.F., Kamhawi, S., Valenzuela, J.G., 2015. A sand fly salivary protein vaccine shows efficacy against vector-transmitted cutaneous leishmaniasis in nonhuman primates. Sci. Transl. Med. 7, 290.

Oliveira, F., Traore, B., Gomes, R., Faye, O., Gilmore, D.C., Keita, S., Traore, P., Teixeira, C., Coulibaly, C.A., Samake, S., Meneses, C., Sissoko, I., Fairhurst, R.M., Fay, M.P., Anderson, J.M., Doumbia, S., Kamhawi, S., Valenzuela, J.G., 2013b. Delayed-type hypersensitivity to sand fly saliva in humans from a leishmaniasis-endemic area of Mali is Th1-mediated and persists to midlife. J. Invest. Dermatol. 133, 452–459.

Olotu, A., Fegan, G., Wambua, J., Nyangweso, G., Leach, A., Lievens, M., Kaslow, D.C., Njuguna, P., Marsh, K., Bejon, P., 2016. Seven-year efficacy of RTS,S/AS01 malaria vaccine among young African children. N. Engl. J. Med. 374, 2519–2529.

Otsuka, A., Nakajima, S., Kubo, M., Egawa, G., Honda, T., Kitoh, A., Nomura, T., Hanakawa, S., Sagita Moniaga, C., Kim, B., Matsuoka, S., Watanabe, T., Miyachi, Y., Kabashima, K., 2013. Basophils are required for the induction of Th2 immunity to haptens and peptide antigens. Nat. Commun. 4, 1739.

Paesen, G.C., Adams, P.L., Nuttall, P.A., Stuart, D.L., 2000. Tick histamine-binding proteins: lipocalins with a second binding cavity. Biochim. Biophys. Acta 1482, 92–101.

Paine, S.H., Kemp, D.H., Allen, J.R., 1983. In vitro feeding of Dermacentor andersoni (Stiles): effects of histamine and other mediators. Parasitology 86 (Pt. 3), 419–428.

Palmer, R.M., Ferrige, A.G., Moncada, S., 1987. Nitric oxide release accounts for the biological activity of endothelium-derived relaxing factor. Nature 327, 524–526.

Palta, S., Saroa, R., Palta, A., 2014. Overview of the coagulation system. Indian J. Anaesth. 58, 515–523.

Paranhos-Silva, M., Oliveira, G.G., Reis, E.A., de Menezes, R.M., Fernandes, O., Sherlock, I., Gomes, R.B., Pontes-de-Carvalho, L.C., dos-Santos, W.L., 2003. A follow-up of Beagle dogs intradermally infected with Leishmania chagasi in the presence or absence of sand fly saliva. Vet. Parasitol 114, 97–111.

Parmar, A., Grewal, A.S., Dhillon, P., 1996. Immunological cross-reactivity between salivary gland proteins of Hyalomma anatolicum anatolicum and Boophilus microplus ticks. Vet. Immunol. Immunopathol. 51, 345–352.

Pasare, C., Medzhitov, R., 2003. Toll-like receptors: balancing host resistance with immune tolerance. Curr. Opin. Immunol. 15, 677–682.

Patramool, S., Choumet, V., Surasombatpattana, P., Sabatier, L., Thomas, F., Thongrungkiat, S., Rabilloud, T., Boulanger, N., Biron, D.G., Misse, D., 2012. Update on the proteomics of major arthropod vectors of human and animal pathogens. Proteomics 12, 3510–3523.

Peterkova, K., Vancova, I., Hajnicka, V., Slovak, M., Simo, L., Nuttall, P.A., 2008. Immunomodulatory arsenal of nymphal ticks. Med. Vet. Entomol. 22, 167–171.

Petrovsky, N., Aguilar, J.C., 2004. Vaccine adjuvants: current state and future trends. Immunol. Cell Biol. 82, 488–496.

Pingen, M., Bryden, S.R., Pondeville, E., Schnettler, E., Kohl, A., Merits, A., Fazakerley, J.K., Graham, G.J., McKimmie, C.S., 2016. Host inflammatory response to mosquito bites enhances the severity of Arbovirus infection. Immunity 44, 1455–1469.

Plotkin, S.A., 2016. Need for a new lyme disease vaccine. N. Engl. J. Med. 375, 911–913.

Poole, N.M., Mamidanna, G., Smith, R.A., Coons, L.B., Cole, J.A., 2013. Prostaglandin E_2 in tick saliva regulates macrophage cell migration and cytokine profile. Parasit. Vectors 6, 261.

Prates, D.B., Araujo-Santos, T., Luz, N.F., Andrade, B.B., Franca-Costa, J., Afonso, L., Clarencio, J., Miranda, J.C., Bozza, P.T., Dosreis, G.A., Brodskyn, C., Barral-Netto, M., Borges, V.M., Barral, A., 2011. Lutzomyia longipalpis saliva drives apoptosis and enhances parasite burden in neutrophils. J. Leukoc. Biol. 90, 575–582.

Preston, S.G., Majtan, J., Kouremenou, C., Rysnik, O., Burger, L.F., Cabezas Cruz, A., Chiong Guzman, M., Nunn, M.A., Paesen, G.C., Nuttall, P.A., Austyn, J.M., 2013. Novel immunomodulators from hard ticks selectively reprogramme human dendritic cell responses. PLoS Pathog. 9, e1003450.

Prevot, P.P., Adam, B., Boudjeltia, K.Z., Brossard, M., Lins, L., Cauchie, P., Brasseur, R., Vanhaeverbeek, M., Vanhamme, L., Godfroid, E., 2006. Anti-hemostatic effects of a serpin from the saliva of the tick Ixodes ricinus. J. Biol. Chem. 281, 26361–26369.

Qureshi, A.A., Asahina, A., Ohnuma, M., Tajima, M., Granstein, R.D., Lerner, E.A., 1996. Immunomodulatory properties of maxadilan, the vasodilator peptide from sand fly salivary gland extracts. Am. J. Trop. Med. Hyg. 54, 665–671.

Radolf, J.D., Caimano, M.J., Stevenson, B., Hu, L.T., 2012. Of ticks, mice and men: understanding the dual-host lifestyle of Lyme disease spirochaetes. Nat. Rev. Microbiol. 10, 87–99.

Radovsky, F.J., 1969. Adaptive radiation in the parasitic mesostigmata. Acarologia 11, 450–483.

Radulovic, Z.M., Kim, T.K., Porter, L.M., Sze, S.H., Lewis, L., Mulenga, A., 2014. A 24-48 h fed Amblyomma americanum tick saliva immuno-proteome. BMC Genomics 15, 518.

Ramachandra, R.N., Wikel, S.K., 1992. Modulation of host-immune responses by ticks (Acari: Ixodidae): effect of salivary gland extracts on host macrophages and lymphocyte cytokine production. J. Med. Entomol. 29, 818–826.

Ramamoorthi, N., Narasimhan, S., Pal, U., Bao, F., Yang, X.F., Fish, D., Anguita, J., Norgard, M.V., Kantor, F.S., Anderson, J.F., Koski, R.A., Fikrig, E., 2005. The Lyme disease agent exploits a tick protein to infect the mammalian host. Nature 436, 573–577.

Rassi Jr., A., Rassi, A., Marin-Neto, J.A., 2010. Chagas disease. Lancet 375, 1388–1402.

Reagan, K.L., Machain-Williams, C., Wang, T., Blair, C.D., 2012. Immunization of mice with recombinant mosquito salivary protein D7 enhances mortality from subsequent West Nile virus infection via mosquito bite. PLoS Negl. Trop. Dis. 6, e1935.

Reddy, V.B., Li, Y., Lerner, E.A., 2008. Maxadilan, the PAC1 receptor, and leishmaniasis. J. Mol. Neurosci. 36, 241–244.

Remoue, F., Cisse, B., Ba, F., Sokhna, C., Herve, J.P., Boulanger, D., Simondon, F., 2006. Evaluation of the antibody response to Anopheles salivary antigens as a potential marker of risk of malaria. Trans. R. Soc. Trop. Med. Hyg. 100, 363–370.

Ribeiro, J.M., 1992. Characterization of a vasodilator from the salivary glands of the yellow fever mosquito Aedes aegypti. J. Exp. Biol. 165, 61–71.

Ribeiro, J.M., 1995. Blood-feeding arthropods: live syringes or invertebrate pharmacologists? Infect. Agents Dis. 4, 143–152.

Ribeiro, J.M., 1996. NAD(P)H-dependent production of oxygen reactive species by the salivary glands of the mosquito Anopheles albimanus. Insect Biochem. Mol. Biol. 26, 715–720.

Ribeiro, J.M., Alarcon-Chaidez, F., Francischetti, I.M., Mans, B.J., Mather, T.N., Valenzuela, J.G., Wikel, S.K., 2006. An annotated catalog of salivary gland transcripts from Ixodes scapularis ticks. Insect Biochem. Mol. Biol. 36, 111–129.

Ribeiro, J.M., Arca, B., Lombardo, F., Calvo, E., Phan, V.M., Chandra, P.K., Wikel, S.K., 2007. An annotated catalogue of salivary gland transcripts in the adult female mosquito, Aedes aegypti. BMC Genomics 8, 6.

Ribeiro, J.M., Charlab, R., Rowton, E.D., Cupp, E.W., 2000. Simulium vittatum (Diptera: Simuliidae) and Lutzomyia longipalpis (Diptera: Psychodidae) salivary gland hyaluronidase activity. J. Med. Entomol. 37, 743–747.

Ribeiro, J.M., Charlab, R., Valenzuela, J.G., 2001. The salivary adenosine deaminase activity of the mosquitoes Culex quinquefasciatus and Aedes aegypti. J. Exp. Biol. 204, 2001–2010.

Ribeiro, J.M., Francischetti, I.M., 2003. Role of arthropod saliva in blood feeding: sialome and post-sialome perspectives. Annu. Rev. Entomol. 48, 73–88.

Ribeiro, J.M., Hazzard, J.M., Nussenzveig, R.H., Champagne, D.E., Walker, F.A., 1993. Reversible binding of nitric oxide by a salivary heme protein from a bloodsucking insect. Science 260, 539–541.

Ribeiro, J.M., Katz, O., Pannell, L.K., Waitumbi, J., Warburg, A., 1999. Salivary glands of the sand fly Phlebotomus papatasi contain pharmacologically active amounts of adenosine and 5'-AMP. J. Exp. Biol. 202, 1551–1559.

Ribeiro, J.M., Makoul, G.T., Levine, J., Robinson, D.R., Spielman, A., 1985. Antihemostatic, antiinflammatory, and immunosuppressive properties of the saliva of a tick, Ixodes dammini. J. Exp. Med. 161, 332–344.

Ribeiro, J.M., Mans, B.J., Arca, B., 2010. An insight into the sialome of blood-feeding Nematocera. Insect Biochem. Mol. Biol. 40, 767–784.

Ribeiro, J.M., Martin-Martin, I., Arca, B., Calvo, E., 2016. A deep insight into the sialome of male and female *Aedes aegypti* mosquitoes. PLoS One 11, e0151400.

Ribeiro, J.M., Schneider, M., Isaias, T., Jurberg, J., Galvao, C., Guimaraes, J.A., 1998. Role of salivary antihemostatic components in blood feeding by triatomine bugs (Heteroptera). J. Med. Entomol. 35, 599–610.

Ribeiro, J.M., Vachereau, A., Modi, G.B., Tesh, R.B., 1989. A novel vasodilatory peptide from the salivary glands of the sand fly *Lutzomyia longipalpis*. Science 243, 212–214.

Ribeiro, J.M., Valenzuela, J.G., 1999. Purification and cloning of the salivary peroxidase/catechol oxidase of the mosquito *Anopheles albimanus*. J. Exp. Biol. 202, 809–816.

Ribeiro, J.M., Weis, J.J., Telford 3rd, S.R., 1990. Saliva of the tick Ixodes dammini inhibits neutrophil function. Exp. Parasitol. 70, 382–388.

Richerson, H.B., Dvorak, H.F., Leskowitz, S., 1970. Cutaneous basophil hypersensitivity. I. A new look at the Jones-Mote reaction, general characteristics. J. Exp. Med. 132, 546–557.

Rizzo, C., Lombardo, F., Ronca, R., Mangano, V., Sirima, S.B., Nebie, I., Fiorentino, G., Modiano, D., Arca, B., 2014. Differential antibody response to the *Anopheles gambiae* gSG6 and cE5 salivary proteins in individuals naturally exposed to bites of malaria vectors. Parasit. Vectors 7, 549.

Roberts, D.R., Andre, R.G., 1994. Insecticide resistance issues in vector-borne disease control. Am. J. Trop. Med. Hyg. 50, 21–34.

Roberts, D.R., Laughlin, L.L., Hsheih, P., Legters, L.J., 1997. DDT, global strategies, and a malaria control crisis in South America. Emerg. Infect. Dis. 3, 295–302.

Rocha, A.C., Braga, E.M., Araujo, M.S., Franklin, B.S., Pimenta, P.F., 2004. Effect of the *Aedes fluviatilis* saliva on the development of *Plasmodium gallinaceum* infection in *Gallus (gallus) domesticus*. Mem. Inst. Oswaldo Cruz 99, 709–715.

Rodriguez, M., Massard, C.L., da Fonseca, A.H., Ramos, N.F., Machado, H., Labarta, V., de la Fuente, J., 1995. Effect of vaccination with a recombinant Bm86 antigen preparation on natural infestations of *Boophilus microplus* in grazing dairy and beef pure and cross-bred cattle in Brazil. Vaccine 13, 1804–1808.

Rodriguez-Mallon, A., 2016. Developing anti-tick vaccines. Methods Mol. Biol. 1404, 243–259.

Rodriguez-Valle, M., Xu, T., Kurscheid, S., Lew-Tabor, A.E., 2015. *Rhipicephalus microplus* serine protease inhibitor family: annotation, expression and functional characterisation assessment. Parasit. Vectors 8, 7.

Roehrig, J.T., Layton, M., Smith, P., Campbell, G.L., Nasci, R., Lanciotti, R.S., 2002. The emergence of West Nile virus in North America: ecology, epidemiology, and surveillance. Curr. Top. Microbiol. Immunol. 267, 223–240.

Rogers, M.E., Chance, M.L., Bates, P.A., 2002. The role of promastigote secretory gel in the origin and transmission of the infective stage of *Leishmania mexicana* by the sandfly *Lutzomyia longipalpis*. Parasitology 124, 495–507.

Rohousova, I., Hostomska, J., Vlkova, M., Kobets, T., Lipoldova, M., Volf, P., 2011. The protective effect against Leishmania infection conferred by sand fly bites is limited to short-term exposure. Int. J. Parasitol. 41, 481–485.

Ronca, R., Kotsyfakis, M., Lombardo, F., Rizzo, C., Curra, C., Ponzi, M., Fiorentino, G., Ribeiro, J.M., Arca, B., 2012. The *Anopheles gambiae* cE5, a tight- and fast-binding thrombin inhibitor with post-transcriptionally regulated salivary-restricted expression. Insect Biochem. Mol. Biol. 42, 610–620.

Rossignol, P.A., Ribeiro, J.M., Spielman, A., 1984. Increased intradermal probing time in sporozoite-infected mosquitoes. Am. J. Trop. Med. Hyg. 33, 17–20.

Russel, R., Otranto, P.D., Wall, L.R., 2013. The Encyclopedia of Medical and Veterinary Entomology. CAB International North America.

Sa-Nunes, A., Bafica, A., Lucas, D.A., Conrads, T.P., Veenstra, T.D., Andersen, J.F., Mather, T.N., Ribeiro, J.M., Francischetti, I.M., 2007. Prostaglandin E2 is a major inhibitor of dendritic cell maturation and function in *Ixodes scapularis* saliva. J. Immunol. 179, 1497–1505.

Sakhon, O.S., Severo, M.S., Kotsyfakis, M., Pedra, J.H., 2013. A Nod to disease vectors: mitigation of pathogen sensing by arthropod saliva. Front. Microbiol. 4, 308.

Samuelson, J., Lerner, E., Tesh, R., Titus, R., 1991. A mouse model of *Leishmania braziliensis braziliensis* infection produced by coinjection with sand fly saliva. J. Exp. Med 173, 49–54.

Sawynok, J., 1998. Adenosine receptor activation and nociception. Eur. J. Pharmacol. 347, 1–11.

Schneider, B.S., Mathieu, C., Peronet, R., Mecheri, S., 2011. *Anopheles stephensi* saliva enhances progression of cerebral malaria in a murine model. Vector Borne Zoonotic Dis. 11, 423–432.

Schneider, B.S., McGee, C.E., Jordan, J.M., Stevenson, H.L., Soong, L., Higgs, S., 2007. Prior exposure to uninfected mosquitoes enhances mortality in naturally-transmitted West Nile virus infection. PLoS One 2, e1171.

Schneider, B.S., Soong, L., Girard, Y.A., Campbell, G., Mason, P., Higgs, S., 2006. Potentiation of West Nile encephalitis by mosquito feeding. Viral Immunol. 19, 74–82.

Schneider, B.S., Soong, L., Zeidner, N.S., Higgs, S., 2004. *Aedes aegypti* salivary gland extracts modulate anti-viral and TH1/TH2 cytokine responses to sindbis virus infection. Viral Immunol 17, 565–573.

Schoeler, G.B., Manweiler, S.A., Wikel, S.K., 1999. *Ixodes scapularis*: effects of repeated infestations with pathogen-free nymphs on macrophage and T lymphocyte cytokine responses of BALB/c and C3H/HeN mice. Exp. Parasitol. 92, 239–248.

Schofield, C.J., Galvao, C., 2009. Classification, evolution, and species groups within the Triatominae. Acta Trop. 110, 88–100.

Schofield, L., Villaquiran, J., Ferreira, A., Schellekens, H., Nussenzweig, R., Nussenzweig, V., 1987. Gamma interferon, CD8⁺ T cells and antibodies required for immunity to malaria sporozoites. Nature 330, 664–666.

Scholl, D.C., Embers, M.E., Caskey, J.R., Kaushal, D., Mather, T.N., Buck, W.R., Morici, L.A., Philipp, M.T., 2016. Immunomodulatory effects of tick saliva on dermal cells exposed to Borrelia burgdorferi, the agent of Lyme disease. Parasit. Vectors 9, 394.

Schroeder, H., Daix, V., Gillet, L., Renauld, J.C., Vanderplasschen, A., 2007. The paralogous salivary anti-complement proteins IRAC I and IRAC II encoded by *Ixodes ricinus* ticks have broad and complementary inhibitory activities against the complement of different host species. Microbes Infect. 9, 247–250.

Schuijt, T.J., Bakhtiari, K., Daffre, S., Deponte, K., Wielders, S.J., Marquart, J.A., Hovius, J.W., van der Poll, T., Fikrig, E., Bunce, M.W., Camire, R.M., Nicolaes, G.A., Meijers, J.C., van't Veer, C., 2013. Factor Xa activation of factor V is of paramount importance in initiating the coagulation system: lessons from a tick salivary protein. Circulation 128, 254–266.

Schuijt, T.J., Coumou, J., Narasimhan, S., Dai, J., Deponte, K., Wouters, D., Brouwer, M., Oei, A., Roelofs, J.J., van Dam, A.P., van der Poll, T., Van't Veer, C., Hovius, J.W., Fikrig, E., 2011a. A tick mannose-binding lectin inhibitor interferes with the vertebrate complement cascade to enhance transmission of the lyme disease agent. Cell Host Microbe 10, 136–146.

Schuijt, T.J., Narasimhan, S., Daffre, S., DePonte, K., Hovius, J.W., Van't Veer, C., van der Poll, T., Bakhtiari, K., Meijers, J.C., Boder, E.T., van Dam, A.P., Fikrig, E., 2011b. Identification and characterization of *Ixodes scapularis* antigens that elicit tick immunity using yeast surface display. PLoS One 6, e15926.

Schwarz, A., Cabezas-Cruz, A., Kopecky, J., Valdes, J.J., 2014. Understanding the evolutionary structural variability and target specificity of tick salivary Kunitz peptides using next generation transcriptome data. BMC Evol. Biol. 14, 4.

Schwarz, A., Medrano-Mercado, N., Billingsley, P.F., Schaub, G.A., Sternberg, J.M., 2010. IgM-antibody responses of chickens to salivary antigens of *Triatoma infestans* as early biomarkers for low-level infestation of triatomines. Int. J. Parasitol. 40, 1295–1302.

Seppo, A., Tiemeyer, M., 2000. Function and structure of *Drosophila glycans*. Glycobiology 10, 751–760.

Shen, H.D., Chen, C.C., Chang, H.N., Chang, L.Y., Tu, W.C., Han, S.H., 1989. Human IgE and IgG antibodies to mosquito proteins detected by the immunoblot technique. Ann. Allergy 63, 143–146.

Shim, W.S., Oh, U., 2008. Histamine-induced itch and its relationship with pain. Mol. Pain 4, 29.

Sinnis, P., Zavala, F., 2012. The skin: where malaria infection and the host immune response begin. Semin. Immunopathol. 34, 787–792.

Siracusa, M.C., Kim, B.S., Spergel, J.M., Artis, D., 2013. Basophils and allergic inflammation. J. Allergy Clin. Immunol. 132, 789–801 quiz 788.

Soares, M.B., Titus, R.G., Shoemaker, C.B., David, J.R., Bozza, M., 1998. The vasoactive peptide maxadilan from sand fly saliva inhibits TNF-alpha and induces IL-6 by mouse macrophages through interaction with the pituitary adenylate cyclase-activating polypeptide (PACAP) receptor. J. Immunol. 160, 1811–1816.

Sokol, C.L., Medzhitov, R., 2010. Role of basophils in the initiation of Th2 responses. Curr. Opin. Immunol. 22, 73–77.

Somda, M.B., Bengaly, Z., Dama, E., Poinsignon, A., Dayo, G.K., Sidibe, I., Remoue, F., Sanon, A., Bucheton, B., 2013. First insights into the cattle serological response to tsetse salivary antigens: a promising direct biomarker of exposure to tsetse bites. Vet. Parasitol. 197, 332–340.

Sonenshine, D.E., Roe, R.M., 2014. Biology of Ticks, second ed. Oxford University Press, New York.

Soulard, V., Bosson-Vanga, H., Lorthiois, A., Roucher, C., Franetich, J.F., Zanghi, G., Bordessoulles, M., Tefit, M., Thellier, M., Morosan, S., Le Naour, G., Capron, F., Suemizu, H., Snounou, G., Moreno-Sabater, A., Mazier, D., 2015. *Plasmodium falciparum* full life cycle and *Plasmodium ovale* liver stages in humanized mice. Nat. Commun. 6, 7690.

Spencer, L.A., Szela, C.T., Perez, S.A., Kirchhoffer, C.L., Neves, J.S., Radke, A.L., Weller, P.F., 2009. Human eosinophils constitutively express multiple Th1, Th2, and immunoregulatory cytokines that are secreted rapidly and differentially. J. Leukoc. Biol. 85, 117–123.

Stark, K.R., James, A.A., 1998. Isolation and characterization of the gene encoding a novel factor Xa-directed anticoagulant from the yellow fever mosquito, *Aedes aegypti*. J. Biol. Chem. 273, 20802–20809.

Stevens, T.L., Bossie, A., Sanders, V.M., Fernandez-Botran, R., Coffman, R.L., Mosmann, T.R., Vitetta, E.S., 1988. Regulation of antibody isotype secretion by subsets of antigen-specific helper T cells. Nature 334, 255–258.

Struik, S.S., Riley, E.M., 2004. Does malaria suffer from lack of memory? Immunol. Rev. 201, 268–290.

Styer, L.M., Lim, P.Y., Louie, K.L., Albright, R.G., Kramer, L.D., Bernard, K.A., 2011. Mosquito saliva causes enhancement of West Nile virus infection in mice. J. Virol. 85, 1517–1527.

Sutherst, R.W., 2004. Global change and human vulnerability to vector-borne diseases. Clin. Microbiol. Rev. 17, 136–173.

Tavares, N.M., Silva, R.A., Costa, D.J., Pitombo, M.A., Fukutani, K.F., Miranda, J.C., Valenzuela, J.G., Barral, A., de Oliveira, C.I., Barral-Netto, M., Brodskyn, C., 2011. *Lutzomyia longipalpis* saliva or salivary protein LJM19 protects against *Leishmania braziliensis* and the saliva of its vector, *Lutzomyia intermedia*. PLoS Negl. Trop. Dis. 5, e1169.

Tellam, R.L., Eisemann, C.H., Vuocolo, T., Casu, R., Jarmey, J., Bowles, V., Pearson, R., 2001. Role of oligosaccharides in the immune response of sheep vaccinated with *Lucilia cuprina* larval glycoprotein, peritrophin-95. Int. J. Parasitol. 31, 798–809.

Thanassi, W.T., Schoen, R.T., 2000. The Lyme disease vaccine: conception, development, and implementation. Ann. Intern. Med. 132, 661–668.

Thangamani, S., Higgs, S., Ziegler, S., Vanlandingham, D., Tesh, R., Wikel, S., 2010. Host immune response to mosquito-transmitted chikungunya virus differs from that elicited by needle inoculated virus. PLoS One 5, e12137.

Theodos, C.M., Povinelli, L., Molina, R., Sherry, B., Titus, R.G., 1991a. Role of tumor necrosis factor in macrophage leishmanicidal activity in vitro and resistance to cutaneous leishmaniasis in vivo. Infect. Immun. 59, 2839–2842.

Theodos, C.M., Ribeiro, J.M., Titus, R.G., 1991b. Analysis of enhancing effect of sand fly saliva on Leishmania infection in mice. Infect. Immun. 59, 1592–1598.

Theodos, C.M., Titus, R.G., 1993. Salivary gland material from the sand fly *Lutzomyia longipalpis* has an inhibitory effect on macrophage function in vitro. Parasite Immunol 15, 481–487.

Thiakaki, M., Rohousova, I., Volfova, V., Volf, P., Chang, K.P., Soteriadou, K., 2005. Sand fly specificity of saliva-mediated protective immunity in *Leishmania amazonensis*-BALB/c mouse model. Microbe. Infect 7, 760–766.

Tirloni, L., Kim, T.K., Coutinho, M.L., Ali, A., Seixas, A., Termignoni, C., Mulenga, A., da Silva Vaz Jr., I., 2016. The putative role of Rhipicephalus microplus salivary serpins in the tick-host relationship. Insect Biochem. Mol. Biol. 71, 12–28.

Titus, R.G., Bishop, J.V., Mejia, J.S., 2006. The immunomodulatory factors of arthropod saliva and the potential for these factors to serve as vaccine targets to prevent pathogen transmission. Parasite Immunol. 28, 131–141.

Titus, R.G., Ribeiro, J.M., 1988. Salivary gland lysates from the sand fly *Lutzomyia longipalpis* enhance Leishmania infectivity. Science 239, 1306–1308.

Trager, W., 1939. Accquired immunity to ticks. J. Parasitol. 25, 57–81.

Trimnell, A.R., Davies, G.M., Lissina, O., Hails, R.S., Nuttall, P.A., 2005. A cross-reactive tick cement antigen is a candidate broad-spectrum tick vaccine. Vaccine 23, 4329–4341.

Trimnell, A.R., Hails, R.S., Nuttall, P.A., 2002. Dual action ectoparasite vaccine targeting 'exposed' and 'concealed' antigens. Vaccine 20, 3560–3568.

Triplett, D.A., 2000. Coagulation and bleeding disorders: review and update. Clin. Chem. 46, 1260–1269.

Trung, H.D., Van Bortel, W., Sochantha, T., Keokenchanh, K., Quang, N.T., Cong, L.D., Coosemans, M., 2004. Malaria transmission and major malaria vectors in different geographical areas of Southeast Asia. Trop. Med. Int. Health 9, 230–237.

Tucker, T., Marra, M., Friedman, J.M., 2009. Massively parallel sequencing: the next big thing in genetic medicine. Am. J. Hum. Genet. 85, 142–154.

Tyson, K.R., Elkins, C., de Silva, A.M., 2008. A novel mechanism of complement inhibition unmasked by a tick salivary protein that binds to properdin. J. Immunol. 180, 3964–3968.

Valenzuela, J.G., Belkaid, Y., Garfield, M.K., Mendez, S., Kamhawi, S., Rowton, E.D., Sacks, D.L., Ribeiro, J.M., 2001a. Toward a defined anti-Leishmania vaccine targeting vector antigens: characterization of a protective salivary protein. J. Exp. Med. 194, 331–342.

Valenzuela, J.G., Belkaid, Y., Rowton, E., Ribeiro, J.M., 2001b. The salivary apyrase of the blood-sucking sand fly *Phlebotomus papatasi* belongs to the novel Cimex family of apyrases. J. Exp. Biol. 204, 229–237.

Valenzuela, J.G., Charlab, R., Gonzalez, E.C., de Miranda-Santos, I.K., Marinotti, O., Francischetti, I.M., Ribeiro, J.M., 2002a. The D7 family of salivary proteins in blood sucking diptera. Insect Mol. Biol. 11, 149–155.

Valenzuela, J.G., Charlab, R., Mather, T.N., Ribeiro, J.M., 2000. Purification, cloning, and expression of a novel salivary anticomplement protein from the tick, *Ixodes scapularis*. J. Biol. Chem. 275, 18717–18723.

Valenzuela, J.G., Francischetti, I.M., Pham, V.M., Garfield, M.K., Mather, T.N., Ribeiro, J.M., 2002b. Exploring the sialome of the tick *Ixodes scapularis*. J. Exp. Biol. 205, 2843–2864.

Valenzuela, J.G., Francischetti, I.M., Pham, V.M., Garfield, M.K., Ribeiro, J.M., 2003. Exploring the salivary gland transcriptome and proteome of the *Anopheles stephensi* mosquito. Insect Biochem. Mol. Biol. 33, 717–732.

Valenzuela, J.G., Ribeiro, J.M., 1998. Purification and cloning of the salivary nitrophorin from the hemipteran *Cimex lectularius*. J. Exp. Biol. 201, 2659–2664.

Vancova, I., Hajnicka, V., Slovak, M., Kocakova, P., Paesen, G.C., Nuttall, P.A., 2010a. Evasin-3-like anti-chemokine activity in salivary gland extracts of ixodid ticks during blood-feeding: a new target for tick control. Parasite Immunol. 32, 460–463.

Vancova, I., Hajnicka, V., Slovak, M., Nuttall, P.A., 2010b. Anti-chemokine activities of ixodid ticks depend on tick species, developmental stage, and duration of feeding. Vet. Parasitol. 167, 274–278.

Vaughan, A.M., Mikolajczak, S.A., Wilson, E.M., Grompe, M., Kaushansky, A., Camargo, N., Bial, J., Ploss, A., Kappe, S.H., 2012. Complete *Plasmodium falciparum* liver-stage development in liver-chimeric mice. J. Clin. Invest. 122, 3618–3628.

Venter, J.C., Adams, M.D., Myers, E.W., Li, P.W., Mural, R.J., Sutton, G.G., Smith, H.O., Yandell, M., Evans, C.A., Holt, R.A., Gocayne, J.D., Amanatides, P., Ballew, R.M., Huson, D.H., Wortman, J.R., Zhang, Q., Kodira, C.D., Zheng, X.H., Chen, L., Skupski, M., Subramanian, G., Thomas, P.D., Zhang, J., Gabor Miklos, G.L., Nelson, C., Broder, S., Clark, A.G., Nadeau, J., McKusick, V.A., Zinder, N., Levine, A.J., Roberts, R.J., Simon, M., Slayman, C., Hunkapiller, M., Bolanos, R., Delcher, A., Dew, I., Fasulo, D., Flanigan, M., Florea, L., Halpern, A., Hannenhalli, S., Kravitz, S., Levy, S., Mobarry, C., Reinert, K., Remington, K., Abu-Threideh, J., Beasley, E., Biddick, K., Bonazzi, V., Brandon, R., Cargill, M., Chandramouliswaran, I., Charlab, R., Chaturvedi, K., Deng, Z., Di Francesco, V., Dunn, P., Eilbeck, K., Evangelista, C., Gabrielian, A.E., Gan, W., Ge, W., Gong, F., Gu, Z., Guan, P., Heiman, T.J., Higgins, M.E., Ji, R.R., Ke, Z., Ketchum, K.A., Lai, Z., Lei, Y., Li, Z., Li, J., Liang, Y., Lin, X., Lu, F., Merkulov, G.V., Milshina, N., Moore, H.M., Naik, A.K., Narayan, V.A., Neelam, B., Nusskern, D., Rusch, D.B., Salzberg, S., Shao, W., Shue, B., Sun, J., Wang, Z., Wang, A., Wang, X., Wang, J., Wei, M., Wides, R., Xiao, C., Yan, C., Yao, A., Ye, J., Zhan, M., Zhang, W., Zhang, H., Zhao, Q., Zheng, L., Zhong, F., Zhong, W., Zhu, S., Zhao, S., Gilbert, D., Baumhueter, S., Spier, G., Carter, C., Cravchik, A., Woodage, T., Ali, F., An, H., Awe, A., Baldwin, D., Baden, H., Barnstead, M., Barrow, I., Beeson, K., Busam, D., Carver, A., Center, A., Cheng, M.L., Curry, L., Danaher, S., Davenport, L., Desilets, R., Dietz, S., Dodson, K., Doup, L., Ferriera, S., Garg, N., Gluecksmann, A., Hart, B., Haynes, J., Haynes, C., Heiner, C., Hladun, S., Hostin, D., Houck, J., Howland, T., Ibegwam, C., Johnson, J., Kalush, F., Kline, L., Koduru, S., Love, A., Mann, F., May, D., McCawley, S., McIntosh, T., McMullen, I., Moy, M., Moy, L., Murphy, B., Nelson, K., Pfannkoch, C., Pratts, E., Puri, V., Qureshi, H., Reardon, M., Rodriguez, R., Rogers, Y.H., Romblad, D., Ruhfel, B., Scott, R., Sitter, C., Smallwood, M., Stewart, E., Strong, R., Suh, E., Thomas, R., Tint, N.N., Tse, S., Vech, C., Wang, G., Wetter, J., Williams, S., Williams, M., Windsor, S., Winn-Deen, E., Wolfe, K., Zaveri, J., Zaveri, K., Abril, J.F., Guigo, R., Campbell, M.J., Sjolander, K.V., Karlak, B., Kejariwal, A., Mi, H., Lazareva, B., Hatton, T., Narechania, A., Diemer, K., Muruganujan, A., Guo, N., Sato, S., Bafna, V., Istrail, S., Lippert, R., Schwartz, R., Walenz, B., Yooseph, S., Allen, D., Basu, A., Baxendale, J., Blick, L., Caminha, M., Carnes-Stine, J., Caulk, P., Chiang, Y.H., Coyne, M., Dahlke, C., Mays, A., Dombroski, M., Donnelly, M., Ely, D., Esparham, S., Fosler, C., Gire, H., Glanowski, S., Glasser, K., Glodek, A., Gorokhov, M., Graham, K., Gropman, B., Harris, M., Heil, J., Henderson, S., Hoover, J., Jennings, D., Jordan, C., Jordan, J., Kasha, J., Kagan, L., Kraft, C., Levitsky, A., Lewis, M., Liu, X., Lopez, J., Ma, D., Majoros, W., McDaniel, J., Murphy, S., Newman, M., Nguyen, T., Nguyen, N., Nodell, M., Pan, S., Peck, J., Peterson, M., Rowe, W., Sanders, R., Scott, J., Simpson, M., Smith, T., Sprague, A., Stockwell, T., Turner, R., Venter, E., Wang, M., Wen, M., Wu, D., Wu, M., Xia, A., Zandieh, A., Zhu, X., 2001. The sequence of the human genome. Science 291, 1304–1351.

Verhamme, P., Hoylaerts, M.F., 2009. Hemostasis and inflammation: two of a kind? Thromb. J. 7, 15.

Versteeg, H.H., Heemskerk, J.W., Levi, M., Reitsma, P.H., 2013. New fundamentals in hemostasis. Physiol. Rev. 93, 327–358.

Vijay, S., Rawat, M., Sharma, A., 2014. Mass spectrometry based proteomic analysis of salivary glands of urban malaria vector *Anopheles stephensi*. Biomed. Res. Int. 2014, 686319.

Vinhas, V., Andrade, B.B., Paes, F., Bomura, A., Clarencio, J., Miranda, J.C., Bafica, A., Barral, A., Barral-Netto, M., 2007. Human anti-saliva immune response following experimental exposure to the visceral leishmaniasis vector, *Lutzomyia longipalpis*. Eur. J. Immunol. 37, 3111–3121.

Vora, N., 2008. Impact of anthropogenic environmental alterations on vector-borne diseases. Medscape J. Med. 10, 238.

Wada, T., Ishiwata, K., Koseki, H., Ishikura, T., Ugajin, T., Ohnuma, N., Obata, K., Ishikawa, R., Yoshikawa, S., Mukai, K., Kawano, Y., Minegishi, Y., Yokozeki, H., Watanabe, N., Karasuyama, H., 2010. Selective ablation of basophils in mice reveals their nonredundant role in acquired immunity against ticks. J. Clin. Invest. 120, 2867–2875.

Waisberg, M., Molina-Cruz, A., Mizurini, D.M., Gera, N., Sousa, B.C., Ma, D., Leal, A.C., Gomes, T., Kotsyfakis, M., Ribeiro, J.M., Lukszo, J., Reiter, K., Porcella, S.F., Oliveira, C.J., Monteiro, R.Q., Barillas-Mury, C., Pierce, S.K., Francischetti, I.M., 2014. *Plasmodium falciparum* infection induces expression of a mosquito salivary protein (Agaphelin) that targets neutrophil function and inhibits thrombosis without impairing hemostasis. PLoS Pathog. 10, e1004338.

Walter, E.D., Proctor, C.H., 2013. Mites: Ecology, Evolution & Behaviour, second ed. Springer.

Wang, J., Zhang, Y., Zhao, Y.O., Li, M.W., Zhang, L., Dragovic, S., Abraham, N.M., Fikrig, E., 2013. *Anopheles gambiae* circumsporozoite protein-binding protein facilitates *plasmodium* infection of mosquito salivary glands. J. Infect Dis. 208, 1161–1169.

Wang, X., Shaw, D.K., Sakhon, O.S., Snyder, G.A., Sundberg, E.J., Santambrogio, L., Sutterwala, F.S., Dumler, J.S., Shirey, K.A., Perkins, D.J., Richard, K., Chagas, A.C., Calvo, E., Kopecky, J., Kotsyfakis, M., Pedra, J.H., 2016. The tick protein sialostatin L2 binds to annexin A2 and inhibits NLRC4-mediated inflammasome activation. Infect. Immun. 84, 1796–1805.

Warburg, A., Saraiva, E., Lanzaro, G.C., Titus, R.G., Neva, F., 1994. Saliva of *Lutzomyia longipalpis* sibling species differs in its composition and capacity to enhance leishmaniasis. Philos. Trans. R. Soc. Lond. B Biol. Sci. 345, 223–230.

Ware, J., Suva, L.J., 2011. Platelets to hemostasis and beyond. Blood 117, 3703–3704.

Wasserman, H.A., Singh, S., Champagne, D.E., 2004. Saliva of the Yellow Fever mosquito, *Aedes aegypti*, modulates murine lymphocyte function. Parasite Immunol. 26, 295–306.

Watanabe, R.M., Tanaka-Azevedo, A.M., Araujo, M.S., Juliano, M.A., Tanaka, A.S., 2011. Characterization of thrombin inhibitory mechanism of rAaTI, a Kazal-type inhibitor from *Aedes aegypti* with anticoagulant activity. Biochimie 93, 618–623.

Weaver, S.C., Costa, F., Garcia-Blanco, M.A., Ko, A.I., Ribeiro, G.S., Saade, G., Shi, P.Y., Vasilakis, N., 2016. Zika virus: history, emergence, biology, and prospects for control. Antiviral Res. 130, 69–80.

Weichsel, A., Andersen, J.F., Champagne, D.E., Walker, F.A., Montfort, W.R., 1998. Crystal structures of a nitric oxide transport protein from a blood-sucking insect. Nat. Struct. Biol. 5, 304–309.

Weichsel, A., Maes, E.M., Andersen, J.F., Valenzuela, J.G., Shokhireva, T., Walker, F.A., Montfort, W.R., 2005. Heme-assisted S-nitrosation of a proximal thiolate in a nitric oxide transport protein. Proc. Natl. Acad. Sci. U. S. A. 102, 594–599.

Weiss, W.R., Sedegah, M., Beaudoin, R.L., Miller, L.H., Good, M.F., 1988. CD8+ T cells (cytotoxic/suppressors) are required for protection in mice immunized with malaria sporozoites. Proc. Natl. Acad. Sci. U. S. A. 85, 573–576.

Wengelnik, K., Spaccapelo, R., Naitza, S., Robson, K.J., Janse, C.J., Bistoni, F., Waters, A.P., Crisanti, A., 1999. The A-domain and the thrombospondin-related motif of *Plasmodium falciparum* TRAP are implicated in the invasion process of mosquito salivary glands. EMBO J. 18, 5195–5204.

White, M.T., Verity, R., Griffin, J.T., Asante, K.P., Owusu-Agyei, S., Greenwood, B., Drakeley, C., Gesase, S., Lusingu, J., Ansong, D., Adjei, S., Agbenyega, T., Ogutu, B., Otieno, L., Otieno, W., Agnandji, S.T., Lell, B., Kremsner, P., Hoffman, I., Martinson, F., Kamthunzu, P., Tinto, H., Valea, I., Sorgho, H., Oneko, M., Otieno, K., Hamel, M.J., Salim, N., Mtoro, A., Abdulla, S., Aide, P., Sacarlal, J., Aponte, J.J., Njuguna, P., Marsh, K., Bejon, P., Riley, E.M., Ghani, A.C., 2015. Immunogenicity of the RTS,S/AS01 malaria vaccine and implications for duration of vaccine efficacy: secondary analysis of data from a phase 3 randomised controlled trial. Lancet Infect. Dis. 15, 1450–1458.

Wikel, S., 2013. Ticks and tick-borne pathogens at the cutaneous interface: host defenses, tick countermeasures, and a suitable environment for pathogen establishment. Front. Microbiol. 4, 337.

Wikel, S.K., 1979. Acquired resistance to ticks: expression of resistance by C4-deficient guinea pigs. Am. J. Trop. Med. Hyg. 28, 586–590.

Wikel, S.K., 1996. Host immunity to ticks. Annu. Rev. Entomol. 41, 1–22.

Wikel, S.K., 1999. Tick modulation of host immunity: an important factor in pathogen transmission. Int. J. Parasitol. 29, 851–859.

Wikel, S.K., Allen, J.R., 1976. Acquired resistance to ticks. I. Passive transfer of resistance. Immunology 30, 311–316.

Wikel, S.K., Ramachandra, R.N., Bergman, D.K., Burkot, T.R., Piesman, J., 1997. Infestation with pathogen-free nymphs of the tick *Ixodes scapularis* induces host resistance to transmission of *Borrelia burgdorferi* by ticks. Infect. Immun. 65, 335–338.

Wilkinson, C.F., 1976. Insecticide Biochemistry and Physiology. Plenum Press, New York.

Willadsen, P., 2004. Anti-tick vaccines. Parasitology (129 Suppl.), S367–S387.

Willadsen, P., Riding, G.A., McKenna, R.V., Kemp, D.H., Tellam, R.L., Nielsen, J.N., Lahnstein, J., Cobon, G.S., Gough, J.M., 1989. Immunologic control of a parasitic arthropod. Identification of a protective antigen from *Boophilus microplus*. J. Immunol. 143 (4), 1346–1351.

Xu, T., Lew-Tabor, A., Rodriguez-Valle, M., 2016. Effective inhibition of thrombin by *Rhipicephalus microplus* serpin-15 (RmS-15) obtained in the yeast *Pichia pastoris*. Ticks Tick Borne Dis. 7, 180–187.

Yang, L., Murota, H., Serada, S., Fujimoto, M., Kudo, A., Naka, T., Katayama, I., 2014. Histamine contributes to tissue remodeling via periostin expression. J. Invest. Dermatol. 134, 2105–2113.

Yates, J.R., Ruse, C.I., Nakorchevsky, A., 2009. Proteomics by mass spectrometry: approaches, advances, and applications. Annu. Rev. Biomed. Eng. 11, 49–79.

Yilmaz, B., Portugal, S., Tran, T.M., Gozzelino, R., Ramos, S., Gomes, J., Regalado, A., Cowan, P.J., d'Apice, A.J., Chong, A.S., Doumbo, O.K., Traore, B., Crompton, P.D., Silveira, H., Soares, M.P., 2014. Gut microbiota elicits a protective immune response against malaria transmission. Cell 159, 1277–1289.

Yoshida, N., Nussenzweig, R.S., Potocnjak, P., Nussenzweig, V., Aikawa, M., 1980. Hybridoma produces protective antibodies directed against the sporozoite stage of malaria parasite. Science 207, 71–73.

Zhu, K., Bowman, A.S., Brigham, D.L., Essenberg, R.C., Dillwith, J.W., Sauer, J.R., 1997. Isolation and characterization of americanin, a specific inhibitor of thrombin, from the salivary glands of the lone star tick *Amblyomma americanum* (L.). Exp. Parasitol. 87, 30–38.

Zivkovic, Z., Esteves, E., Almazan, C., Daffre, S., Nijhof, A.M., Kocan, K.M., Jongejan, F., de la Fuente, J., 2010. Differential expression of genes in salivary glands of male *Rhipicephalus (Boophilus) microplus* in response to infection with *Anaplasma marginale*. BMC Genomics 11, 186.

Considerations for the Translation of Vector Biology Research

Adriana Costero-Saint Denis, Wolfgang W. Leitner, Tonu Wali,
Randall Kincaid

National Institute of Allergy and Infectious Diseases, NIH Rockville, Maryland, United States

The power of basic research lies not only in its contribution to knowledge, but also in its potential for translation, adapting and developing work done in the laboratory to create innovative, novel approaches to prevent and control public health threats. The path for translation of basic research findings is a highly iterative and interdisciplinary approach that is often unfamiliar to those in academic environments. The resources and perspectives required to bring a new idea into the realm of products involve collaborative relationships and learning that are, necessarily, elements of the business world. These include elements of establishing investment priorities, marketing, regulatory requirements, and, of course, sustainability of products.

In the area of vector biology, remarkable advances in basic knowledge have occurred over the last 20 years or so, as represented in the many chapters of this two volume work. Some of these areas clearly have translational potential that may impact public health, both as products to improve the health of individuals (therapeutics, vaccines, personal repellents, etc.) and solutions to reduce environmental burden of vectorborne pathogens. However, we must appreciate that not all good ideas can become good products. The objective of this chapter is to provide some basic principles and guidance for investigators who have identified such translational possibilities in their research.

Translational strategy and product development are generally not the topics that investigators in basic science have experienced firsthand. Therefore, when laboratory-based research evolves to the point where a potential product is identified, the laboratory researcher needs to understand the implications and unique requirements of moving it through the translational process. Their involvement in this process is undoubtedly crucial, but they will need to decide how best to become involved and to what extent. There are likely to be challenges in finding a balance between contributing essential scientific insights and the learning of new principles that are largely outside the purview of academia.

Laboratory researchers lack the financial means to translate their research into a product; however, there are numerous entities with interest in supporting the development of novel

Arthropod Vector: Controller of Disease Transmission, Volume 2
http://dx.doi.org/10.1016/B978-0-12-805360-7.00015-0

products for public health. These include an extraordinary spectrum of interests—government and nongovernmental organizations, venture capitalists, and established commercial entities looking for products that will enhance their portfolios. They may be domestic interests and those from institutions with more global needs. So how can scientists connect to these groups, if they decide they want to translate a discovery from their laboratory? Is there a general process for determining which partnerships are needed and at what times engagement is warranted? The answer is not simple, and one should consider the basic value of establishing proof of concept and of risk reduction, prior to seeking commitments of financial support. These principles are shared, to different degrees, by all who may consider investments in novel ideas.

Based on the output of two meetings that we organized in 2014 and 2015: *Arthropod Vectors and Disease Transmission: Translational Aspects* (Leitner et al., 2015) and *Translational Considerations of Novel Vector Management Approaches* (Costero-Saint Denis et al., 2017), as well as our recent experiences in supporting investigators interested in translating discoveries from their laboratories, we have compiled below some basic considerations for investigators who may be contemplating translation of their basic research. The intent of this chapter is to provide a primer of ideas for investigators to consider when thinking about taking their discoveries down the translational path. At the end of the chapter we have provided a list of resources that investigators may find useful.

THE PIPELINE

The product development pipeline can be viewed schematically as a series of steps and phases from discovery in the laboratory to a marketable product. It is important to recognize that there are many early decision points in a process that may act to narrow the product possibilities. This is often stressful for investigators who consider many potential uses for their ideas; nonetheless, the failure to appropriately define the most tractable product opportunities is a major reason for never getting past the first steps. It is useful to start with two elementary questions: Who is likely to be my customer? What elements of the product are its biggest selling points? Once these questions have been considered, the design of the product will be tailored to meet the expectations and to accentuate the value of the innovation.

Each step in the following figure can be broken down into additional levels, depending on the type of product. The types of input for decision-making along this path may also vary for many reasons, including the changes in customer needs, the dynamics of competition, and financial constraints. Product development is sometimes a paradox that requires highly disciplined planning while, at the same time, maintaining flexibility and alternate scenarios (plan A, B, etc.).

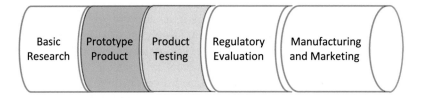

During the Basic Research and Prototype Product phases, the investigator tests the effectiveness of the product under laboratory conditions. If the resulting data are encouraging and reproducible, then the product moves to the next step that involves refinement of the prototype to a working "candidate" and testing under conditions that are relevant to its ultimate use; such testing is essential to confirm effectiveness and safety prior to advancing the product for larger trials.

The Product Testing phase can be a very long process, requiring access to various types of models (animal models, simulated efficacy testing, etc.) and often to human specimens that may require specialized procedures and permissions. However, the value added by these iterative testing efforts acts to "derisk" the proposed development in the eyes of potential investors and to inform additional testing that may be required.

If the testing suggests that a tenable product is likely, it is highly recommended that the investigator considers seeking *patent* protection for the product at this stage. This will allow the investigator to guard the intellectual property during the subsequent phases of product development and assure potential investors that their product cannot be easily infringed. Academic institutions normally house Technology Transfer offices that can provide guidance and help in filing patents, trademarks, and the like. The existence of intellectual property cannot be assumed and there must be due diligence to support claims for invention, which must both be "nonobvious" to those in the field and realizable, in terms of reduction to practice.

For most products having public health implications, studies of safety and efficacy are required to warrant a marketing decision by a regulatory agency; these most often require trials conducted with human subjects. Before a trial outside the laboratory can be done, the appropriate regulatory agency must be consulted at length to define the "intended use" of the product, provide guidance on appropriate trial design, and set broad expectations for the types of data that must be provided in regulatory submission. Ultimately, these elements will be required to grant approval for marketing of the product. We recommend contacting the appropriate regulatory agency early in the process so that the investigator can understand what data will be needed. The extent and types of data required may be considerable, and this often is a first real shock the investigator experiences about the reality of creating products, both in terms of time and cost. Such guidance from regulators is essential for the investigator to properly document and plan future steps for the development of a final product.

PRODUCT DEVELOPMENT PLAN

A product development plan (PDP) is a strategic document that contains all information relevant to a proposed new product. It provides historical and scientific precedent to justify the need for the product, describes its attributes in the context of any current products, and clarifies the major risks and opportunities that may exist. As such, it is akin to a traditional business plan and it serves a very similar purpose. It demonstrates to a potential financial backer (e.g., government program, industry partner, or venture capitalist) that you have done your homework on what it will take to realize a product and ensure its success. It is important to be forthcoming with all information that is available since failure to disclose information is just as damaging as information that may suggest risks or competitive products. If carefully done, creation of the PDP will help an investigator create a "roadmap" for his/her product that allows everyone to have realistic expectations; those expectations can also contribute to the types of development "milestones" commonly associated with funded relationships. There are consulting firms that can be hired to develop PDPs for investigators interested in having this document as a guiding principle for translation of their product.

The content of the PDP should be specific to a product and its intended use. In this regard, it may be restricted to a rather narrow collection of opportunities although one can create multiple PDP documents for other potential uses. It is very helpful to assemble a team of individuals who have different perspectives on the product when preparing this planning document. Ideally, each member of the team can contribute insights that are of value (e.g., scientific, commercialization, funding, regulatory, marketing, etc.). To more easily visualize the various stages of PDP preparation and follow-up, we have included a diagrammatic view of the process below, which outlines the specific areas and types of information that are collected.

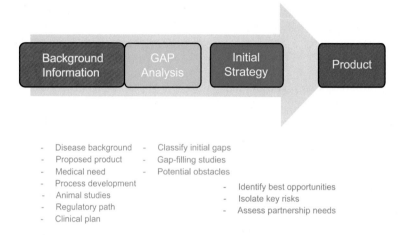

- Disease background
- Proposed product
- Medical need
- Process development
- Animal studies
- Regulatory path
- Clinical plan

- Classify initial gaps
- Gap-filling studies
- Potential obstacles

- Identify best opportunities
- Isolate key risks
- Assess partnership needs

Often, a PDP starts by defining the preferred characteristics of the product and its use—the target product profile. This collection of elements usually includes the target population, the dosing and route of administration of the product, efficacy end points, safety tolerance, stability, etc. and is used to guide selection of an optimal candidate. A background section is then prepared that describes the biology and public health importance of the pathogen being targeted by the product, the current approaches for preventing and treating the pathogen, its importance of preventing transmission of the particular pathogen, and specific products that already are available for this purpose. Rigorous analysis of available products and how they may compare to the proposed product can be very useful to address strengths and weaknesses of the product. This comparison is also useful in considering the marketing niche that might be filled by the new product. The PDP should also include all data on the product, e.g., efficacy, formulation, stability, etc., as well as the facilities and testing that may be required for regulatory approval. The types of data needed will depend on the regulatory agency and the classification of the product (e.g., biological pesticide) and the specific use (personal protectant, environmental use, etc.). Additional data that will inform the process may include relevant preclinical data (e.g., animal studies), the maturity of the current manufacturing process, and the potential need for GMP (Good Manufacturing Practice) and/or GLP (Good Laboratory Practice) standards for production. Not all of these considerations may be important and will depend on those who are expected to fund its further development and understand the needs and requirements of the end users.

After collection of the background information has been completed, a detailed risk–benefit analysis is carried out. Obviously, it is critical that the PDP has access to as much specific information as possible to enable a meaningful "gap analysis." The types of unknowns and shortcomings of the product are considered, and potential solutions for these are put forward to allow them to be classified in terms of solution. Some of these gaps may be easily addressed, whereas others may be very problematic and difficult to solve (e.g., poorly characterized methods to measure product performance, lack of regulatory precedent, shortage of qualified suppliers of key ingredients for the product). Nonetheless, the analysis is critical to making a decision to move ahead, seeking alignments/partnerships that will support further product testing and development, and, ultimately, anticipating the concerns of the investor. The final stages of planning are the creation of strategies that best suit the product's key advantages and that allow for manageable risk.

The above analysis creates an early roadmap for the investigator to fairly assess the risk and to find the best opportunities for their innovation. Just like a good business plan, a good PDP can go a long way to convincing investors to consider funding the product. Failure to have carried out such in-depth self-examination is likely to prove costly.

THE REGULATORY PROCESS

As indicated above, after an investigator identifies a promising product concept in the laboratory and a decision is made to move it forward, an essential goal is to generate the data necessary for regulatory approval (marketing approval, registration of the product). This may require establishing a business partnership with entities having experience in the area, or it may be possible to acquire the expertise through subject matter experts (i.e., consulting firms). Regardless, the investigator should contact the appropriate regulatory agency as early in the process as possible to obtain guidance on their process of evaluation and the data needed to evaluate the product's efficacy and safety. Depending on the product's specific use, the safety aspect may pertain not only to vertebrate animals, but also to other organisms that may be affected. Therefore, it is not unusual that a series of data packets may be required to cover vertebrate as well as environment toxicity.

In the United States, the regulatory agency that oversees the approval of vector control products for use is the Environmental Protection Agency (EPA). Within this agency, the following offices and divisions have oversight for the different classes of vector control products:

EPA—Office of Pesticide Programs
Biopesticides and Pollution Prevention Division (includes biologicals: *Bacillus thuringiensis* var. *israelensis* (Bti), *Bacillus sphaericus* (Bs), *Wolbachia*, fungi)
Registration Division (includes insecticides and repellents)

For more information, please access the EPA website link listed in the Resources section of this chapter.

The requirements for a product are usually based on precedent, and general examples of the types of data can often be obtained through analysis of public databases. However, the amounts and detail of information (and thus the time and expense) can be substantial, requiring at least 2 years of effort to complete even the most basic data package. There may be some room for discussion, particularly if there are scientifically sound reasons that draw upon previously used approaches, but it is unlikely that many of the tests can be avoided. In product areas that are novel, such as the development of transgenic organisms (vectors), there may be additional needs that must be satisfied. In such cases, it is wise to consult with organizations that have close ties to these newly emerging areas to anticipate the types of concerns that are likely to arise.

TO TRANSLATE OR NOT: THAT IS THE QUESTION

The decision on whether or not to begin moving a laboratory discovery through the translational pipeline is made by the investigator and should be done in consultation with the institution's Technical Transfer office and with the sponsor of the research that resulted in the discovery. There is often disagreement on the product potential for a new idea from the research institution's perspective, ranging from the costs of pursuing patent protection to a perceived lack of licensing interest. The investigator may still wish to consider pursuing the idea and he/she has the right to do so. Here, it is important to emphasize that there is an ethical and contractual obligation to disclose potential public health discoveries to the sponsor of the research as well as the investigator's institution. If the investigator's institution decides not to pursue the further development of a discovery, the institution sponsoring the research may have an interest in doing so after the investigator has obtained a patent.

There are governmental and not-for-profit organizations that can provide advice and help to investigators with translatable discoveries. It is the investigator's responsibility to patent a discovery and then contact the appropriate organization(s) that can invest in moving it forward.

Given that research scientists are typically not trained and knowledgeable in product development, it is advisable to partner with an organization that knows how to do this to ensure the maximum possibility of success. The process of product development is time-consuming and involves many regulatory and legal aspects that are very specialized.

DETERMINING THE PUBLIC HEALTH IMPACT OF A PRODUCT

After an invention has been patented and approved by regulatory authorities, it can then be tested in a field setting to determine its public health impact. For novel vector control interventions, this would mean performing epidemiological studies and clinical trials to determine if the product has an impact on transmission and morbidity/mortality because of a vectorborne pathogen.

The following aspects must be considered when thinking about testing a promising vector control product:

1. *Trial site requirements*: Choosing an appropriate field site to test the efficacy and safety of a product is essential. Site selection is not only important to assess the entomologic and epidemiologic impact of a novel vector approach. It is also important from the legal, social, and political perspective that will influence whether the product can be tested or not (Brown et al., 2014). Additional factors in selecting a field site include entomologic and epidemiologic base line data, disease endemicity and periodicity, training of local personnel to help monitor before and after implementation of the novel product, etc.

2. *Regulatory considerations*: This can be considered as part of the trial site selection as a credible regulatory structure/system should be in place. In addition to a well-defined and structured regulatory system, biosafety and ethical oversight should also exist (Ramsey et al., 2014). This framework will ensure that the testing of the product will be done within national and international accepted standards (Entomology, 2004; WHO, 2014).

3. *Risk assessment and management*: Concerns about vector control interventions have centered on safety for humans and the environment. A risk management assessment can identify potentially adverse effects and also takes into account social and economic risks. An example of risk assessment is the Convention on Biological Diversity guidance (https://bch.cbd.int/protocol/guidance_risk_assessment/). For any project receiving US federal government funding, an evaluation of environmental impact must be completed and approved by the National Environmental Policy Act (NEPA), at the NIH, the NEPA office (https://nems.nih.gov/Pages/nepa.aspx).

4. *Social and ethical considerations*: This is perhaps the most relevant item in preparing for field site testing of a vector control intervention. Even if all other parameters are met, if the local population is not in agreement with the testing of the product, the entire project can fail. Lack of understanding of the social, cultural, and ethical issues of populations in sites where products are to be tested can delay or even prevent the evaluation, let alone the implementation of a product (Kolopack et al., 2015).

We hope that these thoughts and ideas may be useful in providing an awareness of the complexities of translating basic science discoveries to the field. We have tried to provide general guidance, particularly about the importance of detailed planning and the importance of anticipating partnerships to advance your ideas. As more novel vector control interventions emerge, the paths on how to develop them into useful public health tools will become better defined. Ultimately, being able to prevent transmission of vectorborne diseases and the burden they impose on people is a goal we should not shy of even if the path to get there is currently not well defined and therefore will require extra effort and time. It is also worthwhile to remember that teamwork is crucial to success: solving complex problems require partnerships.

Resources of Interest

WHO Guidance framework for GMM. http://www.who.int/tdr/publications/year/2014/guide-fmrk-gm-mosquit/en/.
A tool-kit for integrated vector management in Sub-Saharan Africa WHO. http://www.who.int/neglected_diseases/resources/9789241549653/en/.
NIAID preclinical resources website. https://www.niaid.nih.gov/research/tools-datasets-and-services.

IVCC website. http://www.ivcc.com/.
EPA – Office of Pesticide Programs. https://www.epa.gov/pesticide-contacts.
EPA – Office of Biopesticides. https://www.epa.gov/pesticide-contacts/contacts-office-pesticide-programs-biopesticides-and-pollution-prevention.
EPA – Registration Division. https://www.epa.gov/pesticide-contacts/contacts-office-pesticide-programs-registration-division.
EPA – Pesticide Registration. https://www.epa.gov/pesticide-registration/registration-requirements-and-guidance.
NIH NEPA. https://nems.nih.gov/Pages/nepa.aspx.
Convention on Biological Diversity. https://bch.cbd.int/protocol/guidance_risk_assessment/.
The IR-4 Project – Public Health Pesticides. http://ir4.rutgers.edu/publichealth.html.
Deployed War Fighter Protection Program (DoD). http://www.acq.osd.mil/eie/afpmb/dwfp.html.
American Mosquito Control Association. http://www.mosquito.org/.

References

Brown, D.M., Alphey, L.S., McKemey, A., Beech, C., James, A.A., 2014. Criteria for identifying and evaluating candidate sites for open-field trials of genetically engineered mosquitoes. Vector Borne Zoonotic Dis. 14, 291–299.

Costero-Saint Denis, A., Leitner, W.W., Wali, T., James, S., August 11, 2016. Translational considerations of novel vector management approaches. PLoS Negl. Trop. Dis. 10 (8), e0004800. http://dx.doi.org/10.1371/journal.pntd.0004800.

Entomology, A.C.O.M., 2004. Arthropod containment levels. Vector Borne Zoonotic Dis. 3, 75–90.

Kolopack, P.A., Parsons, J.A., Lavery, J.V., 2015. What makes community engagement effective?: Lessons from the Eliminate Dengue Program in Queensland Australia. PLoS Negl. Trop. Dis. 9, e0003713.

Leitner, W.W., Wali, T., Kincaid, R., Costero-Saint Denis, A., 2015. Arthropod vectors and disease transmission: translational aspects. PLoS Negl. Trop. Dis. 9, e0004107.

Ramsey, J.M., Bond, J.G., Macotela, M.E., Facchinelli, L., Valerio, L., Brown, D.M., Scott, T.W., James, A.A., 2014. A regulatory structure for working with genetically modified mosquitoes: lessons from Mexico. PLoS Negl. Trop. Dis. 8, e2623.

WHO, 2014. The guidance framework for testing genetically modified mosquitoes. http://www.who.int/tdr/publications/year/2014/guide-fmrk-gm-mosquit/en/.

Index

'*Note*: Page numbers followed by "f" indicate figures and "t" indicate tables.'

A. aegypti salivary gland,
   40–41
adaptive immune responses, 43
host-vector interactions, 40
neutrophils, 41
pool feeder disruption, 40
ticks, 42

third phase, proliferation and
   vectors, 43–44
tissue remodeling, 35
vector arthropod feeding,
   32–33
vector arthropod modulation,
   35–36

biting flies and ticks, 36
   head lice, 36
Wound repair, 102–103

Y
Yellow proteins, 58–59

Printed in the United States
By Bookmasters